Advances in Lignocellulosic Biofuel Production Systems

Applied Biotechnology Reviews Series

Microbial Biotechnology in Food and Health
Ramesh C. Ray

Sustainable Biofuels, Opportunities and Challenges
Ramesh C. Ray

Lactic Acid Bacteria in Food Biotechnology: Innovations and Functional Aspects
Ramesh C. Ray, Spiros Paramithiotis, Vasco Ariston De Carvalho Azevedo, and Didier Montet

Advances and Developments in Biobutanol Production
Juan Gabriel Segovia-Hernandez, Shuvashish Behera, and Eduardo Sanchez-Ramirez

Lactic Acid Bacteria as Cell Factories
Didier Montet, Ramesh C. Ray, Vasco Ariston De Carvalho Azevedo, and Spiros Paramithiotis

Applied Biotechnology Reviews Series

Advances in Lignocellulosic Biofuel Production Systems

Edited by

Preshanthan Moodley

SARCHI Chair Waste and Climate Change, School of Engineering, University of KwaZulu-Natal, Durban, South Africa

Ramesh C. Ray

Centre for Food Biology and Environment Studies, Bhubaneswar, Odisha, India

Evariste B. Gueguim Kana

School of Life Sciences, University of KwaZulu-Natal, Pietermaritzburg, South Africa

Series Editor

Ramesh C. Ray

ELSEVIER

WP
WOODHEAD PUBLISHING
An imprint of Elsevier

Woodhead Publishing is an imprint of Elsevier
50 Hampshire Street, 5th Floor, Cambridge, MA 02139, United States
The Boulevard, Langford Lane, Kidlington, OX5 1GB, United Kingdom

ISBN: 978-0-323-91192-4 (print)

ISBN: 978-0-323-91344-7 (online)

For information on all Woodhead Publishing publications
visit our website at https://www.elsevier.com/books-and-journals

Publisher: Megan Ball
Acquisitions Editor: Peter Adamson
Editorial Project Manager: Zsereena Rose Mampusti
Production Project Manager: Prem Kumar Kaliamoorthi
Cover Designer: Miles Hitchen

Typeset by MPS Limited, Chennai, India

Contents

Section IV Advances in Modeling and Development

List of contributors

Noé Aguilar-Rivera Faculty of Biological and Agricultural Sciences, Universidad Veracruzana, Córdoba, Veracruz, Mexico

Kalyani Ananthakrishnan Department of Chemical Engineering, SUNY College of Environmental Science and Forestry, Syracuse, NY, United States

Jaya Arora Laboratory of Biomolecular Technology, Department of Botany, Mohanlal Sukhadia University, Udaipur, Rajasthan, India

Gabriel S. Aruwajoye School of Life Sciences, University of KwaZulu-Natal, Pietermaritzburg, South Africa

Mukesh Kumar Awasthi College of Natural Resources and Environment, Northwest A&F University, Yangling, Shaanxi Province, P. R. China

Sudhanshu S. Behera Department of Biotechnology, National Institute of Technology Raipur, Raipur, Chhattisgarh, India

María Mercedes González Bernal Universidad Industrial de Santander, Bucaramanga, Colombia

Anthea N. David School of Life Sciences, University of KwaZulu-Natal, Pietermaritzburg, South Africa

Andrea Dell'Orto SARCHI Chair Waste and Climate Change, School of Engineering, University of KwaZulu-Natal, Durban, South Africa

Evariste B. Gueguim Kana School of Life Sciences, University of KwaZulu-Natal, Pietermaritzburg, South Africa

Md Shahadat Hossain Department of Chemical Engineering, SUNY College of Environmental Science and Forestry, Syracuse, NY, United States

Priyanka Jha Department of Biotechnology, Lovely Faculty of Technology and Sciences, Lovely Professional University, Phagwara, Punjab, India

Abhishek Joshi Laboratory of Biomolecular Technology, Department of Botany, Mohanlal Sukhadia University, Udaipur, Rajasthan, India

Ankita Juneja Department of Biology, Syracuse University, Syracuse, NY, United States

Sonampreet Kaur Enzyme Biotechnology Laboratory, Department of Biotechnology, Punjabi University, Patiala, Punjab, India

Deepak Kumar Department of Chemical Engineering, SUNY College of Environmental Science and Forestry, Syracuse, NY, United States

Prasun Kumar Department of Chemical Engineering, Yeungnam University, Gyeongsan, Republic of Korea

Vijay Kumar Plant Biotechnology Lab, Division of Research and Development, Lovely Professional University, Phagwara, Punjab, India

Vinay Kumar Department of Physiology and Cell Biology, Ohio State University Wexner Medical Centre, Columbus, OH, United States

Milesh Laltha School of Life Sciences, University of KwaZulu-Natal, Pietermaritzburg, South Africa

Edzai Magowo School of Chemical and Metallurgical Engineering, University of the Witwatersrand, Johannesburg, South Africa

Manpreet Kaur Mann Enzyme Biotechnology Laboratory, Department of Biotechnology, Punjabi University, Patiala, Punjab, India

Edson L. Meyer Fort Hare Institute of Technology, University of Fort Hare, Alice, South Africa

Christian Michel-Cuello Multidisciplinary Academic Unit Central Zone, Autonomous University of San Luis Potosí, Rioverde, San Luis Potosí, México

Ashutosh Mittal National Renewable Energy Laboratory, Golden, CO, United States

Preshanthan Moodley SARCHI Chair Waste and Climate Change, School of Engineering, University of KwaZulu-Natal, Durban, South Africa

Gopu Raveendran Nair Department of Agricultural and Biological Engineering, University of Illinois at Urbana-Champaign, Urbana, IL, United States

Luis Alberto Olvera-Vargas Center for Research and Assistance in Technology and Design of the State of Jalisco (CIATEJ) Guadalajara, Jalisco, Mexico

Avirup Panja Amity Institute of Biotechnology, Amity University, Kolkata, West Bengal, India

Karthik Rajendran Department of Environmental Science and Engineering, School of Engineering and Sciences, SRM University-AP, Amaravati, Andhra Pradesh, India

Vishnu D. Rajput Academy of Biology and Biotechnology, Southern Federal University, Rostov-on-Don, Russia

Ramesh C. Ray Centre for Food Biology and Environment Studies, Bhubaneswar, Odisha, India

Daneal C.S. Rorke School of Life Sciences, University of KwaZulu-Natal, Pietermaritzburg, South Africa

Karl Rumbold School of Molecular and Cell Biology, University of the Witwatersrand, Johannesburg, South Africa

Isaac A. Sanusi Fort Hare Institute of Technology, University of Fort Hare, Alice, South Africa

Yeshona Sewsynker-Sukai Fort Hare Institute of Technology, University of Fort Hare, Alice, South Africa

Craig Sheridan School of Chemical and Metallurgical Engineering, University of the Witwatersrand, Johannesburg, South Africa; School Geography, Archaeology and Environmental Studies, University of the Witwatersrand, Johannesburg, South Africa

Julia Sibiya School of Agriculture, Earth and Environmental Sciences, University of KwaZulu-Natal, Scottsville, Pietermaritzburg, South Africa

Ashutosh Singh School of Engineering, University of Guelph, Guelph, ON, Canada

Balwinder Singh Sooch Enzyme Biotechnology Laboratory, Department of Biotechnology, Punjabi University, Patiala, Punjab, India

Manas R. Swain DBT-IOC Centre for Advance Bio-energy Research, Indian Oil Corporation, Faridabad, Haryana, India

Cristina Trois SARCHI Chair Waste and Climate Change, School of Engineering, University of KwaZulu-Natal, Durban, South Africa

Krishan K. Verma Key Laboratory of Sugarcane Biotechnology and Genetic Improvement (Guangxi), Ministry of Agriculture and Rural Affairs/Guangxi Key Laboratory of Sugarcane Genetic improvement/Sugarcane Research Institute, Guangxi Academy of Agricultural Sciences, Nanning, Guangxi, P. R. China

Preface

Population growth, increasing urbanization, the depletion of fossil fuel sources, the necessity for energy security, and the incredible amount of waste produced are some of the inherent problems linked to the current fossil-based linear economy. Exploring and identifying alternative technologies, methods, and feedstocks that can transition to a circular economy is urgently needed, given the trajectory of the current scenario. The world's economy is progressively shifting from one that is dependent on fossil fuels to one that is circular and sustainable. Alternative and sustainable methods for producing liquid and gaseous biofuels are crucial to aid in this paradigm shift. The production of biofuels utilizing residues as a feedstock has emerged as a method for generating sustainable energy. Biofuels are a temporary and natural by-product of numerous biochemical processes powered by microorganisms, supporting environment-friendly and sustainable manufacturing. Climate change and the enormous amounts of waste being produced are concerns that cannot be overlooked. The massive amounts of biogenic waste that are produced may be viewed as a potential fuel for the bio-based economy's structure.

This book is divided into five sections and it is a collection of expertly curated chapters that provide a snapshot of the lignocellulosic biofuel landscape, with a specific focus on current advances. Section I is the introductory chapter which provides an overview of lignocellulosic biofuel processes with a brief look at the contents of each chapter in the book. Section II deals with feedstock and processing. The first chapter in this section provides a detailed account of lignocellulosic biomass as a feedstock to support the circular economy; this is followed by a chapter on genetically engineered feedstocks to enhance biofuel yields. Subsequent chapters deal with pretreatment technologies, application of microwave processing, and role of enzymes for enhanced saccharification. It also presents a case study examining acid mine drainage as a potential pretreatment agent for lignocellulosic biomass. Section III details recent trends in bioprocessing with the first chapter focusing metabolic engineering of microorganisms followed by chapters on insight into microbial factories and then cell immobilization strategies to enhance process yields. Section IV contains chapters dealing with advances in modeling and development, specifically artificial intelligence, integrated biorefineries, two-stage anaerobic digestion, farm to biorefinery, and life cycle assessment. The last section (V) details public policy for a circular economy with a specific focus on Mexico as a cast study.

The editors of this book would like to express their sincere thanks to all the authors for their high-quality contributions. The successful completion of this book has been the result of the cooperation of many people and we would therefore like to express our sincere thanks and gratitude to all of them.

We genuinely hope that the present discourse on biofuels research will significantly advance readers' understanding of the cutting-edge technical potential in this area as well as to inspire others to build upon the ground-breaking research that has been laid.

Preshanthan Moodley
Ramesh C. Ray
Evariste B. Gueguim Kana

Section I

Introduction

Current status of lignocellulosic biofuel production system—an overview

1

Preshanthan Moodley and Cristina Trois
SARCHI Chair Waste and Climate Change, School of Engineering, University of KwaZulu-Natal, Durban, South Africa

1.1 Introduction

Due to the energy industry's growing environmental, economic, and social consequences, experts and researchers are motivated to explore biofuels. According to model forecasts in recent research, the value of the global biofuels market is expected to approach US$201.21 billion by 2030, rising at a compound annual growth rate of 8.3% between 2021 and 2030. There has been a fundamental movement among nations toward higher usage of biofuel to promote a circular economy and aid in the fight against climate change brought on by excessive carbon emissions. The United Nations issued 17 sustainable development objectives (goals), addressing all countries globally. Among these, the aspirational objectives of achieving clean, inexpensive energy and combating climate change provide a bold plan for a better world. The necessity for sustainable development has been acknowledged by national governments worldwide. Countries like the United States, Brazil, Europe, France, India, and China are among those preparing to reduce greenhouse gas (GHG) emissions by boosting the use of biofuels (Wei et al., 2021). The International Air Transport Association has set high goals for its members, including encouraging the construction of carbon-neutral infrastructure and reducing CO_2 emissions by half by 2050. The development of outstanding, cost-effective biofuel production technologies may be ensured by significant infrastructure investment and improved government policy.

It is essential to employ cutting-edge techniques to produce biofuels in an environmentally responsible, cost-competitive, and sustainable way. Biofuels made from biomass resources using ecologically friendly processes are attracting growing attention from scientists and academics. Many liquid and gaseous biofuels may be created from biomass, including ethanol, biodiesel, methane, methanol, biohydrogen, and sustainable aviation fuel (SAF). With the help of these biofuels, future energy security, and sustainability may be attained. Furthermore, biofuels lessen total pollution load and environmental effects by lowering GHG emissions, primarily CO_2, in addition to reducing air pollutants during combustion. Most renewable energy-generating alternatives rely on resources that can provide electrical energy and displace fossil fuels, including wind, solar, tidal, hydropower, and geothermal.

Advances in Lignocellulosic Biofuel Production Systems. DOI: https://doi.org/10.1016/B978-0-323-91192-4.00015-8

However, the platform for energy consumption is still dominated by liquid fuels, and the infrastructure and technologies needed to convey these renewables are still in their infancy. For this reason, many experts consider it a viable choice to gradually replace fossil fuels with liquid biofuels, including ethanol, methanol, butanol, and biokerosene (Kazemi Shariat Panahi et al., 2019).

With a strategic growth in biofuel production from about 16 billion liters in 2000 to 143 billion liters in 2016, which includes a production growth of 65% for bioethanol, 25% for biodiesel, and 10% for other biofuels, the use of liquid biofuels has increased from 1% of the world's total biomass-derived bioenergy supply in 2000 to 6% in 2016 (Tursi, 2019). These biofuels are often combined with gasoline at a specific ratio to fuel automobiles. For example, India aims to mix 20% of ethanol with gasoline by 2025 (Sarwal et al., 2021), compared to Brazil's requirement of 27% (BRAZIL: FUELS: BIOFUELS, 2021).

It will take several million years to rebuild the depleted oil and natural gas supplies. Thus it is vital to turn to carbon-based materials that can produce energy quickly and continuously. One such feedstock that satisfies the criteria is lignocellulosic biomass (LCB) (Popp et al., 2021). All types of plant and animal materials may be classified as biomass, with plant biomass being the most prevalent (Woiciechowski et al., 2020). Although there are many choices for using biomass to produce energy, it's critical to identify those biomasses that are both economical and provide a sufficient and reliable quantity of energy to be able to replace energy from fossil fuels. One of the primary issues in the production of energy and transportation is cost (Sarkar et al., 2021).

Waste biomass and residues are byproducts of the processing and consumption of raw materials, as opposed to biomass that is generated particularly for energy generation. Using waste biomass in energy production adds value to biomass leftovers and increases the profitability of the energy production process since waste biomass, such as agricultural and agro-industrial wastes, is plentiful and, if improperly disposed of, causes environmental difficulties. Agro-biomass, also known as agricultural lignocellulose waste biomass has recently attracted attention since it is produced in vast quantities each year and has a high calorific value (Abdeshahian et al., 2020). Lignocellulosic plant biomass, which is widely accessible, is thought to be an upcoming feedstock for the production of biofuels (Parakh et al., 2020). Plants are comprised of lignocellulose, which contains cellulose, hemicellulose, and lignin as their primary constituents and forms a complex structure that is highly recalcitrant (Deshavath et al., 2019). Agricultural LCB production relies on several variables, including seasonal availability, composition, pretreatment, and fermentation techniques.

1.2 Lignocellulosic biomass: an ideal candidate feedstock for biofuels

The imbalance between global supply and demand for energy resources has raised concerns on a global scale. Utilizing LCB as an energy source might be a great choice in this situation. However, different kinds of biomass have not been fully used for energy recovery. Despite its widespread availability, waste biomass is one kind of LCB that has not been used for energy recovery. On the other hand, improper management of waste biomass disposal led to a wide range of environmental issues. Despite the billion tons of agricultural leftovers that are produced yearly across the globe, most of them are either burnt in open fields or dumped as garbage. Depending on their composition, waste biomasses may be divided into woody and nonwoody biomass (Roy et al., 2022).

The three primary components of LCB are cellulose, hemicellulose, and lignin (Zadeh et al., 2020). Fig. 1.1 provides a broad representation of the LCB structure where lignin encases the sugar polymers while hemicellulose is woven between the cellulose chains. More details on the composition of LCB are given in Chapter 2 of this book.

Cellulose is a linear polymer made up of chains of β-(1bis4)-glucopyranose units. It is an unbranched polymer. Cellulose crystallinity has an impact on the enzymatic digestibility of LCB because highly organized regions are physically more compact and inherently resistant to enzymes, acids, and swelling underwater. Studies have demonstrated that amorphous cellulose might hydrolyze more rapidly

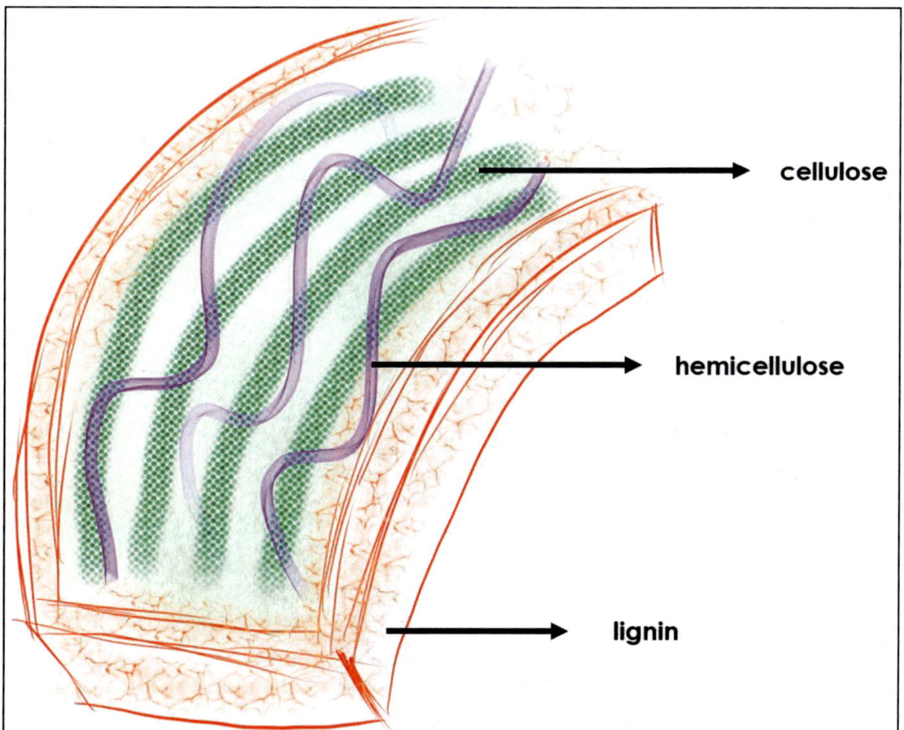

Figure 1.1 General lignocellulosic structure of biomass.

and at a faster rate than crystalline cellulose (Ravindran and Jaiswal, 2016). The degree of polymerization of cellulose also reveals the biomass's resistance. The rate of cellulose hydrolysis increases with the length of the cellulose chain. One of the first stages involved in the conversion of biomass to biofuels is pretreatment.

1.2.1 Pretreatment

Owing to the recalcitrant nature of LCB, pretreatment has become synonymous with lignocellulosic processes. The primary goal of pretreatment is to destabilize and disrupt the complex matrix, removing lignin and enhancing access to the sugar-rich polymers of cellulose and hemicellulose. Pretreatment can be broadly categorized into either physical, chemical, or biological strategies while a combination of these options has routinely been employed as well, as illustrated in Fig. 1.2. Chemical methods such as acid and alkali are the most commonly employed, albeit coupled with many disadvantages. These include high cost and toxicity, environmental concerns, production of inhibitors, and degradation of sugars. For this reason, the development of newer, more environmentally-friendly technologies is being pursued, as illustrated in Table 1.1. Given the integral role that pretreatment plays in the biofuel production process, selecting and optimizing a pretreatment is critical to the overall outlook of the process (Moodley et al., 2020).

1.2.2 Bioethanol

The world's biggest producer of gasoline-ethanol is the United States. In 2020, it was projected that the United States would generate 13.8 billion gallons of ethanol.

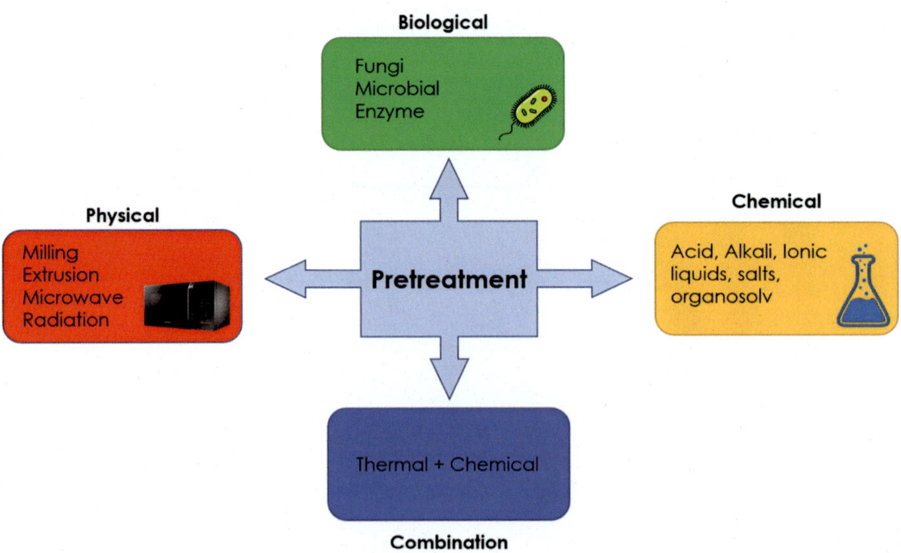

Figure 1.2 Overview of commonly considered pretreatment methods.

Table 1.1 Recent developments in pretreatment strategies.

Pretreatment strategy	Feedstock	Key finding	References
Kraft waste	Corn cobs	Glucose yield of 0.85 g/g	David et al. (2021)
Surfactant-assisted green liquor dregs	Paper waste	16.38 g/L reducing sugar obtained	Rorke et al. (2021)
Microwave-assisted alkali	Bamboo and corn cobs	0.512 g/g reducing sugar yield. High recyclability	Laltha et al. (2021)
Sequential alkalic salt-dilute acid	Corn cobs	56.17% delignification Reducing sugar of 0.99 g/g observed	Sewsynker-Sukai et al. (2018)
Microwave-assisted inorganic salt	Sugarcane leaf waste	Reducing sugar yield of 0.406 g/g observed	Moodley and Gueguim Kana (2017)
Inorganic salt hydrate	Cellulose	Severe physical damage to the structure. High reusability	Sun et al. (2022)
Molten salt hydrate	Corn stover	88.4% cellulose recovery, 75.3% glucose yield	Guan et al. (2021)

In conventional cars, ethanol or grain alcohol may be mixed up to 10% with gasoline (Sonnichsen, 2021). India Glycols Ltd. in Kashipur, Uttarakhand, has constructed a second-generation ethanol demonstration facility for the first time using a range of feedstocks, such as cotton stalk, sugarcane bagasse, and rice straw. On the National Institute for Interdisciplinary Science and Technology campus, Thiruvananthapuram, India, a prototype plant with a feeding capacity of 50 kg/day was built as part of continuing efforts to create second-generation ethanol (Sukumaran et al., 2010). China was Asia's biggest ethanol producer, with 2.1 billion liters of gasoline-ethanol produced in 2012. China has long developed ethanol for use as a vehicle fuel. To decrease its reliance on oil, China plans to promote ethanol-based fuel and establish a new market for its excess grain. In China, corn, cassava, and rice make up the majority of the sources of grain-based bioethanol, making up around 80% of the nation's overall output.

Most of the ethanol produced in the United States is made from starch-based crops that are either processed in dry or wet mills. It was estimated that by 2020 there would be 208 ethanol production facilities in the United States, with an installed capacity of 17.44 billion gallons annually. The total output in 2020 was 13.8 billion gallons, a 12.7% decrease from the record-breaking 15.8 billion gallons produced the year before (Renewable Fuels Association, 2021). In 2020, Canada was ranked as the sixth-largest bioethanol producer worldwide and contributed 1.6% to the world's ethanol production. Corn and wheat were the two most significant feedstocks for manufacturing bioethanol, each contributing 1534.3 million liters and 360.7 million liters, respectively (Renewables Fuels Association, 2020). In 2020, Brazil's ethanol output was anticipated to make up 26.7% of global ethanol production. On the other hand, Argentina produced 1.0% of the world's ethanol production in 2020, placing it ninth in the globe (New Energy Blue, 2021). Colombia produced 0.44% of the total ethanol produced worldwide in 2020, ranking it 13th in the world (Renewables Fuels Association, 2022). The entire amount of ethanol produced in the EU in 2020 made up 4.8% of global production. Due to the country's expanding economy, Vietnam has recently used a substantial volume of gasoline to manufacture biofuels. The United States Department of Agriculture reports that Vietnam's gasoline consumption has increased by 4%–5% in recent years. In addition to its output, Vietnam is anticipated to have spent $2.5 billion in the first quarter of 2020 on crude oil and petroleum products, including bioethanol. Among the top ethanol producers in the world, India is placed in the sixth position. The country's average ethanol blending rate in gasoline was anticipated to reach 7.5% in 2021 due to increased government initiatives to divert more raw materials toward ethanol. Therefore India's ethanol production was projected to reach 3.17 billion liters in 2021, up 7% from 2020, due to excess sugarcane production (India Biofuels Annual Report, 2020).

While there have been some great strides toward developing second-generation bioethanol processes, there are still several bottlenecks facing the commercialization of bioethanol production processes from LCB. Some of these include identifying the best feedstock, efficient and cost-effective pretreatment strategy, employing highly efficient enzymes to hydrolyze the feedstock, selecting a suitable microbial strain, optimizing the fermentation process, and furthering the downstream processing. Once these shortfalls are addressed in an integrated manner, the process in its entirety can potentially transition towards a commercial scale.

1.2.3 Biohydrogen

By using biological processes, hydrogen generation is fundamentally environmentally beneficial and sustainable. Hydrogen may be produced from a wide variety of feedstocks, including household trash, industrial effluents, agricultural residue, municipal solid waste, and even water. Direct biophotolysis, indirect biophotolysis, photo-fermentation, dark fermentation, and microbial electrolysis are the typical methods of producing biohydrogen.

Dark fermentation is one of the most promising clean technologies for producing biohydrogen since it can convert a variety of organic waste into biohydrogen under moderate fermentation conditions. However, its poor biohydrogen yields limit dark fermentation's commercial use. Only a portion of the substrates may be converted to biohydrogen throughout the dark fermentation process, and the majority of them (60%–70%) are still present as volatile fatty acids and alcohols. The highest biohydrogen output, where acetic acid is the sole byproduct, is 4 mol/mol glucose. According to reports, the production of biohydrogen by dark fermentation was often only 1–3 mol/mol glucose (Sekoai et al., 2020). So, in recent years, research has been conducted to improve biohydrogen generation in the dark fermentation process using various cutting-edge technologies, such as innovative microbial culture selection, genetic engineering, cell immobilization, and nanotechnology. Consequently, there is a huge opportunity to research the expansion of these areas.

Briefly, investigations into microbial selection continue to be extensively studied to enhance biohydrogen yields. For instance, a study by Nizzy et al. (2020) discovered that it is possible to produce biohydrogen from sago industrial wastewater using a newly isolated pure culture of *Enterobacter cloacae* NASGE 02 and *Clostridium sartagoforme* NASGE01. With a maximal biohydrogen production of 1.26 mol H_2/mol glucose, *C. sartagoforme* NASGE 01 was capable of degrading up to 56.7% of the substrate. Additionally, novel microbial species (*Bacillus coagulans* MO11 and *Clostridium beijerinckii* CN) with efficient biohydrogen production capacity from molasses and ethanol refinery effluent were successfully screened by Lertsriwong and Glinwong (2020). In a recent study, two pure *Bacillus cereus* strains (*B. cereus* RTUA and RTUB strains) capable of producing several enzymes were isolated from an anaerobic digester and shown to be prospective candidates for biohydrogen generation from various substrates (Saleem et al., 2020). However, it is challenging to maintain a pure culture free of contamination because of different toxins from biowaste and wastewater.

Nanotechnology is another field that has shown immense promise in biohydrogen production. According to Seelert et al. (2015), *Clostridium beijerinckii* NCIMB8052 was immobilized on magnetite nanoparticles, which decreased the lag period of microbial growth and later increased the generation yield of biohydrogen. When nanoparticles are combined with carbon-based materials like activated carbon and biochar, the activity of microorganisms and enzymes may be enhanced synergistically (Yang and Wang, 2019). Due to their complementary roles and the release of additional Fe^{2+} from the Fe^0 nanoparticle-biochar microelectrolysis, biochar and Fe^0 nanoparticles together had the predicted synergetic impact on the improvement of biohydrogen generation from grass fermentation (Yang and Wang, 2019).

1.2.4 Sustainable aviation fuel

Lower carbon alternatives may be produced from LCB sources (agricultural residues, waste organics, etc.) and nonfossil derived CO_2 materials as diverse feedstocks for creating biokerosene, sustainable aviation fuel (SAF) has lately gained a lot of interest. Common organic chemical components of jet fuel include paraffin, aromatics, olefins, and naphthenes. Triglyceride-based sources such as vegetable oil used cooking oil, and animal fats are promising feedstocks to manufacture diesel equivalent fuel since commercial-level SAF production is often constrained. The SAF meets the requirements set out by the American Society for Testing and Materials (ASTM) for jet fuel standards to be approved as a drop-in and distillate fuel. For instance, D1655 is a jet range fuel having synthetic and conventional mix components, while D7566 is an aircraft turbine fuel. Furthermore, according to the Sustainable Development Scenario of the International Energy Agency (IEA), biofuel production should be able to cover the energy needs of 10% of aviation fuel by 2030 and 20% by 2040. Catalytic hydrothermolysis, the sixth process for SAF production as certified by the ASTM in 2020, is one of the most effective techniques found. However, Biokerosene is often produced as the primary feedstock for aircraft turbine fuel by the conversion of a range of biogenic wastes and hydrotreated vegetable oil (Eswaran et al., 2021). Finding alternate feedstock materials and different biorefining processes has received a lot of interest due to the difficulties associated with adopting the food versus fuel idea (Ghatalam, 2020). The two probable methods for producing biokerosene from lignocellulose are as follows:

1. Hydrolyzing the feedstock directly to produce sugars, followed by the production of biokerosene.
2. Thermochemically treating the feedstock to produce intermediate products, which are subsequently upgraded to biokerosene.

However, compared to traditional fossil fuels, the high cost of biokerosene further hinders product development for large-scale commercial use (Thilo, 2018). These problems might be resolved by establishing effective intergovernmental agreements to lower tariffs related to the supply of biofuels and by creating alternative pathways for the manufacture of biokerosene, coupled with useful byproducts that can result in practical affordability.

1.2.5 Biogas

Anaerobic processing of LCB can lead to the production of biogas. However, hydrolysis is a relatively slow process for lignocellulosic materials and is affected by the type and composition of the feedstock. Hydrolysis is the rate-limiting phase for lignocellulosic material that breaks down the complex organic polymer components during anaerobic digestion (AD) (Dahadha et al., 2017). In recent years, there has been a lot of research done on the biogas produced by the AD of lignocellulosic materials. For instance, Khan and Ahring (2019) employed the leftovers from the enzymatic hydrolysis and wet explosion pretreatment steps that remove carbohydrates. The findings demonstrated that lignin from biorefineries might be a suitable substrate for biogas generation, with the greatest severity pretreatment producing the maximum methane output. In a relatively recent paper, modeling, and experiments were employed for the first time to suggest a potential cleaner thermochemical conversion of LCB to biomethane (Yun et al., 2019). Predicted values of CH_4, H_2, and plant thermal efficiency were equivalent to modeled efficiencies reported in the literature for wood gasification based on the autothermal plant's simulation findings. A study exploring the potential for converting waste products from bioethanol manufacturing into gaseous biofuels in a biorefinery setting demonstrated that waste from bioethanol has a high energetic value and may help boost the total energy produced by biomass (Rocha-Meneses et al., 2019).

The variability and low density of lignocellulosic materials further hinder the AD process by causing a floating layer to develop on the surface of AD reactors (Tian et al., 2015). As a result, the microorganisms have difficulty accessing the substrate, which worsens mass and heat transmission and lowers the methane output (Wang et al., 2018). High carbon-to-nitrogen (C/N) ratios have been shown to restrict the amount of LCB that can be digested (Sawatdeenarunat et al., 2015). However, this may be overcome by adding additional nitrogen sources or co-digesting with the substrate that contains more nitrogen. Increased biogas output may offset these additional expenses, making LCB a profitable AD feed for methane production.

1.3 Biorefineries

A single definition of the phrase “biorefinery” is challenging to come up with since it encompasses a wide range of industries, including those in the transportation, chemical, energy, agricultural, and forestry sectors. Biorefining, for example, is described by the International Energy Agency (IEA) as “the sustainable processing of biomass into a spectrum of marketable bio-based products and bioenergy/biofuels, as an innovative and efficient approach to use available biomass resources for the synergistic co-production of power, heat, and biofuels alongside food and feed ingredients, pharmaceuticals, chemicals, materials, minerals, and short-cycle CO_2” (Lindorfer et al., 2019). The concept of a biorefinery is not new, and there are now several industrial facilities in use that may be seen as exemplary examples of actual biorefineries. Most of them are associated with food, pulp, and paper and likely constitute the first examples of technical applications to transform biomass into valuable goods. In the so-called “product-driven” biorefineries, these facilities are concentrated on the manufacture of relatively limited quantities of higher value-added bio-based goods, and they typically generate energy (bioenergy) as a byproduct (ETIP, 2017). However, in recent years the idea of a biorefinery has expanded to include a broader range of feedstocks, processing methods, and end products to maximize the sustainable and efficient use of all biomass components. Utilizing a broad variety of biomasses (including woody and herbaceous energy crops, byproducts from forestry and agricultural sectors, urban garbage, and aquaculture) to create a variety of products and energy is a crucial component of the advanced biorefinery method. Depending on the feedstock and product needed, LCB may go through several processing methods in a biorefinery. Numerous technologies must often be used simultaneously or successively to create the desired result. Thermochemical, biochemical, chemical, and mechanical/physical processes can be used to transform LCB into a wide range of products, including materials, chemicals, organic acids, polymers and resins, and biopolymers, as well as energy-related products (gaseous, solid, and liquid biofuels). Cellulosic ethanol is one of the most significant examples of the advancements made in this field. Currently, it is close to full commercialization. Over the past few decades, a significant portion of the research and development effort carried out in the field of LCB has been focused on fuels and other energy-related products. However, it is also recognized that its manufacturing costs are still too high to be competitive, necessitating the search for creative solutions that lower costs. Recent developments in biocatalysts and microbial strains used in the conversion pathway, applied through the use of tools like high-throughput screening, metagenomics techniques, metabolic

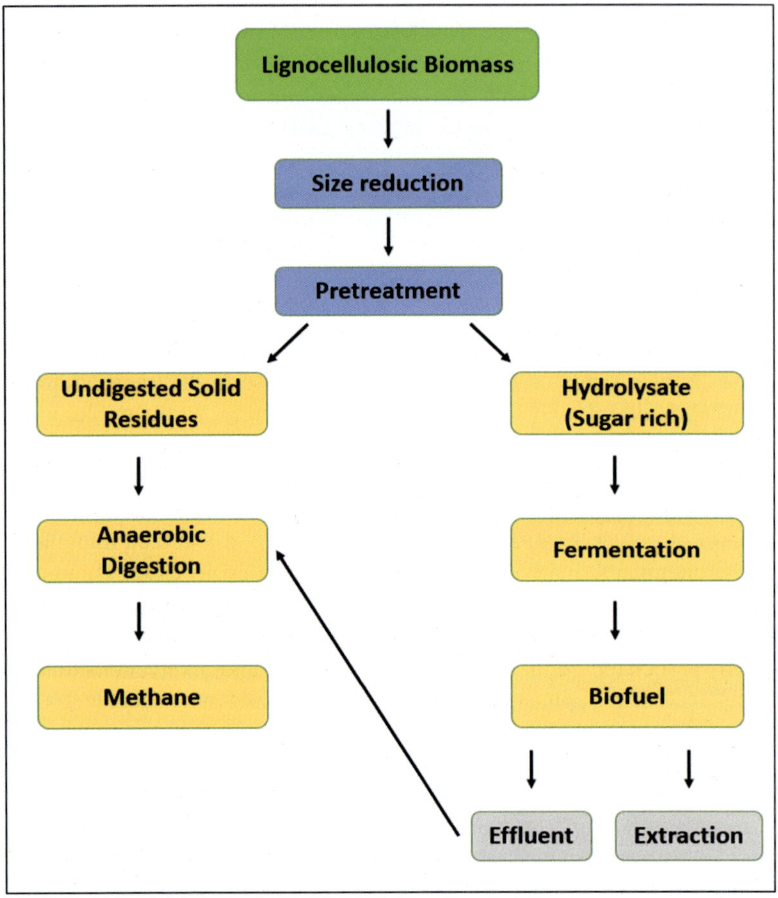

Figure 1.3 The general flow of bioconversion of lignocellulosic biomass.

engineering, and synthetic biology, may enhance the potential for increased production and contribute to reaching a more competitive product from an economic point of view. Such cutting-edge technology has also made it possible to enhance the manufacturing of bio-based goods from lignocellulosic feedstocks meant to replace their equivalents produced by fossil fuels (Duque et al., 2021). A general LCB is illustrated in Fig. 1.3.

1.4 Genetic engineering of feedstocks and fermenting microorganisms

Major contributors to the generation of biofuel are microorganisms. However, the native strains' inadequate yield of the product makes it essential to modify and enhance them using the strategies of metabolic engineering and genetic engineering. To maximize levels of productivity and energy value at a lower cost of production, recent research has concentrated on using metabolic engineering to analyze strain development. There is a good chance that database mining will provide more distinct metabolic pathways for the synthesis of biofuels shortly. Therefore any obstacles to using LCBs as renewable fermentation feedstock may be removed by implementing these routes in commercial fermentation hosts. By inserting useful genes into the genome or deleting obstructive ones, metabolic engineers need to take advantage of the most cutting-edge technologies currently available, such as the omic technologies and CRISPER/Cas9 system, to design and create novel strains of microbial species with the enhanced ability to produce biofuels from a variety of feedstocks (Adegboye et al., 2021).

Through the use of diverse breeding techniques, progress has also been achieved in addressing the engineering of lignocellulosic feedstocks, specifically cell wall recalcitrance, improving biomass output and the quality of bioenergy crops. The methods primarily aim to modify the composition of plant cell walls. Various biofuel crops have been improved using a variety of plant breeding strategies, including molecular and marker-assisted selection, genetic engineering, and resistance to biotic and abiotic challenges. There have been produced crops that can withstand a variety of biotic and abiotic challenges, such as pests, weeds, diseases, drought, heat, and salt (Wang et al., 2016). Recombinant DNA technology has been a key technique in genetic engineering's alteration of plant cell walls to overcome biomass recalcitrance, leading to improved saccharification and ultimately higher yields of bioethanol. To increase feedstock output and quality in various plants, genetic modification efforts have focused on changing the structure of enzymes and overexpressing or silencing genes involved in various processes. Gene discovery will continue to be a crucial task for genetic modification using both conventional breeding methods and genetic engineering since several genes are known to affect various processes in the production, breakdown, and regulation of the plant cell wall. Modern biotechnological technologies that enable comparative genomic investigations of the transfer of genes across distantly related species may also be used for gene discovery (Wang et al., 2021).

1.5 Artificial intelligence in biofuel production

Artificial intelligence (AI) technology has advanced in recent years to the point where researchers are starting to pay more attention to it, especially in the field of bioenergy and biofuel research. Given the significance of AI in this particular area of bioenergy, many AI technologies for improving current systems and solving issues in bioenergy production have been discussed. For effective use of resources, environmental awareness, and efficient production from natural resources, AI offers numerous sophisticated mechanisms in bioenergy production technologies

(Williams et al., 2015). Similar to other sectors, bioenergy might use AI to assist its long-term goals. AI offers a variety of optimization tools, like the ASPEN Plus simulator (Okolie et al., 2021) and artificial neural network (ANN)-genetic algorithm (GA)-based modeling (Liao and Yao, 2021), which makes it simple to examine the economics of different types of bio-oil and syngas production based plants. The user of AI-based optimization tools may also benefit from assistance with economic forecasting and the large-scale commercialization of bioenergy conversion technology. With the use of AI, risks or disruptive events that may potentially occur across the supply chain can be quickly assessed and more effectively minimized. A bioenergy production facility's whole supply chain management is heavily reliant on resource evaluation, logistical planning, and power plant design. By keeping a multiscale feedstock database with information on the characteristics, availability, demand, and logistics of feedstocks, among other things, AI provides various significant applications in the management of supply chains.

1.6 Bioreactor configuration for enhanced biofuel processes

A favorable environment for microbial growth, biocatalysis, the generation of microbial metabolites, and energy conversion is provided by bioreactors. Due to its potential for simple operation, improved reaction parameter control, sustainability, low input of raw materials and energy cost, and maximum carbon footprint conversion, bioreactor technology has attracted a lot of interest in the process development of microbial cultivations and biofuel conversion. For fermentation to proceed efficiently, sugar must be transformed into a useful product. As a result, a crucial factor that depends on the kind of substrate that will be fermented into the desired product is microbial selection. These bacteria convert C6 or C5 carbohydrates into biofuels in a bioreactor under regulated conditions.

For the large-scale production of renewable and sustainable biofuels, energy-efficient reactors that provide a stable and regulated biological environment are crucial, and their optimization focuses on lowering energy consumption and waste gas emissions. Photobioreactors were created for the cultivation of algae or microalgae, and the bioreactors might be either anaerobic or aerobic (Xu et al., 2018). Various modeling techniques for bioreactor design have been developed due to the high expense of manufacturing large volumes of bioreactors. The development of super microbial factories with maximum productivity and metabolic pathway flux is also essential to the development of the optimal biofuel reactor (Keasling et al., 2021). Every biological process requires the use of bioreactors, and they are even thought of as the center of bioprocesses. The structures in which a reaction or biological process takes place are called bioreactors. To ensure a high output of the bioprocess and to meet the needs of the biological system (enzyme, microbe, or cell), the bioreactor must offer the ideal conditions.

1.7 Current status of global energy recovery from lignocelluloses

Biomass has historically been a significant source of energy. The biomass sector has substantial, regional, and international impacts on society and the environment. Numerous disputes and problems have been linked to the use of biomass as an energy source. There is a lot of promise for balancing global trade imbalances, assuring global energy security, and ensuring global food security using a modern biofuel manufacturing technique that extracts energy from biomass. Several efforts are now underway globally to transition the present energy supply system from traditional to contemporary biomass conversion. Renewable Energy Statistics (2021) states that the capacity and output of renewable energy for 2020 was 2799094 MW. Hydropower (1331889 MW) was the most significant energy source, followed by solid biofuels and renewable trash (102852 MW). The leading nations in terms of biofuel capacity are North America (623.000 barrels of oil equivalent per day), South and Central America (446,000 barrels of oil equivalent per day), Asia Pacific (310,000 barrels of oil equivalent per day), and Europe (295,000 barrels of oil equivalent per day). Between 2007 and 2017, the output of biofuels grew by 11.4% worldwide. The IEA reports that the output of biofuels globally hit a record 154 billion liters in 2018. Forecasts indicate that it will expand by 25% by 2024, with an average annual growth rate of 3%. In addition to financial and political concerns, the bioenergy industry faces several obstacles and uncertainties, particularly concerning changes in the price of crude oil. Additionally, it has been challenging to commercialize advanced biofuels due to technical challenges. Despite these obstacles, the biofuel business is still expanding, which adds to the rise in world energy demand. About 2.8 million employment opportunities have been made possible by bioenergy, suggesting its significance for job creation. All of these indicate that lignocellulosic biofuel systems have been advancing at a precipitous rate; however, more research is required to bridge the gap between experimentation and commercialization.

1.8 Conclusion and future perspectives

The potential for producing biofuels effectively has been established by recent advances in lignocellulose bioconversion processes. With the use of contemporary synthetic biology technology, it is possible to synthesize biofuel feedstock that is not naturally generated. In any case, the use of genetic and metabolic engineering will be regarded as an important and cutting-edge strategy that modifies and examines metabolic pathways to improve the production of biofuels. Future research should concentrate on low-cost pretreatment methods, metabolic engineering approaches, and modeling for large-scale commercial lignocellulose biofuel production. Additionally, more effective methods must be created to completely remove the refractory characteristic and maximize the usage of lignocellulose components in fuel production. All of these mentioned fields as well as other areas for improvement are presented and discussed in further detail in this book.

References

Abdeshahian, P., Kadier, A., Rai, P.K., da Silva, S.S., 2020. Lignocellulose as a renewable carbon source for microbial synthesis of different enzymes. In: A. P. Ingle, A.K. Chandel And S. S. Da Silva (Eds.), Lignocellulosic Biorefining Technologies, Wiley, pp. 185–202.

Adegboye, M.F., Ojuederie, O.B., Talia, P.M., et al., 2021. Bioprospecting of microbial strains for biofuel production: metabolic engineering, applications, and challenges. Biotechnol. Biofuels 14, 5. Available from: https://doi.org/10.1186/s13068-020-01853-2.

BRAZIL: FUELS: BIOFUELS, 2021. Transportpolicy.net. <https://www.transportpolicy.net/standard/brazil-fuels-biofuels/> (accessed 10.11.21).

Dahadha, S., Amin, Z., Bazyar Lakeh, A.A., Elbeshbishy, E., 2017. Evaluation of different pretreatment processes of lignocellulosic biomass for enhanced biomethane production. Energy Fuels 31, 10335–10347.

David, A.N., Sewsynker-Sukai, Y., Kana, E.G., 2021. Development of Kraft waste-based pretreatment strategies for enhanced sugar recovery from lignocellulosic waste. Ind.l Crop. Prod. 174, 114222.

Deshavath, N.N., Veeranki, V.D., Goud, V.V., 2019. Chapter 1 - lignocellulosic feedstocks for the production of bioethanol: availability, structure, and composition. Sustain. Bioenergy 1–19.

Duque, A., Álvarez, C., Doménech, P., Manzanares, P., Moreno, A.D., 2021. Advanced bioethanol production: from novel raw materials to integrated biorefineries. Processes 9 (2), 206. Available from: https://doi.org/10.3390/pr9020206.

Eswaran, S., Subramaniam, S., Geleynse, S., Brandt, K., Wolcott, M., Zhang, X., 2021. Techno-economic analysis of catalytic hydrothermolysis pathway for jet fuel production. Renew. Sustain. Energy Rev. 151, 111516. Available from: https://doi.org/10.1016/j.rser.2021.111516.

ETIP, 2017. Biorefinery Concepts; European Technology and Innovation Platform Bioenergy: Gülzow, Germany. <https://etipbioenergy.eu/images/ETIP_Bioenergy_Factsheet_Biorefinery_concepts.pdf> (accessed 16.05.22).

Ghatalam, F., 2020. Sustainable aviation fuel policy in the United States: a pragmatic way forward. Atlantic Council, Global Energy Center.

Guan, M., Liu, Q., Xin, H., Jiang, E., Ma, Q., 2021. Enhanced glucose production from cellulose and corn stover hydrolysis by molten salt hydrates pretreatment. Fuel Process. Technol. 215, 106739.

India Biofuels Annual Report,, 2020. Global agricultural information network; GAIN report number IN2021-0072. Foreign Agricultural Service, U.S. Department of Agriculture. Washington, DC, USA.

Kazemi Shariat Panahi, H., Dehhaghi, M., Kinder, J.E., Ezeji, T.C., 2019. A review on green liquid fuels for the transportation sector: a prospect of microbial solutions to climate change. Biofuel Res. J. 6, 995–1024.

Keasling, J., Martin, H.G., Lee, T.S., Mukhopadhyay, A., Singer, S.W., Sundstrom, E., 2021. Microbial production of advanced biofuels. Nat. Rev. Microbiol. 19, 701–715.

Khan, M.U., Ahring, B.K., 2019. Anaerobic digestion of biorefinery lignin: effect of different wet explosion pretreatment conditions. Bioresour. Technol. 298, 122537.

Laltha, M., Sewsynker-Sukai, Y., EB, G.K., 2021. Development of microwave-assisted alkaline pretreatment methods for enhanced sugar recovery from bamboo and corn cobs: process optimization, chemical recyclability and kinetics of bioethanol production. Ind. Crop. Prod. 174, 114166.

Lertsriwong, S., Glinwong, C., 2020. Newly-isolated hydrogen-producing bacteria and biohydrogen production by *Bacillus coagulans* MO11 and *Clostridium beijerinckii* CN on molasses and agricultural wastewater. Int. J. Hydrog. Energy 45, 26812–26821.

Liao, M., Yao, Y., 2021. Applications of artificial intelligence-based modeling for bioenergy systems: A review. GCB Bioenergy 13, 774–802.

Lindorfer, J.; Lettner, M.; Fazeni, K.; Rosenfeld, D.; Annevelink, B.; Mandl, M. 2019. Technical, economic and environmental assessment of biorefinery concepts; IEA Bioenergy: Paris, France. <https://www.ieabioenergy.com/wp-content/uploads/2019/07/TEE_assessment_report_final_20190704-1.pdf> (accessed 15.1121).

Moodley, P., Kana, E.G., 2017. Microwave-assisted inorganic salt pretreatment of sugarcane leaf waste: effect on physiochemical structure and enzymatic saccharification. Bioresour. Technol. 235, 35–42.

Moodley, P., Sewsynker-Sukai, Y., Kana, E.G., 2020. Progress in the development of alkali and metal salt catalysed lignocellulosic pretreatment regimes: potential for bioethanol production. Biores. Technol. 310, 123372. Available from: https://doi.org/10.1016/j.biortech.2020.123372. Epub 2020 Apr 13.

New Energy Blue to Construct Cellulosic Biorefinery in Iowa. <http://biomassmagazine.com/articles/18180/new-energy-blue-to-construct-cellulosic-biorefinery-in-iowa/> (accessed 12.12.21).

Nizzy, A.M., Kannan, S., Anand, S.B., 2020. Identification of hydrogen gas producing anaerobic bacteria isolated from sago industrial effluent. Curr. Microbiol. 77, 2544–2553.

Okolie, J.A., Nanda, S., Dalai, A.K., Kozinski, J.A., 2021. Techno-economic evaluation and sensitivity analysis of a conceptual design for supercritical water gasification of soybean straw to produce hydrogen. Biores. Technol. 331, 125005.

Parakh, P.D., Nanda, S., Kozinski, J.A., 2020. Eco-friendly transformation of waste biomass to biofuels. Curr. Biochem. Eng 6, 120–134.

Popp, J., Kovács, S., Oláh, J., Divéki, Z., Balázs, E., 2021. Bioeconomy: biomass and biomass-based energy supply and demand. N. Biotechnol 60, 76–84.

Ravindran, R., Jaiswal, A.K., 2016. A comprehensive review on pretreatment strategy for lignocellulosic food industry waste: challenges and opportunities. Bioresour. Technol. 199, 92–102.

Renewable Fuels Association. 2020. Ethanol Industry Outlook. <https://ethanolrfa.org/file/21/2020-OutlookFinal-for-Website.pdf/> (accessed 24.05.22).

Renewable Fuels Association. 2021. Ethanol Industry Outlook. <https://ethanolrfa.org/wp-content/uploads/2021/02/RFA_Outlook_2021_fin_low.pdf/> (accessed 15.03.22).

Renewable Fuels Association Analysis of Public and Private Data. <https://ethanolrfa.org/markets-andstatistics/annual-ethanol-production> (accessed 24.04.22).

Rocha-Meneses, L., Ivanova, A., Atouguia, G., Ávila, I., Raud, M., Orupõld, K., et al., 2019. The effect of flue gas explosive decompression pretreatment on methane recovery from bioethanol production waste. Ind. Crop Prod. 127, 66–72.

Rorke, D.C., Lekha, P., Kana, G.E., Sithole, B.B., 2021. Surfactant-assisted green liquor dregs pretreatment to enhance the digestibility of paper mill sludge. Int. J. Hydrog. Energy 46, 21359–21371.

Roy, R., Debnath, D., Ray, S., 2022. Comprehensive assessment of various lignocellulosic biomasses for energy recovery in a hybrid energy system. Arab. J. Sci. Eng. 47, 5935–5948.

Saleem, A., Umar, H., Shah, T.A., Tabassum, R., 2020. Innovation fermentation of simple and complex substrates to biohydrogen using pure *Bacillus cereus* RTUA and RTUB strains. Environ. Technol. 18, 100704.

Sarkar, B., Mridha, B., Pareek, S., Sarkar, M., Thangavelu, L., 2021. A flexible biofuel and bioenergy production system with transportation disruption under a sustainable supply chain network. J. Clean. Prod. 317, 128079.

Sarwal, R., Kumar, S., Mehta, A., Varadan, A., Singh, S.K., Ramakumar, S.S.V., et al., 2021. Roadmap for ethanol blending in India 2020-25. OSF Prepr 1–72. Available from: https://doi.org/10.31219/osf.io/825mz.

Sawatdeenarunat, C., Surendra, K.C., Takara, D., Oechsner, H., Khanal, S.K., 2015. Anaerobic digestion of lignocellulosic biomass: challenges and opportunities. Bioresour. Technol 178, 178–186.

Sekoai, P.T., Daramola, M.O., Mogwase, B., Engelbrecht, N., Yoro, K.O., du Preez, S.P., et al., 2020. Revising the dark fermentative H_2 research and development scenario—an overview of the recent advances and emerging technological approaches. Biomass Bioenerg 140, 105673.

Seelert, T., Ghosh, D., Yargeau, V., 2015. Improving biohydrogen production using *Clostridium beijerinckii* immobilized with magnetite nanoparticles. Appl. Microbiol. Biot 99, 4107–4116.

Sewsynker-Sukai, Y., Suinyuy, T.N., Kana, E.G., 2018. Development of a sequential alkalic salt and dilute acid pretreatment for enhanced sugar recovery from corn cobs. Energ. Convers. Manage. 160, 22–30.

Sönnichsen, N., 2021. Fuel ethanol production worldwide in 2020, by country. <https://www.statista.com/statistics/281606/ethanol-production-in-selected-countries/>.

Sukumaran, R.K., Surender, V.J., Sindhu, R., Binod, P., Janu, K.U., Sajna, K.V., et al., 2010. Lignocellulosic ethanol in India: prospects, challenges and feedstock availability. Biores. Technol. 101, 4826–4833.

Sun, L., Han, J., Wu, J., Huang, W., Li, Y., Mao, Y., et al., 2022. Cellulose pretreatment with inorganic salt hydrate: dissolution, regeneration, structure and morphology. Ind. Crop. Products 180, 114722.

Thilo, Z., 2018. New oil plants and their potential as feedstock for biokerosene production. In: Kaltschmitt, M., Neuling, U. (Eds.), Biokerosene: Status and Prospects. Springer Nature, Berlin, Heidelberg.

Tian, L., Zou, D., Yuan, H., Wang, L., Zhang, X., Li, X., 2015. Identifying proper agitation interval to prevent floating layers formation of corn stover and improve biogas production in anaerobic digestion. Bioresour. Technol. 186, 1–7.

Tursi, A., 2019. A review on biomass: importance, chemistry, classification, and conversion. Biofuel Res. J. 6, 962–979.

Wang, Y., Fan, C., Hu, H., Li, Y., Sun, D., Wang, Y., et al., 2016. Genetic modification of plant cell walls to enhance biomass yield and biofuel production in bioenergy crops. Biotechnol. Adv. 34, 997–1017.

Wang, D., Shen, F., Yang, G., Zhang, Y., Deng, S., Zhang, J., et al., 2018. Can hydrothermal pretreatment improve anaerobic digestion for biogas from lignocellulosic biomass? Bioresour. Technol. 249, 117–124.

Wang, Y., Liu, P., Zhang, G., Yang, Q., Lu, J., Xia, T., et al., 2021. Cascading of engineered bioenergy plants and fungi sustainable for low-cost bioethanol and high-value biomaterials under green-like biomass processing. Renew. Sustain. Energy Rev. 137, 110586.

Wei, T., Wu, J., Chen, S., 2021. Keeping track of greenhouse gas emission reduction progress and targets in 167 cities worldwide. Front. Sustain. Cities 3, 696381. Available from: https://doi.org/10.3389/frsc.2021.696381.

Williams, C.L., Dahiya, A., Porter, P., 2015. In: Dahiya, A. (Ed.), Bioenergy: Biomass to Biofuels. Academic Press, pp. 5–36.

Woiciechowski, A.L., Neto, C.J., de Souza Vandenberghe, L.P., de Carvalho Neto, D.P., Sydney, A.C., Letti, L.A., et al., 2020. Lignocellulosic biomass: acid and alkaline pretreatments and their effects on biomass recalcitrance–conventional processing and recent advances. Bioresour. Technol. 304, 122848.

Xu, K., Lv, B., Huo, Y.-X., Li, C., 2018. Toward the lowest energy consumption and emission in biofuel production: combination of ideal reactors and robust hosts. Curr. Opin. Biotechnol. 50, 19–24.

Yang, G., Wang, J., 2019. Synergistic enhancement of biohydrogen production from grass fermentation using biochar combined with zero-valent iron nanoparticles. Fuels 251, 420–427.

Yun, H.A.H., Ramírez-Solís, S., Dupont, V., 2019. Bio-CH_4 from palm empty fruit bunch *via* pyrolysis-direct methanation: full plant model and experiments with bio-oil surrogate. J. Clean. Prod. 2019, 1–12.

Zadeh, Z.E., Abdulkhani, A., Aboelazayem, O., Saha, B., 2020. Recent insights into lignocellulosic biomass pyrolysis: a critical review on pretreatment, characterization, and products upgrading. Process 8 (7), 799. Available from: https://doi.org/10.3390/pr8070799.

Section II

Feedstock & processing

Lignocellulosic biomass: A feedstock to support the circular economy

2

Balwinder Singh Sooch, Manpreet Kaur Mann and Sonampreet Kaur
Enzyme Biotechnology Laboratory, Department of Biotechnology, Punjabi University, Patiala, Punjab, India

2.1 Introduction

Lignocellulosic biomass (LCB) is the world's widely available sugar-enriched sustainable feedstock and has an extensive role in producing biofuels. The processing of sustainable biomass is a global challenge that meets the economy's and the environment's demands. In the era of massive depletion of fossil fuels, the production of biofuels through any sustainable approach as an alternative to renewable resources is extensively challenging (Ray, 2021). The continuous production or the supply of biomass through the biological routes for saccharification leads to a high yield of sugars (Singhvi and Gokhale, 2019). The supply and the use of liquid biofuels increased from 1% (in the year 2000) to 6% in 2016. Biofuel production was 16 billion liters in the year 2000, which has increased to 143 billion in 2016, with 65% production of bioethanol, 25% production of biodiesel, and 10% production of other biofuels (Tursi, 2019). According to the Biofuels Renewables 2021 Analysis-IEA, biodiesel, renewable fuel, and biojet will all experience growth by 2021 which is much more than that of 2019, despite biojet having a modest starting point. The total market for these fuels is predicted to rise by 15%, or 7 billion liters, in 2021 compared to 2019 (Kumar and Kumar, 2022). India is assuming a 20% upsurge in ethanol demand by 2025 (Sarwal et al., 2021). Agrobiomass, also known as agricultural lignocellulose waste biomass, has recently gained popularity due to its high calorific value and large annual production (Hahn-Hägerdal et al., 2006; Abdeshahian et al., 2020). Lignocellulosic plant biomass, widely accessible worldwide, is thought to be a future source for biofuel production (Parakh et al., 2020). The building block of plants is lignocellulose, consisting primarily of cellulose, hemicellulose, and lignin in a complex structure that resists deterioration (Deshavath et al., 2019). Agricultural LCB production is influenced by several variables, including seasonality, availability, makeup, pretreatment methods, and fermentation techniques. The maximum yield of bioethanol production from biomass of rice, corn, and wheat has been reported by several researchers in the past (Ray and Ramachandran, 2018; Panahi et al., 2020). The present chapter focuses on producing biofuels using lignocellulosic feedstock through a circular economy to

Advances in Lignocellulosic Biofuel Production Systems. DOI: https://doi.org/10.1016/B978-0-323-91192-4.00009-2

achieve sustainability. A schematic illustration of circular economy through the conversion of lignocellulosic waste materials into biofuels is presented in Fig. 2.1.

2.2 Types and composition of lignocellulosic biomass

An understanding of the chemical and structural makeup of various lignocellulosic materials and their constituent parts is critical before moving forward with biofuel production from these materials. Numerous factors impede the formation of monomeric sugars from LCB and their subsequent conversion into bioethanol. Cellulose, hemicellulose, and lignin make up the structure of LCB. Additionally, LCB includes extractives, pectin, protein, and percentages of metal or ash (Marriott et al., 2016). The composition of the LCB is illustrated in Fig. 2.2.

Polymeric carbohydrates like cellulose and hemicellulose are hydrolyzed to various hexose and pentose sugars. In contrast, lignin, a three-dimensional methoxylated polyphenolic molecule, covers cellulose and hemicellulose predominantly (Zhou et al., 2011). Therefore, during the hydrolysis of LCB, lignin limits the depolymerization of cellulose and hemicellulose. Agricultural feedstock contains more hemicellulose than hardwood biomass, but the percentages of these LCB constituents vary by species. For example, hardwood biomass contains more cellulose than softwood biomass, whereas softwood biomass contains more hemicellulose (Saini et al., 2014). The abundances of cellulose, hemicelluloses, and lignin vary significantly between the various types of LCB materials as shown in Table 2.1. Genetic and environmental variables in plants affect greatly the composition of LCB. The

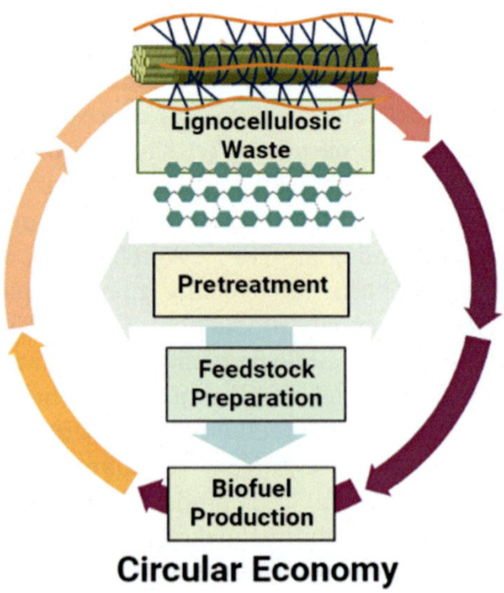

Figure 2.1 Schematic representation of circular economy through the conversion of lignocellulosic waste materials into biofuels.

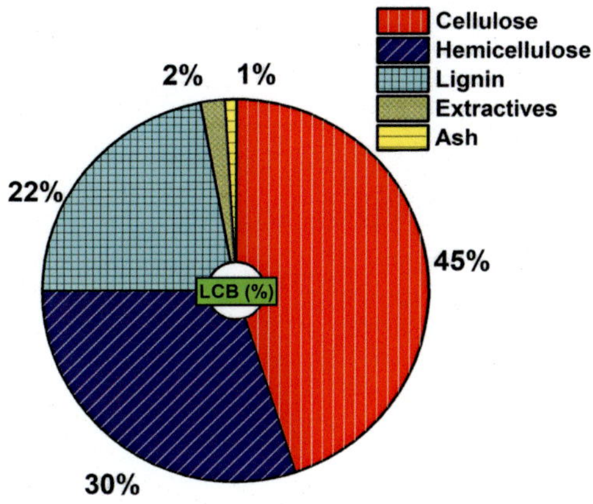

Figure 2.2 Composition of lignocellulosic materials.

varying genetic and environmental characteristics of plants impact the composition of LCB. A homopolysaccharide, cellulose $(C_6H_{10}O_5)n$, is made up of linear chains of D-glucose units connected by 1,4-glycosidic linkages. Strong hydrogen bonds connecting these chains cause the cellulose to crystallize naturally into microfibrils. Cellulose fibers are created by bundling these microfibrils. Cellulose is comprised of an amorphous (poorly ordered) portion that can be easily degraded and a crystalline (organized) part that is resistant to degradation. The cellulose fibers are in an amorphous matrix of lignin, pectin, and hemicelluloses. Lignin and hemicellulose are present between cellulose microfibrils in primary, secondary, and intermediate lamellae of cell walls (Hendriks and Zeeman, 2009; Singhania et al., 2010; Agbor et al., 2011). Fig. 2.3 elucidates the structural components of lignocellulosic biomass depicting cellulose, hemicellulose and lignin. Hexose sugars (D-mannose, D-glucose, and D-galactose) and pentose sugars (D-xylose and L-arabinose) make up branching heteropolymers known as hemicelluloses, with xylose being the most prevalent (Kumar et al., 2008; Juturu and Wu, 2012). The primary heteropolymers that make up hemicelluloses are xylan, mannan, arabinan, and galactan. The primary structural element of plant hemicelluloses is xylan, which is the second most prevalent renewable polysaccharide after cellulose in nature. Xylan is a complex polysaccharide with traces of L-arabinose and a backbone made of xylose residues linked together by 1,4-glycosidic linkage. A third of the world's renewable organic carbon is represented by xylan (Collins et al., 2005). The xylan layer may be crucial for preserving the integrity of cellulose in situ and safeguarding the cellulosic material due to its covalent interaction with lignin and noncovalent coupling with cellulose fibers resistant to cellulases degradation (Uffen, 1997). An aromatic polymer called lignin comprises phenyl propane units arranged in a substantial three-dimensional network structure. P-coumaryl alcohol, coniferyl alcohol, and sinapyl alcohol are connected by C−O−C and C−C links in the phenyl propane unit. The side chains of lignin also include methoxyl, phenolic, hydroxyl, and terminal aldehyde groups. Lignin fills the surrounding spaces between cellulose and hemicelluloses in

Table 2.1 Composition of different lignocellulosic biomass from agrowaste.

Lignocellulosic material	Lignin (%)	Hemicellulose (%)	Cellulose (%)	References
Banana waste	14	14.8	13.2	Sánchez (2009)
Barley straw	14–15	24–29	31–34	Naik et al. (2010)
Black gram residue	23.14	32.48	26.8	Ilyas et al. (2012)
Bamboo	21–31	15–26	26–43	Sánchez (2009)
Corn cob	4.5–6.6	38–42	35	Okeke and Obi (1994)
Coconut coir	18	26	48	Raveendran et al. (1995)
Corn stover	18	20.5	41.7	Merino and Cherry (2007)
Coastal Bermuda grass	19.4	13.3	47.8	Howard et al. (2003)
Cotton gin waste	–	16	78	Raveendran et al. (1995)
Elephant grass	24	24	22	Sánchez (2009)
Flax straw	22	27	29	Naik et al. (2010)
Hardwood stem	18–25	24–40	40–55	Malherbe and Cloete (2002); Howard et al. (2003); Sánchez (2009)
Millet husk	14	27	33	Raveendran et al. (1995)
Newspaper	18–30	25–40	40–55	Howard et al. (2003); Prasad et al. (2007)
Oat straw	16–19	27–38	31–37	Rowell (1992); Sánchez (2009)
Pinewood	20	24	39	Ballesteros et al. (2004)
Rice husk	14	24	31	Raveendran et al. (1995)
Rice straw	18	24	32.1	Howard et al. (2003); Prasad et al. (2007); Li et al. (2009); Sánchez (2009)
Sabai grass	20.88	23.72	49.9	Tang et al. (2013)
Sugarcane bagasse	18.4	26.9	38.1	Lee (1997)
Sugarcane tops	36.1	24.2	33.3	Ferreira-Leitão et al. (2010)
Sunhemp residue	17.4–18.4	11.9–13	43.4–48	Mansoer et al. (1997)

(*Continued*)

Table 2.1 (Continued)

Lignocellulosic material	Lignin (%)	Hemicellulose (%)	Cellulose (%)	References
Sweet Sorghum bagasse	18	25	45	Ballesteros et al. (2004)
Switchgrass	18	22	31	Merino and Cherry (2007)
Timothy grass	18	30	34	Nanda et al. (2013)
Wheat bran	10	22	11	Kabel et al. (2006)
Wheat straw	17	23	33	Merino and Cherry (2007)

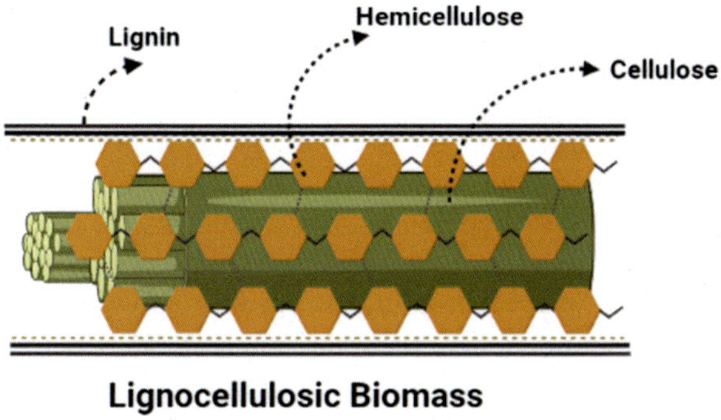

Figure 2.3 Structural components of lignocellulosic biomass presenting cellulose, hemicellulose, and lignin.

LCB, acting as glue to bind them together. An amorphous heteropolymer called lignin renders the cell wall impermeable and resistant to oxidative and microbial attack. Lignin content is higher in softwoods than in hardwoods, but holocellulose content is higher in hardwoods. The release of monomer sugars from holocellulose is challenged by the lignin present in LCB (Nanda et al., 2013).

Agriculture or agrowaste is the best source for producing biofuels and is very beneficial for the economic cycle. It includes different agriculture and forestry products that produce a large amount of LCB each year, with production ranging between 1.7 and 1011 tonnes (Elumalai et al., 2018). In addition, of the 1.2 billion tonnes of biomass produced from different crops in 2011, 38 million tonnes were utilized for biofuels in the European Union (Scarlat et al., 2015). Additionally, it is estimated that by 2030, the United States will have 450 million dry tonnes of LCB available each year, producing 67 million gallons of ethanol annually (Valdivia et al., 2016). Fruits and vegetables generate the maximum waste since they contain leftovers from grains (I9%), roots and tubers (20%), peels, pomace, seeds (25%–30%), and nutritious seeds (Galanakis, 2013). These also have a high concentration of cellulose and hemicellulose, an efficient biofuel source and a bioactive compound with therapeutic applications. Transforming them into high-value goods may be used for sustainable development instead of discarded as waste material. Almost all industries and sectors, including 42% of the domestic, 39% of the industrial, and 9% of the servicing and distribution sectors, all contribute to food waste through processing techniques (Mirabella et al., 2014). The compositions of different types of agricultural wastes comprising lignocellulosic materials are exposed in Table 2.1.

2.3 Pretreatment strategies for the lignocellulosic biomass conversion as a feedstock for biofuel production

To make more cellulose available to the enzymes that transform polysaccharides into fermentable forms of sugar, a variety of pretreatment procedures are employed for this purpose. The strategy used in these approaches is a crucial instrument to effectively modify cellulosic biomass. Pretreatment is the most expensive phase of all the different techniques, accounting for at least 20% of the total manufacturing cost (Amin et al., 2017). The pretreatment process breaks the complex and compact lignocellulosic structure to expose the cellulose fibers for further degradation into fermentable sugars. Pretreatment techniques typically aim to break down cellulose, lignin, and hemicellulose into small pieces so that they may more easily participate in the biomass refinery process and maximize the output of valuable products after enzymatic hydrolysis (Baruah et al., 2018). Nevertheless, there are several pros and downsides to the various pretreatment methods.

Industrially relevant chemicals are produced from LCB under solid-state and submerged fermentation conditions. LCB is used in solid-state fermentation either directly without any preparation or after a few physical, chemical, or biological pretreatments. To make LCB acceptable for fermentation, a pretreatment step is needed since these are resistant to enzymatic attacks. Sugars generated from LCB are a financially appealing feedstock for the large-scale fermentation of various compounds. Through submerged fermentation, sugars released from cellulose and hemicelluloses through hydrolyzation are transformed into a variety of commercial products, including ethanol, butanol, glycerol, organic acids, and bioactive polysaccharides (Tse et al., 2021). A schematic representation of the different pretreatment strategies for the degradation of LCB as a feedstock for biofuel production has been presented in Fig. 2.4 and discussed as follows.

2.3.1 *Physical pretreatment*

Physical pretreatment methods include irradiation, mechanical, hydrothermal, Gamma-ray, pyrolysis, and ultrasound. An overview of the different physical pretreatment methods employed for the LCB pretreatment is elaborated in Table 2.2.

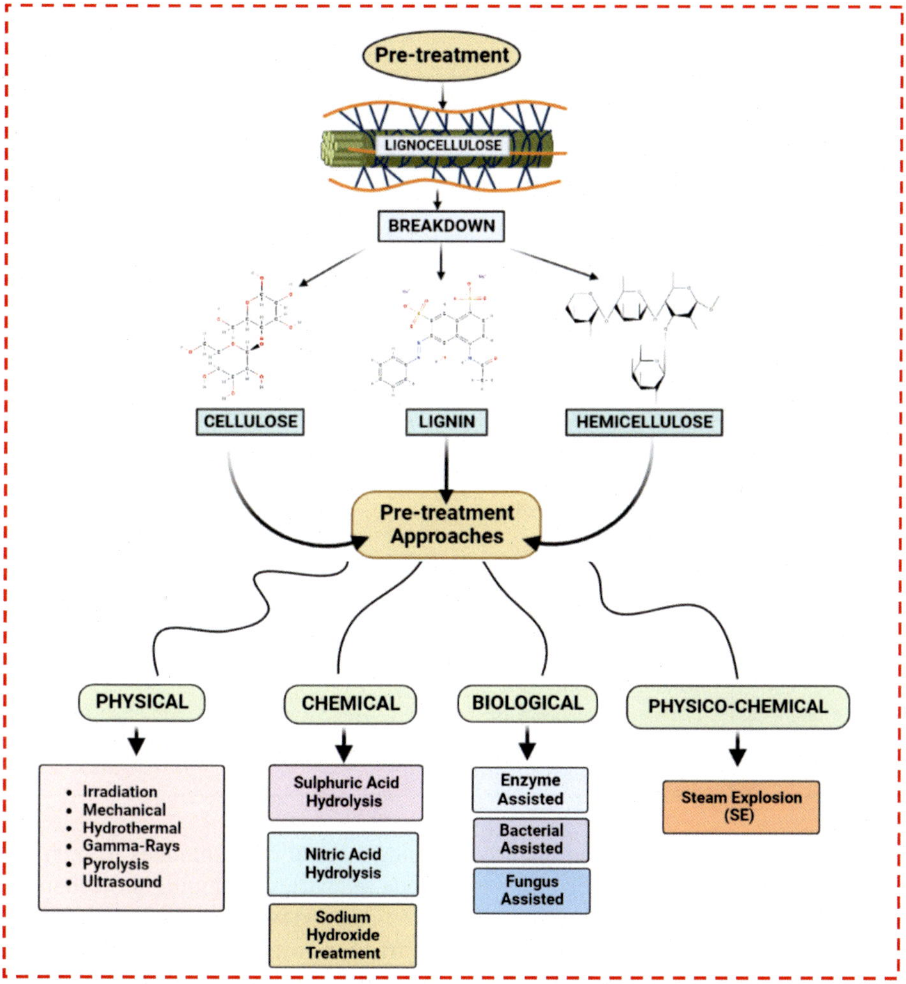

Figure 2.4 Schematic representation of the different pretreatment strategies for the degradation of lignocellulosic biomass.

Dry crushing, grinding, ball mining, chipping, and compression are typical mechanical pretreatment methods crucial in rupturing the carbohydrate polymer structures in the cell wall and also play an important role in feedstock preparation from lignocellulosic materials. Mechanical methods also expose the material for efficient chemical or other types of treatments, thus increasing the effectiveness of the respective pretreatment process. Ultrasonic pretreatment involves the production of high-energy vibrations that permeate and break down the crystal structure of lignocellulose. The high frequency, intensity, and duration of ultrasonic vibrations have been employed to remove 80–100% of the lignin from coffee waste, cabbage, and other fibrous materials (Subhedar et al., 2018; Mankar et al., 2021).

Table 2.2 Pretreatment methods employed for the lignocellulosic biomass pretreatment.

Technique	Effect	Application	References
Irradiation	Effect of lignocellulose irradiation treatment on hydrolysis for production of high levels of fermentable sugars	Glucose production	Liu et al. (2017)
Mechanical method	Impact of mechanical pretreatment on the methane generation of reed stems and wheat straw	Biomethane production	Dell'Omo and Spena (2020)
Hydrothermal	Effect of hydrothermal and nonthermal pretreatment on biodegradability and methane generation rate	Anaerobic digestor for biomethane production	Lee and Park (2020); Sarker et al. (2021)
Gamma ray	The effect of chronic gamma radiation on enzymatic hydrolysis and lignocellulosic composition	Increase the yield of glucose	Kim et al. (2015)
Pyrolysis	Flexible pyrolysis for lignocellulosic pretreatment to a wide range of products and high feed	Biooil production	Alvarez-Chavez et al. (2019); Puig-Arnavat et al. (2019)
Ultrasound	Ultrasonic assistance in pretreating LCB to generate fermentable sugar forms	Bioethanol production	Subhedar and Gogate (2016)

Recent research has focused heavily on using irradiation techniques to treat biomass and its commercial applications in various industries, including biotechnology and fuel generation. The generation of glucose and xylose from LCB utilizing the irradiation technique for the hemicellulose fraction was around 21 and 55 g/L, respectively, in a two-stage hydrolysis process (dilute enzymatic and thermal). On the other hand, the cellulose fraction released 235 and 16 g/L, respectively, of glucose and xylose (Liu et al., 2017). Biopolymer and lignin-based biorefineries may benefit from the irradiation of lignin with high-energy ionizing radiation. In the current study, kraft lignin was exposed to alpha radiation (5.2 MeV) up to a dose of 1200 mGy and gamma radiation (1.17 and 1.33 MeV) up to a dose of 200 kGy. The radiation causes the irradiated material to create free radicals, which cause chemical processes and alter the physicochemical characteristics of the lignin (Kumar and Kumar, 2022).

Gamma-ray pretreatment is another method for pretreating LCB, which uses ionizing radiation produced by Cesium-70 or Cobalt-60 radioisotopes. According to Hassan et al. (2018), ionizing radiation affects cellulose by changing the lignin, permeating the crystalline lignocellulosic structure, and fracturing the crystal areas. Ionizing radiation causes the cellulose molecules to absorb energy, which results in the creation of short or long-lived radicals. The radicals produced actively through a series of chemical processes, contribute to the further breakdown of the LCB (Bhutto et al., 2017; Hassan et al., 2018).

The cavitation effect-based fundamental idea behind ultrasound-assisted pretreatment is used for rupturing polysaccharides and lignin fractions in the biomass. Ultrasound radiation helps in breaking the O-4 and O-4 lignin connections, causing cavitation bubbles to develop (Shirkavand et al., 2016; Kumar and Sharma, 2017). The bubbles reach a critical size and then violently collapse; when the temperature and pressure increase to 2000–5000K and 1800 atm, respectively (Kumar et al., 2020). The effectiveness and performance of the process can be improved by analyzing the conversion of cellulose to fermentation and fermentable sugars utilizing a combined pretreatment strategy with ultrasound and alkali (Subhedar and Gogate, 2016). However, combining ultrasound with other technologies using chemicals and enzymes results in an increase in yield with fewer residues and byproducts that can restrict production.

Pyrolysis is a thermochemical process to convert LCB into chemical compounds and energy. The three main categories of biomass pyrolysis (conventional, flash, and fast pyrolysis) are typically used to categorize this process based on variables like time and heating rate for boosting the yield (Kan et al., 2016). The ability to transform multiple biogenic feedstocks into products, including ash, high-value char products, power, heat, and transportation fuels, makes pyrolysis an intriguing pretreatment method (Puig-Arnavat et al., 2019). Bai et al. (2018) evaluated the efficiency of pyrolysis utilizing rod-milling and hammer-milling pretreatment in wheat straw. They discovered that rod-milling for 60 min resulted in a considerable size reduction and a decrease in crystallinity.

2.3.2 Chemical pretreatment

A wide range of chemicals is used to treat biomass feedstock to make it capable of producing bioproducts. Several chemical pretreatment methods such as acidic pretreatment, alkaline pretreatment, etc., are used for the conversion of LCB into a feedstock for biofuel production. The usage of an acidic pretreatment strategy may upsurge the solubility of cellulose in enzymes. Acid pretreatment procedures are carried out in concentrated acid conditions (up to 70%) at 40°C or in diluted acid conditions (0.1%) at elevated temperatures like 230°C. Some of the acids used for acidic pretreatment are H_2SO_4, HNO_3, HCl, and H_3PO_4, with H_2SO_4 being the most popular acid. Some other acids are better for certain categories of biomass due to their high kPa value and solution pH. These acids also have the benefits of being less tolerant of yeast, odorless, having a wide pH and temperature range for hydrolysis, and not disrupting glycolysis (Zhang and Bao, 2020). The delignification process, which hydrolyzes hemicellulose and cellulose, requires enzymes to remove lignin; hence this pretreatment method is often recommended for biomass with less lignin (Kumar et al., 2020). Crude palm oil can be converted into biodiesel in two steps: first, through an esterification reaction with an H_2SO_4 catalyst, and second, through a transesterification reaction with a CaO catalyst made from the blood calm (*Anadara granosa*) shells that have been calcined at temperatures of 800°C and 900°C. The highest biodiesel yield in this investigation was 87.17%, and it was achieved by transesterification by employing CaO as a catalyst that had been calcined at 900°C for 10 h during the optimal esterification process. This, however, took place for 3 h at 65°C. This biodiesel outcome was approximately 3% higher than biodiesel generated using an efficient single-stage procedure (transesterification) (Nurhayati et al., 2017). The sugarcane biomass was pretreated with phosphoric acid by Morais et al. (2020), producing feedstock with 98% glucose in the lignocellulosic materials hydrolysate. Fitria et al. (2019) compared two pretreatment techniques such as dilute acid and hot water to evaluate the optimal wheat straw pretreatment conditions. Using the optimal dilute acid pretreatment conditions, the 30 wheat types' straw sugar yields were compared. Wheat straw variants exhibited considerable variability in total sugar recovery, which ranged between 0.2 and 0.401 g dry matter/g of raw biomass. Researchers have discovered that pretreatment of eucalyptus chips, hollow palm fruit bunch, rice bran, and forest residues with diluted H_3PO_4 and H_2SO_4 results in up to 60% sugar recovery (Gonzales et al., 2016).

Alkaline pretreatment includes removing lignin and hemicellulose from LCB using the alkaline substances $Ca(OH)_2$, NH_3, and NaOH. A high base concentration is needed to function for a long time and at low temperatures. Certain alkalis are needed to be transformed into inorganic salts before proceeding with the pretreatment. It was reported that pretreating the pinewood feedstock with NaOH results in increased char output by 2.49% (Wang et al., 2017). According to Baruah et al. (2018), pretreating the biomass with 1% NaOH for 1 h improved the methane output from rice straw by 34%. In related work, Safari et al. (2017) discovered that NaOH pretreatment is efficient enough for the anaerobic digestion (AD). However, bamboo pretreated with dilute acid had better catalytic activity than bamboo pretreated with alkali (Zhu et al., 2019). Enhanced biomethanation was achieved by optimizing the alkaline pretreatment conditions of sugarcane bagasse and filter mud, where 86% of lignin was removed using 1% NaOH at 100°C for 3 h, and 82% more methane was produced (Talha et al., 2016). Shen et al. (2016) showed that NaOH pretreatment improves the AD process. They adjusted the pretreatment conditions on vinegar residue and discovered an increased methane yield of 206 mL/g at a concentration of 3% NaOH, which was 54% greater than the untreated vinegar residue. Charcoal is used to remove the colored phenolic compounds, which interfere with the fermentation process from the feedstock. This is another very important strategy for pretreatment used to remove the toxins and colors from the treated hydrolysate produced during chemical treatments (Mussatto and Roberto, 2002).

2.3.3 Biological pretreatment

The biological pretreatment of LCB is carried out with the help of microbes or enzymes to specifically decompose hemicelluloses and lignin. As a consequence, the production of bioethanol and biomass saccharification is boosted (Kumar et al., 2020).

This can be performed under relatively controlled circumstances, such as at or near ambient pressure and temperature. As a result, this process requires less energy and chemical inputs and generates fewer inhibitors. However, this approach still has to be improved in key areas, such as overall effectiveness and incubation time, to make it a viable alternative to the present pretreatment techniques (Vasco-Correa et al., 2016). It uses less energy and milder processing temperatures than other chemical or physical procedures. Although, biological pretreatment is a slower process than other pretreatment techniques, but is more environmentally friendly. For instance, pretreatments based on fungi and ensiling might take weeks or months. Contrarily, chemical pretreatments might take anywhere between 5 and 120 min. Delignification through biological pretreatment is able to expose cellulose and hemicelluloses effectively for further degradation into sugars (Shrestha et al., 2017). AD works well with this pretreatment technique because this technique produces no harmful byproducts that might interfere with AD (Wagner et al., 2018). A detailed functionality of the biological pretreatment method for LCB has been elucidated in Table 2.3.

2.3.4 Physicochemical pretreatment

The pretreatment method for lignocellulosic materials including steam explosion is considered an efficient technique (Bhutto et al., 2017). It was demonstrated that by applying high pressure and a specific temperature to the LCB, the steam explosion approach causes the cell structure to explode. The procedure consists of two stages: an auto-hydrolysis stage and an explosion stage. Due to its low energy usage and quick deposit time, SE is regarded as the most inexpensive pretreatment method. Steam explosion is a commonly used and economically viable pretreatment method for LCB. It is possible to increase yield, nevertheless, by making changes to the reactor's design, operating method, and catalyst choice (Duque et al., 2016). Even though hybrid pretreatment methods like those that include microwaves, ultrasound, and fungal pretreatment show promising results, they need a significant change in design and implementation to be acceptable for large-scale applications (Vasco-Correa and Shah, 2019). Studies have recommended that a combination of chemical and microwave pretreatments can cost-effectively boost the effectiveness of the treatment of LCB (Hassan et al., 2018). It was discovered that steam explosion is an excellent pretreatment method for increasing the microbial digestion of pretreated corn stover (Zhao et al., 2018). Dilute acidic pretreatment is the most suitable for producing biofuels, with an economic value of $39.2 million per year, contrary to what economic feasibility assessments of several pretreatment methods on maize stover for biofuel generation imply (Saini et al., 2020).

2.3.5 Advanced methods in feedstock pretreatment

Several advanced strategies in feedstock pretreatment include hybrid pretreatment methods. The hybrid pretreatment methods using a combination of microwave, ultrasound, and fungal pretreatment, show encouraging results after modification in design for its implementation in large-scale applications (Vasco-Correa and Shah, 2019). It

Table 2.3 A detailed functionality of the different biological pretreatment methods for the conversion of lignocellulosic biomass as a feedstock for bioethanol production.

Technique	Process	Treatment time	References
Enzyme based pretreatment	Pretreatment of corncob using enzymatic and hydrodynamic cavitation combined	Hours to days	Thangavelu et al. (2018)
Bacterial-based pretreatment	Employing *Bacillus subtilis* as a microaerobic pretreatment for maize straw digestion	Few days	Xu et al. (2018)
	Delignification and detoxification of wheat straw		
	Rapeseed preparation with bacteria using rumen fluid from a cow	Few days	Baba et al. (2017)
	Utilizing the bacterial strain *Pandoraea sp.* B-6 alone	Few days	Zhuo et al. (2018)
Fungal based method	Aerobic preparation of cow dung and crop materials including wheat, triticale, barley, and rye using *Trametes versicolor*	Weeks to months	Akyol et al. (2019)
	White-rot fungal strains for corn stover pretreatment	Weeks to months	Saha et al. (2016)
	Pretreated four agricultural wastes-wheat straw, corncobs, barley straw, and corn stover-with the white-rot fungus *Irpex lacteu*	Weeks to months	García-Torreiro et al. (2016)

is proved that chemical and microwave pretreatments can be combined to boost competence cost-effectively (Hassan et al., 2018). Despite what assessments of the economic viability of several pretreatment systems on maize stover for the manufacture of biofuels show, the most effective method for creating biofuels is diluted acidic pretreatment, with an economic value of $39.2 million annually (Saini et al., 2020).

Another advanced treatment is called organosolv pretreatment, where the resistant polymer structures of hemicellulose and lignin are broken down utilizing carbon-based solvents such as methanol, acetic acid, ethanol, formic acid, acetone, and others. Paper and bioethanol production frequently use the pretreatment method known as organosolv method (Roy et al., 2020). Due to their amazing constituents, organic solvents have the benefit of being repurposed and even recycled. The lignin has been effectively removed from biomass by the organosolv pretreatment method (~90%). In a technoeconomic analysis, the modeling of the organosolv pretreatment process resulted in mass and energy balances that are compared to the most established and widely used pretreatment methods involving diluted acids. The organosolv pretreatment led to a higher ethanol concentration, 11.1%, instead of 5.4%. with an energy consumption of 578.1 MW instead of 213.8 MW for dilute acid pretreatment (Rodrigues Gurgel da Silva et al., 2018). Organosolv pretreatment was employed to remove lignin and hydrolyze the polysaccharides, which produced 74.09% of cellulose and 73.17% and 46.62% of solubilized lignin and hemicellulose, respectively (Tsegaye et al., 2020).

The anaerobic codigestion method adds energy-dense organic waste products to dairy or wastewater digesters with extra capacities, such as fats, oils, grease (FOG), and food scraps. These high-energy materials possess at least three times methane generation potential (e.g., biogas) of biosolids and manure, in addition to diverting food waste and FOG from landfills and public sewage systems. Co-digestion employs existing infrastructure and knowledge to redirect food waste and FOG for the goal of producing biogas, which is one of its major advantages. Anaerobic digesters are used in about 140 wastewater treatment plants in California alone, with a 15%–30% surplus capacity. Reductions in greenhouse gas emissions, economic gains, and chances for diversion are further advantages. The organic fraction of municipal solid waste is a viable source for the production of biogas and is frequently utilized as a feedstock for anaerobic digestion (AD) (van Lier et al., 2001; Abudi et al., 2016). Without oxygen, AD is a natural process that converts food into digestate and renewable fuel (in this case, biogas). CH_4 (50%–70%), CO_2 (30%–50%), and a few significant contaminants, including NH_3, H_2S, siloxane, halides, and water vapor, make up the biogas from AD. The biogas produced by AD of organic fraction of municipal solid waste (OFMSW) can be used directly in combined heat and power units for cooking or purified for transportation (Parthiba Karthikeyan et al., 2018). Anaerobic treatment of digestion run-off can result in a sustainable residue that can be used to enrich the soil or rehabilitate a particular type of soil (Castilo M et al., 2006).

2.4 Current insights into the conversion of lignocellulosic biomass as a feedstock for biofuel production

Amongst many biofuels, two biofuels namely bioethanol and biodiesel are widely in demand nowadays. Bioethanol is produced mainly from carbohydrates like cellulose biomass. Although this may differ geographically, coarse grain and sugarcane are two of the most often used ethanol feedstock. Oils and fats are further used to produce biodiesel. One of the most often utilized feedstocks in biodiesel manufacturing is vegetable oil, however, non-agricultural feedstock like trash is gaining popularity in places like the United States and Europe. Nanoparticles are valuable additions to biofuel because of their distinctive qualities, including structure, stability, higher surface area to volume ratio, catalytic activity, and reusability. In terms of selectivity, energy efficiency, and time management, nanoparticles are one of the most satisfactory answers to the problems with biomass usage, especially given their low cost and low participation. Several of these techniques have been recently applied, and numerous nanoparticles, including metal, magnetic, and metal oxides, are currently employed to increase biofuel production. Nanomaterials, including carbon nanotubes, nanofibers, and nanosheets, have also been discovered to be affordable and stable catalysts for enzyme immobilization, enhancing the production of biofuels (Arya et al., 2021). Increased production of biofuel is significantly hampered by the inability of lignocellulosic material and carbohydrate content to be fully digested due to the presence of inhibitory components, especially lignin. LCB has to be processed to improve fragmentation, boost surface area and solubility, reduce cellulose crystallinity, and elevate lignin content for sustainable biorefinery. The disadvantages of traditional pretreatment techniques include high operational costs, equipment corrosion, and the creation of harmful effluents and byproducts. To mitigate the negative effects of these restrictions on the production of biofuels, Kumar et al. (2022) examined and critically analyzed a wide range of environmental friendly methods for the efficient conversion of biomass to ensure high yields of biofuels as a commercial solution. Additionally, several organisms and enzymes that effectively utilize lignocellulosic materials have been discovered. Malode et al. (2022) suggested that by using an anaerobic digester, the biogas can be produced and can be further purified for its practical applications. Biogas from anaerobic digesters can be used as a biomethane or as a long-lasting fuel with a greater concentration of methane. Anto et al. (2020) extensively reviewed the production of biofuels from algal biomass using several methods for growing the algae, such as raceway ponds and photobioreactors, and their bottlenecks. The study addresses the evolution of biofuel feedstock, from food oils to algae, to offer insights into the several generations of biofuel. Dell'Omo and Spena (2020) elucidated that mechanical pretreatment could be performed for biomethane production. The study resulted in the outcome with both LCB wheat straw and reed stem, which significantly increased their methane output compared to the raw materials, with a hike of 49.1% and 137%, respectively. Additionally, wheat straw and reed stem usage increased the overall electric energy production to 41% and 118%, respectively. The numerous pretreatment methods used on maize stover for the generation of biofuels demonstrate that diluted acidic pretreatment is the most suitable method, with studies estimating its economic worth to be $39.2 million annually. Saini et al. (2020) extensively reviewed technoeconomic research on LCB and concluded that the complexity of the LCB-based economy will determine how it can be scaled up. Access to biomass, transportation, and conversion costs are other elements vital to commercialization. Thermochemical and biological processes might be used to convert the wastes from oilseed crops into biofuels such as ethanol, methane (biogas), bio-oil, butanol, and hydrogen. The refractory structure of oilseed crop leftovers might be disrupted by pretreatment techniques such as alkali, acid, deep eutectic solvent, hydrothermal, and steam explosion, which would further improve biofuel production. Improvements to ethanol production from oilseed crop leftovers have been proposed, including simultaneous saccharification and fermentation before hydrolysis, the use of thermotolerant yeasts, and co-fermentation employing xylose and pentose-fermenting yeasts. By preserving an ideal carbon-to-nitrogen ratio during AD, codigestion of oilseed crop leftovers and animal manure has been identified as the essential technique to increase methane generation (Yong and Wu, 2022). An engineered strain of *Rhodosporidium toruloides* can be cultivated using the ensiled sorghum in combination with an ionic liquid-based deconstruction process to produce bisabolene, a potential intermediary for biofuels and other bioproducts from the monomeric sugars in biofuel biorefinery (Magurudeniya et al., 2021).

Bioethanol is frequently combined with gasoline at ratios to enhance the properties of the fuel and engine performance. Although the bioethanol business has been in operation for a while, research is currently ongoing to develop a suitable microbial agent for the effective conversion of lignocelluloses. Engineered bacteria may be among the attractive microbiological possibilities for producing ethanol while also displaying other desirable properties like a high tolerance for ethanol and a thermophilic nature. To increase the ethanol yield, production rate, titer, and tolerance in bacterial hosts, some genes have been introduced, overexpressed, and deleted through research efforts in the past (Kazemi Shariat Panahi et al., 2022). Bacteria and fungi that worked well together demonstrated excellent lignocellulose breakdown abilities. Pretreatment of lignocelluloses employing this functional microbial consortium revealed that wheat straw had improved degradability and significant amounts of organic components like volatile oils and carbohydrates can be utilized to synthesize biogas. In comparison to the procedure based on the untreated system, a subsequent anaerobic fermentation with the use of the pretreatment system demonstrated an increase in total biogas and methane yields of 39.24% and 80.34%, respectively, in a 20-day production process (Zhong et al., 2016). Subhedar and Gogate (2016) preferred the ultrasound method for treating LCB to produce bioethanol, in which fewer chemicals/enzymes are used and the treatment time is also shorter. The procedure is also regarded as unique and eco-friendly. *Saccharomyces cerevisiae* is the most frequently used microorganism for bioethanol synthesis due to the extensive description of its kinetics and metabolism (Behera et al., 2010). *S. cerevisiae* is well-known for its ability to ferment glucose into ethanol and other volatile chemicals. Nonetheless, this is also one of its biggest flaws. LCB frequently produces glucose, xylose, and trace amounts of other sugar molecules; thus, all sugar components must be fermented to successfully valorize lignocellulose. Due to their capacity to metabolize both pentose and hexose carbohydrates, yeasts such as *Pichia stipitis* or genetically modified strains of *S. cerevisiae* are increasingly being explored (Song et al., 2019). Another study utilized *Chlorella* sp. that had been hydrothermally processed to make biohydrogen and biomethane (Wu et al., 2020).

2.5 Link of lignocellulosic biomass with circular economy

LCB is a viable and sustainable alternative to chemicals and fuels obtained from petroleum to produce biofuels and bio-based goods. Despite its abundant availability, biomass decomposition is a significant barrier. There is an urgent need to integrate several unit processes, including biochemical, thermochemical, physical, and catalytic conversion, to manufacture a wide range of bio-based products. Furthermore, the yield can be enhanced while speeding up the reaction time, which results in the integration of different processes that are economically viable. Integration of processes may substantially result in several results that drive the circular economy's direction (Velvizhi et al., 2022). In addition to its potential availability, biomass mobilization is critical as it is currently left on the ground or burned outside. Numerous agricultural and forestry byproducts and waste streams go untapped. Large biomass can still be mobilized if biomass owners are encouraged to release and make their biomass available. These biomass products must be accessible to a broad consumers in various industries, be sustainably certified, and be transportable over extensive distances. Farmers might modify their labor techniques to maximize the amount of biomass that can be made available for the bioeconomy, taking into account the anticipated sales of residues. A market analysis also projects that the biofuel business will be valued at more than $218 billion by 2022. By 2025, the increase is expected to reach approximately $240 billion. These projections are viewed as an opportunity by investors. In addition to being a fuel of the second generation, biofuel has other applications. It is used to address different health issues and vitamin A deficiency; it is a vegetable oil and cosmetics component. These factors explain the predicted growth and profitability of the sector. Biofuel is anticipated to contribute to economic recovery due to its health benefits and appeal to proponents of renewable energy. Due to the notion that biofuels may enhance energy security and reduce greenhouse gas emissions in essential industries, the policies that encourage the use and production of these fuels have been promoted and widely publicized. Other factors like blending requirements, sustainability standards, gasoline quality rules, and import levies also influence the biofuel sector. Biofuels are advantageous because they have a smaller impact on the environment than fossil fuels and utilize waste that would otherwise be disposed of (Awogbemi and Kallon, 2022).

2.6 Conclusions and future prospects

LCB is a potential source for producing bioproducts, which is the solution to the troubles associated with the depletion of fossil fuels. This results in the generation of biofuels, biooils, and biogas, all of which affect the economy. These technological measures are eco-friendly in terms of environmental pollution, helping to reduce pollution and preserve an eco-friendly environment. As we lose fossil fuels, these technological improvements become increasingly life-saving; hence, LCB is the optimal feedstock for manufacturing numerous renewable resources. Bioethanol production has been linked to the potential availability of lignocellulosic materials, particularly agricultural crop waste. The use of bioethanol as a fuel source reduces greenhouse gas emissions, improves energy security, and reduces reliance on crude oil imports. India, the United States, Brazil, China, and the EU-28 have already regulated the use of E5%–15% ethanol blends in automobiles. These ethanol mixtures are compatible with common motorcycle and car engines. However, unlike bioethanol derived from edible resources (corn starch and sugarcane juice), bioethanol derived from LCB has not yet achieved widespread acceptance.

Present techniques for the production of these alternative fuels are still in the experimental phase. Biorefining facilities utilizing lignocellulose derivatives are projected to utilize a broader range of organic raw materials. This process may lead to integrating a practical approach and catalytic layout to strengthen the biofuel production performance of a certain biofuel process. However, the development of this technology is hampered by obstacles such as the absence of cost-effective pretreatment methods. Lignocellulosic biorefineries have contributed to a circular and environmentally friendly economy. But, additional research efforts must be inculcated to enhance its functionality and make it economically viable. More attention should be focused on conversion technologies of raw materials, where the cost will play a significant role in biofuel biorefinery.

Abbreviations

AD anaerobic digestion
FOG fats, oils, grease
LCB lignocellulosic biomass

References

Abdeshahian, P., Kadier, A., Rai, P.K., Silva, S.S., 2020. Lignocellulose as a renewable carbon source for microbial synthesis of different enzymes. In: Ingle, A.P., Chandel, A.K., da Silva, S.S. (Eds.), Lignocellulosic Biorefining Technologies. Wiley, pp. 185–210.

Abudi, Z., Hu, Z., Sun, N., Xiao, B., Rajaa, N., Liu, C., et al., 2016. Batch anaerobic co-digestion of OFMSW (organic fraction of municipal solid waste), TWAS (thickened waste activated sludge) and RS (rice straw): influence of TWAS and RS pretreatment and mixing ratio. Energy 107, 131–140. Available from: https://doi.org/10.1016/j.energy.2016.03.141.

Agbor, V.B., Cicek, N., Sparling, R., Berlin, A., Levin, D.B., 2011. Biomass pretreatment: fundamentals toward application. Biotechnol. Adv. 29, 675–685. Available from: https://doi.org/10.1016/j.biotechadv.2011.05.005.

Akyol, Ç., Ince, O., Bozan, M., Ozbayram, E.G., Ince, B., 2019. Biological pretreatment with *Trametes versicolor* to enhance methane production from lignocellulosic biomass: a metagenomic approach. Ind. Crop. Prod. 140, 111659. Available from: https://doi.org/10.1016/j.indcrop.2019.111659.

Alvarez-Chavez, B.J., Godbout, S., Palacios-Rios, J.H., Le Roux, É., Raghavan, V., 2019. Physical, chemical, thermal and biological pre-treatment technologies in fast pyrolysis to maximize bio-oil quality: a critical review. Biomass Bioenergy 128, 105333. Available from: https://doi.org/10.1016/j.biombioe.2019.105333.

Amin, F.R., Khalid, H., Zhang, H., Rahman, S.U., Zhang, R., Liu, G., et al., 2017. Pretreatment methods of lignocellulosic biomass for anaerobic digestion. AMB. Express 7, 72. Available from: https://doi.org/10.1186/s13568-017-0375-4.

Anto, S., Mukherjee, S.S., Muthappa, R., Mathimani, T., Deviram, G., Kumar, S.S., et al., 2020. Algae as green energy reserve: technological outlook on biofuel production. Chemosphere 242, 125079. Available from: https://doi.org/10.1016/J.CHEMOSPHERE.2019.125079.

Arya, I., Poona, A., Dikshit, P.K., Pandit, S., Kumar, J., Singh, H.N., et al., 2021. Current trends and future prospects of nanotechnology in biofuel production. Catalysts 11 (11), 1308. Available from: https://doi.org/10.3390/catal11111308.

Awogbemi, O., Kallon, D.V.V., 2022. Pretreatment techniques for agricultural waste. Case study. Chem. Env. Eng. 6, 100229. Available from: https://doi.org/10.1016/J.CSCEE.2022.100229.

Baba, Y., Matsuki, Y., Mori, Y., Suyama, Y., Tada, C., Fukuda, Y., et al., 2017. Pretreatment of lignocellulosic biomass by cattle rumen fluid for methane production: bacterial flora and enzyme activity analysis. J. Biosci. Bioeng. 123, 489–496. Available from: https://doi.org/10.1016/j.jbiosc.2016.11.008.

Bai, X., Wang, G., Yu, Y., Wang, D., Wang, Z., 2018. Changes in the physicochemical structure and pyrolysis characteristics of wheat straw after rod-milling pretreatment. Bioresour. Technol. 250, 770–776. Available from: https://doi.org/10.1016/J.BIORTECH.2017.11.085.

Ballesteros, M., Oliva, J.M., Negro, M.J., Manzanares, P., Ballesteros, I., 2004. Ethanol from lignocellulosic materials by a simultaneous saccharification and fermentation process (SFS) with *Kluyveromyces marxianus* CECT 10875. Process. Biochem. 39, 1843–1848.

Baruah, J., Nath, B.K., Sharma, R., Kumar, S., Deka, R.C., Baruah, D.C., et al., 2018. Recent trends in the pretreatment of lignocellulosic biomass for value-added products. Front. Energy Res. 6. Available from: https://doi.org/10.3389/fenrg.2018.00141.

Behera, S., Mohanty. R.C., Ray, R.C., 2010. Comparative study of bio-ethanol production from mahula (*Madhuca latifolia* L.) flowers by *Saccharomyces cerevisiae and Zymomonas*.

Bhutto, A.W., Qureshi, K., Harijan, K., Abro, R., Abbas, T., Bazmi, A.A., et al., 2017. Insight into progress in pre-treatment of lignocellulosic biomass. Energy 122, 724–745. Available from: https://doi.org/10.1016/j.energy.2017.01.005.

Castilo, M.E.F., Cristancho, D.E., Arellano, A.V., 2006. Study of the operational conditions for anaerobic digestion of urban solid wastes. Waste Manag. 26, 546–556. Available from: https://doi.org/10.1016/j.wasman.2005.06.003.

Collins, T., Gerday, C., Feller, G., 2005. Xylanases, xylanase families and extremophilic xylanases. FEMS Microbiol. Rev. 29, 3–23. Available from: https://doi.org/10.1016/j.femsre.2004.06.005.

Dell'Omo, P.P., Spena, V.A., 2020. Mechanical pretreatment of lignocellulosic biomass to improve biogas production: comparison of results for giant reed and wheat straw. Energy 203, 117798. Available from: https://doi.org/10.1016/j.energy.2020.117798.

Deshavath, N., Dasu, V., Goud, V., 2019. Lignocellulosic feedstocks for the production of bioethanol: availability, structure, and composition. In: Rai, M., Ingle, A. (Eds.), Sustainable Bio-energy: Advances and Impacts. Elsevier, pp. 1–19.

Duque, A., Manzanares, P., Ballesteros, I., Ballesteros, M., 2016. Steam explosion as lignocellulosic biomass pretreatment. In: Mussatto, S.I. (Ed.), Biomass fractionation Technology for a Lignocellulosic Based Biorefinery. Elsevier, Amsterdam, pp. 349–368.

Elumalai, S., Agarwal, B., Runge, T.M., Sangwan, R.S., 2018. Advances in the transformation of lignocellulosic biomass to carbohydrate-derived fuel precursors. In: Kumar, S., Sani, R.K. (Eds.), Biorefining of Biomass to Biofuels: Opportunities and Perception. Cham, Springer, pp. 87–116.

Ferreira-Leitão, V., Perrone, C.C., Rodrigues, J., Franke, A.P.M., Macrelli, S., Zacchi, G., 2010. An approach to the utilization of CO_2 as an impregnating agent in steam pretreatment of sugar cane bagasse and leaves for ethanol production. Biotechnol. Biofuels 3, 7. Available from: https://doi.org/10.1186/1754-6834-3-7.

Fitria, Ruan H., Fransen, S.C., Carter, A.H., Tao, H., Yang, B., 2019. Selecting winter wheat straw for cellulosic ethanol production in the Pacific Northwest, U.S.A. Biomass Bioenergy 123, 59–69. Available from: https://doi.org/10.1016/J.BIOMBIOE.2019.02.012.

Galanakis, C.M., 2013. Emerging technologies for the production of nutraceuticals from agricultural by-products: a viewpoint of opportunities and challenges. Food Bioprod. Process. 91, 575–579. Available from: https://doi.org/10.1016/j.fbp.2013.01.004.

García-Torreiro, M., López-Abelairas, M., Lu-Chau, T.A., Lema, J.M., 2016. Fungal pretreatment of agricultural residues for bioethanol production. Ind. Crop Prod. 89, 486–492. Available from: https://doi.org/10.1016/j.indcrop.2016.05.036.

Gonzales, R.R., Sivagurunathan, P., Kim, S.H., 2016. Effect of severity on dilute acid pretreatment of lignocellulosic biomass and the following hydrogen fermentation. Int. J. Hydrog. Energy 41, 21678–21684. Available from: https://doi.org/10.1016/J.IJHYDENE.2016.06.198.

Hahn-Hägerdal, B., Galbe, M., Gorwa-Grauslund, M.F., Lidén, G., Zacchi, G., 2006. Bio-ethanol–the fuel of tomorrow from the residues of today. Trends Biotechnol. 24, 549–556. Available from: https://doi.org/10.1016/j.tibtech.2006.10.004.

Hassan, S.S., Williams, G.A., Jaiswal, A.K., 2018. Emerging technologies for the pretreatment of lignocellulosic biomass. Bioresour. Technol. 262, 310–318. Available from: https://doi.org/10.1016/j.biortech.2018.04.099.

Hendriks, A.T.W.M., Zeeman, G., 2009. Pretreatments to enhance the digestibility of lignocellulosic biomass. Bioresour. Technol. 100, 10–18. Available from: https://doi.org/10.1016/j.biortech.2008.05.027.

Howard, R.L., Abotsi, E., Van Rensburg, E.L.J., Howard, S., 2003. Lignocellulose biotechnology: issues of bioconversion and enzyme production. African J. Biotechnol. 2, 702–733. Available from: https://doi.org/10.5897/ajb2003.000-1115.

Ilyas, U., Ahmed, S., Majeed, A., Nadeem, M.T., 2012. Exploration of indigenous agrowastes for cellulase production by *Aspergillus niger*. Afr. J. Biotechnol. 11, 9276–9279.

Juturu, V., Wu, J.C., 2012. Microbial xylanases: engineering, production, and industrial applications. Biotechnol. Adv. 30, 1219–1227. Available from: https://doi.org/10.1016/j.biotechadv.2011.11.006.

Kabel, M.A., Van der Maarel, M.J.E.C., Klip, G., Voragen, A.G.J., Schols, H.A., 2006. Standard assays do not predict the efficiency of commercial cellulase preparations towards plant materials. Biotechnol. Bioeng. 93, 56–63. Available from: https://doi.org/10.1002/bit.20685.

Kan, T., Strezov, V., Evans, T.J., 2016. Lignocellulosic biomass pyrolysis: a review of product properties and effects of pyrolysis parameters. Renew. Sustain. Energy Rev. 57, 1126–1140. Available from: https://doi.org/10.1016/j.rser.2015.12.185.

Karthikeyan, O.P., Trably, E., Mehariya, S., Bernet, N., Wong, J.W.C., Carrere, H., 2018. Pretreatment of food waste for methane and hydrogen recovery: a review. Bioresour. Technol. 249, 1025–1039.

Kazemi Shariat Panahi, H., Dehhaghi, M., Dehhaghi, S., Guillemin, G.J., Lam, S.S., Aghbashlo, M., et al., 2022. Engineered bacteria for valorizing lignocellulosic biomass into bioethanol. Bioresour. Technol. 344, 126212. Available from: https://doi.org/10.1016/j.biortech.2021.126212.

Kim, J.Y., Na, C.S., Kim, D.S., Kim, J.-B., Seo, Y.W., 2015. The effect of chronic gamma ray irradiation on lignocellulose of *Brachypodium distachyon*. Cellulose 22, 2419–2430. Available from: https://doi.org/10.1007/s10570-015-0687-y.

Kumar, A., Kumar, G.A., 2022. Modification of lignin properties using alpha particles and gamma-rays for diverse applications. Radiat. Phys. Chem. 110562. Available from: https://doi.org/10.1016/j.radphyschem.2022.110562.

Kumar, A.K., Sharma, S., 2017. Recent updates on different methods of pretreatment of lignocellulosic feedstocks: a review. Bioresour. Bioprocess. 4, 7. Available from: https://doi.org/10.1186/s40643-017-0137-9.

Kumar, B., Bhardwaj, N., Agrawal, K., Chaturvedi, V., Verma, P., 2020. Current perspective on pretreatment technologies using lignocellulosic biomass: an emerging biorefinery concept. Fuel Process. Technol. 199, 106244.

Kumar, R., Singh, S., Singh, O.V., 2008. Bioconversion of lignocellulosic biomass: biochemical and molecular perspectives. J. Ind. Microbiol. Biotechnol. 35, 377–391. Available from: https://doi.org/10.1007/s10295-008-0327-8.

Kumar, R., Kim, T.H., Basak, B., Patil, S.M., Kim, H.H., Ahn, Y., et al., 2022. Emerging approaches in lignocellulosic biomass pretreatment and anaerobic bioprocesses for sustainable biofuels production. J. Clean Prod. 333, 130180. Available from: https://doi.org/10.1016/j.jclepro.2021.130180.

Lee, J., 1997. Biological conversion of lignocellulosic biomass to ethanol. J. Biotechnol. 56, 1–24. Available from: https://doi.org/10.1016/S0168-1656(97)00073-4.

Lee, J., Park, K.Y., 2020. Impact of hydrothermal pretreatment on anaerobic digestion efficiency for lignocellulosic biomass: influence of pretreatment temperature on the formation of biomass-degrading byproducts. Chemosphere 256, 127116. Available from: https://doi.org/10.1016/j.chemosphere.2020.127116.

Li, H., Kim, N.-J., Jiang, M., Kang, J.W., Chang, H.N., 2009. Simultaneous saccharification and fermentation of lignocellulosic residues pretreated with phosphoric acid–acetone for bioethanol production. Bioresour. Technol. 100, 3245–3251. Available from: https://doi.org/10.1016/j.biortech.2009.01.021.

Liu, Y., Guo, L., Wang, L., Zhan, W., Zhou, H., 2017. Irradiation pretreatment facilitates the achievement of high total sugars concentration from lignocellulose biomass. Bioresour. Technol. 232, 270–277. Available from: https://doi.org/10.1016/j.biortech.2017.01.061.

Magurudeniya, H.D., Baral, N.R., Rodriguez, A., Scown, C.D., Dahlberg, J., Putnam, D., et al., 2021. Use of ensiled biomass sorghum increases ionic liquid pretreatment efficiency and reduces the biofuel production cost and carbon footprint. Green. Chem. 23, 3127–3140. Available from: https://doi.org/10.1039/D0GC03260C.

Malherbe, S., Cloete, T.E., 2002. Lignocellulose biodegradation: fundamentals and applications. Rev. Env. Sci. Biotechnol. 1, 105–114. Available from: https://doi.org/10.1023/A:1020858910646.

Malode, S.J., Gaddi, S.A.M., Kamble, P.J., Nalwad, A.A., Muddapur, U.M., Shetti, N.P., 2022. Recent evolutionary trends in the production of biofuels. Mater. Sci. Energy Technol. 5, 262–277. Available from: https://doi.org/10.1016/J.MSET.2022.04.001.

Mankar, A.R., Pandey, A., Modak, A., Pant, K.K., 2021. Pretreatment of lignocellulosic biomass: a review on recent advances. Bioresour. Technol. 334, 125235. Available from: https://doi.org/10.1016/J.BIORTECH.2021.125235.

Mansoer, Z., Reeves, D.W., Wood, C.W., 1997. Suitability of sunn hemp as an alternative late-summer legume cover crop. Soil Sci. Soc. Am. J. 61, 246–253. Available from: https://doi.org/10.2136/sssaj1997.03615995006100010034x.

Marriott, P.E., Gomez, L.D., McQueen-Mason, S.J., 2016. Unlocking the potential of lignocellulosic biomass through plant science. N. Phytol. 209 (4), 1366–1381.

Merino S.T., Cherry J. 2007. Progress and challenges in enzyme development for biomass utilization In: L. Olsson (ed), Biofuels, Springer, Berlin, Heidelberg, pp. 95–120.

Mirabella, N., Castellani, V., Sala, S., 2014. Current options for the valorization of food manufacturing waste: a review. J. Clean. Prod. 65, 28–41. Available from: https://doi.org/10.1016/J.JCLEPRO.2013.10.051.

Morais, W.G., Pacheco, T.F., Corrêa, P.S., Martins, A.A., Mata, T.M., Caetano, N.S., 2020. Acid pretreatment of sugarcane biomass to obtain hemicellulosic hydrolysate rich in fermentable sugar. Energy Rep. 6, 18–23. Available from: https://doi.org/10.1016/J.EGYR.2020.10.015.

Mussatto, S.I., Roberto, I.C., 2002. Xilitol: edulcorante com efeitos benéficos para a saúde humana. Rev. Bras. Cienc. Farm. 38, 401–413.

Naik, S., Goud, V.V., Rout, P.K., Jacobson, K., Dalai, A.K., 2010. Characterization of Canadian biomass for alternative renewable biofuel. Renew. Energy 35, 1624–1631. Available from: https://doi.org/10.1016/j.renene.2009.08.033.

Nanda, S., Mohanty, P., Pant, K.K., Naik, S., Kozinski, J.A., Dalai, A.K., 2013. Characterization of North American lignocellulosic biomass and biochars in terms of their candidacy for alternate renewable fuels. BioEnergy Res. 6, 663–677. Available from: https://doi.org/10.1007/s12155-012-9281-4.

Nurhayati, N., Anita, S., Amri, T., Linggawati, A., 2017. Esterification of crude palm oil using H_2SO_4 and transesterification using CaO catalyst derived from *Anadara granosa*. Indones. J. Chem. 17, 309. Available from: https://doi.org/10.22146/ijc.24909.

Okeke, B.C., Obi, S.K.C., 1994. Lignocellulose and sugar compositions of some agro-waste materials. Bioresour. Technol. 47, 283–284. Available from: https://doi.org/10.1016/0960-8524(94)90192-9.

Panahi, H.K.S., Dehhaghi, M., Aghbashlo, M., Karimi, K., Tabatabaei, M., 2020. Conversion of residues from agro-food industry into bioethanol in Iran: an under-valued biofuel additive to phase out MTBE in gasoline. Renew. Energy 145, 699–710.

Parakh, P.D., Nanda, S., Kozinski, J.A., 2020. Eco-friendly transformation of waste biomass to biofuels. Curr. Biochem. Eng. 120–134.

Prasad, S., Singh, A., Joshi, H.C., 2007. Ethanol as an alternative fuel from agricultural, industrial and urban residues. Resour. Conserv. Recycl. 50, 1–39. Available from: https://doi.org/10.1016/j.resconrec.2006.05.007.

Puig-Arnavat, M., Thomsen, T.P., Ravenni, G., Clausen, L.R., Sárossy, Z., Ahrenfeldt, J., 2019. Pyrolysis and gasification of lignocellulosic biomass. In: Bastidas-Oyanedel, J.-R., Schmidt, J.E. (Eds.), Biorefinery: Integrated Sustainable Processes for Biomass Conversion to Biomaterials, Biofuels, and Fertilizers. Cham, Springer, pp. 79–110.

Raveendran, K., Ganesh, A., Khilar, K.C., 1995. Influence of mineral matter on biomass pyrolysis characteristics. Fuel 74, 1812–1822. Available from: https://doi.org/10.1016/0016-2361(95)80013-.

Ray, R.C., Ramachandran, S. (Eds.), 2018. Bioethanol Production from Food Crops: Sustainable Sources, Interventions and Challenges. Academin Press, London, p. 460.

Ray, R.C. (Ed.), 2021. Sustainable Biofuels: Challenges and Opportunities. Academic Press, London, p. 385.

Rodrigues Gurgel da Silva, A., Errico, M., Rong, B.-G., 2018. Techno-economic analysis of organosolv pretreatment process from lignocellulosic biomass. Clean. Technol. Env. Policy 20, 1401–1412. Available from: https://doi.org/10.1007/s10098-017-1389-y.

Rowell, R.M., Schultz, T., Narayan, R. (Eds.), 1992. Opportunities for lignocellulosic materials and composites. In: Emerging Technologies for Materials and Chemicals from Biomass. American Chemical Society, pp. 2–12. Available from: https://doi.org/10.1021/BK-1992-0476, Corpus ID: 92860463.

Roy, R., Rahman, M.S., Raynie, D.E., 2020. Recent advances of greener pretreatment technologies of lignocellulose. Curr. Res. Green. Sustain. Chem. 3, 100035. Available from: https://doi.org/10.1016/j.crgsc.2020.100035.

Safari, A., Karimi, K., Shafiei, M., 2017. Dilute alkali pretreatment of softwood pine: a biorefinery approach. Bioresour. Technol. 234, 67–76. Available from: https://doi.org/10.1016/j.biortech.2017.03.030.

Saha, B.C., Qureshi, N., Kennedy, G.J., Cotta, M.A., 2016. Biological pretreatment of corn stover with white-rot fungus for improved enzymatic hydrolysis. Int. Biodeterior. Biodegrad. 109, 29–35. Available from: https://doi.org/10.1016/j.ibiod.2015.12.020.

Saini, J.K., Saini, R., Tewari, L., 2014. Lignocellulosic agriculture wastes as biomass feedstocks for second-generation bioethanol production: concepts and recent developments. 3 Biotech. 5, 337–353.

Saini, R., Osorio-Gonzalez, C.S., Hegde, K., Brar, S.K., Magdouli, S., Vezina, P., et al., 2020. Lignocellulosic biomass-based biorefinery: an insight into commercialization and economic standout. Curr. Sustain. Energy Rep. 7, 122–136. Available from: https://doi.org/10.1007/s40518-020-00157-1.

Sánchez, C., 2009. Lignocellulosic residues: biodegradation and bioconversion by fungi. Biotechnol. Adv. 27, 185–194. Available from: https://doi.org/10.1016/J.BIOTECHADV.2008.11.001.

Sarker, T.R., Pattnaik, F., Nanda, S., Dalai, A.K., Meda, V., Naik, S., 2021. Hydrothermal pretreatment technologies for lignocellulosic biomass: a review of steam explosion and subcritical water hydrolysis. Chemosphere 284, 131372. Available from: https://doi.org/10.1016/j.chemosphere.2021.131372.

Sarwal, R., Kumar, S., Mehta, A., Varadan, A., Singh, S.K., Ramakumar, S., Mathai, R. 2021. Roadmap for ethanol blending in India 2020-25: Report of the Expert Committee-NITI Aayog. OSF Preprints. June 8. Available from: https://doi.org/10.31219/osf.io/825mz.

Scarlat, N., Dallemand, J.F., Monforti-Ferrario, F., Nita, V., 2015. The role of biomass and bioenergy in a future bioeconomy: policies and facts. Env. Dev. 15, 3–34.

Shen, J., Zhao, C., Liu, G., Chen, C., 2016. Enhancing the performance on anaerobic digestion of vinegar residue by sodium hydroxide pretreatment. Waste Biomass Valoriz. 2016 84, 1119–1126. Available from: https://doi.org/10.1007/S12649-016-9666-2.

Shirkavand, E., Baroutian, S., Gapes, D.J., Young, B.R., 2016. Combination of fungal and physicochemical processes for lignocellulosic biomass pretreatment – a review. Renew. Sustain. Energy Rev. 54, 217–234. Available from: https://doi.org/10.1016/j.rser.2015.10.003.

Shrestha, S., Fonoll, X., Khanal, S.K., Raskin, L., 2017. Biological strategies for enhanced hydrolysis of lignocellulosic biomass during anaerobic digestion: current status and future perspectives. Bioresour. Technol. 245, 1245–1257. Available from: https://doi.org/10.1016/j.biortech.2017.08.089.

Singhania, R.R., Sukumaran, R.K., Patel, A.K., Larroche, C., Pandey, A., 2010. Advancement and comparative profiles in the production technologies using solid-state and submerged fermentation for microbial cellulases. Enzyme Microb. Technol. 46, 541–549. Available from: https://doi.org/10.1016/j.enzmictec.2010.03.010.

Singhvi, M.S., Gokhale, D.V., 2019. Lignocellulosic biomass: hurdles and challenges in its valorization. Appl. Microbiol. Biotechnol. 103, 9305–9320. Available from: https://doi.org/10.1007/s00253-019-10212-7.

Song, S., Zhang, J., Gözaydın, G., Yan, N., 2019. Production of terephthalic acid from corn stover lignin. Angew. Chem. 131, 4988–4991. Available from: https://doi.org/10.1002/ANGE.201814284.

Subhedar, P.B., Gogate, P.R., 2016. Chapter 6 - Use of ultrasound for pretreatment of biomass and subsequent hydrolysis and fermentation. In: Mussatto, S.I. (Ed.), Biomass fractionation Technology for a Lignocellosic Based Biorefinry. Elsevier, Amsterdam, pp. 127–149.

Subhedar, P.B., Ray, P., Gogate, P.R., 2018. Intensification of delignification and subsequent hydrolysis for the fermentable sugar production from lignocellulosic biomass using ultrasonic irradiation. Ultrason. Sonochem. 40, 140–150. Available from: https://doi.org/10.1016/J.ULTSONCH.2017.01.030.

Talha, Z., Ding, W., Mehryar, E., Hassan, M., Bi, J., 2016. Alkaline pretreatment of sugarcane bagasse and filter mud codigested to improve biomethane production. Biomed. Res. Int. 2016. Available from: https://doi.org/10.1155/2016/8650597.

Tang, J., Chen, K., Huang, F., xu, J., Li, J., 2013. Characterization of the pretreatment liquor of biomass from the perennial grass, *Eulaliopsis binata*, for the production of dissolving pulp. Bioresour. Technol. 129, 548–552. Available from: https://doi.org/10.1016/j.biortech.2012.11.096.

Thangavelu, K., Desikan, R., Taran, O.P., Uthandi, S., 2018. Delignification of corncob via combined hydrodynamic cavitation and enzymatic pretreatment: process optimization by response surface methodology. Biotechnol. Biofuels 11, 203. Available from: https://doi.org/10.1186/s13068-018-1204-y.

Tse, T.J., Wiens, D.J., Reaney, M.J.T., 2021. Production of bioethanol—a review of factors affecting ethanol yield. Fermentation 7 (4), 268. Available from: https://doi.org/10.3390/fermentation7040268.

Tsegaye, B., Balomajumder, C., Roy, P., 2020. Organosolv pretreatments of rice straw followed by microbial hydrolysis for efficient biofuel production. Renew. Energy 148, 923–934. Available from: https://doi.org/10.1016/J.RENENE.2019.10.176.

Tursi, A., 2019. A review on biomass: importance, chemistry, classification, conversion. Biofuel Res. J. 6, 962–979. Available from: https://doi.org/10.18331/BRJ2019.6.2.3.

Uffen, R.L., 1997. Xylan degradation: a glimpse at microbial diversity. J. Ind. Microbiol. Biotechnol. 19, 1–6. Available from: https://doi.org/10.1038/sj.jim.2900417.

Valdivia, M., Galan, J.L., Laffarga, J., Ramos, J.-L., 2016. Biofuels 2020: biorefineries based on lignocellulosic materials. Microb. Biotechnol. 9, 585–594. Available from: https://doi.org/10.1111/1751-7915.12387.

van Lier, J.B., Tilche, A., Ahring, B.K., Macarie, H., Moletta, R., Dohanyos, M., et al., 2001. New perspectives in anaerobic digestion. Water Sci. Technol. J. Int. Assoc. Water Pollut. Res. 43, 1–18.

Vasco-Correa, J., Ge, X., Li, Y., 2016. Biological pretreatment of lignocellulosic biomass. In: Mussatto, S.I. (Ed.), Biomass Fractionatopn Technology for a Lignocellulosic Feedstocks Based Biorefinery. Elsevier, pp. 561–585.

Vasco-Correa, J., Shah, A., 2019. Techno-economic bottlenecks of the fungal pretreatment of lignocellulosic biomass. Fermentation 5 (2), 30. Available from: https://doi.org/10.3390/fermentation5020030.

Velvizhi, G., Balakumar, K., Shetti, N.P., Ahmad, E., Kishore Pant, K., Aminabhavi, T.M., 2022. Integrated biorefinery processes for conversion of lignocellulosic biomass to value added materials: paving a path towards circular economy. Bioresour. Technol. 343, 126151. Available from: https://doi.org/10.1016/j.biortech.2021.126151.

Wagner, A.O., Lackner, N., Mutschlechner, M., Prem, E.M., Markt, R., Illmer, P., 2018. Biological pretreatment strategies for second-generation lignocellulosic resources to enhance biogas production. Energ 11 (7), 1797. Available from: https://doi.org/10.3390/en11071797.

Wang, S., Dai, G., Yang, H., Luo, Z., 2017. Lignocellulosic biomass pyrolysis mechanism: a state-of-the-art review. Prog. Energy Combust. Sci. 62, 33–86. Available from: https://doi.org/10.1016/J.PECS.2017.05.004.

Wu, H., Li, J., Wang, C., Liao, Q., Fu, Q., Liu, Z., 2020. Sequent production of proteins and biogas from *Chlorella* sp. via CO_2 assisted hydrothermal treatment and anaerobic digestion. J. Clean. Prod. 277, 123563. Available from: https://doi.org/10.1016/J.JCLEPRO.2020.123563.

Xu, W., Fu, S., Yang, Z., Lu, J., Guo, R., 2018. Improved methane production from corn straw by microaerobic pretreatment with a pure bacteria system. Bioresour. Technol. 259, 18–23. Available from: https://doi.org/10.1016/j.biortech.2018.02.046.

Yong, K.J., Wu, T.Y., 2022. Second-generation bioenergy from oilseed crop residues: recent technologies, techno-economic assessments, and policies. Energy Convers. Manag. 267, 115869. Available from: https://doi.org/10.1016/j.enconman.2022.115869.

Zhang, J., Bao, J., 2020. Lignocellulose pretreatment using acid as catalyst. In: Bisaria, V. (Ed.), Handbook of Biorefinery Research and Technology. Springer, Dordrecht, pp. 1–14.

Zhao, S., Li, G., Zheng, N., Wang, J., Yu, Z., 2018. Steam explosion enhances digestibility and fermentation of corn stover by facilitating ruminal microbial colonization. Bioresour. Technol. 253, 244–251. Available from: https://doi.org/10.1016/J.BIORTECH.2018.01.024.

Zhong, C., Wang, C., Wang, F., Jia, H., Wei, P., Zhao, Y., 2016. Enhanced biogas production from wheat straw with the application of synergistic microbial consortium pretreatment. RSC Adv. 6, 60187–60195. Available from: https://doi.org/10.1039/C5RA27393E.

Zhou, C.-H., Xia, X., Lin, C.-X., Tong, D.-S., Beltramini, J., 2011. Catalytic conversion of lignocellulosic biomass to fine chemicals and fuels. Chem. Soc. Rev. 40, 5588–5617. Available from: https://doi.org/10.1039/c1cs15124j.

Zhu, Y., Huang, J., Sun, S.-L., Wu, A., Li, H., 2019. Effect of dilute acid and alkali pretreatments on the catalytic performance of bamboo-derived carbonaceous magnetic solid acid. Catalysts 9, 245. Available from: https://doi.org/10.3390/catal9030245.

Zhuo, S., Yan, X., Liu, D., Si, M., Zhang, K., Liu, M., et al., 2018. Use of bacteria for improving the lignocellulose biorefinery process: the importance of pre-erosion. Biotechnol. Biofuels 11, 146. Available from: https://doi.org/10.1186/s13068-018-1146-4.

Genetically engineered lignocellulosic feedstocks for enhanced biofuel yields

3

Julia Sibiya
School of Agriculture, Earth and Environmental Sciences, University of KwaZulu-Natal, Scottsville, Pietermaritzburg, South Africa

3.1 Introduction

Lignocellulosic biomass constitutes the largest, low-cost renewable source of organic carbon available for conversion into biofuels and biomaterials (Farrell et al., 2006). These lignocellulosic biofuels are produced from nonedible residues of cereal crops, including maize (cob, husk, stover), sugarcane bagasse, rice (straw, husk), sorghum (straw, stalk), and straw of wheat, barley, millet, oats, and rye (Yuan et al., 2008; Xie and Peng, 2011; Mitchell et al., 2016; Bhatia et al., 2017; Rocha-Meneses et al., 2020). In addition, other dedicated bioenergy crops, which are nonfood crops and are common in tropical and subtropical climates, include C4 grasses such as *Miscanthus* species, switchgrass (*Panicum virgatum*), and elephant or Napier grass (*Pennisetum purpureum*) (Xie and Peng, 2011; Mitchell et al., 2016; Bhatia et al., 2017; Brandon and Scheller, 2020). The C4 grasses are favored for their low input use and high water and nutrient use efficiency, resulting in increased biomass per unit of land and carbon intensity compared to the traditional crops (Sanderson and Adler, 2008; Carroll and Somerville, 2009; Xie and Peng 2011; Feltus and Vandenbrink, 2012; Bhatia et al., 2017; Clifton-Brown et al., 2017). In addition, short-rotation woody trees, which include poplar, aspen, and willow, have also been utilized, especially in temperate regions (Heaton et al., 2008; Brereton et al., 2010; Ye et al., 2011; Guidi et al., 2013). These woody species grow fast; have a low land occupation, disease resistance, and low management cost (Wang et al., 2016).

The presence of these diverse sources of feedstocks implies that a sustainable and secure biomass supply for biofuel production is possible. However, the major drawback in lignocellulosic ethanol production is the low ethanol yields, due to cell wall recalcitrance that affects enzymatic saccharification (sugar release) (Phitsuwan et al., 2013; Markel et al., 2020). Consequently, efforts to improve the feedstocks through various approaches, including crop genetics and agronomic practices, are important. Nevertheless, plant breeders experience many challenges in the improvement of energy crops. These range from the long-life cycles of some species to the

Advances in Lignocellulosic Biofuel Production Systems. DOI: https://doi.org/10.1016/B978-0-323-91192-4.00007-9

mode of reproduction of others, such as high percentages of outcrossing that limit the implementation of classical genetic studies through inbred line production; in addition to the vast and complex genomes of most of these crops (Rae et al., 2009; López-Bellido et al., 2014).

However, during the last decades, diverse plant breeding technologies, including genetic engineering, have been used for some of the edible and nonedible biofuel crops to improve them for different traits, including biofuel yield and quality and tolerance to various biotic and abiotic stresses (Sexton et al., 2009; Salehi et al., 2018). For example, studies to understand cell wall biosynthesis and degradation mechanisms have been conducted to identify plants with reduced cell wall recalcitrance. In addition, gene expressions important for traits such as biotic stresses, tolerance to herbicides, salt, drought, heat, enhancement of grain quality, plant growth, and biomass production have been identified (Enríquez-Obregón et al., 1998; Zhang et al., 1999, Zhang et al., 2006; Lam et al., 2009; Harrison et al., 2011; Hotta et al., 2011; Mathur et al., 2017; Rocha-Meneses et al., 2020). Other genetic engineering strategies have targeted modification of the structure of enzymes and overexpression or silencing of genes involved in different pathways to improve feedstock yield and quality in different plants (Torres et al., 2015; Loqué et al., 2015; Biswal et al., 2015, 2018; Kalluri and Keller, 2010; Wang et al., 2016; Bhatia et al., 2017; Mathur et al., 2017; Donev et al., 2018; Salehi et al., 2018; Li et al., 2019b; Brandon and Scheller, 2020).

Genetic approaches based on recombinant DNA technologies developed in the 1970s enabled researchers to directly transfer or remove desired genes from the plant cell and develop new biological functions (Regina and Taherzadeh, 2019; Ozsoz et al., 2019; Rocha-Meneses et al., 2020). Genetic engineering is a technique that uses recombinant DNA technology, enabling the introduction of a broader range of traits not available through conventional breeding, making it possible to make changes directly in the organism. It includes cisgenesis, which uses a natural gene from sexually compatible plants, and transgenesis, using one or more genes from nonplant organisms or sexually incompatible plants (Khan et al., 2016; Rocha-Meneses et al., 2020). This technology has been used to optimize the productivity of lignocellulosic feedstocks and modify crops to enhance their conversion to fuels. Other recent technologies also relying on recombinant DNA are genome editing tools. These include zinc finger nucleases (ZFN), transcription activator-like effector nucleases (TALENS), and the clustered regularly interspaced short palindromic repeats (CRISPR) technique, which uses an RNA-guided nuclease (Cas9) (Ozsoz et al., 2019). These techniques have resulted in more efficient manipulation of genomes with the potential to solve complex problems, including creating mutant and transgenic plants (Song et al., 2016; Kamburova et al., 2017; Ozsoz et al., 2019). This chapter aims to provide an overview of the progress made in applying recombinant DNA technologies, namely genetic engineering, to produce transgenic biofuel feedstocks and the use of CRISPR-Cas9 technology. It highlights different strategies employed, limiting the discussion mainly to lignocellulosic feedstocks, challenges faced, and future perspectives of genetic engineering technology for sustainable biofuel production.

3.2 Lignocellulose ethanol production

Approximately 60%–80% of dry matter yields in grasses is lignocellulosic biomass, which consists of three components of secondary cell walls (SCWs). These are cellulose (25%–55%), a crystalline 6-carbon (C6) sugar (β-1,4-glucan polymer); hemicelluloses (20%–50%), branched polymers, mainly xylose and other five-carbon sugars; and lignans (10%–35%), noncarbohydrates complex cross-linked polymer built from aromatic monomers (Pauly and Keegstra 2008, 2010; Xie and Peng 2011; Bhatia et al., 2017). The SCWs are responsible for structural support and mechanical strength, thus protecting the cell against physical and biochemical degradation by microorganisms (Himmel et al., 2007; Lynd et al., 2008; Wang et al., 2016; Bhatia et al., 2017).

Similarities in the polysaccharide composition of the cell wall provide a standard method for the conversion of lignocellulosic feedstocks to bioethanol, mainly involving pretreatment to reduce lignin, enzyme hydrolysis of cell wall polymers to sugars, yeast fermentation, and distillation (Olofsson et al., 2008; Rubin, 2008; Li et al., 2014; Wang et al., 2016; Brandon and Scheller 2020; Rocha-Meneses et al., 2020). However, efficient conversion of the feedstocks to fermentable sugars is limited by the cell walls, which are naturally recalcitrant to degradation and fermentation processes (Mathur et al., 2017; Markel et al., 2020). In addition, lignin and hemicellulose, respectively, play a significant role in the inhibition of saccharification (sugar release) yields in woody and monocot biomass crops, resulting in low ethanol yields (DeMartini et al., 2013; Raud et al., 2015; Escamez et al., 2017; Mathur et al., 2017; Rocha-Meneses et al., 2020; Markel et al., 2020).

Therefore, to enhance biomass yield and quality of bioenergy crops, modification of plant cell wall structure is crucial in addressing the biomass recalcitrance (Chen and Dixon, 2007; Torney et al., 2007; Demura and Ye, 2010; Vega-Sánchez and Ronald, 2010; Xie and Peng, 2011; Wang et al., 2014; Wang et al., 2016). With more than 1000 genes involved in plant cell wall biosynthesis, degradation, and regulations (Torney et al., 2007; Vega-Sánchez and Ronald, 2010), gene discovery is important for genetic engineering modification. Due to the diversity in plant cell wall structure and the complex functions, genetic modification might have adverse effects on plant cell growth and development (Torney et al., 2007; Vega-Sánchez and Ronald, 2010; Xie and Peng, 2011). Gene discovery can also take advantage of modern biotechnological tools that allow comparative genomic studies to transfer genes between distantly related species (Xie and Peng, 2011; López-Bellido et al., 2014). The process that can be adopted for genome discovery is shown in Fig. 3.1.

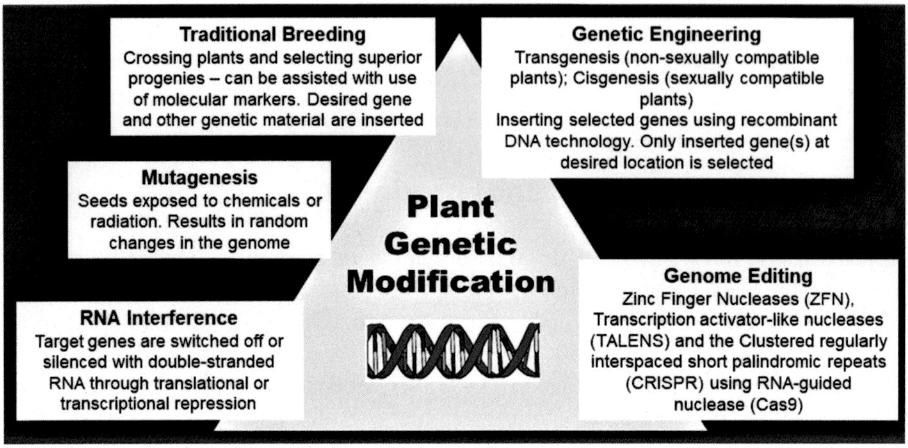

Figure 3.1 Genetic modification approaches.

3.3 Key traits to increasing lignocellulosic biomass production and yield

The main objectives in bioenergy crop breeding for lignocellulosic biofuels are biomass yield, quality enhancement, and maintaining biomass supply with minimum agricultural inputs (Chen and Peng, 2013; Wang et al., 2016). Therefore essential traits are those that improve yields while reducing agronomic input use and maintaining other priority traits such as pest and disease resistance, drought, heat and salinity tolerance, and nutrient and water use efficiencies (Spiertz and Ewert, 2009; Dubois, 2011; Jaradat, 2010; Ray et al., 2012; López-Bellido et al., 2014; Allwright and Taylor 2015). In addition, bioenergy-specific traits, including feedstock composition and low requirements for conversion efficiency, need to be enhanced (Jaradat, 2010; Ray et al., 2012; Allwright and Taylor, 2015). Table 3.1 lists traits that are valuable for improving biomass yield and quality, tolerance to stresses, resource use efficiency, and ecosystem services provision. For C4 grasses, traits related to photosynthesis pathways such as plant architecture may be improved to capture more light, thus increasing carbon fixation efficiency and biomass yield (Jaradat, 2010).

Improving traits for biomass yield and enzymatic digestibility is important for bioenergy food crops such as rice, wheat, and maize, which provide major agricultural lignocellulose residues. For example, increasing arabinose (Ara) levels in transgenic rice mutants altered hemicellulose composition and structure, improved lodging resistance, increased biomass yield, and exhibited efficient lignocellulose enzymatic saccharification (Li et al., 2015a; Wang et al., 2016). Important variations for high biomass and enzymatic saccharification observed in wheat and maize varieties make transgenic plant selection possible (Wu et al., 2013, 2014; Jia et al., 2014). Sweet sorghum is increasingly becoming an important source of biofuels

Table 3.1 Genes, and phenotypic effect due to genetic engineering on various cell wall targets.

Species	Target gene	Cell wall target	Phenotype/effect on biomass yield	References
Arabidopsis	*AtCesA1,3*	Cellulose biosynthesis	Decrease in cellulose and crystalline index, no significant impact on biomass yield, positive impact on enzymatic digestibility	Harris et al. (2012), Handakumbura et al. (2013)
	AtCesA2,5	Cellulose synthesis	Increase cellulose	Demura et al. (2015)
	AtCOBL2	Cellulose synthesis	Reduce cellulose crystallinity index	Takahashi et al. (2009)
	PttCel9A1	Cellulose degradation	Decrease in cellulose and crystalline index, positive impact on biomass yield	Takahashi et al. (2009)
	PtGH9B5, C2	Cellulose degradation	Increase in cellulose, negative impact on biomass yield	Glass et al. (2015), Guo et al. (2014)
	AtGH9C2	Cellulose degradation	Decrease in cellulose and crystalline index, positive impact on biomass yield	Glass et al. (2015), Guo et al. (2014)
	AtKORRIGAN1	Cellulose degradation	Reduce primary cellulose, increase cellulose crystallinity index, defective plant growth	Zhang et al. (2009)
	AtTBL129	Hemicellulose synthesis	Decrease in acetate content,	Xiong et al. (2013)
	AtESK1 (DUF231)	Hemicellulose synthesis	Decrease in acetylated xylan, negative impact on biomass yield	Yuan et al. (2013), Takahashi et al. (2009)
	AtIRX10, 10 L (GT47)	Hemicellulose synthesis	Reduce glucuronoxylan (GX), defective plant growth	Rennie and Scheller (2014)
	AtIRX14, 14 L (GT43)	Hemicellulose synthesis	Reduce Xyl, GX; defective plant growth	Brown et al. (2009)
	AtIRX15, 15 L (DUF579)	Hemicellulose synthesis	Reduce xylan, Xyl	Handakumbura et al. (2013)
	AnAXE1 (GT43)	Hemicellulose synthesis	Decrease in xylan content, no significant impact on biomass yield	Pawar et al. (2015) Gui et al. (2011)

(Continued)

Table 3.1 (Continued)

Species	Target gene	Cell wall target	Phenotype/effect on biomass yield	References
	AtMUR3/AtMURUS3 (GT47)	Hemicellulose synthesis	Decrease in xyloglucan, negative impact on biomass yield	Kong et al. (2015), Chen et al. (2013)
	AtAXY9(TBL)	Hemicellulose synthesis	Decrease in xyloglucan, negative impact on biomass yield	Schultink et al. (2015)
	IRX8, PARVUS (GT8); FRA8, FH8 (GT47)	Hemicellulose biosynthesis	Synthesis of a unique β-d-Xyl-(1−3)-α-1-Rha-(1−2)-α-d-GalA-(1−4)-d-Xyl tetrasaccharide commonly found at the reducing end of glucuronoxylan (GX). This tetrasaccharide may act as a primer sequence for the initiation of short xylan chains which presumably splice together to form longer xylan polymers	Brown et al. (2007), Pena et al. (2007), York and O'neill (2008), Lee et al. (2009)
	IRX9, 9 L (GT43)	Hemicellulose biosynthesis	Elongation of GX oligomeric backbones	Brown et al. (2009), Wu et al. (2009), Keppler and Showalter (2010), Wu et al. (2010)
	GUX-1, GUX-2 (GT8)	Hemicellulose biosynthesis	Addition of glucuronic acid and 4-Omethylglucuronic acid side branches on GX backbone	Mortimer et al. (2010), Bromley et al. (2013)
	AtCSE-1	Lignin synthesis	Decreased lignin content and S unit, increase in H and G unit, no significant effect on biomass yield, positive impact on enzymatic digestion	Vanholme et al. (2013), Wu et al. (2014)
	AtREF4/ AtRFR1	Lignin regulation	Increase in G unit, reduction of S unit and S/G ratio, negative impact on enzymatic digestibility	Bonawitz et al. (2014)
	AMOMT	Lignin synthesis	Reduce lignin, G, S	Gao et al. (2013)
	AtQUA2−1	Pectin synthesis	Decrease of GalA and de-methyl-esterified HG, increase Gal and Xyl, negative impact on biomass yield, positive impact on enzymatic digestion	Francocci et al. (2013), Li et al. (2010a)

	PcPL1	Pectin degradation	Increase Glc; affect growth, negative impact on biomass yield, positive impact on enzymatic digestion	Tomassetti et al. (2015), Alvira et al. (2010)
	AtPMEI	Pectin synthesis	Reduce dimethyl-esterified HG	Rubin (2008)
	AtPME3	Pectin synthesis	Reduce de-methyl-esterified HG, increase DM	Li et al. (2010)
	AnPG	Pectin degradation	Reduce GalA and de-methyl-esterified HG	Alvira et al. (2010)
Tobacco	*PsnSuSy1/2*		Increase cellulose, decrease lignin	Wei et al. (2015), Li et al. (2019a), Brandon and Scheller (2020)
	AcCel5		Increase saccharification efficiency	Brunecky et al. (2011)
	TrCel5		Decrease in biomass fresh weight and cellulose	Klose et al. (2015)
	CIEXPA1/2		Increase in biomass fresh weight and cellulose	Wang et al. (2011)
	AnPGA2		Increase in saccharification efficiency and decrease in biomass fresh weight	Lionetti et al. (2010)
	AtCESA3	Cellulose synthesis	No significant effect on cell wall, negative impact on biomass yield, positive impact on enzymatic digestion	Sahoo et al. (2013)
Barley	*HvCesA4,8*	Cellulose synthesis	Decrease in cellulose and cellulose crystalline index, negative impact on biomass yield	Tan et al. (2015)
Rice	*OsMYB103L*	Cellulose regulation	Increased secondary wall thickness, no significant impact on biomass yield	Yang et al. (2014)
	OsMYB61; NAc29, 31	Cellulose regulation	Increased secondary wall thickness, no significant impact on biomass yield	Huang et al. (2015)
	OsXAX1 (GT61)	Hemicellulose synthesis	Decrease in Xyl, ferulic, coumaric acid, negative impact on biomass yield, positive impact on enzymatic digestion	Chiniquy et al. (2012), Shen et al. (2012)
	OsIRX10 (GT47)	Hemicellulose synthesis	Reduced ratio of xylose and arabinose, affect growth, negative impact on biomass yield, positive impact on enzymatic digestion	Chen et al. (2012), Zhang et al. (2012)

(Continued)

Table 3.1 (Continued)

Species	Target gene	Cell wall target	Phenotype/effect on biomass yield	References
	Os4CL3	Lignin synthesis	Lignin reduction	Vanholme et al. (2010)
	OsSuSy3		Increase in cellulose and hemicellulolose	Fan et al. (2017, 2019)
	OsGH9B1, OsGH9B3		Increase in saccharification efficiency	Huang et al. (2019)
	OsARAF1, OsARAF3		Decrease in arabimose and cellulose	Sumiyoshi et al. (2013)
	OsAT10		Increase in cellulose and saccharification efficiency	Bartley et al. (2013)
Poplar	*PtGAUT12.1, 12.2 (GT8)*	Hemicellulose synthesis	Reduce xylan content, positive impact on biomass yield, positive impact on enzymatic digestion	Biswal et al. (2015), Chiniquy et al. (2012)
	AtKNAT7	Hemicellulose degradation	Increased expression of *IRX8, IRX9* and *FRA8* negative impact on biomass yield, no significant impact on enzymatic digestion	Li et al. (2012)
	PtxtXyn10A (GT10)	Hemicellulose degradation	Increased biomass yield	Im Kim et al. (2014)
	PtCCR	Lignin synthesis	Decrease in lignin content, increase in G unit and hemicellulose, negative impact on biomass yield, positive impact on enzymatic digestion	Van Acker et al. (2014), Baxter et al. (2014)
	AsFMT	Lignin interlinking	Increased ferulic acid conjugates, ester bonds and Degradation of β-ether bonds, positive impact on enzymatic digestion	Wilkerson et al. (2014)
	PtGAUT12	Pectin synthesis	Decreased Xyl, GalA, HG, RG; increased Man and Gal, positive impact on biomass yield, positive impact on enzymatic digestion	Biswal et al. (2015), Chiniquy et al. (2012)
	PtdCesA8	Cellulose synthesis	Reduce cellulose, defective plant growth	Tan et al. (2015)
	PtxtPL1−27	Pectin degradation	Increase pectin and xylan's solubility	Ma et al. (2009)

	GhSuSy		(increase in cellulose and cellulose crystallinity index)	Coleman et al. (2009
	PdDUF266A		increase in biomass dry weight, cellulose, cellulose degree of polymerization and saccharification efficiency	Yang et al. (2017a)
	AnAXE1		increase in saccharification efficiency	Pawar et al. (2016, 2017)
	PdDUF231A		Increase in cellulose and decrease in lignin.	Yang et al. (2017b)
	AaXEG2		(increase in biomass dry weight, and increase in saccharification efficiency)	Kaida et al. (2009), Park et al. (2004)
	PtPL1		Increase in saccharification efficiency	Biswal et al. (2014)
Alfalfa	*MsHCT*	Lignin synthesis	Decreased lignin content; increased H, negative impact on biomass yield	Gallego-Giraldo et al. (2014)
Wheat	*TaXAT1, 2 (GT61)*	Hemicellulose synthesis	Increase Xyl/Ara	Guillaumie et al. (2008)
	TaGT43–4, TaGT47–13, TaGT75–4	Hemicellulose biosynthesis	Presumably involved in glucuronoarabinoxylan (GAX) biosynthesis, although specific functions are yet undefined. *TaGT43–4* and *TaGT47–13* are, respectively, orthologous to Arabidopsis *IRX14* and *IRX10*	Zeng et al. (2010)
Maize	*XynB, ThiXnyB (GT11)*	Hemicellulose degradation	Enhanced biomass enzymatic digestibility	Reddy et al. (2016)
	ZmCOMT	Lignin synthesis	Reduce lignin and S	Sathitsuksanoh et al. (2015)
	GRMZM2G100143, GRMZM2G059825 (GT47) (homologous to AtIRX10, 10 L)	Hemicellulose synthesis	biosynthesis of (glucurono) arabinoxylan (GAX)	Bosch et al. (2011), Torres et al. (2015)
Sorghum	*SbCOMT*	Lignin synthesis		Guerriero et al. (2015)

(*Continued*)

Table 3.1 (Continued)

Species	Target gene	Cell wall target	Phenotype/effect on biomass yield	References
Switchgrass	*Pv4CL1*	Lignin synthesis	Reduce lignin, G and increase H, S/G	Mester et al. (2013)
	PvCOMT	Lignin synthesis	Lignin reduction	Baxter et al. (2014)
	PvCesA4, PvCesA6		Decreased biomass dry weight, decreased cellulose and increased xylose	Mazarei et al. (2018)
	OsAT10		Increased cellulose	Li et al. (2018)
Sugarcane	*SbCOMT*	Lignin synthesis	Decreased lignin content, negative impact on biomass yield, positive impact on enzymatic digestion	Jung et al. (2012)
	CsCESA		Increased cellulose, saccharification efficiency, sucrose, noncellulose glucose, galactose and galacturonic acid	Ndimande (2013)
Brassica distachyon	*BdCesA4, 7*	Cellulose synthesis	Decrease in cellulose and cellulose crystallinity index, negative impact on biomass yield	Handakumbura et al. (2013), Yuan et al. (2013)
	BdCAD1	Lignin synthesis	Decrease in lignin content and S unit, increase in H and G, no significant impact on biomass yield, positive impact on enzymatic digestion	Bouvier d'Yvoire et al. (2013)

Notes: Syringyl (S), Guaiacyl (G), P-hydroxy phenol (H); Glucuronoxylan (GX) in dicots and (glucurono) arabinoxylan (GAX) in grasses, Xyl, xylose, Arabinose (Ara), Degree of polymerization (DP), galacturonic acid (GalA), Homogalacturonan (HG), Xylogalacturonan (XGA), Apiogalacturonan (AGA), Rhamnogalacturonan-I (RGI) and Rhamnogalacturonan- II (RGII).

due to the large amounts of fermentable soluble sugars in the stalks and degradable lignocellulose as bagasse (Zegada-Lizarazu and Monti, 2012; Zegada-Lizarazu et al., 2012; Wang et al., 2016). In sugarcane, transgenic plants have been improved in herbicide tolerance, drought tolerance, pest and disease resistance, flowering inhibition, and expression of cellulosic enzymes in leaves (Enríquez-Obregón et al., 1998; Zhang et al., 1999; Zhang et al., 2006; Lam et al., 2009; Harrison et al., 2011; Hotta et al., 2011). These genotypes can be utilized in biofuel production.

3.4 Genetic engineering strategies to modify plant biomass properties

Genetic engineering approaches that can be employed to enhance biomass yield and lignocellulose enzymatic digestibility should not negatively affect plant growth and development (Casler et al., 2002; Pedersen et al., 2005; Vega-Sánchez and Ronald, 2010; Xie and Peng, 2011; Li et al., 2015a; Wang et al., 2016). Given that more than one thousand genes are associated with cell wall biosynthesis, degradation, and modification, identifying appropriate genes, promoters, and effective systems for gene transformation and genetic mutagenesis is essential (Guo et al., 2014; Wang et al., 2016). The major objective of genetic modification of plant cell walls is to reduce biomass recalcitrance for efficient bioethanol production; therefore, genes targeting cell wall polymer synthesis, degradation, and regulation have been used (Wang et al., 2016; Rocha-Meneses et al., 2020).

3.4.1 Changing the structure and content of lignin in the cell

Lignin is the leading cause of cell wall recalcitrance in the thermochemical pretreatment step, and it is essential to change its structure and amount in the cell (Gomez et al., 2008; Sainz, 2011; Allwright and Taylor, 2015; Rocha-Meneses et al., 2020). Considerable work has gone into lignin transgenics and using RNA interference (RNAi)-mediated gene suppression to target genes involved in the lignin pathway and cell wall biosynthesis (Bottcher et al., 2013). Both Arabidopsis (*Arabidopsis thaliana*) as a model organism and the bioenergy crops have been used in the studies on cell walls through downregulated or altered lignin biosynthesis (Khan et al., 2011). As a result, a reduction in recalcitrance increased ethanol yield, and improved conversion efficiency has been achieved but at the expense of yield in the field and plant vigor (Allwright and Taylor, 2015).

Modifying plant lignin using manipulation of three hydroxycinnamic alcohol monolignols involved in its synthesis with separate but interconnected pathways has been used (Ralph et al., 2004; Ragauskas et al., 2006; Rocha-Meneses et al., 2020). In alfalfa (*Medicago sativa*), downregulation of lignin synthesis pathway enzymes, 4-cumarate 3- hydroxylase, and cinnamyl alcohol dehydrogenase (CAD) altered lignin structure, improving its digestibility, but lignin amount remained unchanged (Baucher et al., 1999). In *Populus* sp. CAD downregulation improved pretreatment through enhanced delignification (Pilate et al., 2002; Rocha-Meneses et al., 2020). Lignin modification was observed with Caffeic acid 3-O-methyltransferase (COMT) downregulation in transgenic maize (Piquemal et al., 2002) and sorghum O-methyltransferase (OMT) antisense (He et al., 2003; Rocha-Meneses et al., 2020).

Another strategy reduces lignin accumulation through 3-dehydroshikimate dehydratase (QsuB) from *Corynebacterium glutamicum*. Reduction in lignin accumulation and increased saccharification yield were observed in transgenic *A. thaliana* plastids that expressed *QsuB* (Eudes et al., 2015; Markel et al., 2020). Some brown midrib (bm) maize and sorghum mutants characterized as CAD or COMT mutations reduced lignin and increased digestibility (Sattler et al., 2010; Li et al., 2014). The transgenic sweet sorghum lines expressing caffeoyl CoA-O-methyltransferase (CCoAOMT) altered lignin content (Casas et al., 1993; Mathur et al., 2017). In addition, 14 *CAD* genes were identified on the sorghum genome, with *SbCAD2* and the target gene in brown midrib6 mutants demonstrated to have a significant role in lignification (Saballos et al., 2009; Li et al., 2014; Li et al., 2015b). Four site mutations of COMT in sorghum *SbCOMT*s showed low levels of lignin and better biomass digestibility (Sattler et al., 2012; Wang et al., 2016). Saccharification efficiency was increased in sugarcane COMT RNAi (RNA interference) transgenic plants (Jung et al., 2012, 2013; Li et al., 2014). Lignin content was reduced in switchgrass lines downregulating COMT by RNAi (Fu et al., 2011; Li et al., 2014). RNAi silencing of the enzyme gene *Pv4CL1* in transgenic switchgrass reduced lignin levels resulting in improved biomass digestibility (Xu et al., 2011).

3.4.2 The increasing cellulose content in biomass

Cellulose constitutes 40%–50% of plant SCW and is the chief component utilized for bioethanol synthesis due to cellulose digesting commercially available enzyme mixtures (Sticklen, 2008; Li et al., 2014; Rocha-Meneses et al., 2020). Therefore, some genetic modifications to increase the amount of cellulose have focused on the overexpression of sucrose synthase in the fiber cells (Pauly and Keegstra, 2010). Fan et al. (2020) showed that transgenic rice lines overexpressing *OsSUS3* genes reduced cellulose crystallinity, improved biomass saccharification, and biotic stress resistance. Overexpressing *EGII* gene also in transgenic rice (Li et al., 2019b) increased hexose yields resulting in higher saccharification and bioethanol yields (Rocha-Meneses et al., 2020).

Another approach is to divert plant carbon resources from lignin production; for example, downregulation of 4-coumarate CoA ligase in aspen (*Populus tremuloides*) reduced lignin and increased cellulose content. Coniferaldehyde 5-hydroxylase suppression in aspen also decreased lignin content and improved cellulose (Chabannes et al., 2001; Rocha-Meneses et al., 2020).

Some indirect approaches, such as increasing gibberellin biosynthesis or delaying flowering enhanced growth rate and indirectly improved biomass yield (Eriksson et al., 2000; Rocha-Meneses et al., 2020). Plants overexpressing a floral repressor "flowering locus C" (*FLC*) gene had extended vegetative growth and increased biomass yield (Sheldon et al., 1999; Michaels and Amasino, 2000; Salehi et al., 2005).

Genes for cellulose biosynthesis characterized in cotton (the *CesA* and *Csl* superfamilies) have been identified in rice, maize, and other crops (Xie and Peng, 2011). In addition, non-*CesA* genes involved in cell wall synthesis, including *Korrigan (KOR)*, a membrane-bound b-1,4-glucanase, *Cobra* (*COB*), and *COB*-like genes (*COBLs*), encoding glycosylphosphatidylinositol (GPI) anchored proteins and *Kobito*, can be utilized for genetic modification for cellulose production (Pagant et al., 2002; Bhandari et al., 2006; Xie and Peng, 2011; Liu et al., 2013). Overexpression of *GhCOBL9A*, a COBL gene from cotton (*Gossypium hirsutum*), increased cellulose content and total biomass in transgenic *Arabidopsis* by promoting cellulose deposition, resulting in more robust seedlings, longer roots, and taller plants (Niu et al., 2018). Of the 33 *GhCOBL* genes, *GhCOBL9A* was preferentially expressed during SCW development and was coexpressed with *CesA* genes (*GhCESA4, GhCESA7*, and *GhCESA8* clades) (Niu et al., 2018; Brandon and Scheller, 2020)

Another candidate gene is sucrose synthase (*SuSy*). Some of its isoforms were observed to interact with the CesA complex, channeling UDP-Glc to cellulose biosynthesis (Fujii et al., 2010; Stein and Granot, 2019; Brandon and Scheller, 2020). SuSy overexpression in poplar increased cellulose without negatively affecting plant growth (Coleman et al., 2009; Xie and Peng, 2011). Studies using transgenic rice (*Oryza sativa*) lines overexpressing OsSuSy3 showed enhanced Lodging Index, implying vigorous growth and high-yielding plants (Fan et al., 2017) and resistance to diseases and pests (Fan et al., 2019) due to increased infection-induced callose accumulation, which contributes to increased saccharification through b-(1,3)-Glc bonds in callose. However, some species observed adverse plant growth effects due to the *SuSy* gene expression (Brandon and Scheller, 2020). Cotton (*G. hirsutum*) *SuSy* gene in hybrid poplar (*Populus alba* × *grandidentata*) increased cellulose slightly (2%–6%) while increasing cellulose crystallinity (CrI), which contributes to biomass recalcitrance (Coleman et al., 2009). Although biomass increased in switchgrass due to overexpression of *PvSUS1*, lignin also increased (Poovaiah et al., 2015). According to Brandon and Scheller (2020), this implies that the selection of promoters is crucial in crops' overexpression of SCW *CesA*s.

3.4.3 Hemicellulose biosynthesis and engineering

Hemicelluloses are removed to increase the efficiency of enzymatic cellulose hydrolysis step (Li et al., 2014). Therefore, genetic engineering has targeted the reduction of xylan [consisting of β-1,4-linked xylose (a pentose sugar)], the most important hemicellulose causing biomass recalcitrance in most bioenergy crops. Approaches have included decreasing C5 sugars like xylose by increasing hexose/pentose (Rocha-Meneses et al., 2020). However, reduction in xylan biosynthesis in mutants of model species caused xylem vessels to collapse, thus interfering with water and nutrient movement and, consequently, adverse effects on plant growth (Brown et al., 2005, 2009; Lee et al., 2007; Wu et al., 2010). (Brandon et al., 2020; Brandon and Scheller, 2020). This demonstrates the importance of xylan in strengthening vessels and fiber SCWs, despite its effects on biomass recalcitrance (Brandon and Scheller, 2020).

The hexose polysaccharide present in grass cell walls is mixed-linkage glucan. Overexpression of cellulose synthase-like F6 from rice (*OsCslF6*) in transgenic *Arabidopsis* resulted in increased amounts of mixed-linkage glucans and saccharification efficiency (Vega-Sánchez et al., 2015). In barley, *Arabidopsis* and tobacco, however; overexpression of *CslF6* with the most commonly used (constitutive) or SCW-specific promoters negatively affected plant growth, thus, emphasizing the importance of promoter selection (Burton et al., 2011; Vega-Sánchez et al., 2015; Brandon and Scheller, 2020). Acetyl xylan esterase from *Aspergillus niger* (AnAXE1) was overexpressed in *Arabidopsis* to decrease xylan acetylation without detrimental effects on plant growth (Pawar et al., 2016). The same gene was overexpressed in hybrid aspen (*Populus tremula* L. × *P. tremuloides* Michx), effectively reducing the average xylan degree of polymerization (DP) and increasing saccharification efficiency (Pawar et al., 2017). Modifying arabinoxylan, a major hemicellulose in monocots, resulted in improved saccharification efficiency, as demonstrated in transgenic rice that overexpressed *OsARAF1* and *OsARAF3* genes from native rice (Sumiyoshi et al., 2013; Brandon and Scheller, 2020).

3.4.4 Cellulase enzymes for enzymatic hydrolysis

The process of enzymatic hydrolysis is important in converting lignocellulose into monomer sugars for yeast fermentation into bioethanol (Balat, 2011; Singh et al., 2015; Wang et al., 2016). Cellulase enzymes are produced by a variety of microorganisms, including both fungi and bacteria, to digest cellulosic materials when they grow. Genera widely studied for cellulase enzyme production are *Clostridium, Cellulomonas, Thermomonospora, Trichoderma*, and *Aspergillus* (Kuhad et al., 1999; Magrey et al., 2018). Genetic engineering of these microbes focuses on increasing the secretion and productivity of enzymes, efficient enzyme purification and stable storage, supply of enzyme cofactors, and inhibitor reduction in enzyme activity (Yang and Fang, 2014; Wang et al., 2016). Successful expression of cellulases and hemicellulases has been achieved in crops such as maize, rice, and barley (Sainz, 2011; Sticklen, 2008; Rocha-Meneses et al., 2020). Genetic modification of feedstock plants to express these enzymes can be achieved by altering codons in the coding region, as most of them are of bacterial origin (Sticklen, 2008).

Different glycosyl hydrolases (GH) can also be expressed in plants (Barnes and Anderson, 2018). The GH9 b-1,4-endoglucanase in plants is assumed to contribute to cellulose remodeling and biosynthesis (Hayashi et al., 2005). Transgenic rice that overexpressed two GH9Bs genes (*OsGH9B1* and *OsGH9B3*) significantly improved biomass quality decreased cellulose DP and CrI and increased reducing sugars with no negative effects on plant growth (Huang et al., 2019). Plants expressing genes encoding GH enzymes from lignocellulose-degrading fungi can also be developed. A thermophilic endoglucanase gene from *Acidothermus cellulolyticus, AcCel5A* enhanced enzymatic digestibility by affecting microfibril crystallinity, resulting in high sugar yield (Brunecky et al., 2011; Donohoe et al., 2017; Brandon and Scheller, 2020)

3.4.5 Pectin biosynthesis and modification

Pectin, a galacturonic acid (GalA)-rich linear polymer, is a major constituent of primary cell walls in dicots and nongraminaceous monocots but is also found in SCWs (in the middle lamella) and grasses (Mohnen, 2008; Li et al., 2014). Pectin biosynthesis and modification influence plant growth and stress response (Ridley and O'Neill, 2001; Caffall and Mohnen, 2009; Wolf et al., 2009). High levels of cell wall galactose were observed after overexpression of *AtUGE2* and *GalS1* at the same time as overexpressing either one on its own, resulting in an improvement of the biomass composition with no effect on plant growth (Gondolf et al., 2014).

Increased enzymatic saccharification was shown in transgenic *Arabidopsis*, tobacco, and wheat leaves that overexpressed the fungal polygalacturonase (PG) or the *Arabidopsis* pectin methylesterase inhibitor (PMEI) due to reduced de-methyl esterified homogalacturonan (HG) (Capodicasa et al., 2004; Lionetti et al., 2010; Wang et al., 2016). Biomass production and saccharification could also be improved through the downregulation of pectin methylesterase (PME) in transgenic *Arabidopsis* (Chen and Peng, 2013; Wang et al., 2016). In poplar, overexpression of pectate lyase (that degrades HG) resulted in greater release of pentoses and hexoses and, subsequently, enhanced saccharification (Biswal et al., 2014; Li et al., 2014).

An increase in saccharification efficiency was observed in transgenic *Arabidopsis* and tobacco overexpressing *A. niger* polygalacturonase (*AnPGA2*) gene, but plants were short, and biomass was reduced (Capodicasa et al., 2004; Lionetti et al., 2010; Brandon and Scheller 2020).

3.4.6 Yeast fermentation step

The final step in bioethanol synthesis is yeast fermentation, utilizing soluble sugars from the enzymatic hydrolysis of biomass. The commonly used organism is *Saccharomyces cerevisiae*, which ferments only hexose sugars (Li et al., 2010; Wang et al., 2016). Developing yeast strains through genetic engineering has improved hexose fermentation and converted xylose to bioethanol through the pentose fermentation genes (*XYL1*, *XYL2*, and *XKS1*) (Gonçalves et al., 2014; Wang et al., 2016). Other bacterial species used in this process include *Thermoanaerobacter mathranii*, *T. brockii*, *T. ethanolicus*, *Clostridium thermohydrosulfuricum*, *C. thermosaccharolyticum*, and *Zymomonas mobilis*. Most microbial strains used in bioethanol production have been improved using the latest genome editing technology (Javed et al., 2019).

3.5 Genetic modification through CRISPR-Cas9 technology

CRISPR is a technique based on the natural defense system of bacteria and archaea, which uses an RNA-guided nuclease (Cas9) from the bacteria *Streptococcus pyogenes* to cut DNA (Xu et al., 2016; Javed et al., 2019; Kamburova et al., 2017).

It, therefore allows researchers to change DNA sequences easily and gene functions. In addition, the technology eliminates the isolation of a gene, in vitro engineering, and transfer of the gene to the host cell as is done in conventional genetic engineering (Ulaganathan et al., 2017; Javed et al., 2019).

The CRISPR-Cas9 has been recently used for strain improvement in biofuel synthesis (Xu et al., 2016; Javed et al., 2019), targeting traits like increased tolerance of the microbes to toxic compounds that would inhibit the fermentation process and microbial growth (Tkalec et al., 2014). For example, traits including b-cyclodextrin glycosyltransferase production and resistance to spore formation in *Bacillus subtilis* were improved using CRISPR-Cas9. In addition, the genome of *S. cerevisiae*, which is important in the fermentation process, was modified to enhance mevalonate or (R-R)-2,3-butanediol production and xylose use (Vervoort et al., 2017; Javed et al., 2019).

Gene editing resulting in a single amino acid change in four different genes, *ADH3* (G416 T, T966 G, T201 A), *SKS1* (G821 T, C617 A), *GIS4* (G1322C, G295 A), and *ASG1* (G1248 A, G1248 T, A1979 G, G2881C) at different loci on the genome improved tolerance of *S. cerevisiae* to acetic acid (González-Ramos et al., 2016; Javed et al., 2019). Cellulases in different microorganisms were improved for heat tolerance through one or more amino acid changes in different organisms (Table 3.2) (Ulaganathan et al., 2017; Javed et al., 2019). The hemicellulose enzyme, endo-1, 4-β-xylanase, responsible for the breakdown of the β-1,4-xylosidic links in xylan, was also modified through amino acid changes at various locations (Gallardo et al., 2010). Improved xylanases activity from *Thermomyces lanuginosus* and *Geobacillus stearothermophilus* and thermotolerance in xylanases from *Aspergillus usamii*, *B. subtilis*, *A. niger*, and *Yarrowia lipolytica* were all achieved through amino acid substitutions using CRISPR/Cas9 genome editing (Ulaganathan et al., 2017).

Besides strain improvement, CRISPR-Cas9 has been successfully tested in other crops utilized as lignocellulosic feedstocks (Miladinovic et al., 2021). For example, in barley CRISPR-Cas9 was used to modify the lignin biosynthetic gene, caffeic acid O-methyltransferase 1 (*HvCOMT1*), resulting in lignin content reduction (Lee et al., 2021). This increased both bioethanol concentration and yield from the barley mutant biomass. Park et al. (2017) employed a CRISPR/Cas9 genome editing system in switchgrass targeting one of the key enzymes, 4-Coumarate genes: coenzyme A ligase (*Pv4CL1*) involved in the monolignol biosynthesis. The resulting *Pv4CL1* knockout plants, when compared to the wild type, had thinner cell walls, reduced total lignin, and released higher amounts of glucose and xylose. The same enzymes, 4CLs, were targeted using woody perennial *Populus tremula* × *P. alba* CRISPR-Cas9 resulting in *4CL1* reduction in lignification and *4CL2* mutants, chlorogenic acid production in leaves (Zhou et al., 2015). In another study, Matsumoto (2018) increased saccharification efficiency through lignification alteration in rice by creating *CAD2* double mutants deficient in CAD gene, which encodes a key enzyme in lignin. Table 3.2 shows some of the traits and gene targets and crops.

Table 3.2 Candidate genes targeted in genomic editing of bioenergy crops using CRISPR-Cas to improve different traits.

Plant/crop	Target gene	Trait description	Reference
Rice	*OsCAld5H1*	Alteration of S/G subunit in lignin composition of rice cell wall	Takeda et al. (2019)
	OsMYB108.	Lignin-enriched transgenic rice	Miyamoto et al. (2019)
	CAD2	Altered lignification in rice and synergistic increase in saccharification efficiency	Matsumoto (2017)
	OsAT3, OsAT4	Suppressions on pectin methyltransferase activity	
Switchgrass	*4CL*	Produced plants with thinner cell walls, reduced lignin content by 30% and increased glucose and xylose release	Park et al. (2017)
	mutations at three gene loci, *teosinte branched 1(tb1) a, b* and phosphoglycerate mutase (*PGM*),	Plants with *tb1* gene mutations had increased tiller numbers, which is a useful genetic material for breeding cultivars with high biomass yield	Liu et al. (2018)
Sorghum	*CAD* and phytoene desaturase (*PDS*)	Edited gene passed down to next generation	Liu et al. (2019)
Woody perennial *P. tremula* × *alba*	three *4-coumarate: CoA ligase genes* (*4CL1, 4CL2* and *4CL5*)	Lignin biosynthesis *via* phenylpropanoid metabolism, reduction in lignification (*4CL1*), whereas *4CL2* gene proved to be involved in chlorogenic acid production in leaves, *4CL* gene revealed a 20% reduction in lignin content and a 30% decrease in S/G ratio	Zhou et al. (2015)
Poplar	*PtoMYB156*, *PtoMYB115* and *PtoMYB170*)	Negative regulatory role of some MYB transcription factors	Wan et al. (2017)
	MYB115-OE and *myb115* mutant plants *PtoMYB156* and *PtoMYB57*	Increased proanthocyanidin biosynthesis resistance to the fungal pathogen *Dothiorella gregaria*	Wang et al. (2017), Xu et al. (2017), Yang et al. (2017)

(*Continued*)

Table 3.2 (Continued)

Plant/crop	Target gene	Trait description	Reference
Arabidopsis	*PtoMYB170*	Lignification, drought tolerance	Xu et al. (2017), Yang et al. (2017)
	PtrMYB57 and *PtoMYB115*	Flavonoid accumulation	Wan et al. (2017), Wang et al. (2017), Xu et al. (2017), Yang et al. (2017)
	PtoDWF4 gen	Decreased biomass production	Shen et al. (2018)
	PcBRC1−1 and *PcBRC2−1*	(Increased branch numbers - altered shoot architecture and increased bud outgrowth)	Muhr et al. (2018)
	hydroxycinnamoyl transferase	Reduce lignification in xylem cells	Liang et al. (2019)
rice	*C3′H*	Suppression on lignin composition, and on the assembly of other cell wall components in rice	Takeda et al. (2018)
Tobacco	Miscanthus (*MYB* transcription factors) *MsSCM1* and *MsMYB103*, *MsSND1*	Act as regulators of lignin biosynthesis leading to specific lignin qualities, represent interesting targets for lignin content manipulation and composition towards tailored biomass (enhanced biomass digestibility and saccharification)	Golfier et al. (2019)

3.6 Conclusions and future perspectives

Advances have been made in addressing cell wall recalcitrance, thus enhancing biomass yield and quality of bioenergy crops using various breeding approaches. The approaches have mainly targeted the modification of plant cell wall structure. Diverse plant breeding technologies, such as molecular and marker-assisted selection and genetic engineering, have been used for various biofuel crops to improve them for different biofuel-specific traits such as tolerance to biotic and abiotic stresses. As a result, crops with tolerances to various biotic and abiotic stresses (pests, weeds, diseases, drought, heat, and salinity) have been developed. Genetic engineering using recombinant DNA technology has been a major tool in modifying the plant cell wall to overcome biomass recalcitrance, thus improving saccharification and bioethanol yields. The genetic modification strategies have targeted alteration of the structure of enzymes and overexpression or silencing of genes involved in different pathways to enhance feedstock yield and quality in the various plants. With many genes known to control different pathways in the plant cell wall biosynthesis, degradation and regulations, gene discovery will continue to be an important undertaking for modification using genetic engineering and traditional breeding techniques. Gene discovery can also take advantage of modern biotechnological tools that allow comparative genomic studies to transfer genes between distantly related species.

More recent developments such as next generation sequencing, genotyping-by-sequencing, genome-wide association studies, and genomic selection, are encouraging rapid development of bioenergy crops using molecular marker approaches. They have also reduced the cost of whole-genome sequencing, making it possible to sequence large and complex genomes like most bioenergy crops. In addition, the different "omics" technologies (genomics, transcriptomics, metabolomics, and metagenomics), including progress in genome mapping, DNA microarray, expression profiling, RNAi (RNA interference), and data mining tools, are some of the promising solutions to improve different traits in bioenergy crops. The development of new markers and their incorporation in genetic maps will accelerate breeding programs leading to the development of an approved version of sugarcane. Furthermore, the discovery of the CRISPR-Cas9 technology, a simple yet powerful tool for editing genomes, promises to accelerate and completely revolutionize the breeding process. It allows researchers to alter DNA sequences easily and modify gene function without gene isolation, in vitro engineering and retransferring the gene to the host cell as conventional genetic engineering does. Success has been achieved in microbial cells for enhanced biofuel production and characteristics improvement, inhibitor tolerance, and thermotolerance, as well as modification in cellulose and hemicellulase, which are vital for enhanced biofuel production. Therefore as better tools are developed for gene modification, the prospect of better, high-yielding feedstocks for biofuel also increases.

Abbreviations

4CL	4-coumarate CoA ligase
AnAXE1	acetyl xylan esterase from *Aspergillus niger*
AnPGA2	*Aspergillus niger* polygalacturonase
Ara	Arabinose
C6	6-carbon
CAD	cinnamyl alcohol dehydrogenase
CAS9	CRISPR associated protein 9
CCoAOMT	caffeoyl CoA-O-methyltransferase
CesA	cellulose synthase genes
COB	*Cobra*
COBL	*COBRA-like*
COMT	caffeic acid 3-O-methyltransferase
CrI	cellulose crystallinity
CRISPR	clustered regularly interspaced short palindromic repeats
Csl	cellulose synthase-like
DP	degree of polymerization
EGII	Endoglucanase II
FLC	'flowering locus C'gene
G	guaiacyl
GalA	galacturonic acid
GAX	glucurono arabinoxylan in grasses
GBS	genotyping-by-Sequencing
GH	glycosyl hydrolases
GH9B	GH9 b-1,4-endoglucanase
GPI	glycosyl-phosphatidyl-inositol
gRNA	guide RNA (crispr RNA (crRNA) and trans activating RNA (tracr RNA))
GS	genomic selection
GWAS	genome-wide association studies
GX	glucurono xylan in dicots
H	P-hydroxy phenol
HG	homogalacturonan
KOB	*Kobito*, defect in the synthesis of cellulose
KOR	*Korrigan*, a membrane-bound b-1,4-glucanase
NGS	next generation sequencing
OMT	O-methyltransferase
OsARAF	arabinofuranosidase in rice (*Oryza sativa*)
OsCslF6	cellulose synthase-like F6 from rice (*Oryza sativa*)
PG	polygalacturonase
PME	pectin methylesterase
PMEI	pectin methylesterase inhibitor
PvSUS1	Sucrose synthase from switchgrass (*Panicum virgatum* L.)
QsuB	3-dehydroshikimate dehydratase
RG	rhamnogalacturonan
RNAi	RNA interference
S	syringyl
SCW	secondary cell wall

SuSy	sucrose synthase
TALENS	transcription activator-like effector nucleases
UDP-Glc	uridine 5′-diphosphate-glucose
XGA	xylogalacturonan
Xyl	xylose
ZFN	zinc finger nucleases

References

Allwright, M.R., Taylor, G., 2015. Review molecular breeding for improved second-generation bioenergy crops. Trends Plant Sci. 1348. Available from: https://doi.org/10.1016/j.tplants.2015.10.002.

Alvira, P., Ballesteros, M., Negro, M.J., 2010. Pretreatment technologies for an efficient bioethanol production process based on enzymatic hydrolysis, A review. Bioresour Technol 101, 4851–4861.

Balat, M., 2011. Production of bioethanol from lignocellulosic materials *via* the biochemical pathway: a review. Energy Convers. Manag. 52, 858–875.

Barnes, W.J., Anderson, C.T., 2018. Release, recycle, rebuild: cell-wall remodeling, autodegradation, and sugar salvage for new wall biosynthesis during plant development. Mol. Plant 11, 31–46.

Bartley, L.E., Peck, M.L., Kim, S.-R., Ebert, B., Manisseri, C., Chiniquy, D.M., et al., 2013. Overexpression of a BAHD acyltransferase, OsAt10, alters rice cell wall hydroxycinnamic acid content and saccharification. Plant Physiol 161, 1615–1633.

Baucher, M., Bernard-vailhé, M.A., Chabbert, B., Besle, J.-M., Opsomer, C., Van Montagu, M., et al., 1999. Down-regulation of cinnamyl alcohol dehydrogenase in transgenic alfalfa (*Medicago sativa* L.) and the effect on lignin composition and digestibility. Plant Mol. Biol. 39, 437–447.

Baxter, H.L., Mazarei, M., Labbe, N., Kline, L.M., Cheng, Q., Windham, M.T., Mann, D.G., Fu, C., Ziebell, A., Sykes, R.W., et al., 2014. Two-year field analysis of reduced recalcitrance transgenic switchgrass. Plant Biotechnol. J. 12, 914–924.

Bhandari, S., Fujino, T., Thammanagowda, S., Zhang, D., Xu, F., Joshi, C.P., 2006. Xylem-specific and tension stress-responsive co-expression of KORRIGAN endoglucanase and three secondary wall-associated cellulose synthase genes in aspen trees. Planta 224, 828–837.

Bhatia, R., Gallagher, J.A., Gomez, L.D., Bosch, M., 2017. Genetic engineering of grass cell wall polysaccharides for biorefining. Plant Biotechnol. J. 15, 1071–1092.

Biswal, A.K., Soeno, K., Gandla, M.L., Immerzeel, P., Pattathil, S., Lucenius, J., et al., 2014. Aspen pectate lyase PtxtPL1–27 mobilizes matrix polysaccharides from woody issues and improves saccharification yield. Biotechnol. Biofuels 7, 11.

Biswal, A.K., Hao, Z., Pattathil, S., Yang, X., Winkeler, K., Collins, C., et al., 2015. Downregulation of GAUT12 in *Populus deltoides* by RNA silencing results in reduced recalcitrance, increased growth and reduced xylan and pectin in a woody biofuel feedstock. Biotechnol. Biofuels 8, 41.

Biswal, A.K., Atmodjo, M.A., Li, M., Baxter, H.L., Yoo, C.G., Pu, Y., et al., 2018. Sugar release and growth of biofuel crops are improved by the downregulation of pectin biosynthesis. Nat. Biotechnol. 36, 249–257.

Bonawitz, N.D., Kim, J.I., Tobimatsu, Y., Ciesielski, P.N., Anderson, N.A., Ximenes, E., Maeda, J., Ralph, J., Donohoe, B.S., Ladisch, M., et al., 2014. Disruption of mediator rescues the stunted growth of a lignin-deficient Arabidopsis mutant. Nature 509, 376–380.

Bosch, M., Mayer, C.-D., Cookson, A., Donnison, I.S., 2011. Identification of genes involved in cell wall biogenesis in grasses by differential gene expression profiling of elongating and non-elongating maize internodes. J. Exp. Bot 62, 3545–3561.

Bottcher, A., Cesarino, I., dos Santos, A.B., Vicentini, R., Mayer, J.L., Vanholme, R., et al., 2013. Lignification in sugarcane: biochemical characterization, gene discovery, and expression analysis in two genotypes contrasting for lignin content. Plant. Physiol. 163, 1539–1557.

Brandon, A.G., Scheller, H.V., 2020. Engineering of bioenergy crops: dominant genetic approaches to improve polysaccharide properties and composition in biomass. Front. Plant Sci. 11, 282.

Bouvier d'Yvoire, M., Bouchabke-Coussa, O., Voorend, W., Antelme, S., Cézard, L., Legée, F., Lebris, P., Legay, S., Whitehead, C., McQueen-Mason, S.J., et al., 2013. Disrupting the cinnamyl alcohol dehydrogenase 1 gene (BdCAD1) leads to altered lignification and improved saccharification in *Brachypodium distachyon*. Plant J 73, 496–508.

Brandon, A.G., Birdseye, D.S., Scheller, H.V., 2020. A dominant negative approach to reduce xylan in plants. Plant Biotechnol. J. 18 (1), 5–7.

Brereton, N., Pitre, F., Hanley, S., Ray, M., Karp, A., Murphy, R., 2010. QTL mapping of enzymatic saccharification in short rotation coppice willow and its independence from biomass yield. Bioenerg. Res. 3 (3), 251e61.

Bromley, J.R., Wicher, M.B., Tryfona, T., Mortimer, J.C., Zhang, Z.N., Brown, M.D., et al., 2013. GUX1 and GUX2 glucuronyltransferases decorate distinct domains of glucuronoxylan with different substitution patterns. Plant J 74, 423–434.

Brown, D.M., Goubet, F., Wong, V.W., Goodacre, R., Stephens, E., Dupree, P., et al., 2007. Comparison of five xylan synthesis mutants reveals new insight into the mechanisms of xylan synthesis. Plant J 52, 1154–1168.

Brown, D.M., Zeef, L.A.H., Ellis, J., Goodacre, R., Turner, S.R., 2005. Identification of novel genes in *Arabidopsis* involved in secondary cell wall formation using expression profiling and reverse genetics. Plant Cell 17, 2281–2295.

Brown, D.M., Zhang, Z., Stephens, E., Dupree, P., Turner, S.R., 2009. Characterization of IRX10 and IRX10-like reveals an essential role in glucuronoxylan biosynthesis in Arabidopsis. Plant J. 57, 732–746.

Brunecky, R., Selig, M.J., Vinzant, T.B., Himmel, M.E., Lee, D., Blaylock, M.J., et al., 2011. In planta expression of *A. cellulolyticus* Cel5A endocellulase reduces cell wall recalcitrance in tobacco and maize. Biotechnol. Biofuels 4, 1.

Burton, R.A., Collins, H.M., Kibble, N.A.J., Smith, J.A., Shirley, N.J., Jobling, S.A., et al., 2011. Over-expression of specific HvCslF cellulose synthase-like genes in transgenic barley increases the levels of cell wall (1,3;1,4)-(-d-glucans and alters their fine structure. Plant Biotechnol. J. 9, 117–135.

Caffall, K.H., Mohnen, D., 2009. The structure, function, and biosynthesis of plant cell wall-pectic polysaccharides. Carbohydr. Res. 344, 1879–1900.

Capodicasa, C., Vairo, D., Zabotina, O., McCartney, L., Caprari, C., Mattei, B., et al., 2004. Targeted modification of homogalacturonan by transgenic expression of a fungal polygalacturonase alters plant growth. Plant. Physiol. 135, 1294–1304.

Carroll, A., Somerville, C., 2009. Cellulosic biofuels. Annu. Rev. Plant. Biol. 60, 165–182.

Casas, A.M., Kononowicz, A.K., Zehr, U.B., Tomes, D.T., Axtell, J.D., Butler, L.G., et al., 1993. Transgenic sorghum plants *via* microprojectile bombardment. Proc. Natl. Acad. Sci., USA 90, 11212–11216.

Casler, M.D., Buxton, D.R., Vogel, K.P., 2002. Genetic modification of lignin concentration affects fitness of perennial herbaceous plants. Theor. Appl. Genet. 104, 127–131.

Chabannes, M., Barakate, A., Lapierre, C., Marita, J.M., Ralph, J., Pean, M., et al., 2001. Strong decrease in lignin content without significant alteration of plant development is induced by simultaneous downregulation of cinnamoyl CoA reductase (CCR) and cinnamyl alcohol dehydrogenase (CAD) in tobacco plants. Plant J. 28, 257–270.

Chen, F., Dixon, R.A., 2007. Lignin modification improves fermentable sugar yields forbiofuel production. Nat. Biotechnol. 25, 759.

Chen, P., Peng, L.C., 2013. The diversity of lignocellulosic biomass resources and their evaluation for use as biofuels and chemicals. In: Sun, J.Z., Ding, S.Y., Peterson, J.D. (Eds.), Biological Conversion of Biomass for Fuels and Chemicals: Explorations From Natural Biomass Utilization Systems. Royal Society of Chemistry. pp. 83–109.

Chen, W., VanOpdorp, N., Fitzl, D., Tewari, J., Friedemann, P., Greene, T., Thompson, S., Kumpatla, S., Zheng, P., 2012. Transposon insertion in a cinnamyl alcohol dehydrogenase gene is responsible for a brown midrib1 mutation in maize. Plant Mol Biol 3, 289–297.

Chen, X.W., Vega-Sánchez, M., Verhertbruggen, Y., Chiniquy, D., Canlas, P.E., Fagerström, A., et al., 2013. Inactivation of OsIRX10 leads to decreased xylan content in rice culm cell walls and improved biomass saccharification. Mol. Plant 6, 570–573.

Chiniquy, D., Sharma, V., Schultink, A., Baidoo, E.E., Rautengarten, C., Cheng, K., Carroll, A., Ulvskov, P., Harholt, J., Keasling, J.D., et al., 2012. XAX1 from glycosyltransferase family 61 mediates xylosyltransfer to rice xylan. Proc. Natl. Acad. Sci., USA 109, 17117–17122.

Clifton-Brown, J., Hastings, A., Mos, M., McCalmont, J.P., Ashman, C., Awty Carroll, D., et al., 2017. Progress in upscaling Miscanthus biomass production for the European bioeconomy with seed-based hybrids. GCB Bioenergy 9, 6–17.

Coleman, H.D., Yan, J., Mansfield, S.D., 2009. Sucrose synthase affects carbon partitioning to increase cellulose production and altered cell wall ultrastructure. Proc. Natl. Acad. Sci., USA 106, 13118–13123.

DeMartini, J.D., Pattathil, S., Miller, J.S., Li, H., Hahn, M.G., Wyman, C.E., 2013. Investigating plant cell wall components that affect biomass recalcitrance in poplar and switchgrass. Energy Environ. Sci. 6, 898–909.

Demura, T., Ehrhardt, D.W., Samuels, A.L., Mansfield, S.D., 2015. Visualization of cellulose synthases in Arabidopsis secondary cell walls. Sci 350, 198–203.

Demura, T., Ye, Z., 2010. Regulation of plant biomass production. Curr. Opin. Plant Biol. 13, 299–304.

Donev, E., Gandla, M.L., Jönsson, L.J., Mellerowicz, E.J., 2018. Engineering non-cellulosic polysaccharides of wood for the biorefinery. Front. Plant Sci. 9, 1537.

Donohoe, B.S., Wei, H., Mittal, A., Shollenberger, T., Lunin, V.V., Himmel, M.E., et al., 2017. Towards an understanding of enhanced biomass digestibility by in planta expression of a family 5 glycoside hydrolase. Sci. Rep. 7, 4389.

Dubois, J., 2011. Requirements for the development of a bioeconomy for chemicals. Curr. Opin. Environ. Sustain 3, 11–14.

Enríquez-Obregón, G.A., Vázquez-Padrón, R.I., Prieto-Samsonov, D.L., De La Riva, G.A., Selman-Housein, G., 1998. Herbicide resistant sugarcane (*Saccharum officinarum* L.) plants by *Agrobacterium*-mediated transformation. Planta 206, 20–27.

Eriksson, M.E., Israelsson, M., Olsson, O., Moritz, T., 2000. Increased gibberellin biosynthesis in transgenic trees promotes growth, biomass production and xylem fiber length. Nat. Biotechnol. 18, 784–788.

Escamez, S., Gandla, L.M., Derba-Maceluch, M., Lundqvist, S.O., Mellerowicz, E.J., Jönsson, L.J., et al., 2017. A collection of genetically engineered Populus trees reveals wood biomass traits that predict glucose yield from enzymatic hydrolysis. Sci. Rep. 7, 15798.

Eude, S.A., Sathitsuksanoh, N., Baidoo, E.E.K., George, A., Liang, Y., Yang, F., et al., 2015. Expression of a bacterial 3-dehydroshikimate dehydratase reduces lignin content and improves biomass saccharification efficiency. Plant Biotechnol. J. 13, 1241–1250.

Fan, C., Feng, S., Huang, J., Wang, Y., Wu, L., Li, X., et al., 2017. AtCesA8- driven OsSUS3 expression leads to largely enhanced biomass saccharification and lodging resistance by distinctively altering lignocellulose features in rice. Biotechnol. Biofuels 10, 221.

Fan, C., Wang, G., Wu, L., Liu, P., Huang, J., Jin, X., et al., 2019. Distinct cellulose and callose accumulation for enhanced bioethanol production and biotic stress resistance in OsSUS3 transgenic rice. Carbohydr. Polym. 232, 115448.

Fan, C., Wang, G., Wu, L., Liu, P., Huang, J., Jin, X., Zhang, G., He, Y., Peng, L., Luo, K., Feng, S., 2020. Distinct cellulose and callose accumulation for enhanced bioethanol production and biotic stress resistance in OsSUS3 transgenic rice. Carbohydr. Polym. 232, 115448.

Farrell, A.E., Plevin, R.J., Turner, B.T., Jones, A.D., O'Hare, M., Kammen, D.M., 2006. Ethanol can contribute to energy and environmental goals. Science 311, 506–509.

Feltus, F.A., Vandenbrink, J.P., 2012. Bioenergy grass feedstock: current options and prospects for trait improvement using emerging genetic, genomic, and systems biology toolkits. Biotechnol. Biofuels 5, 80.

Francocci, F., Bastianelli, E., Lionetti, V., Ferrari, S., De Lorenzo, G., Bellincampi, D., Cervone, F., 2013. Analysis of pectin mutants and natural accessions of Arabidopsis highlights the impact of de-methyl-esterified homogalacturonan on tissue saccharification. Biotechnol. Biofuel 6, 163.

Fu, C., Mielenz, J.R., Xiao, X., Ge, Y., Hamilton, C.Y., Rodriguez Jr., M., et al., 2011. Genetic manipulation of lignin reduces recalcitrance and improves ethanol production from switchgrass. Proc. Natl. Acad. Sci., USA 108, 3803–3808.

Fujii, S., Hayashi, T., Mizuno, K., 2010. Sucrose synthase is an integral component of the cellulose synthesis machinery. Plant Cell Physiol. 51, 294–301.

Gallardo, Ó., Pastor, F.J., Polaina, J., Diaz, P., Łysek, R., Vogel, P., et al., 2010. Structural insights into the specificity of Xyn10B from *Paenibacillus barcinonensis* and its improved stability by forced protein evolution. J. Biol. Chem. 285 (4), 2721–2733.

Gallego-Giraldo, L., Bhattarai, K., Pislariu, C., Nakashima, J., Jikumaru, Y., Kamiya, Y., Udvardi, M.K., Monteros, M.J., Dixon, R.A., 2014. Lignin modification leads to increased nodule numbers in alfalfa. Plant Physiol 164, 1139–1150.

Gao, Y., Xu, J., Zhang, Y., Yu, Q., Yuan, Z., et al., 2013. Effects of different pretreatment methods on chemical composition of sugarcane bagasse and enzymatic hydrolysis. Bioresour Technol 144, 396–400.

Glass, M., Barkwill, S., Unda, F., Mansfield, S.D., 2015. Endo-beta-1,4-glucanases impact plant cell wall development by influencing cellulose crystallization. J. Integr. Plant Biol. 57, 396–410.

Golfier, P., Unda, F., Murphy, E.K., Xie, J., He, F., Zhang, W., Mansfield, S.D., Rausch, T., Wolf, S., 2019. Distinct and overlapping functions of Miscanthus sinensis MYB transcription factors SCM1 and MYB103 in lignin biosynthesis. bioRxiv.

Gomez, L.D., Steele-King, C.G., McQueen-Mason, S.J., 2008. Sustainable liquid biofuels from biomass: the writing's on the walls. N. Phytol. 178, 473–485.

Gonçalves, D.L., Matsushika, A., de Sales, B.B., Goshima, T., Bon, E.P., Stambuk, B.U., 2014. Xylose and xylose/glucose co-fermentation by recombinant *Saccharomyces cerevisiae* strains expressing individual hexose transporters. Enz. Microb. Technol. 63, 13–20.

Gondolf, V.M., Stoppel, R., Ebert, B., Rautengarten, C., Liwanag, A.J., Loqué, D., et al., 2014. A gene stacking approach leads to engineered plants with highly increased galactan levels in Arabidopsis. BMC Plant. Biol. 14, 344.

González-Ramos, D., de Vries, A.R.G., Grijseels, S.S., Berkum, M.C., Swinnen, S., Broek, M., et al., 2016. A new laboratory evolution approach to select for constitutive acetic acid tolerance in *Saccharomyces cerevisiae* and identification of causal mutations. Biotechnol. Biofuels 9 (1), 173.

Guerriero, G., Hausman, J., Strauss, J., Ertan, H., Siddiqui, K.S., 2015. Destructuring plant biomass: Focus on fungal and extremophilic cell wall hydrolases. Plant Sci 234, 180–193.

Gui, J., Shen, J., Li, L., 2011. Functional characterization of evolutionarily divergent 4-coumarate: coenzyme a ligases in rice. Plant Physiol 157, 574–586.

Guidi, W.E., Pitre, F., Labrecque, M., 2013. Short-rotation coppice of willows for the production of biomass in Eastern Canada. In: Matovic, M.D. (Ed.), Biomass Now -Sustainable Growth and Use. InTech, London. Available from: http://doi.org/10.5772/51111.

Guillaumie, S., Goffner, D., Barbier, O., Martinant, J., Pichon, M., Barrière, Y., 2008. Expression of cell wall related genes in basal and ear internodes of silking brown-midrib-3, caffeic acid O-methyltransferase (COMT) down-regulated, and normal maize plants. BMC Plant Biol 8, 71.

Guo, K., Zou, W., Feng, Y., Zhang, M., Zhang, J., Tu, F., et al., 2014. An integrated genomic and metabolomic framework for cell wall biology in rice. BMC Genomics 15, 596.

Handakumbura, P.P., Matos, D.A., Osmont, K.S., Harrington, M.J., Heo, K., Kafle, K., Kim, S.H., Baskin, T.I., Hazen, S.P., 2013. Perturbation of Brachypodium distachyon CELLULOSE SYNTHASE A4 or 7 results in abnormalcell walls. BMC Plant Biol 13, 131.

Harris, D.M., Corbin, K., Wang, T., Gutierrez, R., Bertolo, A.L., Petti, C., Smilgies, D.M., Estevez, J.M., Bonetta, D., Urbanowicz, B.R., et al., 2012. Cellulose microfibril crystallinity is reduced by mutating C-terminal transmembrane region residues CESA1A903V and CESA3T942I of cellulose synthase. Proc. Natl. Acad. Sci. USA 109, 4098–4103.

Harrison, M.D., Geijskes, J., Coleman, H.D., Shand, K., Kinkema, M., Palupe, A., et al., 2011. Accumulation of recombinant cellobiohydrolase and endoglucanase in the leaves of mature transgenic sugar cane. Plant Biotechnol. J. 9, 84–896.

Hayashi, T., Yoshida, K., Park, Y.W., Konishi, T., Baba, K., 2005. Cellulose metabolism in plants. Int. Rev. Cytol. 247, 1–34.

He, X., Hall, M.B., Gallo-Meagher, M., Smith, R.L., 2003. Improvement of forage quality by downregulation of maize o-methyltransferase. Crop Sci. 43, 2240–2251.

Heaton, E.A., Dohlemon, F.G., Long, 2008. Meeting US biofuel goals with less land: the potential of Miscanthus. Glob. Change Biol. 14, 2000–2014.

Himmel, M.E., Ding, S.Y., Johnson, D.K., Adney, W.S., Nimlos, M.R., Brady, J.W., et al., 2007. Biomass recalcitrance: engineering plants and enzymes for biofuels production. Science 315, 804–807.

Hotta, C.T., Lembke, C.G., Domingues, D.S., Ochoa, E.A., Cruz, G.M., Melotto-Passarin, D. M., et al., 2011. The biotechnology roadmap for sugarcane improvement. Trop. Plant Biol. 3, 75–87.

Huang, D., Wang, S., Zhang, B., Shang-Guan, K., Shi, Y., Zhang, D., Liu, X., Wu, K., Xu, Z., Fu, X., et al., 2015. A gibberellin-mediated DELLA-NAC signaling cascade regulates cellulose synthesis in rice. Plant Cell 27, 1681–1696.

Huang, J., Xia, T., Li, G., Li, X., Li, Y., Wang, Y., et al., 2019. Overproduction of native endo-(-1,4-glucanases leads to largely enhanced biomass saccharification and bioethanol production by specific modification of cellulose features in transgenic rice. Biotechnol. Biofuels 12, 11.

Im Kim, J., Ciesielski, P.N., Donohoe, B.S., Chapple, C., Li, X., 2014. Chemically induced conditional rescue of the reduced epidermal fluorescence8 mutant of Arabidopsis reveals rapid restoration of growth and selective turnover of secondary metabolite pools. Plant Physiol 164, 584–595.

Jaradat, A.A., 2010. Genetic resources of energy crops: biological systems to combat climate change. Aust. J. Crop. Sci. 4, 309–323.

Javed, M.R., Noman, M., Shahid, M., Ahmed, T., Khurshi, M., Rashid, M.H., et al., 2019. Current situation of biofuel production and its enhancement by CRISPR/Cas9-mediated genome engineering of microbial cells. Microbiol. Res. 219, 1–11.

Jia, J., Yu, B., Wu, L.M., Wang, H.W., Wu, Z.L., Li, M., et al., 2014. Biomass enzymatic saccharification is determined by the non-KOH-extractable wall polymer features that predominately affect cellulose crystallinity in corn. PLoS One 9, e108449.

Jung, J.H., Fouad, W.M., Vermerris, W., Gallo, M., Altpeter, F., 2012. RNAi suppression of lignin biosynthesis in sugarcane reduces recalcitrance for biofuel production from lignocellulosic biomass. Plant Biotechnol. J. 10, 1067–1076.

Jung, J.H., Vermerris, W., Gallo, M., Fedenko, J.R., Erickson, J.E., Altpeter, F., 2013. RNA interference suppression of lignin biosynthesis increases fermentable sugar yields for biofuel production from field-grown sugarcane. Plant Biotechnol. J. 11, 709–716.

Kaida, R., Kaku, T., Baba, K., Oyadomari, M., Watanabe, T., Nishida, K., Kanaya, T., Shani, Z., Shoseyov, O., Hayashi, T., 2009. Loosening xyloglucan accelerates the enzymatic degradation of cellulose in wood. Mol. Plant 2, 904–909.

Kalluri, U.C., Keller, M., 2010. Bioenergy research: a new paradigm in multidisciplinary research. J. R. Soc. Interface 7, 1391–1401.

Kamburova, V.S., Nikitina, E.V., Shermatov, S.E., Buriev, Z.T., Kumpatla, S.P., Emani, C., et al., 2017. Genome editing in plants: an overview of tools and applications. Int. J. Agron. 2017, 7315351.

Keppler, B.D., Showalter, A.M., 2010. IRX14 and IRX14-LIKE, Two glycosyl transferases involved in glucuronoxylan biosynthesis and drought tolerance in Arabidopsis. Mol. Plant 3, 834–841.

Khan, M.S., Ali, S., Iqbal, J., 2011. Developmental and photosynthetic regulation of *Bacillus thuringiensis* δ-endotoxin reveals that engineered sugarcane conferring resistance to 'dead heart' contains no toxins in cane juice. Mol. Biol. Rep. 38, 2359–2369.

Khan, S., Ullah, M.W., Siddique, R., Nabi, G., Manan, S., Yousaf, M., et al., 2016. Role of recombinant DNA technology to improve life. Int. J. Genom. 2016, 2405954.

Klose, H., Günl, M., Usadel, B., Fischer, R., Commandeur, U., 2015. Cell wall modification in tobacco by differential targeting of recombinant endoglucanase from *Trichoderma reesei*. BMC Plant Biol 15, 54.

Kong, Y.Z., Peña, M.J., Renna, L., Avci, U., Pattathil, S., Tuomivaara, S.T., Li, X., Reiter, W.D., Brandizzi, F., Hahn, M.G., et al., 2015. Galactose-depleted xyloglucan is dysfunctional and leads to dwarfism in Arabidopsis. Plant Physiol 167, 1296–1306.

Kuhad, R.C., Manchanda, M., Singh, A., 1999. Hydrolytic potential of extracellular enzymes from a mutant strain of *Fusarium oxysporum*. Bioprocess. Eng. 20, 133–135.

Lam, E., Shine Jr, J., Da Silva, J., Lawton, M., Bonos, S., Calvino, M., et al., 2009. Improving sugarcane for biofuel: engineering for an even better feedstock. GCB Bioenergy 1 (3), 251–255.

Lee, C.H., O'Neill, M.A., Tsumuraya, Y., Darvill, A.G., Ye, Z.H., 2007. The irregular xylem9 mutant is deficient in xylan xylosyltransferase activity. Plant Cell Physiol. 48, 1624–1634.

Lee, C.H., Teng, Q., Huang, W.L., Zhong, R.Q., Ye, Z.H., 2009. The F8H glycosyltransferase is a functional paralog of FRA8 involved in glucuronoxylan biosynthesis in Arabidopsis. Plant Cell Physiol 50, 812–827.

Lee, J.H., Won, H.J., Tran, N.P.H., Lee, S., Kim, H.-Y., Jung, J.H., 2021. Improving lignocellulosic biofuel production by CRISPR/ Cas9-mediated lignin modification in barley. GCB Bioenergy 13, 742–752.

Li, E., Bhargava, A., Qiang, W., Friedmann, M.C., Forneris, N., Savidge, R.A., Johnson, L.A., Mansfield, S.D., Ellis, B.E., Douglas, C.J., 2012. The Class II KNOX gene KNAT7 negatively regulates secondary wall formation in Arabidopsis and is functionally conserved in Populus. New Phytol 194, 102–115.

Li, S.J., Du, J., Sun, J., Galazka, J.M., Glass, N.L., Cate, J.H.D., et al., 2010. Overcoming glucose repression in mixed sugar fermentation by co-expressing a cellobiose transporter and a beta-glucosidase in *Saccharomyces cerevisiae*. Mol. BioSyst. 6, 2129–2132.

Li, G., Jones, K.C., Eudes, A., Pidatala, V.R., Sun, J., Xu, F., et al., 2018. Overexpression of a rice BAHD acyltransferase gene in switchgrass (*Panicum virgatum* L.) enhances saccharification. BMC Biotechnol 18, 54.

Li, Q., Song, J., Peng, S., Wang, J.P., Qu, G.-Z., Sederoff, R.R., et al., 2014. Plant biotechnology for lignocellulosic biofuel production. Plant Biotechnol. J. 1–19.

Li, F., Zhang, M., Guo, K., Hu, Z., Zhang, R., Feng, Y., et al., 2015a. High-level hemicellulosic arabinose predominately affects lignocellulose crystallinity for genetically enhancing both plant lodging resistance and biomass enzymatic digestibility in rice mutants. Plant Biotechnol. J. 13, 514–525.

Li, J., Wang, L., Zhan, Q., Liu, Y., 2015b. Map-based cloning and expression analysis of BMR-6 in sorghum. J. Genet. 94, 445–452.

Li, Y., Sun, H., Fan, C., Hu, H., Wu, L., Jin, X., et al., 2019a. Overproduction of fungal endo-OE≤-1,4-glucanase leads to characteristic lignocellulose modification for considerably enhanced biomass enzymatic saccharification and bioethanol production in transgenic rice straw. Cellulose 26, 8249–8261.

Li, M., Yoo, C.G., Pu, Y., Biswal, A.K., Tolbert, A.K., Mohnen, D., et al., 2019b. Downregulation of pectin biosynthesis gene GAUT4 leads to reduced ferulate and lignin-carbohydrate cross-linking in switchgrass. Commun. Biol. 2, 22.

Liang, Y., Eudes, A., Yogiswara, S., Jing, B., Benites, V.T., Yamanaka, R., Cheng-Yue, C., Baidoo, E.E., et al., 2019. A screening method to identify efficient sgRNAs in Arabidopsis, used in conjunction with cell-specific lignin reduction. Biotechnol Biofuels 12,130.

Lionetti, V., Francocci, F., Ferrari, S., Volpi, C., Bellincampi, D., Galletti, R., et al., 2010. Engineering the cell wall by reducing de-methyl-esterified homogalacturonan improves saccharification of plant tissues for bioconversion. Proc. Natl. Acad. Sci., USA 107, 616–621.

Liu, L., Shang-Guan, K., Zhang, B., Liu, X., Yan, M., Zhang, L., et al., 2013. Brittle Culm1, a COBRA-like protein, functions in cellulose assembly through binding cellulose microfibrils. PLoS Genet. 9, e1003704.

Liu, Y., Merrick, P., Zhang, Z., Ji, C., Yang, B., Fei, S.Z., 2018. Targeted mutagenesis in tetraploid switchgrass (*Panicum virgatum* L.) using CRISPR/Cas9. Plant Biotechnol. J. 16 (2), 381–393.

Liu, G., Li, J., D, I., 2019. Genome editing by CRISPR/Cas9 in sorghum through biolistic bombardment. In: Zhao, Z.-Y., Dahlberg, J. (Eds.), Sorghum: Methods and Protocols, Methods in Molecular Biology, 1931. © Springer Science + Business Media, LLC, pp. 169–183.

López-Bellido, L., Wery, J., López-Bellido, R.J., 2014. Energy crops: prospects in the context of sustainable agriculture. Eur. J. Agron. 60, 1–12.

Loqué, D., Scheller, H.V., Pauly, M., 2015. Engineering of plant cell walls for enhanced biofuel production. Curr. Opin. Plant Biol. 25, 151–161.

Lynd, L.R., Laser, M.S., Brandsby, D., Dale, B.E., Davison, B., Hamilton, R., et al., 2008. How biotech can transform biofuels. Nat. Biotechnol. 26, 169–172.

Magrey, A., Sahay, S., Gothalwal, R., 2018. Cellulases for biofuel: a review. Int. J. Recent Trends Sci. Technol., P-ISSN 2277–2812 E-ISSN 2249–8109, Special Issue, ACAEE.

Ma, H., Liu, W., Chen, X., Wu, Y., Yu, Z., 2009. Enhanced enzymatic saccharification of rice straw by microwave pretreatment. Bioresour Technol 100, 1279–1284.

Markel, K., Belcher, M.S., Shih, P.M., 2020. Defining and engineering bioenergy plant feedstock ideotypes. Curr. Opin. Biotechnol. 62, 196–201.

Mathur, S., Umakanth, A.V., Tonapi, V.A., Sharma, R., Sharma, M.K., 2017. Sweet sorghum as biofuel feedstock: recent advances and available resources. Biotechnol. Biofuels 10, 146.

Matsumoto, N., 2017. Generation and characterization of rice CAD2 CAldMT1 double mutants with altered lignin content and structure. Sustain. Humanosphere 13, 44.

Matsumoto, N., 2018. Generation and characterization of rice CAD2 CAldMT1 double mutants with altered lignin content and structure. Sustaine Humanosphere 13, 44.

Mazarei, M., Baxter, H.L., Li, M., Biswal, A.K., Kim, K., Meng, X., et al., 2018. Functional analysis of cellulose synthase CesA4 and CesA6 Genes in switchgrass (*Panicum virgatum*) by overexpression and RNAi-Mediated gene silencing. Front. Plant Sci. 9,1114.

Mester, T., Varela, E., Tien, M., 2013. Wood Degradation by Brown-Rot and White-Rot Fungi. Genet Biotechnol 2, 355–368.

Michaels, S.D., Amasino, R.M., 2000. Memories of winter: vernalization and the competence to flower. Plant Cell Env. 23, 1145–1153.

Miladinovic, D., Antunes, D., Yildirim, K., Bakhsh, A., Cvejić, S., Kondić-Špika, A., et al., 2021. Targeted plant improvement through genome editing: from laboratory to field. Plant. Cell Rep. 40, 935–951.

Mitchell, R.B., Schmer, M.R., Anderson, W.F., Jin, V., Balkcom, K.S., Kiniry, J., et al., 2016. Dedicated energy crops and crop residues for bioenergy feedstocks in the Central and Eastern USA. Bioenerg. Res. 9, 384–398.

Miyamoto, T., Takada, R., Tobimatsu, Y., Takeda, Y., Suzuki, S., Yamamura, M., Osakabe, K., Osakabe, Y., et al., 2019. OsMYB108 loss-of-function enriches p-coumaroylated and tricin lignin units in rice cell walls. Plant J 98, 975–987.

Mohnen, D., 2008. Pectin structure and biosynthesis. Curr. Opin. Plant Biol. 11, 266–277.

Mortimer, J.C., Miles, G.P., Brown, D.M., Zhang, Z.N., Segura, M.P., Weimar, T., et al., 2010. Absence of branches from xylan in Arabidopsis guxmutants reveals potential for simplification of lignocellulosic biomass. Proc. Natl. Acad. Sci. USA. 107, 17409–17414.

Muhr, M., Paulat, M., Awwanah, M., Brinkkötter, M., Teichmann, T., 2018. CRISPR/Cas9-mediated knockout of Populus BRANCHED1 and BRANCHED2 orthologs reveals a major function in bud outgrowth control. Tree Physiol 38, 1588–1597.

Ndimande, S., 2013. Increasing Cellulosic Biomass in Sugarcane. Ph.D. Thesis, Stellenbosch University, Stellenbosch. South Africa.

Niu, E., Fang, S., Shang, X., Guo, W., 2018. Ectopic expression of GhCOBL9A, a cotton glycosyl-phosphatidyl inositol-anchored protein encoding gene, promotes cell elongation, thickening and increased plant biomass in transgenic Arabidopsis. Mol. Genet. Genomics 293, 1191–1204.

Olofsson, K., Bertilsson, M., Lidén, G., 2008. A short review on SSF – an interesting process option for ethanol production from lignocellulosic feedstocks. Biotechnol. Biofuels 1, 7.

Ozsoz, M., Ibrahim, A.U., Coston, P.P., 2019. Application of CRISPR technology for the generation of biofuels: a review. J. Fundamentals Renew. Energy Appl. 9 (2), 2090–4541.

Pagant, S., Bichet, A., Sugimoto, K., Lerouxel, O., Desprez, T., McCann, M., et al., 2002. KOBITO1 encodes a novel plasma membrane protein necessary for normal synthesis of cellulose during cell expansion in Arabidopsis. Plant Cell 14, 2001–2013.

Park, J.J., Yoo, C.G., Flanagan, A., Pu, Y., Debnath, S., Ge, Y., et al., 2017. Defined tetra-allelic gene disruption of the 4-coumarate:coenzyme A ligase 1 (Pv4CL1) gene by CRISPR/Cas9 in switchgrass results in lignin reduction and improved sugar release. Biotechnol. Biofuels 10, 284.

Park, Y.W., Baba, K., Furuta, Y., Iida, I., Sameshima, K., Arai, M., et al., 2004. Enhancement of growth and cellulose accumulation by overexpression of xyloglucanase in poplar. FEBS Lett 564, 183–187.

Pauly, M., Keegstra, K., 2008. Cell-wall carbohydrates and their modification as a resource for biofuels. Plant J. 54, 559–568.

Pauly, M., Keegstra, K., 2010. Plant cell wall polymers as precursors for biofuels. Curr. Opin. Plant Biol. 13, 304–311.

Pawar, P.M.-A., Derba-Maceluch, M., Chong, S.-L., Gómez, L.D., Miedes, E., Banasiak, A., et al., 2016. Expression of fungal acetyl xylan esterase in *Arabidopsis thaliana* improves saccharification of stem lignocellulose. Plant Biotechnol. J. 14, 387–397.

Pawar, P.M.-A., Derba-Maceluch, M., Chong, S.-L., Gandla, M.L., Bashar, S.S., Sparrman, T., et al., 2017. *In muro* deacetylation of xylan affects lignin properties and improves saccharification of aspen wood. Biotechnol. Biofuels 10, 98.

Pawar, P.M.A., Derba-Maceluch, M., Chong, S.L., Gómez, L.D., Miedes, E., Banasiak, A., Ratke, C., Gaertner, C., Mouille, G., McQueen-Mason, S.J., et al., 2015. Expression of fungal acetyl xylan esterase in Arabidopsis thaliana improves saccharification of stem lignocellulose. Plant Biotechnol. J. 14, 387–397.

Pedersen, J.F., Vogel, K.P., Funnell, D.L., 2005. Impact of reduced lignin on plant fitness. Crop Sci. 45, 812–819.

Pena, M.J., Zhong, R.Q., Zhou, G.K., Richardson, E.A., O'Neill, M.A., Darvill, A.G., et al., 2007. Arabidopsis irregular xylem8 and irregular xylem9: implications for the complexity of glucuronoxylan biosynthesis. Plant Cell 19, 549–563.

Phitsuwan, P., Sakka, K., Ratanakhanokchai, K., 2013. Improvement of lignocellulosic biomass in planta: a review of feedstocks, biomass recalcitrance, and strategic manipulation of ideal plants designed for ethanol production and processability. Biomass Bioenergy 58, 390–405.

Pilate, G., Guiney, E., Holt, K., Petit-Conil, M., Lapierre, C., Leplé, J.-C., et al., 2002. Field and pulping performances of transgenic trees with altered lignification. Nat. Biotechnol. 20, 607–612.

Piquemal, J., Chamayou, S., Nadaud, I., Beckert, M., Barrière, Y., Mila, I., et al., 2002. Down-regulation of caffeic acid o-methyltransferase in maize revisited using a transgenic approach. Plant Physiol. 130, 1675–1685.

Poovaiah, C.R., Mazarei, M., Decker, S.R., Turner, G.B., Sykes, R.W., Davis, M.F., et al., 2015. Transgenic switchgrass (*Panicum virgatum* L.) biomass is increased by overexpression of switchgrass sucrose synthase (PvSUS1). Biotechnol. J. 10, 552–563.

Rae, A.M., Street, N.R., Robenson, K.M., Harris, N., Taylor, G., 2009. Five QTL hotspots for yield in short rotation coppice bioenergy poplar: the poplar biomass loci. BMC Plant Biol. 9, 23.

Ragauskas, A.J., Williams, C.K., Davison, B.H., Britovsek, G., Cairney, J., Eckert, C.A., et al., 2006. The path forward for biofuels and biomaterials. Science 311, 484–489.

Ralph, J., Lundquist, K., Brunow, G., Lu, F., Kim, H., Schatz, P.F., et al., 2004. Lignins: natural polymers from oxidative coupling of 4-hydroxyphenyl- propanoids. Phytochem. Rev. 3, 29–60.

Raud, M., Tutt, M., Olt, J., Kikas, T., 2015. Effect of lignin content of lignocellulosic material on hydrolysis efficiency. Agron. Res. 13, 405–412.

Ray, M.J., Brereton, N.J.B., Shield, I., Karp, A., Murphy, R.J., 2012. Variation in cell wall composition and accessibility in relation to biofuel potential of short rotation coppice willows. Bioenergy Res. 5, 685–698.

Reddy, M.S., Chen, F., Shadle, G., Jackson, L., Aljoe, H., et al., 2016. Targeted down-regulation of cytochrome P450 enzymes for forage quality improvement in alfalfa (*Medicago sativa* L.). Proc Natl Acad Sci USA 102, 16573–16578.

Regina, J., Taherzadeh, M.J., 2019. Fermentation processes for second generation biofuels. In: Basile, A., Dalena, F. (Eds.), Second and Third Generation of Feedstocks: The Evolution of Biofuels. Elsevier B.V, pp. 241–272.

Rennie, E.A., Scheller, H.V., 2014. Xylan biosynthesis. Curr. Opin. Biotechnol. 26, 100–107.

Ridley, B., O'Neill, M.D., 2001. Pectin: structure, biosynthesis, and oligogalacturonide related signalling. Phytochemistry 57, 292-67.

Rocha-Meneses, L., Ferreira, J.A., Mushtaq, M., Karimi, S., Orupõld, K., Kikasa, T., 2020. Genetic modification of cereal plants: a strategy to enhance bioethanol yields from agricultural waste. Ind. Crop Prod. 150, 112408.

Rubin, E.M., 2008. Genomics of cellulosic biofuels. Nature 454, 841–845.

Saballos, A., Ejeta, G., Sanchez, E., Kang, C., Vermerris, W.A., 2009. Genome-wide analysis of the cinnamyl alcohol dehydrogenase family in sorghum (*Sorghum bicolor* (L.) Moench) identifies SbCAD2 as the Brown midrib6 gene. Genetics 181 (2), 783–795.

Sahoo, D.K., Stork, J., DeBolt, S., Maiti, I.B., 2013. Manipulating cellulose biosynthesis by expression of mutant Arabidopsis proM24: CESA3ixr1-2 gene in transgenic tobacco. Plant Biotechnol. J. 11, 362–372.

Sainz, M.B., 2011. Commercial cellulosic ethanol: the role of plant-expressed enzymes. In: Tomes, D., Lakshmanan, P., Songstad, D. (Eds.), Biofuels: Global Impact on Renewable Energy, Production Agriculture, and Technological Advancements. Springer, New York, New York, NY, pp. 237–264.

Salehi, H., Ransom, C.B., Oraby, H.F., Seddighi, Z., Sticklen, M.B., 2005. Delay in lowering and increase in biomass of transgenic tobacco expressing the Arabidopsis floral repressor gene 'Flowering locus C'. J. Plant Physiol. 162, 711–717.

Salehi, J.G., Sharafi, R., Soheilivand, S., 2018. Fuelling the future: plant genetic engineering for sustainable biodiesel production. Biofuel Res. J. 19, 829–845.

Sanderson, M.A., Adler, P.R., 2008. Perennial forages as second generation bioenergy crops. Int. J. Mol. Sci. 9, 768–788.

Sathitsuksanoh, N., Sawant, M., Truong, Q., Tan, J., Canlas, C.G., Sun, N., et al., 2015. How alkyl chain length of alcohols affects lignin fractionation and ionic liquid recycle during lignocellulose pretreatment. Bioenergy Res 8, 973–981.

Sattler, S.E., Funnell-Harris, D.L., Pedersen, J.F., 2010. Brown midrib mutations and their importance to the utilization of maize, sorghum, and pearl millet lignocellulosic tissues. Plant Sci. 178, 229–238.

Sattler, S.E., Palmer, N.A., Saballos, A., Greene, A.M., Xin, Z.G., Sarath, G., et al., 2012. Identification and characterization of four missense mutations in Brown midrib 12 (Bmr12), the caffeic O-methyltranferase (COMT) of sorghum. Bioenergy Res. 5, 855–865.

Schultink, A., Naylor, D., Dama, M., Pauly, M., 2015. The role of the plant-specific ALTERED XYLOGLUCAN9 protein in Arabidopsis cell wall polysaccharide O-acetylation. Plant Physiol 167, 1271–1283.

Sexton, S., Zilberma, D., Rajagopal, D., Hochman, G., 2009. The role of biotechnology in a sustainable biofuel future. AgBioForum 12 (1), 1–11.

Sheldon, C.C., Burn, J.E., Perez, P.P., Metzger, J., Edwards, J.A., Peacock, W.J., et al., 1999. The FLF MADS box gene: a repressor of flowering in Arabidopsis regulated by vernalization and methylation. Plant Cell 11, 445–458.

Shen, B.Z., Sun, X.G., Zou, X., Shilling, T., Aogar, J., Ross, M., et al., 2012. Engineering a thermoregulated intein-modified xylanase into maize for consolidated lignocellulosic biomass processing. Nat. Biotechnol. 30, 1131–1136.

Shen, Y., Li, Y., Xu, D., Yang, C., Li, C., Luo, K., 2018. Molecular cloning and characterization of a brassinosteriod biosynthesis-related gene PtoDWF4 from *Populus tomentosa*. Tree Physiol 38, 1424–1436.

Singh, J., Suhag, M., Dhaka, A., 2015. Augmented digestion of lignocellulose by steam explosion, acid and alkaline pre-treatment methods: a review. Carbohydr. Polym. 117, 624–631.

Song, G., Jia, M., Chen, K., Kong, X., Khattak, B., et al., 2016. CRISPR/Cas9: a powerful tool for crop genome editing. Crop J. 4, 75–82.

Spiertz, J.H.J., Ewert, F., 2009. Crop production and resource use to meet the growing demand for food, feed and fuel: opportunities and constraints. NJAS-Wagen. J. Life. Sci. 56, 281–300.

Stein, O., Granot, D., 2019. An overview of sucrose synthases in plants. Front. Plant Sci. 10, 95.

Sticklen, M.B., 2008. Plant genetic engineering for biofuel production: towards affordable cellulosic ethanol. Nat. Rev. Genet. 9, 433.

Sumiyoshi, M., Nakamura, A., Nakamura, H., Hakata, M., Ichikawa, H., Hirochika, H., et al., 2013. Increase in cellulose accumulation and improvement of saccharification by over-expression of arabinofuranosidase in rice. PLoS One 8, e78269.

Takahashi, J., Rudsander, U.J., Hedenstrom, M., Banasiak, A., Harholt, J., Amelot, N., Immerzeel, P., Ryden, P., Endo, S., Ibatullin, F.M., et al., 2009. KORRIGAN1 and its aspen homolog PttCel9A1 decrease cellulose crystallinity in Arabidopsis stems. Plant Cell Physiol 50, 1099–1115.

Takeda, Y., Suzuki, S., Tobimatsu, Y., Osakabe, K., Osakabe, Y., Ragamustari, S.K., Sakamoto, M., Umezawa, T., 2019. Lignin characterization of rice coniferaldehyde 5-hydroxylase loss-of-function mutants generated with the CRISPR/Cas9 system. Plant J 97, 543–554.

Takeda, Y., Tobimatsu, Y., Karlen, S.D., Koshiba, T., Suzuki, S., Yamamura, M., Murakami, S., Mukai, M., et al., 2018. Downregulation of p-coumaroyl ester 3-hydroxylase in rice leads to altered cell wall structures and improves biomass saccharification. Plant J 95, 796–811.

Tan, H.T., Shirley, N.J., Singh, R.R., Henderson, M., Dhugga, K.S., Mayo, G.M., Fincher, G. B., Burton, R.A., 2015. Powerful regulatory systems and post-transcriptional gene silencing resist increases in cellulose content in cell walls of barley. BMC Plant Biol 15, 62.

Tkalec, M., Štefanić, P.P., Cvjetko, P., Šikić, S., Pavlica, M., Balen, B., 2014. The effects of cadmium-zinc interactions on biochemical responses in tobacco seedlings and adult plants. PLoS One 9 (1), e87582.

Tomassetti, S., Pontiggia, D., Verrascina, I., Reca, I.B., Francocci, F., Salvi, G., Cervone, F., Ferrari, S., 2015. Controlled expression of pectic enzymes in Arabidopsis thaliana enhances biomass conversion without adverse effects on growth. Phytochemistry 112, 221–230.

Torney, F., Moeller, L., Scarpa, A., Wang, K., 2007. Genetic engineering approaches to improve bioethanol production from maize. Curr. Opin. Plant Biol. 18, 193–199.

Torres, A.F., Visser, R.G.F., Trindade, L.M., 2015. Bioethanol from maize cell walls: genes, molecular tools, and breeding prospects. GCB Bioenergy 7, 591–607.

Ulaganathan, K., Goud, S., Reddy, M., Kayalvili, U., 2017. Genome engineering for breaking barriers in lignocellulosic bioethanol production. Renew. Sustain. Energy Rev. 74, 1080–1107.

Vega-Sánchez, M.E., Ronald, P.C., 2010. Genetic and biotechnological approaches for biofuel crop improvement. Curr. Opin. Biotechnol. 21, 218–224.

Van Acker, R., Leplée, J.C., Aerts, D., Stormea, V., Goeminne, G., Ivens, B., Légée, F., Lapierre, C., Piens, K., Van Montagu, M.C., et al., 2014. Improved saccharification and ethanol yield from field-grown transgenic poplar deficient in cinnamoyl-CoA reductase. Proc. Natl. Acad. Sci. USA 111, 845–850.

Vanholme, R., Cesarino, I., Rataj, K., Xiao, Y., Sundin, L., Goeminne, G., Kim, H., Cross, J., Morreel, K., Araujo, P., et al., 2013. Caffeoyl Shikimate Esterase (CSE) is an enzyme in the lignin biosynthetic pathway in Arabidopsis. Science 341, 1103–1106.

Vanholme, R., Demedts, B., Morreel, K., Ralph, J., Boerjan, W., 2010. Lignin biosynthesis and structure. Plant Physiol 153, 895–905.

Vega-Sánchez, M.E., Loqué, D., Lao, J., Catena, M., Verhertbruggen, Y., Herter, T., et al., 2015. Engineering temporal accumulation of a low recalcitrance polysaccharide leads to increased C6 sugar content in plant cell walls. Plant Biotechnol. J. 13, 903–914.

Vervoort, Y., Linares, A.G., Roncoroni, M., Liu, C., Steensels, J., Verstrepen, K.J., 2017. High-throughput system-wide engineering and screening for microbial biotechnology. Curr. Opin. Biotechnol. 46, 120–125.

Wan, S., Li, C., Ma, X., Luo, K., 2017. PtrMYB57 contributes to the negative regulation of anthocyanin and proanthocyanidin biosynthesis in poplar. Plant Cell Rep 3, 1263–1276.

Wang, G., Gao, Y., Wang, J., Yang, L., Song, R., Li, X., et al., 2011. Overexpression of two cambium-abundant Chinese fir (*Cunninghamia lanceolata*) (-expansin genes ClEXPA1 and ClEXPA2 affect growth and development in transgenic tobacco and increase the amount of cellulose in stem cell walls. Plant Biotechnol. J. 9, 486–502.

Wang, Y., Ma, N., Qiu, S., Zou, H., Zang, G., Kang, Z., et al., 2014. Regulation of the a-expansin gene OsEXPA8 expression affects root system architecture in transgenic rice plants. Mol. Breed. 34, 47–57.

Wang, Y., Fan, C., Hu, H., Li, Y., Sun, D., Wang, Y., et al., 2016. Genetic modification of plant cell walls to enhance biomass yield and biofuel production in bioenergy crops. Biotechnol. Adv. 34, 997–1017.

Wang, L., Ran, L., Hou, Y., Tian, Q., Li, C., Liu, R., Fan, D., Luo, K., 2017. The transcription factor MYB115 contributes to the regulation of proanthocyanidin biosynthesis and enhances fungal resistance in poplar. New Phytol 215, 351–367.

Wei, Z., Qu, Z., Zhang, L., Zhao, S., Bi, Z., Ji, X., et al., 2015. Overexpression of poplar xylem sucrose synthase in tobacco leads to a thickened cell wall and increased height. PLoS One 10, e0120669.

Wilkerson, C.G., Mansfield, S.D., Lu, F., Withers, S., Park, J.Y., Karlen, S.D., Gonzales-Vigil, E., Padmakshan, D., Unda, F., Rencoret, J., et al., 2014. Monolignol ferulate transferase introduces chemically labile linkages into the lignin backbone. Science 344, 90–93.

Wolf, S., Mouille, G., Pelloux, J., 2009. Homogalacturonan methylesterification and plant development. Mol. Plant 2, 851–860.

Wu, A.-M., Hörnblad, E., Voxeur, A., Gerber, L., Rihouey, C., Lerouge, P., et al., 2010. Analysis of the Arabidopsis IRX9/IRX9-L and IRX14/IRX14-L pairs of glycosyltransferase genes reveals critical contributions to biosynthesis of the hemicellulose glucuronoxylan. Plant Physiol. 153, 542–554.

Wu, A.M., Rihouey, C., Seveno, M., Hörnblad, E., Singh, S.K., Matsunaga, T., et al., 2009. The Arabidopsis IRX10 and IRX10-LIKE glycosyltransferases are critical for glucuronoxylan biosynthesis during secondary cell wall formation. Plant J 57, 718–731.

Wu, Z., Zhang, M., Wang, L., Tu, Y., Zhang, J., Xie, G., et al., 2013. Biomass digestibility is predominantly affected by three factors of wall polymer features distinctive in wheat accessions and rice mutants. Biotechnol. Biofuels 6, 183.

Wu, Z.L., Hao, H.H., Zahoor, T.Y.Y., Hu, Z., Wei, F., et al., 2014. Diverse cell wall composition and varied biomass digestibility in wheat straw for bioenergy feedstock. Biomass Bioenergy 70, 347–355.

Xie, G., Peng, L., 2011. Genetic engineering of energy crops: a strategy for biofuel production in China. J. Integr. Plant Biol. 53, 143–150.

Xiong, G.Y., Cheng, K., Pauly, M., 2013. Xylan O-acetylation impacts xylem development and enzymatic recalcitrance as indicated by the Arabidopsis mutant tbl29. Mol. Plant 2013, 6, 1373–1375.

Xu, B., Escamilla-Treviño, L.L., Sathitsuksanoh, N., Shen, Z., Shen, H., Zhang, Y.H., et al., 2011. Silencing of 4-coumarate: coenzyme A ligase in switchgrass leads to reduced lignin content and improved fermentable sugar yields for biofuel production. N. Phytol. 192, 611–625.

Xu, C., Fu, X., Liu, R., Guo, L., Ran, L., Li, C., Tian, Q., Jiao, B., Wang, B., Luo, K., 2017. PtoMYB170 positively regulates lignin deposition during wood formation in poplar and confers drought tolerance in transgenic Arabidopsis. Tree Physiol 37, 1713–1726.

Xu, M., Hu, T., Zhao, J., Park, M.-Y., Earley, K.W., Wu, G., et al., 2016. Developmental functions of miR156-regulated squamosa promoter bindingprotein-like (SPL) genes in *Arabidopsis thaliana*. PLoS Genet. 12 (8), e1006263.

Yang, C.H., Li, D.Y., Liu, X., Ji, C.J., Hao, L.L., Zhao, X.F., Li, X.B., Chen, C.Y., Cheng, Z.K., Zhu, L.H., 2014. OsMYB103L, an R2R3MYB transcription factor, influences leaf rolling and mechanical strength in rice (*Oryza sativa* L.). BMC Plant Biol 14, 158.

Yang, C.Y., Fang, T.J., 2014. Combination of ultrasonic irradiation with ionic liquid pretreatment for enzymatic hydrolysis of rice straw. Bioresour. Technol. 164, 198–202.

Yang, L., Zhao, X., Ran, L., Li, C., Fan, D., Luo, K., 2017. PtoMYB156 is involved in negative regulation of phenylpropanoid metabolism and secondary cell wall biosynthesis during wood formation in poplar. Sci Rep 7, 41209.

Yang, Y., Yoo, C.G., Guo, H.-B., Rottmann, W., Winkeler, K.A., Collins, C.M., et al., 2017a. Overexpression of a domain of unknown function 266-containing protein results in high cellulose content, reduced recalcitrance, and enhanced plant growth in the bioenergy crop populus. Biotechnol. Biofuels 10, 74.

Yang, Y., Yoo, C.G., Winkeler, K.A., Collins, C.M., Hinchee, M.A.W., Jawdy, S.S., et al., 2017b. Overexpression of a domain of unknown function 231-containing protein increases O-xylan acetylation and cellulose biosynthesis in populus. Biotechnol. Biofuels 10, 311.

Ye, X., Busov, V., Zhao, N., Meilan, R., McDonnell, L.M., Coleman, H.,D., et al., 2011. Transgenic Populus trees for forest products, bioenergy, and functional genomics. Crit. Rev. Plant Sci. 30 (5), 415e34.

York, W.S., O'neill, M.A., 2008. Biochemical control of xylan biosynthesis—which end is up? Current Opinion in Plant Biology 11, 258–265.

Yuan, J.S., Tiller, K.H., Al-Ahmad, H., Stewart, N.R., Stewart, C.N., 2008. Plants to power: bioenergy to fuel the future. Trends Plant Sci. 13, 421–429.

Yuan, Y.X., Teng, Q., Zhong, R.Q., Ye, Z.H., 2013. The Arabidopsis DUF231 domain-containing protein ESK1 mediates 2-O- and 3-O-acetylation of xylosyl residues in xylan. Plant Cell Physiol 54, 1186–1199.

Zegada-Lizarazu, W., Monti, A., 2012. Are we ready to cultivate sweet sorghum as a bioenergy feedstock? A review on field management practices. Biomass Bioenergy 40, 1–12.

Zegada-Lizarazu, W., Zatta, A., Monti, A., 2012. Water uptake efficiency and above and below ground biomass development of sweet sorghum and maize under different water regimes. Plant Soil 351, 47–60.

Zeng, W., Jiang, N., Nadella, R., Killen, T.L., Nadella, V., et al., 2010. A glucurono (arabino) xylan synthase complex from wheat contains members of the GT43, GT47, and GT75 families and functions cooperatively. Plant Physiol 154, 78–97.

Zhang, B.C., Deng, L.W., Qian, Q., Xiong, G.Y., Zeng, D.L., Li, R., et al., 2009. A missense mutation in the transmembrane domain of CESA4 affects protein abundance in the plasma membrane and results in abnormal cell wall biosynthesis in rice. Plant Mol. Biol. 71, 509–524.

Zhang, K., Bhuiya, M.W., Pazo, J.R., Miao, Y., Kim, H., et al., 2012. An engineered monolignol 4-O-methyltransferase depresses lignin biosynthesis and confers novel metabolic capability in Arabidopsis. The Plant Cell 24, 3135–3152.

Zhang, L., Xu, J., Birch, R.G., 1999. Engineered detoxification confers resistance against a pathogenic bacterium. Nat. Biotechnol. 17, 1021–1024.

Zhang, S.-Z., Yang, B.-P., Feng, C.-L., Chen, R.-K., Luo, J.-P., Cai, W.-W., et al., 2006. Expression of the *Grifola frondosa* trehalose synthase gene and improvement of drought-tolerance in sugarcane (*Saccharum officinarum* L.). J. Integr. Plant Biol. 48, 453–459.

Zhou, X., Jacobs, T.B., Xue, L.J., Harding, S.A., Tsai, C.J., 2015. Exploiting SNPs for biallelic CRISPR mutations in the outcrossing woody perennial Populus reveals 4-coumarate: CoA ligase specificity and redundancy. N. Phytol. 208, 298–301.

Pretreatment technologies for lignocellulosic biomass refineries

4

Ankita Juneja[1], *Deepak Kumar*[2], *Karthik Rajendran*[3] *and Ashutosh Mittal*[4]
[1]Department of Biology, Syracuse University, Syracuse, NY, United States, [2]Department of Chemical Engineering, SUNY College of Environmental Science and Forestry, Syracuse, NY, United States, [3]Department of Environmental Science and Engineering, School of Engineering and Sciences, SRM University-AP, Amaravati, Andhra Pradesh, India, [4]National Renewable Energy Laboratory, Golden, CO, United States

4.1 Introduction

An increase in energy demand due to rising population and industrialization, rapid depletion of fossil resources, and environmental concerns associated with the use of fossil fuels, necessitate the need for alternative and renewable energy sources. The United States is the biggest consumer of oil in the world, and about one-third of the energy is used in the transportation sector, derived mostly from petroleum ($\sim$90%). Biofuel derived from plant-based sources (grain crops, agricultural residues, algae, etc.) would play a critical role in addressing these challenges (Ray and Ramachandran, 2018). Bioethanol is one such important biofuel that provides significant advantages [high blending octane rating, lower greenhouse gas (GHG) emissions, high net energy ratio] compared to gasoline. The United States produces more than 16 billion gallons of bioethanol annually. According to a recent report, the use of ethanol-blend gasoline in the United States resulted in a decrease of approximately 55 million metric tonnes of CO_2-equivalent GHG emissions in the year 2021, equivalent to taking 12 million cars from the road for an entire year (RFA, 2022). However, only a limited number of feedstocks are successful on a commercial scale in biofuel production. For instance, approximately 95% of the bioethanol in the United States is produced using corn starch. Most of the bioethanol in Brazil (second biggest producer of ethanol in the world) is produced from sugarcane. The use of food crops for biofuel creates "food versus fuel" debates, and suffers from the limitation of limited capacity to replace a large amount of gasoline use, intensive agriculture inputs, and the need for arable land. Alternatively, lignocellulosic biomass (LCB), such as agricultural and forestry residues, municipal solid wastes, and dedicated energy crops, are nonfood feedstock for biofuel production due to its ample global availability and low cost (Ray, 2021).

The lignocellulosic feedstock is a hetero-matrix of three major polymers, cellulose, hemicellulose, and lignin. Other than these major components, lignocellulose

Advances in Lignocellulosic Biofuel Production Systems. DOI: https://doi.org/10.1016/B978-0-323-91192-4.00004-3

contains small amounts of proteins, chlorophyll, resins, terpenoids, silicates, sulfates, carbonates, and nitrates, which are not an active part of the biomass structure (Vassilev et al., 2010). The carbohydrate fractions (cellulose and hemicellulose) can be hydrolyzed into sugars that can be subsequently fermented to biofuel or used for biochemical and bioproduct production. However, lignin acts as a severe barrier for the enzymes or microbial access to the carbohydrates to produce the desired product. Therefore a pretreatment step is necessary to break the biomass recalcitrance, remove the lignin barrier, and improve the accessibility of microbes to carbohydrates for efficient bioconversion. The structure of the matrix varies to different degrees and compositions depending on the source of biomass. The relative amounts of cellulose, hemicellulose, or lignin are key factors in determining the optimum conversion route and end product produced (McKendry, 2002). Cellulose is the most abundant polymer in biomass and is a major constituent of the plant cell wall. It is a polymer of β-D-glucopyranose moieties linked with β-1,4 glycosidic bonds (Fig. 4.1A) in its unbranched form. The repeating unit of cellulose is the cellobiose with a degree of polymerization ranging from 10,000 to 15,000. About 200–300 cellulose chains are bundled together to form microfibrils, which are further grouped to form cellulose fibers (Kumar and Murthy, 2013). Most of the cellulose is in crystalline form. However, a small portion of cellulose is in an amorphous and unorganized form, which is more susceptible to degradation with enzymes. Cellulose exhibits low solubility in water at ambient temperature. However, solubility increases with increasing process temperature (Klemm et al., 2005; Roy et al., 2009).

Hemicellulose is the second most abundant polymer in the LCB. Hemicellulose is a heteropolymer with short lateral branches. It contains a variety of

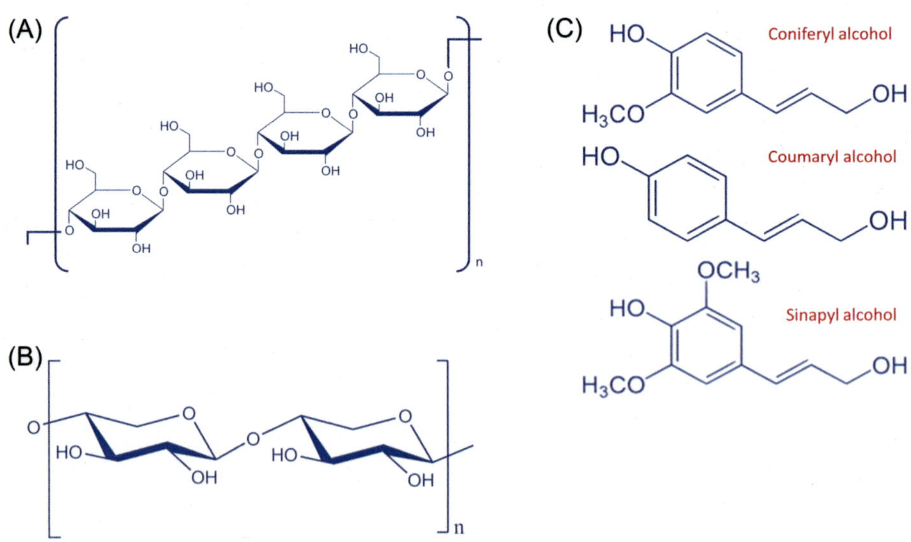

Figure 4.1 Structural backbone of (A) cellulose, (B) hemicellulose, and (C) lignin.

monosaccharides, including hexoses (mannose, galactose, and glucose), pentoses (xylose, arabinose, and rhamnose), and uronic acids (4-O-methyl glucuronic acid, d-glucuronic and d-galacturonic acids) (Fig. 4.1B) (Bajpai, 2016). Hemicellulose is essentially amorphous due to the side groups linked to the linear backbone chain of xylan constructed from β-1→4-linked d-xylopyranosyl residues by β-1→4-glycosidic bonds. Similar to cellulose, hemicellulose also shows low solubility in water at ambient temperature. However, solubility improves with increasing the temperature and lowering the pH (Harmsen et al., 2010). Since hemicelluloses have lower molecular weight and branches with short chains, these are easily hydrolyzed.

Lignin is a complex, amorphous, irregular hetero-structure with cross-linked polymers of phenolic monomers. The three monomers of lignin are coniferyl alcohol, coumaryl alcohol, and sinapyl alcohol (Fig. 4.1C) (Bajpai, 2016), that form p-hydroxyphenyl (H), guaiacyl (G), and syringyl (S) units, respectively. Lignin encompasses cellulose and hemicellulose and forms a resistant and impermeable layer for the attack against enzymes and microorganisms. Lignin is a major restriction to biomass hydrolysis due to its adherence to the cellulose microfibrils (Avgerinos and Wang, 1983; Kumar and Murthy, 2011b). It has been shown that increasing lignin removal can lead to enhancing biomass digestibility (Chang and Holtzapple, 2000). Although lignin is insoluble in water at ambient and neutral conditions, similar to hemicelluloses, solubility is increased in the presence of acids at very high temperatures (above 180°C) (Harmsen et al., 2010).

Each of these components can act as a promising polymer for producing value-added products. Cellulose and hemicellulose are composed of monomeric sugars which have wide applications in the production of biofuels, bioproducts, platform chemicals, bioplastics, and many more. Recently, lignin has been extensively investigated to be recovered in a pure form for producing high-value fuels and chemicals. Valorization of all the components of LCB can lead to an economic biorefinery producing high-value products using low-cost feedstock.

4.2 Bioprocessing schemes of lignocellulosic biomass

The conversion of biomass to desired bioproducts can be attained through biochemical or thermochemical conversions. Biochemical pathways lead to the production of fermentative bioproducts such as bioethanol, polyhydroxyalkanoates, organic acids, and biobutanol, whereas thermochemical methods can be utilized to produce bio-oil, furan-based chemicals, and other organic acids (Al-Battashi et al., 2019; Jang et al., 2012; Yoo et al., 2017b). Possible schemes for processing LCB are shown in Fig. 4.2.

4.3 Pretreatment of lignocellulosic biomass

Pretreatment is a crucial step in the bioprocessing of LCB to break the recalcitrance of the lignocellulosic material and facilitate hydrolysis and fermentation by

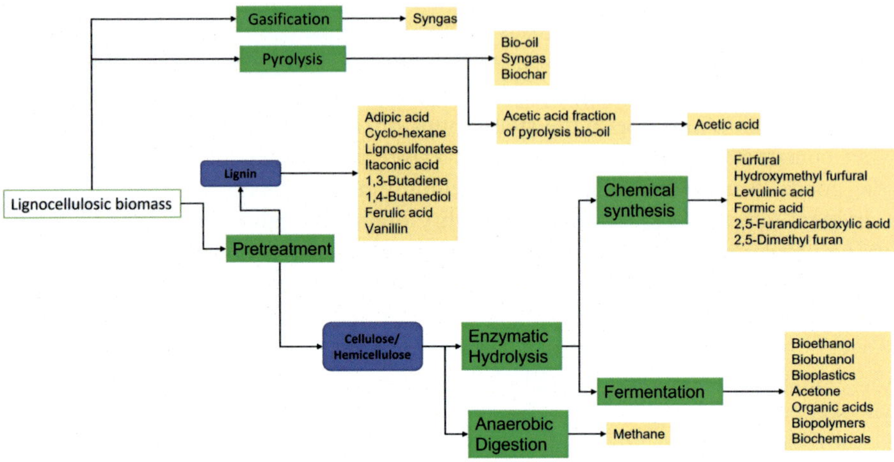

Figure 4.2 Bioprocessing schemes for lignocellulosic biomass.

releasing the cellulose and hemicellulose. Pretreatment also facilitates lignin removal, hemicellulose degradation, reduction of cellulose crystallinity, and increased porosity of the biomass surface, making it more susceptible to chemical reactions and usage across multiple industries. The broad categories that the pretreatment technologies can be categorized into are physical, chemical, physicochemical (combination of physical and chemical technologies), and biological treatments, which have been used to break and solubilize the lignin matrix (Fig. 4.3). All pretreatment methods work with a different mechanism and have advantages and disadvantages over other pretreatment methods, which are summarized in Table 4.1.

4.3.1 Physical pretreatments

Physical pretreatment is performed mainly by reducing the size of the biomass to change the physical structure of the biomass and expose the maximum surface area. Physical pretreatments avoid any use of chemicals, thereby reducing the generation of inhibitors and chemical waste. Mechanical pretreatment is one of the most conventional physical pretreatment methods, reducing the size of the particles by grinding the biomass and increasing the specific surface area and bulk density. Higher surface area helps in further chemical processing due to the elimination of heat transfer limitation and the development of a phase boundary between the chemicals and lignocellulosic material (Khan et al., 2016). On the other hand, higher bulk density improves the handling of biomass during storage and transport.

Gu et al. (2018) investigated the effect of different physical pretreatments, hammer milling, air classifier milling, ball milling, chipping, and mashing, with different residence times (7–30 min) on the production of sugar monomers. The authors found ball milling for 30 min to be the best condition to obtain the highest yield of

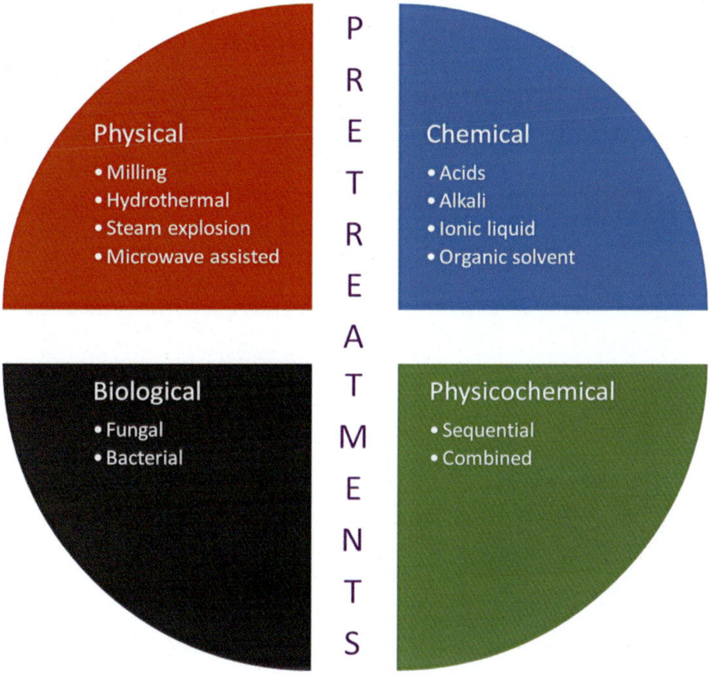

Figure 4.3 Categories of pretreatments.

glucose (59.67%) and xylose/mannose (23.82%). The crystallinity index with this condition also reduced from 40.73% to 11.70% (Gu et al., 2018). Zhang et al. (2018) compared the effectiveness of ball milling to two physico-chemical pretreatments (microwave-assisted alkali and ultrasound-assisted alkali) on the fermentation of rumen-digested residue of rice straw (2.5% and 10% digestate) and observed the highest ethanol yield of 116.65 and 147.42 mg/g for ball mill pretreated 2.5% and 10% digestate residue, respectively (Zhang et al., 2018). The authors also found that for 2.5% digested residue, the energy output for the integrated system involving ball milling was maximum (7010 kJ/kg). Bychkov et al. (2018) pretreated the *Phragmites australis* straw using ball milling at a temperature of 180°C and observed substantial changes in the cell wall structure due to increased mobility of the devitrified lignin (Bychkov et al., 2018). Due to high temperature alongside mechanical treatment under shear action on the biomass, the lignin melts and forms pseudo-lignin accumulations on the surface of the particles (Meng and Ragauskas, 2017; Sannigrahi et al., 2011). Another widely used physical pretreatment is disk milling to finely grind the biomass. Disk mill induces class II size reductions, where biomass size is reduced beyond nanofibrils with the significant breakup of microfibril cross-links and compression-induced internal fibrillation (Juneja et al., 2021). Disk mill has been used as a sole pretreatment (Hideno et al., 2009; Zhang et al., 2013), as well as sequentially with other chemical pretreatments (Cheng et al.,

Table 4.1 Comparison of various pretreatment methods for lignocellulosic biomass.

Category	Pretreatment	Conditions	Mechanism	Advantages	Disadvantages
Physical	Milling (Dry or wet)	Dry biomass or increased moisture biomass	Reduce the particle size of biomass to increase the surface area for enzymatic attack Reduce cellulose crystallinity	Easy size reduction Minimal hazardous process High hydrolysis yield Low feedstock processing	Require high energy Low lignin removal
Chemical	Acid	Up to 5% acid concentration High temperature 5–30 min residence time	Dissolution of hemicellulose Reduce cellulose polymerization	High yield of sugars High hemicellulose removal	Require neutralization Corrosion Formation of inhibitors Chemical waste generation Expensive chemicals
	Alkali		Solubilizes hemicellulose Lignin degradation Decreases polymerization degree Increase of surface area	Increase enzymatic hydrolysis	Require high temperature ($>120°C$) Require neutralization Hydrolysate contains lignin and toxic compounds
	Organic solvent			High extraction efficiency Solvent recovery and reuse are simple	High cost Flammability Volatility of solvents The low boiling point of solvents
	Ionic liquid	High temperature ($\sim100°C$) Residence time ~50 min	Strong hydrogen bond coordination leads to the dissolution of the entire biomass	High solubilization of cellulose and hemicellulose	Require high temperature (100°C) Require high biomass load High-cost investment

physicochemical	Steam explosion	Temperature range: 150°C–240°C Pressure range: 0.7–4.8 MPa pressure is explosively released	Disruption in fiber structure due to sudden pressure change Hemicellulose breakdown at 150°C Lignin deconstruction starts at 180°C	High hydrolysis yield No use of chemicals Hemicellulose solubilization and lignin transformation	Formation of inhibitory compounds Incomplete breakdown of lignin carbohydrate matrix
	Ammonia Fiber Explosion (AFEX)	Liquid ammonia under pressure (200–500 psi) at varying temperatures (80°C–150°C) for a short residence time (5–30 min) pressure is explosively released	Cellulose swelling by liquid ammonia and a phase change in the crystal structure from cellulose I to cellulose III Disruption in fiber structure due to sudden pressure change Cleavage of lignin-carbohydrate complexes	Low use of chemicals High solid recovery (low sugar loss) Minimal inhibitor formation	Challenges with ammonia recycling High cost Alters lignin structure
	Microwave-assisted		The heat generated pressure inside the hydrated and nonhydrated components	low reaction time Uniform heating High yields Low activation energy Minimal inhibitor formation	High energy expenditure The moisture content of biomass High operating cost and capital investment
Biological	Fungi Bacteria	Conditions vary with the microorganism Days to weeks	Microorganisms attack, depolymerize and degrade the polymers in lignocellulosic substrates Degrade lignin by the action of lignin-degrading enzymes such as peroxidases and laccases Hemicellulose and lignin decomposition	Low energy consumption Environmentally friendly No inhibitor generation No catalyst	Low hydrolysis rate Low efficiency Long incubation time The slow rate of delignification It needs a large sterile area

2019; Juneja et al., 2020; Juneja et al., 2021; Kim et al., 2016). Dell'Omo and Spena (2020) observed the effect of dry milling on the anaerobic digestion of giant reed and wheat straw for methane production (Dell'Omo and Spena, 2020). The cumulative biogas production showed a gain of 137% and 49.1% for giant reed and wheat straw, respectively. Pretreatment also significantly decreased the acid-detergent fiber content of processed giant reed. Another study reported an increase of 53% in biogas yield with wheat straw when pretreated at 180°C due to an increase in the cellulose content. Hideno et al. (2009) compared dry ball milling, wet disk milling, and hot compressed water treatment as pretreatments for rice straw before enzymatic hydrolysis (Hideno et al., 2009). Although dry ball mill (at the highest residence time of 60 min) (78.2%) and liquid hot water pretreatment at 160°C (74.1%) gave higher sugar yields, their energy consumption (108 and 5.7 MJ/kg rice straw, respectively) was also higher than the wet disk milling with 10 cycles (67.5% sugar yield and 5.4 MJ/kg rice straw). Wet disk mill leads to reduction in the thickness of fiber and increases the available surface area without reduction of crystallinity.

One of the major limitations that physical methods face is high energy consumption. The amount of energy consumed significantly increases with an increase in the residence time of pretreatment. For example, increasing the pretreatment time from 7 to 30 min of ball milling increased the energy consumption from 0.50 to 2.15 kWh/kg, with the glucose yield after the pretreatment increasing from 24.45% to 59.67%, and xylose/mannose yield from 11.92% to 23.82% (Gu et al., 2018). In the case of only wet disk milling, it is observed that hydrolysis yield increases with the number of disk mill cycles, which increases the energy consumption too. The energy consumption for one cycle of disk milling was calculated as 0.3 kJ/kg rice straw and was observed to significantly increase to 5.4 kJ/kg rice straw with 10 cycles (Hideno et al., 2009). One potential approach to reduce this energy consumption is to use sequential pretreatment (use of mechanical refining after a mild hydrothermal pretreatment), discussed in the later sections of the chapter.

4.3.2 Chemical pretreatments

Chemical pretreatments are generally carried out using acid (dilute sulfuric, nitric, and hydrochloric acids) or alkali (sodium, potassium, calcium, and ammonium hydroxide) chemical methods that demonstrate distinct mechanisms for the destruction of the cell wall. Acid-based pretreatments expose the cellulose for enzymatic digestion by hydrolyzing the hemicellulose component (Silverstein et al., 2007), whereas alkali, during pretreatment, breaks the ether–ester bonds cross-linking the hemicellulose and lignin, swells and partially recrystallizes the cellulose to make the cellulose accessible to the enzymes (Kapoor et al., 2015; Kumar and Murthy, 2011a; Silverstein et al., 2007). Liquid hot water treatment is also considered chemical pretreatment since water acts as a very dilute acid at higher temperatures and pressure due to increased hydronium ion concentration under these conditions (Mood et al., 2013). Two new chemical treatments have been recently investigated: organic solvent and ionic liquid (IL) pretreatment. The attraction of organic solvents comes from their ability to fractionate the three major polymers, cellulose, hemicellulose, and lignin, with high purity and an added advantage of easy solvent recovery and reuse. On the other hand, ILs have gained attention due to the solubility of biomass in these liquids and the high recoveries of monomeric sugars and carbohydrates (Brandt et al., 2013).

4.3.2.1 Acid pretreatment

Dilute acid is the most used chemical for the pretreatment of lignocellulose, where acids can be sulfuric, nitric, phosphoric, acetic, or hydrochloric acid. Concentrated acids can also be utilized but are not preferred due to the high operation and maintenance cost of equipment due to corrosion and the formation of inhibitory compounds (Wyman, 1996). Acids reduce the recalcitrance by degradation of hemicellulose, and with an increase in severity, it can also lead to some cellulose hydrolysis. Acid breaks all lignin–polysaccharide linkages to facilitate the recovery of most of the monomeric sugars (Gonzales et al., 2016). The severe acid pretreatment can lead to monomeric sugar degrading into inhibitors, such as 5-hydroxymethylfurfural (HMF) and furfurals.

The major factors affecting acid pretreatment are the choice of acid, acid concentration, solid loading, residence temperature, and time (Solarte-Toro et al., 2019). Santos et al.(2018) observed that increasing the acid concentration from 5% to 20% w/w at 121°C for 30 min reduced the solid recovery of pretreated elephant grass from 59.9% to 54.3%. This was primarily due to the hydrolysis of hemicellulosic sugar in addition to the partial solubilization of lignin in the LCB. The pretreatment of wheat straw with a sulfuric acid concentration of 0.5% w/w at 160°C for 30 min provided 94.89% sugar yield, compared to 70.64% for 30 min hot water pretreatment (Milano et al., 2018). Increasing the combined severity factor reduces the recovery of solid residue after pretreatment since more sugar is released from xylan and cellulose (Hsu et al., 2010). Temperature, residence time, and acid concentration can be altered to reduce the severity and obtain similar sugar recovery.

Some organic acids, such as acetic acid, citric acid, and oxalic acid, have also been evaluated for the pretreatment of LCB. For example, Amnuaycheewa et al. (2016) used these organic acids for the pretreatment of rice straw. Oxalic acid at 5.01% at 135.91°C for 30.86 min led to the highest sugar recovery (42.68%) after enzymatic saccharification (2.68 times higher than the untreated biomass). In contrast, citric acid was the best pretreatment acid to enhance biogas production (376.04 mL/g volatile solids (VS), which is 7.4 times higher than untreated biomass).

Despite being one of the most commonly used and highly efficient pretreatment, acid pretreatment suffers major limitations in terms of high chemical cost, acidic waste generation, neutralization requirement, reactor corrosion, and high severity.

4.3.2.2 Alkaline pretreatment

Alkaline pretreatment is the most traditional form of treatment used in pulp and paper processing, where cellulose is recovered from the biomass by solubilizing hemicellulose and lignin. Alkaline pretreatment is very efficient in reducing the crystallinity of cellulose and solubilizing lignin and a part of hemicellulose. Hydroxides of sodium, potassium, calcium, and ammonium are the various alkaline reagents used for the process but sodium hydroxide is the most used and effective pretreatment reagent (Gonzales et al., 2016). The mechanism of alkaline pretreatment is the cleavage of the intermolecular ester linkages between hemicelluloses and lignin due to saponification reaction occurring throughout the process, which leads to the solubilization of hemicellulose and lignin in the alkali solution (Sun et al., 2016). Alkali pretreatment also changes the structure via cellulose swelling which reduces crystallinity and degree of polymerization, increasing the internal surface area (Behera et al., 2014).

Talha et al. (2016) optimized the alkaline pretreatment conditions for sugarcane bagasse and filtered mud to enhance methane production. Biomass pretreated with 1% NaOH at 100°C for 3 h showed lignin removal of 86.27% and an increase of 82.2% methane yield. Similarly, 1% NaOH for 3 h at room temperature pretreatment of rice straw showed a significant reduction in hemicellulose and lignin contents, leading to 34% higher methane yield compared to untreated biomass (Shetty et al., 2017). Alkaline pretreatment has effectively increased the methane yield for other LCBs such as vinegar residue (Shen et al., 2017) and corn cob (Shah and Tabassum, 2018). Sugar yields postenzymatic saccharification have also increased with alkaline pretreatment (Rabelo et al., 2011).

Alkali pretreatment is an effective technique in removing lignin and makes carbohydrates more accessible to use for downstream processes. However, the process suffers from some limitations, such as long residence times, high alkali consumption, and the use of acid for neutralization before the hydrolysis and fermentation step.

4.3.2.3 Organic solvent pretreatment

Organic solvent (or organosolv) pretreatment has been investigated for fractionating the LCB in its components with high purity. A significant improvement has been observed in the conversion of biomass into sugars (Shuai and Luterbacher, 2016). Organosolv uses organic or aqueous-organic solvents at temperatures ranging from 100°C to 250°C. A wide range of solvents have been used for organosolv pretreatments, such as alcohol, phenol, esters, propionic acid, acetone, formaldehyde dioxane, and amines, but alcohols (ethanol and methanol) are preferred due to their low cost, and ease of recovery due to low boiling points (Espinoza-Acosta et al., 2014; Zhao et al., 2009). This pretreatment is used to recover high-purity cellulose or high-purity lignin with minimal degradation of the hemicellulose fraction. The separation of the fractions with organosolv occurs in two steps: first, at high temperatures, organic solvent breaks the internal lignin and hemicellulose bonds and breaks them into small molecular weight fragments, that dissolve in the solvent; and second, the lignin is precipitated and recovered, whereas the solvent is sent back for reuse (Zhang et al., 2016).

Ethanol pretreatment is one of the most preferred solvents for the pretreatment of biomass, with or without a catalyst. Ethanol (50%) has been used for pretreating miscanthus (12% solid content at 190°C for 80 min), eucalyptus (10% solid content at 200°C for 60 min), and wheat straw (10% solid content at 210°C for 60 min) without catalyst leading to lignin removal of 66%–75% (Brosse et al., 2010; Wen et al., 2013; Wildschut et al., 2013). Wheat straw has also been reported to be pretreated with organosolv at two conditions 180°C for 40 min with ethanol (Salapa et al., 2017) and 190°C for 1 h (Smit and Huijgen, 2017) and with eutectic solvents at 70°C for 9 h (Zhao, 2018). The delignification rate in each case was observed to be 76.4%, 84.5%, and 71.4%, with 89.2%, 99%, and 79.7% reducing sugar recovery, respectively.

4.3.2.4 Ionic liquid pretreatment

ILs are the greener alternatives to volatile organic solvents with the potential to break down and dissolve LCB entirely and therefore have gained significant interest in recent years (Bhatia et al., 2020; Sun et al., 2017). IL are organic salts with an organic cation and an organic or inorganic anion and are liquid at temperatures below 100°C. They have been identified as thermally and chemically stable, nonvolatile, and nonflammable (Alayoubi et al., 2020). ILs have a modular nature, and structural alterations can be made to the cation core, anion, or substituents. The structure of IL can be modified by altering the branching and length of the alkyl chains attached to the cation. Due to their strong hydrogen bonding coordination, ILs dissolve complete biomass, rather than individual components (Bhatia et al., 2020) or selectively dissolve the components not soluble in the conventional solvent without compromising the polymer. The cations in IL are organic, such as imidazolium, pyridinium, choline, aliphatic ammonium, alkylated phosphonium, and sulfonium ions. In contrast, anions can be both organic and inorganic, such as hexafluorophosphate, tetrafluoroborate, acetate, chloride, bromide, iodide, and methyl sulfate, to name a few (Yoo et al., 2017a). The mechanism of IL interaction with biomass is based on the precise combination of the cations and anions and the specific component of the biomass (cellulose, hemicellulose, or lignin).

For cellulose dissolution, a suitable solvent should contain a strong hydrogen bond acceptor and a moderate hydrogen bond donor. The solvation free energy calculated for various ions and experimental cellulose solubility in IL as $[OAc]^- > [Cl]^- > [BF4]^- \sim [PF6]^-$ suggest a crucial role of anions in the dissolution of cellulose (Payal et al., 2015). Cations also play an important role in cellulose dissolution; however, the role is secondary. Zhao et al. concluded that the cations' heterocyclic structure and alkyl chain length influence the interactions between the anion of the IL and cellulose (Zhao et al., 2012). As for lignin dissolution, Hart et al. reported that hydrogen bonding strength was not a crucial factor, but still, minimum hydrogen bonding basicity was required for solubilizing lignin (Hart et al., 2015). Lignin solubility is largely affected by the nature of anions and the interaction between cations and anions. The IL treatment cleaves the glycosidic, ester, and β-O-4 ether bonds to decompose lignin (Brandt et al., 2015).

Imidazolium-based ILs have been intensively investigated for biomass pretreatment. A variety of LCB have been investigated for dissolution with IL, for instance, wheat straw (Lavarda et al., 2019), corn stover (Papa et al., 2017), and bagasse (Vieira et al., 2020). Raj and coworkers investigated the influence of imidazolium-based IL pretreatment on the mustard stalk at different operating temperatures (Raj et al., 2018). They found that IL with acetate ions improves the surface area accessible to the enzymes and enhances porosity, boosting the formation of reducing sugars. The impact of [Emim][Cl] pretreatment on rye straw at 120°C for 2 h with 1 mL/g rye matter was studied by Smuga-Kogut et al. (Smuga-Kogut et al., 2017). They obtained almost three times the yield of reducing sugars in IL pretreated biomass compared to untreated biomass. It is observed that various factors such as temperature time, biomass, cations, and anions, along with their synergistic effects, affect the IL pretreatment. Below 100°C, 1-butyl-3-methylimidazolium chloride [Bmim][Cl] and 1 allyl-3-methylimidazolium chloride [Amim][Cl] can be effectively used to dissolve cellulose, as up to 10% (w/w) cellulose was dissolved using [Amim][Cl] under the operating condition of 50°C–100°C (Usmani et al., 2020). Yavir et al. (2019) suggested, IL cations exhibit $\pi-\pi$ interactions with aromatic compounds of lignin and IL with p-electrons on the side-chain ([Amim]]Cl]) is efficient in breaking down lignin, which makes IL an excellent solvent for woody biomass also (Yavir et al., 2019).

Despite their remarkable efficacy for biomass pretreatment, ILs suffer several disadvantages, such as slow dissolution rate, inhibitory effects on enzyme activity, toxicity towards microorganisms, the high viscosity of the obtained polymer/IL solution, and extensive water input in the downstream process, high cost of ILs (Mai et al., 2014; Yoo et al., 2017b). One of the ways the high cost and viscosity of the ILs can be subsidized is the use of additives such as polar organic solvents to minimize the use of ILs.

4.3.3 Physicochemical pretreatments

Physicochemical pretreatments use a combination of physical forces and chemical alteration in a single step to break biomass recalcitrance. First, biomass is heated under pressure in the presence or absence of chemicals (e.g., NH_3, H_2SO_4, SO_2), followed by the rapid release of pressure (flashing). The swift change in pressure at elevated temperatures causes an explosive breakdown of the physical and chemical structure and reduces its recalcitrance.

4.3.3.1 Steam explosion

Steam explosion is a well-known technique for physicochemical pretreatment, where biomass is subjected to high-pressure saturated steam at high temperatures (160°C–260°C) and pressure (5–50 atm) for a shorter residence time. When the pressure is suddenly released, the steam within the biomass expands, leading to disruption of the cell wall structure and separation of the fibers (Agbor et al., 2011; Kumar et al., 2009). Many variables, such as particle size, moisture content, temperature, pressure, and residence time, affect the efficacy of steam explosion (Talebnia et al., 2010). The relation between temperature and time can be determined by the composition and particle size of the biomass (Viola et al., 2008).

The severity of pretreatment is a measure used to describe steam pretreatment termed as "severity factor" (log R_o), which is 3.14–3.56 for the steam explosion. Severity factors can be calculated using the following equation:

$$\log(R_0) = \log\left(t * e^{\left(\frac{(T-100)}{14.75}\right)}\right)$$

Where log R_o is the severity factor as a function of time and temperature, T is the temperature in °C with 100°C is the reference temperature at which no solubilization occurs, t is the residence time in minutes, and 14.75 is the activation energy, where the process obeys first-order kinetics and the Arrhenius law (Overend and Chornet, 1987).

The effectiveness of steam explosion pretreatment on the enzymatic hydrolysis of cellulose on a lab scale has previously been observed on rice straw (Moniruzzaman, 1996), eucalyptus wood (Nunes and Pourquie, 1996), pine chips (San Martin et al., 1995), miscanthus (Sørensen et al., 2008) and spent Shiitake mushroom media (Asada et al., 2011). Bhatia et al. (2020) investigated steam explosion on miscanthus biomass to recover xylooligosaccharides (XOS) at the pilot scale (Bhatia et al., 2020). A steam explosion at 200°C; 15 bar; 10 min led to XOS yields of 52% in the hydrolysate. Enzymatic hydrolysis of XOS-rich hydrolysate yielded 74%–90% conversion from XOS to xylobiose. In addition, the glucose yield in the solid residue increased eight to ninefold compared to untreated biomass with the steam explosion.

Generally, steam explosion pretreatment is performed without using any chemicals; however, acid (about 1%) can be added during the process for increased lignin depolymerization. The steam explosion has also been combined with other pretreatments for increased recovery of the sugars. A study investigated combined acid hydrolysis and steam explosion pretreatment for sugar production from olive stone (Padilla-Rascón et al., 2020). The olive stone was sequentially pretreated with 10.5 g acid/100 g olive stones at 33% solid loading and 128°C, followed by steam pretreatment at 195°C for 5 min (optimal condition observed) and provided an overall sugar production yield of 83%. Another study investigated the combined pretreatment of acetosolv (acetic acid organic solvent) followed by the steam explosion to recover lignin from sugarcane bagasse and oil palm (Marques et al., 2021). With the steam explosion, lignin yield increased by 15%, whereas the severity of acetosolv decreased by 75%. The steam explosion was observed to improve the lignin yield without changing its structure. Although autohydrolysis is an efficient method for hydrolyzing LCB, one of its major drawbacks is the generation of inhibitory compounds, which can hamper microbial growth during fermentation.

4.3.3.2 Ammonia fiber explosion

Ammonia fiber explosion or ammonia fiber expansion (AFEX) is another highly investigated physicochemical pretreatment and has been found effective on several feedstocks (Alvira et al., 2010; Brodeur et al., 2011). During the AFEX process, biomass is treated with liquid ammonia under pressure (200–400 psi) at moderate temperatures (80°C–150°C) for a short (5–30 min) residence time. After holding it for a specific time at high pressure, the pressure is explosively released, similar to the case of a steam explosion (Bals et al., 2010). Swelling of cellulose by ammonia and an instant release in pressure causes a synergistic action to disrupt the fibrous structure and cleavage of lignin–carbohydrate complexes, leading to the high accessibility of enzymes during the hydrolysis process. Most of the ammonia is recycled back to reduce operating costs. The residual ammonia (0.5%–1%) in the biomass serves as a nitrogen source for the microorganisms (yeast) during subsequent fermentation. Compared to chemical pretreatments, AFEX is a relatively dry process because most of the cellulose and hemicellulose are retained in the solids' residue (Taherzadeh and Karimi, 2008). The process involves cleavage of lignin–carbohydrate complexes instead of lignin removal, so up to 100%, solid recoveries have been reported after AFEX pretreatment of biomass (Brodeur et al., 2011; Wyman et al., 2005). The primary factors that influence the AFEX parameters process are ammonia loading, solid content, temperature, blowdown pressure, and retention time. Teymouri et al. (2005) investigated the effect of the ammonia-to-biomass ratio (0.5:1, 0.7:1, 1:1, and 1.3:1 g of anhydrous ammonia/g of dry biomass) on the subsequent enzymatic hydrolysis of AFEX-treated corn stover, keeping all other parameters constant (temperature, moisture content). The glucan and xylan conversions increased with increasing ammonia loading and attained a maximum value at a mass ratio of 1:1. A similar trend was observed in the hydrolysis of AFEX-treated switchgrass (Alizadeh et al., 2005). Biomass treated with ammonia loading of 1:1 kg of ammonia: kg of dry matter 20% solid loading at 100°C and 5 min residence time resulted in almost six times hydrolysis yield compared to untreated switchgrass. Pretreatment temperature is another important parameter because it determines the amount of ammonia vaporized during the explosive flash and influences system pressure. Higher temperature also accelerates chemical reactions such as alkaline hydrolysis of hemicellulose (Teymouri et al., 2005). Teymouri et al. (2005) observed that glucan conversion was high for samples treated at high temperatures. However, an increase in temperature beyond 90°C did not show any beneficial effect. More details on the effect of these parameters and their interactions can be found in some comprehensive review papers on this topic (Brodeur et al., 2011; Chundawat et al., 2013, 2020).

4.3.3.3 Microwave-assisted pretreatment

Microwaves are nonionizing electromagnetic radiations that induce explosions within the biomass particles, facilitating the disruption of recalcitrant structures. The mechanism of microwave irradiations affecting biomass recalcitrance has been theorized in multiple ways. According to Budarin et al. (2010), the presence of crystalline ionic and nonionic amorphous regions in the biomass serves as a biological conductor. Continuous blasts inside the cellulosic components due to the presence of polar groups and heat from microwaves accelerate the relocation of the crystalline structure of the biomass (Hu and Wen, 2008). In another theory, it is suggested that microwave heating generates pressure inside the hydrated and nonhydrated components (Palav and Seetharaman, 2007). Microwave pretreatment offers multiple advantages, such as low reaction time, uniform heating, high yields, low activation energy, and minimal inhibitor formation (Li et al., 2016).

The first study on microwave pretreatment reported improved enzymatic hydrolysis of rice straw and bagasse by 1.6 times and 3.2 times with microwave heating at 170°C and 200°C for 5 min, respectively (Ooshima et al., 1984). Since then, microwave pretreatment has seen a lot of development for superior pretreatment of LCB. Microwave pretreatment is generally used in combination with other physical or chemical pretreatment technologies. A study investigated microwave-assisted alkali pretreatment followed by enzymatic hydrolysis of wheat straw and observed more lignin and hemicellulose removal; hence, hydrolysis yields in comparison to conventional alkali pretreatment (Zhu et al., 2006). Microwave-assisted alkali pretreatment at a microwave power of 224 W, NaOH concentration of 1.96 N, and pretreatment time of 2.63 min of pistachio shells led to 92.46% cellulose and 58.35% hemicellulose recovery (Özbek et al., 2021). With the optimal condition pretreatment, the glucose conversion yield of 82.67 mol% was measured during enzymatic hydrolysis, significantly higher than conventional alkali pretreatment (49.97 mol%) and no pretreatment (11.29 mol %).

Another study investigated the efficacy of microwave-assisted deep eutectic solvent (DES) pretreatment on the sugar yields during enzymatic hydrolysis of wheat straw (Isci et al., 2020). Glucose and xylose yields obtained were 99% and 85%, respectively, with the total sugar release (619 mg/g pretreated wheat straw fibers) at 1:3 choline chloride mol ratio, 360 W microwave power, and 8 min retention time being twice as that of conventional DES pretreatment. Microwave pretreatment was combined with ammoniation to obtain higher methane yields during the anaerobic digestion of rice straw (Liu et al., 2021). Microwave ammoniation showed 281.56 mL/g (volatile solids of the substrate) methane production after 18-day digestion, which was 25.43%, 7.96%, and 18.18% higher than that produced with no pretreatment, only microwave pretreatment, and only ammoniation pretreatment, respectively.

Although microwave pretreatment has many advantages over conventional physical methods, it also suffers from some drawbacks, such as high energy expenditure, moisture content of biomass, and scale-up issues (operating costs at high load and high pressure and high capital investment).

4.3.4 Biological pretreatment

With chemical and physical pretreatments suffering from inherent drawbacks of high energy consumption and inhibitory compound generation, biological pretreatment is being investigated as an environmentally friendly method with milder operating conditions, low energy requirement, and low disposal costs. Biological pretreatments use microorganisms such as white rot fungi and bacteria that degrade lignin through the action of lignin-degrading enzymes such as peroxidases and laccases.

Fungi release a family of enzymes called "lignases" extracellularly in an anaerobic environment, which helps in the breakdown of lignin in the LCB (Howard et al., 2003). Phenol oxidase (laccase) and peroxidases (lignin peroxidase, LiP, manganese peroxidase, MnP) are the two families of lignolytic enzymes that are considered to play a significant role in the enzymatic degradation of lignin in LCB (Krause et al., 2003). In addition, the peroxidases produced by rot fungi catalyze the initial depolymerization of lignin (Bonnen et al., 1994). Phenol oxidase belongs to the blue copper oxidase enzyme family, which preferably polymerizes lignin by coupling the phenoxy radicals produced from the oxidation of lignin phenolic groups (Kunamneni et al., 2007).

Biological pretreatment can be carried out in submerged state or solid-state techniques. For better efficiency, the initial solid loading for submerged pretreatment is limited to around 5%. However, solid-state pretreatments can utilize 15%–30% initial solid loading (Wan and Li, 2012). Many factors affect biological pretreatment, broadly categorized as biological factors (strain species, consortia of microorganisms, their interaction, and competition), physical factors (temperature, solid loading, aeration, substrate size, accessible surface area), and chemical factors (pH, culture media, source of carbon and nitrogen, amount of organic and inorganic compound) (Sharma et al., 2019).

Although biological pretreatments are environmentally friendly, long incubation times (days to months) make the process less attractive on a commercial scale. Lignin is efficiently removed using fungal pretreatment, but the loss of holocellulose is a major drawback of this pretreatment. Bacteria use shorter pretreatment time, but lignin removal efficiency is lower than that of fungi. A promising technology for biological pretreatments could be the use of microorganism consortia consisting of fungal-fungal or bacterial–bacterial, or bacterial fungal co-culture. Co-cultures have several advantages, such as relatively lower incubation times, Continuous and stable generation of metabolites required for lignin breakdown, acceleration of organic hydrolysis, and enhanced productivity and biomass uptake. However, individual strains in the consortia might face difficulties in adapting to environmental fluctuations, which might be nonoptimal for their growth and action.

4.4 Recent advancements in the pretreatment

As explained in the chapter, various pretreatments have been used to break down the recalcitrance of the lignocellulosic structure effectively, but efforts have consistently been made to improve the pretreatment efficiency. Combining two or more pretreatments is one step that is being taken to ensure energy efficiency, cost minimization, and a faster process. A combination of pretreatments usually provides better solubility of hemicellulose and breakdown of lignin than the individual pretreatments and helps lower the severity of the drawbacks of a single step. Juneja et al. (2021) investigated the sequential pretreatment of liquid hot water followed by three cycles of disk milling for corn fiber. After hydrolysis of pretreated biomass, the authors obtained sugar yields of 94.9%, 74.2%, and 66.3% for glucose, xylose, and arabinose, respectively (Juneja et al., 2021). Another study investigated the effect of liquid hot water pretreatment as solo pretreatment and sequential pretreatment followed by disk milling for energy cane (Juneja et al., 2020). It was noticed that at lower temperatures and residence times, the disk mill significantly improved the hydrolysis yield. A combination of ball milling and lithium chloride (LiCl)/dimethyl sulfoxide (DMSO) dissolution pretreatment on bagasse was observed to convert cellulose crystal structure from type I to II and provided a yield of 87.8% glucose with 84.2% xylose during enzymatic hydrolysis (Chen et al., 2021). These sequential pretreatments are particularly important for bioprocessing recently developed engineered biomass that accumulates oil and sugars in their stems. A low-temperature chemical-free pretreatment is required to avoid oil quality deterioration and minimize the inhibitor formations. A combination of low severity how water pretreatment and disk refining fulfill these conditions along with high subsequent hydrolysis yields (Cheng and Liu, 2010; Juneja et al., 2021; Wang et al., 2018).

The combination of irradiation and chemicals is another promising pretreatment approach, as discussed previously in the chapter. Microwave-assisted IL pretreatment for rice straw with [Emim][OAc]-DMSO (1:1 volume ratio) was found to be hydrolyzed 22 times faster than that of untreated rice straw (Mai et al., 2014). Lignin removal with microwave-assisted [Emim][OAc]-DMSO pretreatment was 44.1%, which was significantly higher than [Emim][OAc]-DMSO pretreatment independently (27.2%). The combination of the two pretreatments provides the advantage of better lignin extraction, reduced cellulose crystallinity, and lower residual chemicals. Ultrasound-assisted dilute acid pretreatment reduces the amount required and leads to higher sugar recovery with lower acid concentrations. Paramasivan et al. (2021) used ultrasound-assisted diluted pretreatment on four perennial LCB with 1.0% (w/w) H_2SO_4 kept in the ultrasonicator (Power—70% max and 24 kHz) at 60°C for 60 min. They observed no inhibitor formation, which is a primary concern in acid pretreatments, and improved recovery of pretreated biomass. Ultrasound has also been used with an alkaline solution and aqueous DES for improved lignin recovery (Ong et al., 2021). Lignin removal of 47.00% was achieved with ultrasound-assisted NaOH-aqueous DES (choline chloride: urea) of oil palm frond, compared to 30.92% without ultrasound pretreatment. Subsequent hydrolysis after ultrasound-assisted NaOH-aqueous DES achieved a glucose yield of 90.12%.

Although physical and chemical methods have been more studied for pretreatment, some biological methods have also been combined with chemical methods to improve the pretreatment efficiency. Si et al. (2019) investigated the effect of combined biological–chemical pretreatment to enhance the enzymatic digestibility of biomass. The first step in the pretreatment was chemical pretreatment with 70 mL of ethanol–water solution (65:35, v/v) containing 0.5 wt.% NaOH (for alkaline pretreatment) or 0.5 wt.% $FeCl_3$ (for acid pretreatment) and 400 W of microwave irradiation for 10 min, followed by ligninolytic (*Pandoraea* sp.) and saccharolytic (*Acinetobacter* sp.) bacteria at 30°C for 2 days. They observed that ligninolytic bacteria enhanced acid pretreatment (lignin fraction decreased from 8.4%–14.0% to 7.6%–10.4%, and sugar recovery increased to 83.9%–85.9%) and saccharolytic bacteria enhanced alkaline pretreatment (hemicellulose content 19.5%–28.2% from 24.4% to 33.8%, while the lignin content was nearly unchanged).

4.5 Challenges in the commercialization of pretreatment technologies

Even though pretreatment is an important step for the efficient conversion of LCB, it still stays at the cost center of the process. A techno-economic analysis is performed to determine the commercial feasibility of the overall process with pretreatment. The analysis can compare pretreatment methods for the production of desired chemicals/fuel to minimize the minimum selling price of the main product. This can depend on a lot of factors, such as the value of the main product, secondary products generated, the volume of the main and secondary products produced, the efficiency of the individual processes, energy transfer between operations, raw material cost, etc. the capital cost of pretreatment, however, depends on the special reactor material requirement for carrying out pretreatment, temperature, residence time, solid loading, and chemical recovery strategy (Tao et al., 2011). Tao et al. compared six pretreatments, ammonia fiber expansion, dilute acid, lime, liquid hot water, soaking in aqueous ammonia, and sulfur dioxide-impregnated steam explosion in techno-economic analysis (Tao et al., 2011). They reported that the minimum ethanol selling price (MESP) highly depends on the monomer sugar yield. Therefore pretreatments with higher efficacy in releasing sugar monomers lead to higher bioethanol production and lower MESP. The capital cost of the pretreatment was higher for all pretreatments than liquid hot water, primarily due to no chemical recovery system associated with the pretreatment. Lime pretreatment used the highest equipment cost due to the installation of a lime recovery system. High capital costs for dilute acid pretreatment were due to longer residence time and high pretreatment reactor costs (corrosion resistant). For pretreatments requiring high temperatures, heat recovery is one of the significant factors in deciding the capital cost of pretreatments. The number of reactors is determined depending on the operating temperature and residence time.

Baral and Shah (2015) performed a technoeconomic analysis of the IL pretreatment for the production of fermentable sugars. They reported that the cost of IL and IL recovery were two important parameters that can change the cost of sugar production, with the latter having a more predominant effect on the cost. It was shown that even with a 20-fold increase in the IL cost (from base price), the IL recovery increase from 80% to 99.6% reduced the sugar production cost from $47 to $3/kg. However, that high IL recovery might not be practically feasible. At an optimistic recovery value of 97%, the cost of IL needs to be less than $1/kg for the pretreatment to be competitive with dilute acid pretreatment. da Silva et al. (2017) investigated the commercial feasibility of organsolv pretreatment on the LCB. The overall energy requirement for the organosolv pretreatment (578.1 MW) was higher than that for dilute acid pretreatment, but on the other hand, ethanol concentration after biomass fermentation for organosolv (11.1%) was higher than dilute acid. In the current situation, it was concluded that dilute acid pretreatment had a cost advantage over the organosolv technology; however, improvement of the process, such as reduction in usage of organosolv, reduced solvent ratio, and increased solid loading, can lead to a reduction in the pretreatments using organic solvents.

4.6 Conclusion and future perspectives

Pretreatment is the most energy-intensive and expensive process in the LCB bioconversion process. Although process efficiency varies with different pretreatment methods and pretreatment conditions, most of the pretreatment methods deal with high energy requirements, capital and operating cost, inhibitor generation, and the environmental impact of the waste generated. With pretreatment being a prerequisite in the conversion of low-cost lignocellulosic feedstock, necessary efforts should be taken to increase efficiency, lower the operating costs, and intensify the possible recovery and valorization of each component of the biomass. The key factors for biomass pretreatment that should be considered are: (1) the possibility of large-scale feedstock processing; (2) reaching high yields regardless of the type and origin of biomass; (3) minimizing the amount of waste and toxic compounds; (4) compatibility of the pretreatment with further processing; (5) efficient recovery of lignin; and (6) reducing the cost of equipment and energy demand.

References

Agbor, V.B., Cicek, N., Sparling, R., Berlin, A., Levin, D.B., 2011. Biomass pretreatment: fundamentals toward application. Biotechnol. Adv. 29 (6), 675–685.

Alayoubi, R., Mehmood, N., Husson, E., Kouzayha, A., Tabcheh, M., Chaveriat, L., et al., 2020. Low temperature ionic liquid pretreatment of lignocellulosic biomass to enhance bioethanol yield. Renew. Energy 145, 1808–1816.

Al-Battashi, H.S., Annamalai, N., Sivakumar, N., Al-Bahry, S., Tripathi, B.N., Nguyen, Q. D., et al., 2019. Lignocellulosic biomass (LCB): a potential alternative biorefinery feedstock for polyhydroxyalkanoates production. Rev. Environ. Sci. Bio/Technol. 18 (1), 183–205.

Alizadeh, H., Teymouri, F., Gilbert, T.I., Dale, B.E., 2005. Pretreatment of switchgrass by ammonia fiber explosion (AFEX). Appl. Biochem. Biotechnol. 124 (1), 1133–1141.

Alvira, P., Tomás-Pejó, E., Ballesteros, M., Negro, M.J., 2010. Pretreatment technologies for an efficient bioethanol production process based on enzymatic hydrolysis: a review. Bioresour. Technol. 101 (13), 4851–4861.

Amnuaycheewa, P., Hengaroonprasan, R., Rattanaporn, K., Kirdponpattara, S., Cheenkachorn, K., Sriariyanun, M., 2016. Enhancing enzymatic hydrolysis and biogas production from rice straw by pretreatment with organic acids. Ind. Crop. Products 87, 247–254.

Asada, C., Asakawa, A., Sasaki, C., Nakamura, Y., 2011. Characterization of the steam-exploded spent Shiitake mushroom medium and its efficient conversion to ethanol. Bioresour. Technol. 102 (21), 10052–10056.

Avgerinos, G., Wang, D.I.C., 1983. Selective solvent delignification for fermentation enhancement. Biotechnol. Bioeng. 25 (1), 67–83.

Bajpai, P., 2016. Structure of lignocellulosic biomass. Pretreatment of Lignocellulosic Biomass for Biofuel Production. Springer, pp. 7–12.

Bals, B., Murnen, H., Allen, M., Dale, B., 2010. Ammonia fiber expansion (AFEX) treatment of eleven different forages: improvements to fiber digestibility in vitro. Anim. Feed. Sci. Technol. 155 (2–4), 147–155.

Baral, N.R., Shah, A., 2015. Techno-economic analysis of cellulose dissolving ionic liquid pretreatment of lignocellulosic biomass for fermentable sugars production. Biofuels, Bioprod. Biorefin. 10 (1), 70–88.

Behera, S., Arora, R., Nandhagopal, N., Kumar, S., 2014. Importance of chemical pretreatment for bioconversion of lignocellulosic biomass. Renew. Sustain. Energy Rev. 36, 91–106.

Bhatia, S.K., Jagtap, S.S., Bedekar, A.A., Bhatia, R.K., Patel, A.K., Pant, D., et al., 2020. Recent developments in pretreatment technologies on lignocellulosic biomass: effect of key parameters, technological improvements, and challenges. Bioresour. Technol. 300, 122724.

Bonnen, A.M., Anton, L.H., Orth, A.B., 1994. Lignin-degrading enzymes of the commercial button mushroom, *Agaricus bisporus*. Appl. Environ. Microbiol. 60 (3), 960–965.

Brandt, A., Gräsvik, J., Hallett, J.P., Welton, T., 2013. Deconstruction of lignocellulosic biomass with ionic liquids. Green Chem. 15 (3), 550–583.

Brandt, A., Chen, L., van Dongen, B.E., Welton, T., Hallett, J.P., 2015. Structural changes in lignins isolated using an acidic ionic liquid water mixture. Green Chem. 17 (11), 5019–5034.

Brodeur, G., Yau, E., Badal, K., Collier, J., Ramachandran, K., Ramakrishnan, S., 2011. Chemical and physicochemical pretreatment of lignocellulosic biomass: a review. Enzyme Res. 2011.

Brosse, N., Hage, El, Sannigrahi, R., Ragauskas, A., P., 2010. Dilute sulphuric acid and ethanol organosolv pretreatment of Miscanthus x Giganteus. Cellulose Chem. Technol. 44 (1–3), 71–78.

Budarin, V.L., Clark, J.H., Lanigan, B.A., Shuttleworth, P., Macquarrie, D.J., 2010. Microwave-assisted decomposition of cellulose: a new thermochemical route for biomass exploitation. Bioresour. Technol. 101 (10), 3776–3779.

Bychkov, A., Podgorbunskikh, E., Ryabchikova, E., Lomovsky, O., 2018. The role of mechanical action in the process of the thermomechanical isolation of lignin. Cellulose 25 (1), 1–5.

Chang, V.S., Holtzapple, M.T., 2000. Fundamental factors affecting biomass enzymatic reactivity. Twenty-First Symposium on Biotechnology for Fuels and Chemicals. Springer, pp. 5–37.

Chen, Z., Wang, H., Wei, W., Yuan, Z., 2021. Enhancing bagasse enzymatic hydrolysis through combination of ball-milling and LiCl/DMSO dissolution and regeneration. Renew. Energy 171, 994–1001.

Cheng, X.Y., Liu, C.Z., 2010. Enhanced biogas production from herbal-extraction process residues by microwave-assisted alkaline pretreatment. J. Chem. Technol. Biotechnol. 85 (1), 127–131.

Cheng, M.-H., Dien, B.S., Lee, D., Singh, V., 2019. Sugar production from bioenergy sorghum by using pilot scale continuous hydrothermal pretreatment combined with disk refining. Bioresour. Technol. 289, 121663.

Chundawat, S., Bals, B., Campbell, T., Sousa, L., Gao, D., Jin, M., et al., 2013. Primer on ammonia fiber expansion pretreatment. Aqueous Pretreat. Plant Biomass Biol. Chem. Convers. Fuels Chem. 169–200.

Chundawat, S.P., Pal, R.K., Zhao, C., Campbell, T., Teymouri, F., Videto, J., et al., 2020. Ammonia fiber expansion (AFEX) pretreatment of lignocellulosic biomass. JoVE (J. Visual. Exp.) 158, e57488.

da Silva, A.R.G., Errico, M., Rong, B.G., 2017. Techno-economic analysis of organosolv pretreatment 33 process from lignocellulosic biomass. Clean. Techn Env. Policy 20, 1401–1412.

Dell'Omo, P.P., Spena, V.A., 2020. Mechanical pretreatment of lignocellulosic biomass to improve biogas production: comparison of results for giant reed and wheat straw. Energy 203, 117798.

Espinoza-Acosta, J.L., Torres-Chávez, P.I., Carvajal-Millán, E., Ramírez-Wong, B., Bello-Pérez, L.A., Montaño-Leyva, B., 2014. Ionic liquids and organic solvents for recovering lignin from lignocellulosic biomass. BioResources 9 (2), 3660–3687.

Gonzales, R.R., Sivagurunathan, P., Kim, S.-H., 2016. Effect of severity on dilute acid pretreatment of lignocellulosic biomass and the following hydrogen fermentation. Int. J. Hydrog. Energy 41 (46), 21678–21684.

Gu, B.-J., Wang, J., Wolcott, M.P., Ganjyal, G.M., 2018. Increased sugar yield from pre-milled Douglas-fir forest residuals with lower energy consumption by using planetary ball milling. Bioresour. Technol. 251, 93–98.

Harmsen, P.F., Huijgen, W., Bermudez, L., Bakker, R. 2010. Literature review of physical and chemical pretreatment processes for lignocellulosic biomass. Wageningen UR-Food and Biobased Research.

Hart, W.E., Harper, J.B., Aldous, L., 2015. The effect of changing the components of an ionic liquid upon the solubility of lignin. Green Chem. 17 (1), 214–218.

Hideno, A., Inoue, H., Tsukahara, K., Fujimoto, S., Minowa, T., Inoue, S., et al., 2009. Wet disk milling pretreatment without sulfuric acid for enzymatic hydrolysis of rice straw. Bioresour. Technol. 100 (10), 2706–2711.

Howard, R.L., Abotsi, E., Van Rensburg, E.J., Howard, S., 2003. Lignocellulose biotechnology: issues of bioconversion and enzyme production. Afr. J. Biotechnol. 2 (12), 602–619.

Hsu, T.-C., Guo, G.-L., Chen, W.-H., Hwang, W.-S., 2010. Effect of dilute acid pretreatment of rice straw on structural properties and enzymatic hydrolysis. Bioresour. Technol. 101 (13), 4907–4913.

Hu, Z., Wen, Z., 2008. Enhancing enzymatic digestibility of switchgrass by microwave-assisted alkali pretreatment. Biochem. Eng. J. 38 (3), 369–378.

Isci, A., Erdem, G.M., Elmaci, S.B., Sakiyan, O., Lamp, A., Kaltschmitt, M., 2020. Effect of microwave-assisted deep eutectic solvent pretreatment on lignocellulosic structure and bioconversion of wheat straw. Cellulose 27 (15), 8949–8962.

Jang, Y.-S., Malaviya, A., Cho, C., Lee, J., Lee, S.Y., 2012. Butanol production from renewable biomass by clostridia. Bioresour. Technol. 123, 653–663.

Juneja, A., Kumar, D., Singh, V.K., Singh, V., 2020. Chemical free two-step hydrothermal pretreatment to improve sugar yields from energy cane. Energies 13 (21), 5805.

Juneja, A., Noordam, B., Pel, H., Basu, R., Appeldoorn, M., Singh, V., 2021. Optimization of two-stage pretreatment for maximizing ethanol production in 1.5 G technology. Bioresour. Technol. 320, 124380.

Kapoor, M., Raj, T., Vijayaraj, M., Chopra, A., Gupta, R.P., Tuli, D.K., et al., 2015. Structural features of dilute acid, steam exploded, and alkali pretreated mustard stalk and their impact on enzymatic hydrolysis. Carbohydr. Polym. 124, 265–273.

Khan, A.S., Man, Z., Bustam, M.A., Kait, C.F., Khan, M.I., Muhammad, N., et al., 2016. Impact of ball-milling pretreatment on pyrolysis behavior and kinetics of crystalline cellulose. Waste Biomass Valoriz. 7 (3), 571–581.

Kim, S.M., Dien, B.S., Tumbleson, M., Rausch, K.D., Singh, V., 2016. Improvement of sugar yields from corn stover using sequential hot water pretreatment and disk milling. Bioresour. Technol. 216, 706–713.

Klemm, D., Heublein, B., Fink, H.P., Bohn, A., 2005. Cellulose: fascinating biopolymer and sustainable raw material. Angew. Chem. Int. Ed. 44 (22), 3358–3393.

Krause, D.O., Denman, S.E., Mackie, R.I., Morrison, M., Rae, A.L., Attwood, G.T., et al., 2003. Opportunities to improve fiber degradation in the rumen: microbiology, ecology, and genomics. FEMS Microbiol. Rev. 27 (5), 663–693.

Kumar, D., Murthy, G.S., 2011a. Impact of pretreatment and downstream processing technologies on economics and energy in cellulosic ethanol production. Biotechnol. Biofuels 4 (1), 1–19.

Kumar, D., Murthy, G.S., 2011b. Pretreatments and enzymatic hydrolysis of grass straws for ethanol production in the Pacific Northwest US. Biol. Eng. Trans. 3 (2), 97–110.

Kumar, D., Murthy, G.S., 2013. Stochastic molecular model of enzymatic hydrolysis of cellulose for ethanol production. Biotechnol. Biofuels 6 (1), 1–20.

Kumar, P., Barrett, D.M., Delwiche, M.J., Stroeve, P., 2009. Methods for pretreatment of lignocellulosic biomass for efficient hydrolysis and biofuel production. Ind. Eng. Chem. Res. 48 (8), 3713–3729.

Kunamneni, A., Ballesteros, A., Plou, F.J., Alcalde, M., 2007. Fungal laccase—a versatile enzyme for biotechnological applications. Commun. Curr. Res. Educ. Top. Trends Appl. Microbiology 1, 233–245.

Lavarda, G., Morales-delaRosa, S., Centomo, P., Campos-Martin, J.M., Zecca, M., Fierro, J. L., 2019. Gel-type and macroporous cross-linked copolymers functionalized with acid groups for the hydrolysis of wheat straw pretreated with an ionic liquid. Catalysts 9 (8), 675.

Li, H., Qu, Y., Yang, Y., Chang, S., Xu, J., 2016. Microwave irradiation–a green and efficient way to pretreat biomass. Bioresour. Technol. 199, 34–41.

Liu, Q., Pan, S., Zhou, L., Feng, L., Li, Z., Du, L., et al., 2021. Improving the biogas potential of rice straw through microwave-assisted ammoniation pretreatment during anaerobic digestion. BioEnergy Res. 1–11.

Mai, N.L., Ha, S.H., Koo, Y.-M., 2014. Efficient pretreatment of lignocellulose in ionic liquids/co-solvent for enzymatic hydrolysis enhancement into fermentable sugars. Process. Biochem. 49 (7), 1144–1151.

Marques, F.P., Soares, A.K.L., Lomonaco, D., ESilva, L.M.A., Santaella, S.T., de Freitas Rosa, M., et al., 2021. Steam explosion pretreatment improves acetic acid organosolv delignification of oil palm mesocarp fibers and sugarcane bagasse. Int. J. Biol. Macromol. 175, 304–312.

McKendry, P., 2002. Energy production from biomass (part 1): an overview of biomass. Bioresour. Technol. 83 (1), 37–46.

Meng, X., Ragauskas, A.J., 2017. Pseudo-lignin formation during dilute acid pretreatment for cellulosic ethanol. Recent. Adv. Petrochem. Sci. 1 (1).

Milano, J., Ong, H.C., Masjuki, H.H., Silitonga, A.S., Chen, W.-H., Kusumo, F., et al., 2018. Optimization of biodiesel production by microwave irradiation-assisted transesterification for waste cooking oil-Calophyllum inophyllum oil via response surface methodology. Energy Convers. Manag. 158, 400–415.

Moniruzzaman, M., 1996. Saccharification and alcohol fermentation of steam-exploded rice straw. Bioresour. Technol. 55 (2), 111–117.

Mood, S.H., Golfeshan, A.H., Tabatabaei, M., Jouzani, G.S., Najafi, G.H., Gholami, M., et al., 2013. Lignocellulosic biomass to bioethanol, a comprehensive review with a focus on pretreatment. Renew. Sustain. Energy Rev. 27, 77–93.

Nunes, A.P., Pourquie, 1996. Steam explosion pretreatment and enzymatic hydrolysis of eucalyptus wood. Bioresour. Technol. 57, 107–110.

Ong, V.Z., Wu, T.Y., Chu, K.K.L., Sun, W.Y., Shak, K.P.Y., 2021. A combined pretreatment with ultrasound-assisted alkaline solution and aqueous deep eutectic solvent for enhancing delignification and enzymatic hydrolysis from oil palm fronds. Ind. Crop Products 160, 112974.

Ooshima, H., Aso, K., Harano, Y., Yamamoto, T., 1984. Microwave treatment of cellulosic materials for their enzymatic hydrolysis. Biotechnol. Lett. 6 (5), 289–294.

Overend, R.P., Chornet, E., 1987. Fractionation of lignocellulosics by steam-aqueous pretreatments. Philosoph. Trans. Roy. Soc. London. Ser. A, Math. Phys. Sci. 321 (1561), 523–536.

Özbek, H.N., Koçak Yanık, D., Fadıloğlu, S., Göğüş, F., 2021. Effect of microwave-assisted alkali pre-treatment on fractionation of pistachio shell and enzymatic hydrolysis of cellulose-rich residues. J. Chem. Technol. Biotechnol. 96 (2), 521–531.

Padilla-Rascón, C., Ruiz, E., Romero, I., Castro, E., Oliva, J., Ballesteros, I., et al., 2020. Valorisation of olive stone by-product for sugar production using a sequential acid/steam explosion pretreatment. Ind. Crop Products 148, 112279.

Palav, T., Seetharaman, K., 2007. Impact of microwave heating on the physico-chemical properties of a starch–water model system. Carbohydr. Polym. 67 (4), 596–604.

Papa, G., Feldman, T., Sale, K.L., Adani, F., Singh, S., Simmons, B.A., 2017. Parametric study for the optimization of ionic liquid pretreatment of corn stover. Bioresour. Technol. 241, 627–637.

Paramasivan, S., Sankar, S., Velavan, R.S., Krishnakumar, T., Batcha, R.S.I., Muthuvelu, K. S., 2021. Assessing the potential of lignocellulosic energy crops as an alternative resource for bioethanol production using ultrasound assisted dilute acid pretreatment. Mater. Today: Proc. 45, 3279–3285.

Payal, R.S., Bejagam, K.K., Mondal, A., Balasubramanian, S., 2015. Dissolution of cellulose in room temperature ionic liquids: anion dependence. J. Phys. Chem. B 119 (4), 1654–1659.

Rabelo, S., Fonseca, N.A., Andrade, R., Maciel Filho, R., Costa, A., 2011. Ethanol production from enzymatic hydrolysis of sugarcane bagasse pretreated with lime and alkaline hydrogen peroxide. Biomass Bioenergy 35 (7), 2600–2607.

Raj, T., Gaur, R., Lamba, B.Y., Singh, N., Gupta, R.P., Kumar, R., et al., 2018. Characterization of ionic liquid pretreated plant cell wall for improved enzymatic digestibility. Bioresour. Technol. 249, 139–145.

Bioethanol production from food crops: sustainable sources. In: Ray, R.C., Ramachandran, S. (Eds.), Interventions, and Challenges. Academic Press, p. 560.

Ray, R.C. (Ed.), 2021. Sustainable Biofuels: Opportunities and Challenges. Academic Press, p. 502.

RFA, 2022. Zeroing IN on New Opportunities: 2022 Ethanol Industry Outlook. RFA, Washington, DC. Available from: https://ethanolrfa.org/publications/ (accessed 21 August 2022).

Roy, D., Semsarilar, M., Guthrie, J.T., Perrier, S., 2009. Cellulose modification by polymer grafting: a review. Chem. Soc. Rev. 38 (7), 2046–2064.

Salapa, I., Katsimpouras, C., Topakas, E., Sidiras, D., 2017. Organosolv pretreatment of wheat straw for efficient ethanol production using various solvents. Biomass Bioenergy 100, 10–16.

San Martin, R., Perez, C., Briones, R., 1995. Simultaneous production of ethanol and kraft pulp from pine (*Pinus radiata*) using steam explosion. Bioresour. Technol. 53 (3), 217–223.

Sannigrahi, P., Kim, D.H., Jung, S., Ragauskas, A., 2011. Pseudo-lignin and pretreatment chemistry. Energy Environ. Sci. 4 (4), 1306–1310.

Santos, C.C., de Souza, W., Sant'Anna, C., Brienzo, M., 2018. Elephant grass leaves have lower recalcitrance to acid pretreatment than stems, with higher potential for ethanol production. Ind. Crop Products 111, 193–200.

Shah, T.A., Tabassum, R., 2018. Enhancing biogas production from lime soaked corn cob residue. Int. J. Renew. Energy Res. 8 (2), 761–766.

Sharma, H.K., Xu, C., Qin, W., 2019. Biological pretreatment of lignocellulosic biomass for biofuels and bioproducts: an overview. Waste Biomass Valoriz. 10 (2), 235–251.

Shen, J., Zhao, C., Liu, G., Chen, C., 2017. Enhancing the performance on anaerobic digestion of vinegar residue by sodium hydroxide pretreatment. Waste Biomass Valoriz. 8 (4), 1119–1126.

Shetty, D.J., Kshirsagar, P., Tapadia-Maheshwari, S., Lanjekar, V., Singh, S.K., Dhakephalkar, P.K., 2017. Alkali pretreatment at ambient temperature: a promising method to enhance biomethanation of rice straw. Bioresour. Technol. 226, 80–88.

Shuai, L., Luterbacher, J., 2016. Organic solvent effects in biomass conversion reactions. ChemSusChem 9 (ARTICLE), 133–155.

Si, M., Liu, D., Yan, X., Gao, C., Chai, L., Shi, Y., 2019. Complementary effect of combined bacterial-chemical pretreatment to promote enzymatic digestibility of lignocellulose biomass. Bioresour. Technol. 272, 275–280.

Silverstein, R.A., Chen, Y., Sharma-Shivappa, R.R., Boyette, M.D., Osborne, J., 2007. A comparison of chemical pretreatment methods for improving saccharification of cotton stalks. Bioresour. Technol. 98 (16), 3000–3011.

Smit, A., Huijgen, W., 2017. The promotional effect of water-soluble extractives on the enzymatic cellulose hydrolysis of pretreated wheat straw. Bioresour. Technol. 243, 994–999.

Smuga-Kogut, M., Zgórska, K., Kogut, T., Kukiełka, K., Wojdalski, J., Kupczyk, A., et al., 2017. The use of ionic liquid pretreatment of rye straw for bioethanol production. Fuel 191, 266–274.

Solarte-Toro, J.C., Romero-García, J.M., Martínez-Patiño, J.C., Ruiz-Ramos, E., Castro-Galiano, E., Cardona-Alzate, C.A., 2019. Acid pretreatment of lignocellulosic biomass for energy vectors production: a review focused on operational conditions and techno-economic assessment for bioethanol production. Renew. Sustain. Energy Rev. 107, 587–601.

Sørensen, A., Teller, P.J., Hilstrøm, T., Ahring, B.K., 2008. Hydrolysis of Miscanthus for bioethanol production using dilute acid presoaking combined with wet explosion pretreatment and enzymatic treatment. Bioresour. Technol. 99 (14), 6602–6607.

Sun, S., Sun, S., Cao, X., Sun, R., 2016. The role of pretreatment in improving the enzymatic hydrolysis of lignocellulosic materials. Bioresour. Technol. 199, 49–58.

Sun, J., Konda, N.M., Parthasarathi, R., Dutta, T., Valiev, M., Xu, F., et al., 2017. One-pot integrated biofuel production using low-cost biocompatible protic ionic liquids. Green Chem. 19 (13), 3152–3163.

Taherzadeh, M.J., Karimi, K., 2008. Pretreatment of lignocellulosic wastes to improve ethanol and biogas production: a review. Int. J. Mol. Sci. 9 (9), 1621–1651.

Talebnia, F., Karakashev, D., Angelidaki, I., 2010. Production of bioethanol from wheat straw: an overview on pretreatment, hydrolysis and fermentation. Bioresour. Technol. 101 (13), 4744–4753.

Talha, Z., Ding, W., Mehryar, E., Hassan, M., Bi, J., 2016. Alkaline pretreatment of sugarcane bagasse and filter mud codigested to improve biomethane production. BioMed. Res. Int. 2016.

Tao, L., Aden, A., Elander, R.T., Pallapolu, V.R., Lee, Y., Garlock, R.J., et al., 2011. Process and technoeconomic analysis of leading pretreatment technologies for lignocellulosic ethanol production using switchgrass. Bioresour. Technol. 102 (24), 11105–11114.

Teymouri, F., Laureano-Perez, L., Alizadeh, H., Dale, B.E., 2005. Optimization of the ammonia fiber explosion (AFEX) treatment parameters for enzymatic hydrolysis of corn stover. Bioresour. Technol. 96 (18), 2014–2018.

Usmani, Z., Sharma, M., Gupta, P., Karpichev, Y., Gathergood, N., Bhat, R., et al., 2020. Ionic liquid based pretreatment of lignocellulosic biomass for enhanced bioconversion. Bioresour. Technol. 304, 123003.

Vassilev, S.V., Baxter, D., Andersen, L.K., Vassileva, C.G., 2010. An overview of the chemical composition of biomass. Fuel 89 (5), 913–933.

Vieira, S., Barros, M.V., Sydney, A.C.N., Piekarski, C.M., de Francisco, A.C., de Souza Vandenberghe, L.P., et al., 2020. Sustainability of sugarcane lignocellulosic biomass pretreatment for the production of bioethanol. Bioresour. Technol. 299, 122635.

Viola, E., Cardinale, M., Santarcangelo, R., Villone, A., Zimbardi, F., 2008. Ethanol from eel grass via steam explosion and enzymatic hydrolysis. Biomass Bioenergy 32 (7), 613–618.

Wan, C., Li, Y., 2012. Fungal pretreatment of lignocellulosic biomass. Biotechnol. Adv. 30 (6), 1447–1457.

Wang, Z., Dien, B.S., Rausch, K.D., Tumbleson, M., Singh, V., 2018. Fermentation of undetoxified sugarcane bagasse hydrolyzates using a two stage hydrothermal and mechanical refining pretreatment. Bioresour. Aechnol. 261, 313–321.

Wen, J.-L., Sun, S.-L., Yuan, T.-Q., Xu, F., Sun, R.-C., 2013. Structural elucidation of lignin polymers of Eucalyptus chips during organosolv pretreatment and extended delignification. J. Agric. Food Chem. 61 (46), 11067–11075.

Wildschut, J., Smit, A.T., Reith, J.H., Huijgen, W.J., 2013. Ethanol-based organosolv fractionation of wheat straw for the production of lignin and enzymatically digestible cellulose. Bioresour. Technol. 135, 58–66.

Wyman, C., 1996. Handbook on Bioethanol: Production and Utilization. CRC press.

Wyman, C.E., Dale, B.E., Elander, R.T., Holtzapple, M., Ladisch, M.R., Lee, Y., 2005. Comparative sugar recovery data from laboratory scale application of leading pretreatment technologies to corn stover. Bioresour. Technol. 96 (18), 2026–2032.

Yavir, K., Marcinkowski, Ł., Marcinkowska, R., Namieśnik, J., Kloskowski, A., 2019. Analytical applications and physicochemical properties of ionic liquid-based hybrid materials: a review. Anal. Chim. Acta 1054, 1–16.

Yoo, C.G., Pu, Y., Ragauskas, A.J., 2017a. Ionic liquids: promising green solvents for lignocellulosic biomass utilization. Curr. Opin. Green Sustain. Chem. 5, 5–11.

Yoo, C.G., Zhang, S., Pan, X., 2017b. Effective conversion of biomass into bromomethylfurfural, furfural, and depolymerized lignin in lithium bromide molten salt hydrate of a biphasic system. RSC Adv. 7 (1), 300–308.

Zhang, J., Wang, M., Gao, M., Fang, X., Yano, S., Qin, S., et al., 2013. Efficient acetone–butanol–ethanol production from corncob with a new pretreatment technology—wet disk milling. Bioenergy Res. 6 (1), 35–43.

Zhang, K., Pei, Z., Wang, D., 2016. Organic solvent pretreatment of lignocellulosic biomass for biofuels and biochemicals: a review. Bioresour. Technol. 199, 21–33.

Zhang, H., Zhang, P., Ye, J., Wu, Y., Liu, J., Fang, W., et al., 2018. Comparison of various pretreatments for ethanol production enhancement from solid residue after rumen fluid digestion of rice straw. Bioresour. Technol. 247, 147–156.

Zhao, X., Cheng, K., Liu, D., 2009. Organosolv pretreatment of lignocellulosic biomass for enzymatic hydrolysis. Appl. Microbiol. Biotechnol. 82 (5), 815–827.

Zhao, Y., Liu, X., Wang, J., Zhang, S., 2012. Effects of cationic structure on cellulose dissolution in ionic liquids: a molecular dynamics study. ChemPhysChem 13 (13), 3126–3133.

Zhu, S., Wu, Y., Yu, Z., Chen, Q., Wu, G., Yu, F., et al., 2006. Microwave-assisted alkali pre-treatment of wheat straw and its enzymatic hydrolysis. Biosyst. Eng. 94 (3), 437–442.

Application of microwave energy in the processing of lignocellulosic biomass

5

Gopu Raveendran Nair[1], Deepak Kumar[2] and Ashutosh Singh[3]
[1]Department of Agricultural and Biological Engineering, University of Illinois at Urbana-Champaign, Urbana, IL, United States, [2]Department of Chemical Engineering, SUNY College of Environmental Science and Forestry, Syracuse, NY, United States, [3]School of Engineering, University of Guelph, Guelph, ON, Canada

5.1 Introduction

Microwaves (MWs) are a form of electromagnetic energy that exists between 300 and 300,000 MHz on the electromagnetic spectrum in an area that lies between infrared and radio frequencies and corresponds to wavelengths of 1 cm to 1 m. Most MW reactors (typically used for chemical synthesis reactions) and domestic MW ovens run at a frequency of 2.45 GHz or 900 MHz to prevent interference with telecommunications and mobile cellular phone frequencies (Kumar et al., 2014; Nüchter et al., 2004). Dipole rotation and ionic conduction are key mechanisms that efficiently cause rapid heating when MWs come into contact with polar molecules or ions. Polar molecules attempt to realign themselves with the rapidly oscillating electric field of the MW, which results in dipole rotations. Dipole rotations are interactions that occur when polar molecules try to realign with the rapidly oscillating electric field of MW. Through molecular friction and dielectric loss, energy is lost in the form of heat during this phenomenon. When an MW-absorbing material is placed in an alternating electric field, it absorbs energy irreversibly, resulting in quick volumetric heating (Fig. 5.1). The center of the material gets heated to a significantly higher temperature than the outer material, as shown in Fig. 5.1. MW heating is the opposite of conventional heating, in which heat is transferred to a material's surface from an external heat source (Kostas et al., 2017). As electromagnetic energy is converted into heat, MW heating can be described as an energy conversion rather than a form of heating. Because electromagnetic waves can transfer and dissipate energy volumetrically, MW processing can be fast, continuous, compact, and flexible.

Unlike traditional heating methods, the heating effect is instantaneous, and no time is spent heating up and cooling down. Another advantage of MW is that the process can be controlled in terms of energy intensity and temperature, preventing overheating and overprocessing of the material (Remya and Lin, 2011). Not all materials, nevertheless, can absorb MWs. Materials can be classified as (1) MW-

Advances in Lignocellulosic Biofuel Production Systems. DOI: https://doi.org/10.1016/B978-0-323-91192-4.00002-X

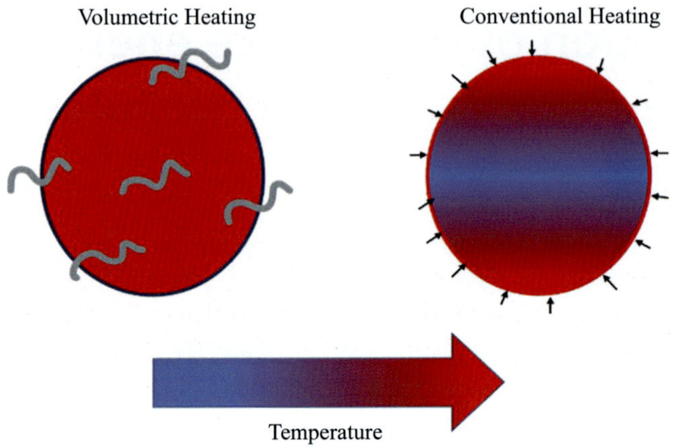

Figure 5.1 Volumetric heating versus conventional heating.

transparent/insulator (e.g., glass and ceramics), (2) MW-conducting (e.g., metals), or (3) MW-absorbing (e.g., polar molecules such as water) where the MWs are absorbed and converted into heat. The relative degree of MW heating can be obtained by "loss tangent (tan δ)," which can be defined by the following equation [Eq. (5.1)]:

$$\tan \delta = \frac{\varepsilon''}{\varepsilon'} \tag{5.1}$$

where ε' is the dielectric constant of the material (the ability of the material to store electromagnetic energy) and ε'' is the dielectric loss factor (the ability of the material to convert the stored electromagnetic energy to heat). MW absorbing materials can be classified as high ($\tan \delta > 0.5$), medium ($0.1 < \tan \delta > 0.5$), and low ($\tan \delta < 0.1$) depending upon their dielectric loss tangent values. Table 5.1 illustrates the dielectric properties of some lignocellulosic feedstocks.

MW energy is not thermal energy by itself. The heating is caused by the electromagnetic energy generated by the dielectric characteristics of the material in combination with the applied electromagnetic field. The interaction between the electric field and the matrices is mainly determined by dielectric characteristics (Buffler, 1993). The rate of conversion of electrical energy into thermal energy in the material is described by (Chen et al., 1993):

$$P = K.f\varepsilon' \mathrm{E}^2 \tan \delta \tag{5.2}$$

where P is the MW power dissipation per volume unit, K is a constant, f is the frequency applied, E is the electric field strength, and tan δ is the dielectric loss tangent.

Table 5.1 Dielectric properties of some lignocellulosic feedstocks.

Feedstock	Frequency	Dielectric constant	Loss factor	Tangent loss	Penetration depth (cm)	References
Oil palm fiber	2.45 GHz	1.99	0.16	0.08	24.8	Salema et al. (2013)
Oil palm shell	2.45 GHz	2.76	0.35	0.12	13.4	Salema et al. (2013)
Pinewood	2.45 GHz	2.7	0.53	–	5.9	Farag et al. (2012)
Switchgrass	2.45 GHz at 150°C	2.2	0.09	0.041	10	Aguilar-Reynosa et al. (2017)
Pulp mill sludge (PS 50–50)[a]	2.45 GHz	5.5	0.4	–	0.64	Namazi et al. (2015)
Pulp mill sludge (PS 70–30)[a]	2.45 GHz	3.5	0.12	–	0.86	Namazi et al. (2015)
Corn Stover	2.45 GHz	2.6	0.8		32	Motasemi et al. (2015)
Sweet Sorghum	2.45 GHz	2.67	0.9	—	48	Fennell et al. (2014)

[a]PS: ratio of primary secondary sludge.

The distribution of the electric field is determined by the geometry and dielectric characteristics of the irradiated sample. The depth of penetration of a wave (D_p) can influence the working frequency choice and is determined by the thickness of the matrix to be treated. The energy absorption inside a solid material produces an electric field that diminishes as one moves away from the substance's surface. The penetration depth (D_p) is the distance from the material surface at which the absorbed electric field (ε) is decreased to $1/\varepsilon$ of the electric field at the surface, resulting in a 37% energy loss (Udalov et al., 2011). The penetration depth is inversely proportional to the frequency and the dielectric properties of the material, as shown by the following equation:

$$Dp = \frac{c}{2\pi f' \sqrt{2\varepsilon'}[\sqrt{1+\tan^2\delta} - 1]^{1/2}} \tag{5.3}$$

where c is the velocity of light in m/s, this equation is approximated by the following equation when $\tan \delta << 1$:

$$d = \frac{\lambda_0 \sqrt{\varepsilon'_r}}{2\pi\varepsilon''_r} \tag{5.4}$$

where λ_0 the wavelength in vacuum, and d is the approximate depth. The depth of penetration varies inversely with the loss factor and even less when the sample is MW-sensitive, as shown by the equation. When the MW penetration depth is significantly smaller than the thickness of the material, only the surface of the material is heated, and the rest of the material is heated by conduction.

The advantages of MW heating are associated with selective and volumetric heating rather than energy consumption. Selective heating occurs when a feedstock is heterogeneous and contains different materials or phases with different tan values. Parts with higher $\tan \delta$ values will reach higher temperatures than those with lower $\tan \delta$ values, resulting in thermal gradients within the biomass that are unique to MW heating and can affect the decomposition pathway and subsequent product chemistry. Volumetric heating will overcome heat transfer limitations, resulting in much higher heating rates, smaller processing equipment, and the ability to process large particle sizes (Motasemi and Afzal, 2013). Volumetric heating can benefit the overall process system because the overall flowsheet can be significantly simplified, with fewer requirements for feedstock size reduction, hot gas handling and recovery systems, and overall equipment size. While the energy requirements are comparable to those of conventional processes, MW heating allows for lower processing system capital costs and the production of products that are not possible with conventional systems (Kostas et al., 2017).

MW-assisted technologies have been developed in various sectors of food and bioprocessing, such as food drying, cooking, extraction, thermos-chemical processing, biomass pretreatment, sterilization, etc. This chapter specifically focuses on the application of MWs in the processing of lignocellulosic biomass for the production of bioenergy, biofuel, and bioproducts. Biomass converts solar energy and carbon dioxide into chemical energy in the form of carbohydrates through photosynthesis. Since the carbon dioxide absorbed during photosynthesis is released after combustion, using biomass as a fuel is a carbon-neutral process. Forestry and agricultural residues, wood, byproducts of material processing, and municipal and organic sludge waste are all examples of biomass. Plants capture around 4000 EJ each year in the form of energy in biomass and food. The worldwide biomass potential, according to Fischer and Schrattenholzer (2001), ranges from 91 to 675 EJ/year from 1990 to 2060. Biomass processing technologies can be broadly classified into thermochemical (combustion, gasification, pyrolysis, and liquefaction) conversion processes and biological [alcoholic fermentation, anaerobic digestion (AD), etc.] conversion processes.

5.2 Microwave-assisted thermochemical conversion—gasification and pyrolysis

While biomass combustion is the most direct and technically simple process, biomass energy generation has low overall efficiency. There are numerous advantages to gasification over combustion. It is capable of converting low-value feedstocks not only into electricity but also into transportation fuels. It will be an amazing development for supplementing the energy demands of the world in the next years. Biomass gasification is the incomplete combustion of biomass which produces combustible gases such as carbon monoxide (CO), hydrogen (H_2), and methane traces (CH_4). Producer gas is the term applied to this mixture. Producer gas can be used to power internal combustion engines (both compression and spark ignition) and a direct heat replacement for furnace oil. Producer gas can be utilized to make methanol—an incredibly appealing chemical that can be used as a heat engine fuel as well as a chemical feedstock for industries—in an economically viable manner. Because any biomass material may be gasified, this approach is far more appealing than ethanol or biogas generation, which requires only certain biomass feedstock. Gasification is considered the heart of biomass conversion. The type and design of the gasifier, the gasification temperature, the flow rates of biomass and oxidizing agents (air or steam), the type and quantity of catalysts, and the biomass type and characteristics are the primary operating parameters of the gasification process. Gasification happens in the presence of an oxidizing agent at a high temperature (also called a gasifying agent). Heat is delivered to the gasifier either directly or indirectly, raising the temperature to 600°C–1000°C for gasification. Air, steam, nitrogen, carbon dioxide, oxygen, or a mix of these are commonly used as oxidizing agents. Since air or oxygen interacts with biomass in a gasifier, they are categorized according to how air or oxygen is supplied into it. Downdraft, Updraft, and Crossdraft are the three types of gasifiers. And as the name implies, an updraft gasifier has air traveling through the biomass from the bottom and flammable gases emerging from the top.

Similarly, in the downdraft gasifier, the air is passed downdraft from the tuyers (Rajvanshi and Goswami, 1986). A study by Xiao et al. proved the significance of MWs as a tool for the gasification of rice straws (Xiao et al., 2015). MW heating was shown to be more effective than traditional heating, lowering the reaction temperature by 200°C while maintaining the same results. Liu et al. (2019) experimented on coal char steam gasification characteristics under MW heating. During the steam gasification, the heating source significantly impacted the CO and CO_2 content more than the H_2 content. MW-assisted gasification (MAG) improved the product gas quality by increasing syngas content and lowering CO_2 concentration during steam gasification.

Furthermore, at the same temperature, the carbon conversation rate of steam gasification under MW-assisted heating was substantially higher than that under conventional heating, particularly in the low-temperature range. Lahijani et al. studied the effect of catalysts in MW-assisted biomass char gasification from oil palm shells (Lahijani et al., 2015). The system included 1150 W MW power, 10% ash-loaded char as a catalyst, and oil palm shell and empty fruit bunch as feedstock. The sample size of 8 g was used in the presence of CO_2 at temperatures of 750°C and 900°C. They achieved a CO yield of 93% when compared to 58% of yield without a catalyst.

Biomass pyrolysis is the thermal degradation of the organic matrix of biomass in nonoxidizing environments, resulting in noncondensable gas products, liquid bio-oil, and solid biochar. Depending on the heating rate and solid residence period, pyrolysis can be divided into slow, fast, and flash pyrolysis (Balat et al., 2009). Traditionally, slow pyrolysis has been used to produce charcoal. Because of the long residence time (hours to days), relatively low temperature (300°C–700°C), and the acceptability of a wide range of particle sizes (5–50 mm), the thermal decomposition of biomass occurs under a very low heating rate with sufficient time allowed for repolymerization reactions to maximize the solid yields. High heating rates ($>$ 10°C–200°C/s) and short residence lengths (0.5–10 s, mainly 2 s) are characteristics of fast pyrolysis. The yield of bio-oil can be as high as 50–70 wt.% (dry basis). The flash pyrolysis process happens at higher heating rates of 10^3–104°C/s and shorter residence times ($<$0.5 s), resulting in extremely high bio-oil yields up to 75–80 wt.% (Kan et al., 2016). Fast and flash pyrolysis is the conditions of choice for maximizing bio-oil yields when bio-oil production is the principal product of interest. Bio-oils have been thoroughly studied as potential combustion fuels for generating electricity and heat in boilers, furnaces, and combustors, as well as diesel engines and gas turbines. Biochar is the main solid product, consisting of unconverted organic solids, carbonaceous wastes, and a mineral fraction produced by the partial or total breakdown of biomass components. Chars have different physical, chemical, and mechanical properties depending on the feedstock and pyrolysis parameters. Slow pyrolysis (typical product yields: bio-oil 30 wt.%, biochar 35 wt.%, and gas 35 wt.%) favors biochar production. The high HHV (higher heating values) (20–36 MJ/kg) makes chars attractive in some fuel applications as substitutes for coal (Kan et al., 2016). Carbon dioxide (CO_2), carbon monoxide (CO), hydrogen (H_2), low carbon number hydrocarbons like methane

(CH_4), ethane (C_2H_6), and ethylene (C_2H_4), and minor amounts of other gases like propane (C_3H_8), ammonia (NH_3), nitrogen oxides (NO_x), sulfur oxides (SO_x), and low carbon number alcohols are among the gases emitted during biomass pyrolysis. Depending on their practical composition, the lower heating values (LHVs) of pyrolytic gases typically vary from 10 to 20 MJ/Nm^3. CO_2 and CO are produced primarily by the breakdown and reformation of carbonyl ($C=O$) and carboxyl (COO) groups as the major products of biomass pyrolysis. Light hydrocarbons (mainly CH_4) are mostly attributed to the secondary breakdown of oxygenated compounds as well as the decomposition of weakly bound methoxyl ($-O-CH_3$) and methylene ($-CH_2-$) groups, whereas H_2 is mostly attributed to the secondary decomposition and reformation of aromatic $C=C$ and $C-H$ groups at high temperatures (Qu et al., 2011; Strezov et al., 2004). In 2014, Wu et al. conducted studies on MW-assisted pyrolysis (MAP) under different heating rates (Wu et al., 2014). They concluded that the release of CH_4 was increased for MAP with an increase in heating power. In contrast, in conventional pyrolysis, there was no release of methane and hydrogen at 310°C. Robinson et al. (2015) researched the effect of MW pyrolysis on converting European larch wood chips by using an MW pyrolysis system with a 6 kW MW generator. The study found that when the sample size was kept within 15 mm, the degree of valorization was 70%, which is equivalent to conventional pyrolysis. Overheating occurred using MW with a larger sample size (50 mm). The quality of the bio-oils generated was substantially better than that of traditional heating techniques, with order-of-magnitude enhancements in levoglucosan and phenolic content. MW- processing has been found to benefit from the use of cold sweep gas and hence does not necessitate the use of a fluidized bed, hot gas handling, or energy recovery devices.

Considering years of development, existing pyrolysis and gasification processes still encounter substantial technical difficulties in terms of product yield, quality, and process energy efficiency. Developing new ways for efficient heating, managing heating settings, and limiting negative effects on product quality is one way to improve the current technologies. Fluidized bed reactors are the most common reactors for large-scale pyrolysis and gasification; however, they suffer from several limitations. Heat is delivered from the heating media, commonly heated sand, to the fine biomass particles in these systems by conduction. Precision control of heating operating parameters during processing is extremely difficult to achieve. MW irradiation is a cost-effective approach to heat biomass for thermochemical conversion. Tech En Ltd. (United Kingdom) in early 1990, was the first to report MAP, although it was not widely used in biomass conversion until the late 2000s. The number of peer-reviewed articles and patents on the topic has been quickly increasing in recent years, indicating a significant increase in interest in the subject. Because of a lack of understanding and control of the heating characteristics and their correlation to product yields, quality, and affordability, MAP or MAG (MW-assisted gasification) of biomass for biofuel production has not seen widespread use. MW heating is more uniform (volumetric heating) and easy to control. It does not require a high degree of feedstock grinding (e.g., huge chunks of wood logs can be utilized) and can manage mixed feedstock.

Because MW-aided techniques do not require biomass powders or rapid agitation and fluidization, the conversion products are cleaner than those produced by traditional procedures. Since the syngas produced is not diluted by the carrying gas, it has a higher heating value. MW heating is a developed technology with a low-cost MW-heating system for biomass conversion. Additional claims are that nonthermal MW effects significantly boost reaction rates (Kappe, 2004). The study of differences in reaction rate between conventional and MW heating when carried out at the same apparent temperature led to this hypothesis of nonthermal effects. Some speculated that the nonthermal effects are caused by dipolar molecules' orientation effects due to the electric field's direct interaction with reacting molecules, causing changes in activation energy. The observed rate improvements were caused by a thermal/kinetic effect because of the high reaction temperatures, according to most scientists in the area. However, the role of MW irradiation in chemical reactions and synthesis is uncertain and even debatable.

5.3 Microwave-assisted biological conversion

During the biological conversion of lignocellulosic biomass to fuel and bioproducts, biomass recalcitrance is a major hurdle in achieving high conversion efficiencies. The structural carbohydrates (cellulose and hemicellulose) and lignin in the biomass form a complex matrix that makes the structure relatively impervious to microbial degradation. A pretreatment step is commonly used that involves the use of chemicals, high temperature, and pressure to open up the biomass structure, remove lignin, and increase the surface area of biomass for efficient microbial conversion (Kumar and Murthy, 2011; Kumar and Singh, 2019). Due to high capital costs and energy-intensive, extensive research efforts have been made to investigate and develop efficient pretreatment technologies. Although a large number of pretreatment technologies have been developed and investigated, these can be broadly classified as physical pretreatment, chemical pretreatment, physicochemical pretreatment, and biological pretreatment. Most of the physical pretreatments (e.g., disk refining) mainly increase the surface area of biomass without altering the chemical composition. Chemical pretreatments that use chemicals under high temperatures and pressure alter the chemical structure of biomass (mostly by removing lignin and hydrolyzing hemicellulose) and increase the accessibility to cellulose polymers for efficient conversion. The mode of alteration and changes in biomass depends on the type of chemical used and other operating conditions (da Costa Sousa et al., 2009). Physicochemical pretreatment, combining chemicals and physical energy, is another promising approach to efficiently reduce biomass recalcitrance. Some common examples of physicochemical pretreatment processes include steam explosion, ammonia fiber expansion, extrusion, ultrasound-assisted pretreatment, MW-assisted pretreatment, and supercritical CO_2 explosion (Hoang et al., 2021; Kumar and Singh, 2020). Although all of these pretreatments have been effective in enhancing the conversion yields of various lignocellulosic feedstocks, chemical, and physicochemical pretreatments are considered the most promising option for commercial-scale applications based on the maturity levels, economics, and environmental impacts (da Costa Sousa et al., 2009). In this section, a brief discussion of MW-assisted pretreatment is provided. To learn details of other pretreatment technologies, the readers are suggested to refer to Chapter 3 of this book, which focuses on the pretreatment technologies for lignocellulosic conversion.

MW-assisted pretreatment is considered a highly promising approach due to lower energy consumption, environmental benefits (no smoke or wastes, no/less use of chemicals), and significantly shorter pretreatment times (up to 10 times) compared to conventional chemical and physicochemical pretreatments (Aguilar-Reynosa et al., 2017; Hoang et al., 2021; Milano et al., 2018). Zhu et al. (2016) investigated MW-assisted acid and alkali pretreatment to enhance sugar recovery from the sugarcane bagasse, one of the high-potential lignocellulosic feedstocks. They observed that using MW energy reduced the pretreatment time by five to seven times compared to the conventional pretreatment and enhanced sugar yields by more than four times. As discussed earlier, MWs heat the entire volume from the inside of biomass, leading to multiple effects, including delignification (the extent depends on the chemical use), hemicellulose degradation, cell wall rupture, and an increase in the surface area that leads to higher yields during enzymatic hydrolysis. Peng et al. reported a 56% and 75% increase in the surface area of microcrystalline cellulose pretreated with 20 min of MW irradiation (800 W) and NaOH-supported MW irradiation, respectively (Peng et al., 2014). Several studies using MW-assisted pretreatment for various lignocellulosic feedstocks have reported delignification due to the thermal effects of the pretreatment that enhances the subsequent conversion efficiencies. For example, Chen et al. reported 46% of lignin removal while retaining most (90%) of the glucans during MW-assisted dilute ammonia pretreatment of sorghum bagasse (Chen et al., 2012). The delignification has been reported to be higher with alkali use than the acid (Zhu et al., 2016). This phenomenon is commonly observed in conventional pretreatments; however, the pretreatment time is usually very long in using alkali. The use of MW energy addresses this challenge and achieves high delignification quickly (Li et al., 2016).

Like conventional pretreatment, chemical concentrations and holding times are important factors when designing MW-assisted pretreatment. The sugar yields are found to increase up to a point with an increase in holding time during the pretreatment and start dropping at high holding times. This is observed mainly due to the degradation of sugars, especially the degradation of xylose to furfural, formic acid, formaldehyde, lactic acid, etc. (Fan et al., 2013; Zhu et al., 2016). Using salts during MW-assisted pretreatment could help achieve the required temperature in even shorter times (Li et al., 2016). Li and Xu. (2013) investigated using calcium chloride during MW-assisted pretreatment of corn stover. Under optimum pretreatment conditions, they observed 85.9% hemicellulose degradation and a 169% increase in the specific surface area, which led to a 90.66% enzymatic hydrolysis ratio (Li and Xu, 2013). Although most studies used MW irradiation in combination with a chemical catalyst to achieve synergistic effects, some recent studies have investigated the use of MW in combination with other irradiation technologies.

Bussemaker et al. (2013) investigated the combined MW and ultrasound pretreatment to enhance the bioconversion efficiency of wheat straw. The combined pretreatment resulted in a 3.6% higher delignification compared to only MW pretreatment. Several other approaches, such as combining physical pretreatments (e.g., extrusion, ball milling, etc.) with MW pretreatment, use of high boiling solvents, and various designs of pretreatment reactors, etc., have been investigated. Details of these approaches and their results have been discussed in detail in a recent review papers (Aguilar-Reynosa et al., 2017; Ethaib et al., 2015; Laghari et al., 2014; Li et al., 2016).

Similar to MW-assisted pretreatment used to enhance sugar yields during enzymatic hydrolysis, the approach has also been investigated to improve the biogas productivity during AD of lignocellulosic biomass. Pretreatment can increase the biodigestibility of waste and increase overall biogas yield. Passos et al. (2013) investigated MW pretreatment during AD of microalgae-bacterial biomass cultivated in high-rate algal ponds for wastewater treatment. The results showed an increase in biomass solubilization concerning MW-specific energy. A linear correlation was established between biomass solubilization and biogas yield, which proved the effect of MW treatment in biogas production from microalgae. Their experiments obtained a cumulative biogas yield of 307.11 mL/g VS (volatile solids) at applied specific energy of 65400 kJ. kg. TS (total solid) from MW pretreated samples with 900 W applied power and at 98°C (Table. 5.2). Another study conducted by Kuglarz et al. (2013) about the effect of MW pretreatment on the increase in biogas potential of secondary sludge from wastewater treatment plants. The sludge was pretreated with MW irradiation within the temperature range of 30 to 100. The sludge pretreated at 60°C–70°C by MWs of 900 W was the most efficient method for optimal biogas production, which had 35% more methane when compared to untreated sludge. MW pretreatment along with sodium hydroxide soaking was performed on paddy straw to evaluate the biogas production rate by Kaur and Phutela. (2016). As per their investigation, 54.7% enhancement in biogas production was reported in 4% NaOH—30 min MW (720 W, 180°C treated paddy straw), which resulted from a 65% reduction in lignin due to the pretreatment. The NaOH pretreatment was performed by submerging the paddy straws in 200 mL 4% NaOH for 24 h at 28°C before MW treatment for 30 min. The MW pretreatment method was applied to the food industry waste too. Experiments on the application of ozone-MW pretreatments on biogas production from canned maize production sludge were conducted by Beszédes et al. (2009).

Ozone treatment was performed in continuously mixed solutions diluted to 6% dry matter content with an ozone contact time of 30 min with a bubbling gas flow rate of 1 dm^3/min. MW treatment was performed with a power of 250 W and 2.45 GHz frequency for 5 min for maximum biogas production of 9.52 cm^3/g/day. Cheng and Liu. (2010) used MW-assisted-alkali pretreatment to improve the biogas production from herbal extraction process residues. Microorganisms do not readily decompose herbal extraction process residues (HPR) due to their physical shielding by nondigestible lignin. Maximum biogas output of 1477 mL and a biogas yield of 443 mL/g were obtained when MW-assisted-alkali pretreatment hydrolyzed HPR

Table 5.2 Microwave pretreatment to enhance biogas production during anaerobic digestion of various lignocellulosic feedstocks.

Feedstock	Pretreatment type	Process conditions	Major results	References
Microalgae	MW	Specific energy: 21,800–65,400 kJ/kg TS, MW power: 300–900 W	12%–78% increase in biogas; Specific energy was major influential factor irrespective of MW power and time	Passos et al. (2013)
Waste activated sludge	MW	MW power: 1250 W, time: 5 min, Temp: 96°C, solid loading: 5.5%	211% increase in biogas production after 23 days of AD	Eskicioglu et al. (2006)
Wastewater sludge	MW	MW power: 900, Temp: 60°C–70°C	35% increase in methane production when compared to control	Kuglarz et al. (2013)
Paddy Straw	MW- NaOH	Paddy straw soaked in NaOH (2%–10%) for 24 h at 28°C; MW pretreatment at 720 W and 180°C for 30 min	65% decrease in lignin content; 88.7% in silica content; 54.7% increase in biogas production volumes	Kaur and Phutela (2016)
Maize production sludge	Ozone- MW	Ozone treatment with an ozone contact time of 30 min with a bubbling gas flow rate of 1 dm^3/min MW pretreatment at 250 W and 2.45 GHz frequency for 5 min	96% biodegradability compared to 24% in control; Maximum biogas production of 9.52 cm^3/g/day	Beszédes et al. (2009)
Herbal extraction residues	MW- NaOH	10% solid loading; 0.02–0.16 g- NaOH/g biomass; MW power: 700 W; Time: 15 min	48% decrease in lignin content; 66% increase in specific surface area; 94% increase in methane yield (1286 mL vs 663 mL)	Cheng and Liu (2010)
Food waste	MW	MW power: 1400 W; Time: 30 min; Solid loading: 7.5%	6.43% increase in biogas yield compared to untreated feedstock (control)	Deepanraj et al. (2017)

for 30 min (MW power of 700 W) with an alkali loading (10 g of grounded HPR mixed with 100 mL of NaOH) of 0.12 g NaOH/g HPR with a gross weight loss of HPR after fermentation of biogas reached 72.1%. Deepanraj et al. (2017) analyzed the effect of MW pretreatment on the conversion of food waste-chicken manure combination into biogas. MW irradiation was conducted with a 1450 W MW apparatus with a frequency of 2450 MHz. The sample volume was 2 L at a temperature of 180°C for 30 min. The total solid removal efficiency of the reactor after MW pretreatment was noted as 60.9%, whereas the TS removal efficiency of untreated samples was 58.8%. Maximum biogas yield was observed as 9287 mL when compared to untreated samples of 8921 mL which showed that MW pretreatment enhanced the biogas yield in food waste.

5.4 Microwave-assisted extraction of high-value compounds

Most of the lignocellulosic biorefinery concepts involve the extraction of high-value compounds at the front or back end of the primary conversion process. The conventional extraction processes use selective solvents at various temperature and time combinations. MW-assisted extraction (MAE) differs from traditional approaches such as solid and liquid extraction because the extraction happens because of changes in cell structures caused by electromagnetic waves. In traditional extractions, mass is transferred from the interior to the outside, while heat is transferred from outside to inside the substrate. The high extraction yield and process acceleration in MAE could be due to a synergistic combination of two transport phenomena: heat and mass gradients moving in the same direction. In traditional extractions, on the other hand, mass is transferred from the interior to the outside, while heat is transferred from the outside to the inside of the substrate (Fig. 5.2).

The rate of extract recovery is not a linear function of time throughout the extraction process: the concentration of solute inside the solid changes, resulting in a nonstationary or unstable condition. During the period of interaction between the solid-containing particle and the solvent effectuating the separation, a sequence of phenomenological processes must occur, including:

1. Penetration of the solvent into the solid matrix,
2. Solubilization and breakdown of components,
3. Movement of the solute out of the solid matrix,
4. Transfer of the extracted solute into the bulk solution from the solid's exterior surface and separation of the extract and the solid (Veggi et al., 2012).

As a result, the solvent diffuses into the solid matrix, and the solute dissolves until it reaches a concentration limited by the properties of the solid. Effective diffusion transports the solute-containing solution to the surface. Finally, the solution is transported from the surface to the bulk solution via natural or forced convection.

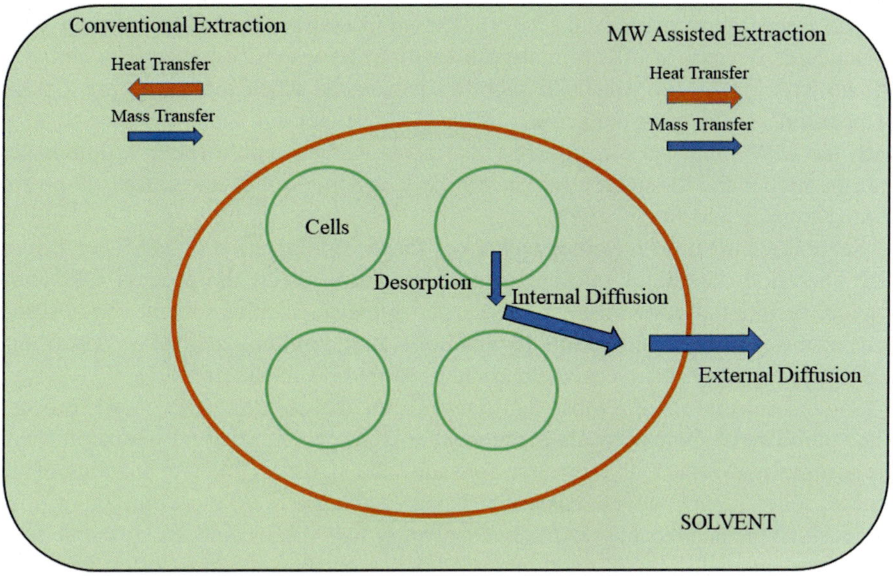

Figure 5.2 Heat and mass transfer mechanism of conventional and MW-assisted extraction of natural products (Périno-Issartier et al., 2011).
Source: Adapted from Périno-Issartier, S., Abert-Vian, M., Chemat, F. 2011. Solvent free microwave-assisted extraction of antioxidants from sea buckthorn (Hippophae rhamnoides) food by-products. Food Bioprocess. Technol., 4(6), 1020–1028.

The basic property of MW heating is energy transfer. The energy is transmitted to the material by convection, conduction, and radiation processes through the external material surface in the presence of thermal gradients in the traditional heat transfer process. In MAE, on the other hand, MW energy is delivered directly to materials via molecular interactions with the electromagnetic field and subsequent conversions of electromagnetic energy to thermal energy (Thostenson and Chou, 1999). The solvent used is the first factor to consider while choosing MW physical constants. It is crucial to pick a solvent with a lot of extraction capacity and interaction with the matrix and the analyte. Because of their permanent dipole moment, polar molecules and ionic solutions (usually acids) absorb much MW radiation. On the other hand, Hexane and other nonpolar solvents do not heat up when exposed to MWs.

The degree of MW absorption usually increases with the dielectric constant. Because water has a larger dielectric constant, it has a lower dissipation factor, meaning the system receives more MW radiation than it can dissipate. This is known as superheating, which occurs when there is water in the matrix. Because of the significant absorption, the temperature inside the sample rises, causing the cells to rupture due to in situ water. In certain situations, it can cause the target compound to degrade or cause a solvent "explosion." In others, it can enhance the target chemical's diffusivity in the matrix. As a result, the MW power must be high enough to reach the boiling point of the water or other solvent, establishing the separation temperature (Jain et al., 2009). The solid matrix is the second factor to be considered. Because it affects molecule rotation, its viscosity impacts its ability to absorb MW radiation. Molecular mobility is reduced when molecules are "locked in position" as viscous molecules, making it difficult for the molecules to align with the MW field. As a result, the heat generated by dipole rotation diminishes, and the higher the dissipation factor (δ), the faster the heat is transferred to the solvent (Kingston and Jassie, 1988).

Several applications have been explored for the optimization of MAE conditions. The operation conditions chosen have a direct impact on the process efficiency. Typical factors that may influence MAE performance, such as solvent composition, solvent-to-feed ratio, extraction temperature and duration, and MW power and matrix characteristics such as water content, should be considered.

Solvent selection is the most significant factor that influences the MAE process. The solubility of the chemicals of interest, solvent penetration and interaction with the sample matrix and its dielectric constant, and the mass transfer kinetics of the process all influence solvent selection (Chen et al., 2008). Another important point to consider is that the best extraction solvents for MAE could be different from those used in traditional extractions: it all relies on the solvent's ability to absorb MW radiation and, as a result, heat up (Chan et al., 2011). When MWs are used, solvents that are transparent to MWs do not heat up. A classic example of MW-transparent solvent is hexane.

In contrast, ethanol is a popular MW-absorbing solvent. MAE may employ both polar and nonpolar solvents, with solvents like ethanol, methanol, and water being sufficiently polar to be heated by MW radiation. When various solvents are mixed, the characteristics of the solvent may be changed, allowing for variable solvent selectivity for different molecules. According to various studies, small quantities of water in the extraction solvent allow water to diffuse into the matrix cells, resulting in improved heating and promoting the transport of chemicals into the solvent at greater mass transfer rates. Hexane is commonly employed in the extraction of volatile oils. A solvent-free MAE (SFMAE) method has also been developed for aromatic plants rich in volatile oils, where the moisture content of the plants itself serves for the extraction. According to studies, adding ethanol or water to weak MW absorbers like hexane can increase extraction efficiency. Hexane-acetone is one of the most often used solvent combinations, and just a small quantity of water ($\sim 10\%$) is required to enhance the heating rate in nonpolar solvents such as hexane xylene or toluene.

Similar to conventional extraction procedures, the solvent-to-solid-feed (S/F) ratio is another important parameter during MAE. While using a matrix that will enlarge during the extraction, the solvent volume must be adequate to ensure that the whole sample is submerged in the solvent throughout the entire irradiation process. The use of high quantities of solvent in traditional extractions enhances extraction yield. Conversely, MAE may yield poorer recoveries due to nonuniform distribution and MW exposure (Eskilsson et al., 1999). According to studies, the extraction solution should not include more than 30%–34% (w/v) of water (Pan et al., 2003).

As discussed in other sections also, heating duration is an important parameter affecting process efficiency. Compared to traditional methods, MAE durations are relatively short, minimizing potential thermal degradation and oxidation, which is especially essential for heat-sensitive high-value compounds (Chan et al., 2011). When a longer extraction period is needed, the samples are sometimes extracted in numerous phases utilizing sequential extraction cycles. This is an example of employing a greater volume of solvent and a longer MW application duration. In this scenario, a new solvent is supplied to the residue, and the procedure is repeated to ensure that the matrix has been exhausted. According to Li et al. (2010), three 7-min cycles were acceptable for the MAE of triterpene saponins from the yellow horn.

In contrast, 5-min cycles were suggested for MAE of triterpenoid saponins from the *Ganoderma atrum*. The heating duration will also be dependent on the MW power used in the process. The solvent power rises at high temperatures due to decreased viscosity and surface tension, making it easier for the solvent to solubilize solutes and increasing matrix wetting and penetration. Furthermore, when MAE is performed in closed containers, the temperature can rise much above the solvent's boiling point, resulting in improved extraction efficiency due to solute desorption. Routray and Orsat, (2012), on the other hand, claim that efficiency increases as temperature rises until an optimum temperature is reached, after which it begins to decline as the temperature increases further: this is because the ideal extraction temperature is directly linked to the stability and, thus, the yield of the target compound. In addition, high MW power produces localized heating in the sample, which causes MAE to break down the plant matrix, allowing the solute to diffuse out and dissolve in the solvent. As a result, raising the power will enhance the extraction yield while reducing the extraction time. However, a very high MW power might result in a low extraction yield due to the destruction of thermally sensitive compounds.

Furthermore, when applying greater power, a fast rupture of the cell wall occurs, and as a result, contaminants can be leached out into the solvent with the desired solute. As a result, it is critical to choose the MAE power carefully to reduce the time taken to achieve the desired temperature and avoid temperature "bumping" during the extraction. Furthermore, overexposure to MW radiation, even at low temperatures or low operating power, was shown to reduce extraction yield due to the loss of the chemical structure of the active compounds.

5.5 Factors affecting efficiency of microwave-assisted biomass processing

As mentioned earlier, the dielectric property of the material to be treated controls the extent of heating. Materials with high dielectric loss factors possess no penetration, and materials with low dielectric loss factors possess high penetration. For MW treatment, materials with a medium dielectric loss factor are to be considered, which are called MW absorbers (Fig. 5.3). Lignocellulosic feedstocks are carbon-

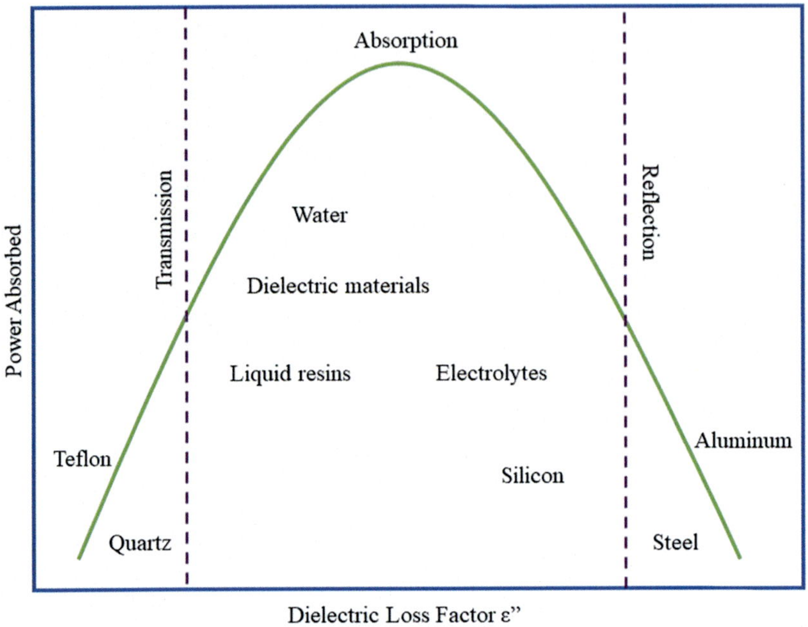

Figure 5.3 Interaction of the electromagnetic field with different materials.

rich, and all carbon-rich materials except coal and carbon foam are good MW absorbers. It is favorable to add activated carbon and silicon carbide to minimize biomass loss while MAP (Xie et al., 2014).

Even while water in the biomass promotes MW absorption, adding more water takes more energy to evaporate the water, despite the higher absorption rate. When using water to increase MW absorption rate, there is a problem: too much free water in the biomass reduces the positive temperature effect on tan δ because the loss factor of free water decreases rapidly with an increase in the temperature. Because the dielectric properties of materials are temperature-dependent, the nature of that dependence is a function of the dielectric relaxation processes operating under the existing conditions. The frequency of the MW being used and the temperature of the MW-assisted biomass conversion process must be optimized. Apart from the properties of biomass (carbon content/moisture level), process temperature, and frequency of MW used, some other important factors also should be considered for MW-assisted biomass treatments:

1. *MW source and cavity*: Magnetrons or solid-state oscillators are utilized as MW sources for MW-assisted processing. Magnetrons release MW power in a wide range of frequencies that can not be controlled. Solid-state MW oscillators, on the other end, use a semiconductor transistor to generate single-defined MW frequencies in a controlled way, or the emitted frequency can be adjusted, and are undoubtedly superior to the magnetrons in many ways; frequency control, power control, energy efficiency and size (Barham et al., 2019).

2. *Waveguide position*: It is critical to understand how MW energy enters the MW cavity from an MW-assisted biomass processing perspective. When the electromagnetic energy enters the MW cavity from the side, it seems to scatter distinctively around the oven's horizontal plane, but when it enters from the bottom, it appears to disperse heavily around the vertical plane. A unique pattern of the hot and cold spots in a single pellet was found compared to a bottom-fed waveguide MW oven. Halim and Swithenbank. (2019) investigated wood pellets in an MW cavity with two distinct wave guide locations, As per their findings, when the MW energy is excited from the side, the wood pellet reached approximately 420°C after 300 s, but when the energy is excited from the bottom, the wood pellet reached just 80°C which made the side entry superior to bottom entry MW ovens.
3. *Effect of a stirrer*: When the hot and cold spots inside the cavity do not move, heating uniformity is nearly impossible to achieve. Mode stirrers or turntables are used to induce relative motion between the electric field and the load. Without the use of a mode stirrer or turntable, heating will be uneven, which is quite undesirable.
4. *Loading height of biomass*: Within the MW cavity, biomass has an ideal loading height where the highest MW energy is absorbed. The highest absorption could be due to a high electric field intensity at such a height. The absorption rate decreases as the height of the biomass bed rise after the peak (Salema and Afzal, 2015).
5. *Effect of biomass blending with an adsorbing agent:* The addition of MW absorbers to a sample during MW processing has increased heat dispersion in the sample. Generally, using char as an MW-absorbent agent helps quick heating of low MW-absorbing materials (such as biomass), resulting in a more uniform heating distribution throughout the sample. However, char material (adsorbing agent) has a stronger interaction with MW radiation than raw biomass material due to superior dielectric characteristics. As a result, biomass interacts poorly with MW energy, making rapid heating of the reaction media nearly impossible. Blending raw biomass with char, on the other hand, can help solve the problem because char can absorb a lot of MW radiation and dissipate it to the surrounding area.

5.6 Summary and conclusion

MW energy is not the same as thermal energy. The heating is triggered by electromagnetic energy created by the dielectric properties of the material in combination with the applied electromagnetic field. The interaction of the electric field with the matrices is mainly governed by dielectric properties. When MW-assisted heating is used in biomass char gasification, the conversion rate is higher than when conventional heating is used. In MAP, the release of CH_4 is increased with an increase in heating power; in contrast, there is no release of CH_4 in conventional pyrolysis. The quality of bio-oil generated is substantially better than that of traditional heating techniques. MW-assisted pretreatments result in an increase in sugar yield during enzymatic hydrolysis; the strategy has also been explored and shown to enhance biogas productivity during AD. In the extraction of high-value compounds, it is critical to choose the MAE power carefully to reduce the time taken to achieve the desired temperature and avoid temperature hikes during the extraction. As a pretreatment, MW is useful in a wide variety of lignocellulose conversion processes.

However, the hefty initial investment and scalability issues are the main reasons for its lack of appeal.

Abbreviations

AD anaerobic digestion
CH_4 methane
CO carbon monoxide
D_p the penetration depth
H_2 hydrogen
HHV higher heating values
LHV lower heating values
MAE MW-assisted extraction
MAG MW-assisted gasification
MAP MW-assisted pyrolysis
MW microwave

References

Aguilar-Reynosa, A., Romani, A., Rodriguez-Jasso, R.M., Aguilar, C.N., Garrote, G., Ruiz, H.A., 2017. Microwave heating processing as an alternative of pretreatment in second-generation biorefinery: an overview. Energy Convers. Manag. 136, 50–65.

Balat, M., Balat, M., Kırtay, E., Balat, H., 2009. Main routes for the thermo-conversion of biomass into fuels and chemicals. Part 1: pyrolysis systems. Energy Convers. Manag. 50 (12), 3147–3157.

Barham, J.P., Koyama, E., Norikane, Y., Ohneda, N., Yoshimura, T., 2019. Microwave flow: a perspective on reactor and microwave configurations and the emergence of tunable single-mode heating toward large-scale applications. Chem. Rec. 19 (1), 188–203.

Beszédes, S., Kertész, S., László, Z., Szabo, G., Hodur, C., 2009. Biogas production of ozone and/or microwave-pretreated canned maize production sludge. Ozone: Sci. Eng. 31 (3), 257–261.

Buffler, C.R., 1993. Microwave Cooking and Processing. Van Nostrand Reinhold.

Bussemaker, M.J., Mu, X., Zhang, D., 2013. Ultrasonic pretreatment of wheat straw in oxidative and nonoxidative conditions aided with microwave heating. Ind. Eng. Chem. Res. 52 (35), 12514–12522.

Chan, C.-H., Yusoff, R., Ngoh, G.-C., Kung, F.W.-L., 2011. Microwave-assisted extractions of active ingredients from plants. J. Chromatogr. A 1218 (37), 6213–6225.

Chen, C., Boldor, D., Aita, G., Walker, M., 2012. Ethanol production from sorghum by a microwave-assisted dilute ammonia pretreatment. Bioresour. Technol. 110, 190–197.

Chen, L., Song, D., Tian, Y., Ding, L., Yu, A., Zhang, H., 2008. Application of on-line microwave sample preparation techniques. TrAC. Trends Anal. Chem. 27 (2), 151–159.

Chen, M., Siochi, E., Ward, T., McGrath, J., 1993. Basic ideas of microwave processing of polymers. Polym. Eng. Sci. 33 (17), 1092–1109.

Cheng, X.Y., Liu, C.Z., 2010. Enhanced biogas production from herbal-extraction process residues by microwave-assisted alkaline pretreatment. J. Chem. Technol. Biotechnol. 85 (1), 127–131.

da Costa Sousa, L., Chundawat, S.P., Balan, V., Dale, B.E., 2009. Cradle-to-grave assessment of existing lignocellulose pretreatment technologies. Curr. Opin. Biotechnol. 20 (3), 339–347.

Deepanraj, B., Sivasubramanian, V., Jayaraj, S., 2017. Effect of substrate pretreatment on biogas production through anaerobic digestion of food waste. Int. J. Hydrog. Energy 42 (42), 26522–26528.

Eskicioglu, C., Kennedy, K.J., Droste, R.L., 2006. Characterization of soluble organic matter of waste activated sludge before and after thermal pretreatment. Water Res. 40 (20), 3725–3736.

Eskilsson, C.S., Björklund, E., Mathiasson, L., Karlsson, L., Torstensson, A., 1999. Microwave-assisted extraction of felodipine tablets. J. Chromatogr. A 840 (1), 59–70.

Ethaib, S., Omar, R., Kamal, S.M., Biak, D.A., 2015. Microwave-assisted pretreatment of lignocellulosic biomass: a review. J. Eng. Sci. Technol. 10, 97–109.

Fan, J., Zhu, Z., Budarin, V., Gronnow, M., Gomez, L.D., Macquarrie, D., et al., 2013. Microwave-enhanced formation of glucose from cellulosic waste. Chem. Eng. Process.: Process. Intensif. 71, 37–42.

Farag, S., Sobhy, A., Akyel, C., Doucet, J., Chaouki, J., 2012. Temperature profile prediction within selected materials heated by microwaves at 2.45 GHz. Appl. Therm. Eng. 36, 360–369.

Fennell, Laura, Picou, Boldor, Dorin, 2014. Dielectric and thermal properties of sweet sorghum biomass. J. Microw. Power Electromag. Energy 48, (no. 4), 244–260.

Fischer, G., Schrattenholzer, L., 2001. Global bioenergy potentials through 2050. Biomass Bioenergy 20 (3), 151–159.

Halim, S.A., Swithenbank, J., 2019. Simulation study of parameters influencing microwave heating of biomass. J. Energy Inst. 92 (4), 1191–1212.

Hoang, A.T., Nižetić, S., Ong, H.C., Mofijur, M., Ahmed, S., Ashok, B., et al., 2021. Insight into the recent advances of microwave pretreatment technologies for the conversion of lignocellulosic biomass into sustainable biofuel. Chemosphere 281, 130878.

Jain, T., Jain, V., Pandey, R., Vyas, A., Shukla, S., 2009. Microwave assisted extraction for phytoconstituents—an overview. Asian J. Res. Chem. 2 (1), 19–25.

Kan, T., Strezov, V., Evans, T.J., 2016. Lignocellulosic biomass pyrolysis: a review of product properties and effects of pyrolysis parameters. Renew. Sustain. Energy Rev. 57, 1126–1140.

Kappe, C., Oliver., 2004. Controlled microwave heating in modern organic synthesis. Angewandte Chemie International Edition 43 (46), 6250–6284.

Kaur, K., Phutela, U.G., 2016. Enhancement of paddy straw digestibility and biogas production by sodium hydroxide-microwave pretreatment. Renew. Energy 92, 178–184.

Kingston, H.M., Jassie, L.B., 1988. Introduction to Microwave Sample Preparation. American Chemical Society.

Kostas, E.T., Beneroso, D., Robinson, J.P., 2017. The application of microwave heating in bioenergy: a review on the microwave pre-treatment and upgrading technologies for biomass. Renew. Sustain. Energy Rev. 77, 12–27.

Kuglarz, M., Karakashev, D., Angelidaki, I., 2013. Microwave and thermal pretreatment as methods for increasing the biogas potential of secondary sludge from municipal wastewater treatment plants. Bioresour. Technol. 134, 290–297.

Kumar, D., Murthy, G.S., 2011. Impact of pretreatment and downstream processing technologies on economics and energy in cellulosic ethanol production. Biotechnol. Biofuels 4 (1), 1–19.

Kumar, D., Murthy, G.S., Tiwari, G., 2014. Microwave and radio frequency heating of food products: fundamentals, modeling, and applications. In: Sahu, J.K. (Ed.), Introduction to Advanced Food Process Engineering. CRC Press, pp. 3–30.

Kumar, D., Singh, V., 2019. Bioethanol production from corn. Sergio Serna-Saldivar (Ed.), Corn: Chemistry and Technology. Elsevier, pp. 615–631.

Kumar, D., Singh, V., 2020. Bioconversion of processing waste from agro-food industries to bioethanol: creating a sustainable and circular economy. Waste Valorisation: Waste Streams a Circular Economy 161–181.

Laghari, S.M., Isa, M.H., Saleem, H., 2014. Microwave individual and combined pretreatments on lignocellulosic biomasses. IOSR J. Eng. 4 (2), 14–28.

Lahijani, P., Mohammadi, M., Zainal, Z.A., Mohamed, A.R., 2015. Improvement of biomass char-CO_2 gasification reactivity using microwave irradiation and natural catalyst. Thermochim. Acta 604, 61–66.

Li, H., Qu, Y., Yang, Y., Chang, S., Xu, J., 2016. Microwave irradiation–a green and efficient way to pretreat biomass. Bioresour. Technol. 199, 34–41.

Li, H., Xu, J., 2013. Optimization of microwave-assisted calcium chloride pretreatment of corn stover. Bioresour. Technol. 127, 112–118.

Li, J., Zu, Y.-G., Fu, Y.-J., Yang, Y.-C., Li, S.-M., Li, Z.-N., et al., 2010. Optimization of microwave-assisted extraction of triterpene saponins from defatted residue of yellow horn (*Xanthoceras sorbifolia* Bunge.) kernel and evaluation of its antioxidant activity. Innov. Food Sci. Emerg. Technol. 11 (4), 637–643.

Liu, Q., He, H., Li, H., Jia, J., Huang, G., Xing, B., et al., 2019. Characteristics and kinetics of coal char steam gasification under microwave heating. Fuel 256, 115899.

Milano, J., Ong, H.C., Masjuki, H.H., Silitonga, A.S., Chen, W.-H., Kusumo, F., et al., 2018. Optimization of biodiesel production by microwave irradiation-assisted transesterification for waste cooking oil-*Calophyllum inophyllum* oil via response surface methodology. Energy Convers. Manag. 158, 400–415.

Motasemi, F., Afzal, M.T., 2013. A review on the microwave-assisted pyrolysis technique. Renew. Sustain. Energy Rev. 28, 317–330.

Motasemi, F., Salema, A.A., Afzal, M.T., 2015. Dielectric characterization of corn stover for microwave processing technology. Fuel Process. Technol. 131, 370–375.

Namazi, A.B., Allen, D.G., Jia, C.Q., 2015. Probing microwave heating of lignocellulosic biomasses. J. Anal. Appl. Pyrolysis 112, 121–128.

Nüchter, M., Ondruschka, B., Bonrath, W., Gum, A., 2004. Microwave assisted synthesis–a critical technology overview. Green. Chem. 6 (3), 128–141.

Pan, X., Niu, G., Liu, H., 2003. Microwave-assisted extraction of tea polyphenols and tea caffeine from green tea leaves. Chem. Eng. Process.: Process. Intensif. 42 (2), 129–133.

Passos, F., Solé, M., García, J., Ferrer, I., 2013. Biogas production from microalgae grown in wastewater: effect of microwave pretreatment. Appl. Energy 108, 168–175.

Peng, H., Chen, H., Qu, Y., Li, H., Xu, J., 2014. Bioconversion of different sizes of microcrystalline cellulose pretreated by microwave irradiation with/without NaOH. Appl. Energy 117, 142–148.

Périno-Issartier, S., Abert-Vian, M., Chemat, F., 2011. Solvent free microwave-assisted extraction of antioxidants from sea buckthorn (*Hippophae rhamnoides*) food by-products. Food Bioprocess. Technol. 4 (6), 1020–1028.

Qu, T., Guo, W., Shen, L., Xiao, J., Zhao, K., 2011. Experimental study of biomass pyrolysis based on three major components: hemicellulose, cellulose, and lignin. Ind. Eng. Chem. Res. 50 (18), 10424–10433.

Rajvanshi, A.K., Goswami, D.Y., 1986. Biomass gasification. In: Goswami, D.Y. (Ed.), Alternative Energy in Agriculture. Vol. II, CRC Press, Boca Raton, pp. 83–102.

Remya, N., Lin, J.-G., 2011. Current status of microwave application in wastewater treatment—a review. Chem. Eng. J. 166 (3), 797–813.

Robinson, J., Dodds, C., Stavrinides, A., Kingman, S., Katrib, J., Wu, Z., et al., 2015. Microwave pyrolysis of biomass: control of process parameters for high pyrolysis oil yields and enhanced oil quality. Energy Fuels 29 (3), 1701–1709.

Routray, W., Orsat, V., 2012. Microwave-assisted extraction of flavonoids: a review. Food Bioprocess. Technol. 5 (2), 409–424.

Salema, A.A., Afzal, M.T., 2015. Numerical simulation of heating behaviour in biomass bed and pellets under multimode microwave system. Int. J. Therm. Sci. 91, 12–24.

Salema, A.A., Yeow, Y.K., Ishaque, K., Ani, F.N., Afzal, M.T., Hassan, A., 2013. Dielectric properties and microwave heating of oil palm biomass and biochar. Ind. Crop Products 50, 366–374.

Strezov, V., Moghtaderi, B., Lucas, J.A., 2004. Computational calorimetric investigation of the reactions during thermal conversion of wood biomass. Biomass Bioenergy 27 (5), 459–465.

Thostenson, E., Chou, T.-W., 1999. Microwave processing: fundamentals and applications. Compos. A Appl. Sci. Manuf. 30 (9), 1055–1071.

Udalov, E., Bolotov, V., Tanashev, Y.Y., Chernousov, Y.D., Parmon, V., 2011. Pyrolysis of liquid hexadecane with selective microwave heating of the catalyst. Theor. Exp. Chem. 46 (6), 384–392.

Veggi, P.C., Martinez, J., Meireles, M.A.A., 2012. Fundamentals of microwave extraction. Microwave-assisted Extraction for Bioactive Compounds. Springer, pp. 15–52.

Wu, C., Budarin, V.L., Gronnow, M.J., De Bruyn, M., Onwudili, J.A., Clark, J.H., et al., 2014. Conventional and microwave-assisted pyrolysis of biomass under different heating rates. J. Anal. Appl. Pyrol. 107, 276–283.

Xiao, N., Luo, H., Wei, W., Tang, Z., Hu, B., Kong, L., et al., 2015. Microwave-assisted gasification of rice straw pyrolytic biochar promoted by alkali and alkaline earth metals. J. Anal. Appl. Pyrol. 112, 173–179.

Xie, Q., Borges, F.C., Cheng, Y., Wan, Y., Li, Y., Lin, X., et al., 2014. Fast microwave-assisted catalytic gasification of biomass for syngas production and tar removal. Bioresour. Technol. 156, 291–296.

Zhu, Z., Rezende, C.A., Simister, R., McQueen-Mason, S.J., Macquarrie, D.J., Polikarpov, I., et al., 2016. Efficient sugar production from sugarcane bagasse by microwave assisted acid and alkali pretreatment. Biomass Bioenergy 93, 269–278.

Cellulosic-based enzymes for enhanced saccharification for biofuel production

6

Avirup Panja[1], *Vijay Kumar*[2] *and Priyanka Jha*[3]
[1]Amity Institute of Biotechnology, Amity University, Kolkata, West Bengal, India, [2]Plant Biotechnology Lab, Division of Research and Development, Lovely Professional University, Phagwara, Punjab, India, [3]Department of Biotechnology, Lovely Faculty of Technology and Sciences, Lovely Professional University, Phagwara, Punjab, India

6.1 Introduction

As fossil fuels are depleting, the whole world is now more focused on a more sustainable and renewable source of fuels. Lignocellulosic biomass is currently an attractive renewable resource for liquid fuel. Furthermore, the majority of the biofuels produced from lignocellulosic biomass are advantageous in having reduced greenhouse gas (GHG) emissions ($\sim$80%) as compared to various other feedstocks (corn stover, straw, food waste, waste wood), which leads to term lignocellulose-based biofuels as advanced biofuels (Hahn-Hägerdal et al., 2006).

The term "lignocellulose" includes primary constituents in plants such as cellulose, hemicellulose, and lignin. Cellulose is the primary component and the most abundant natural carbohydrate source. It is abundant in the plant cell wall and classified as a glucan polysaccharide with potential biofuel production from biomass. Cellulose from biomass is one of the significant products of photosynthesis and is the most abundant renewable source of biofuel. Cellulose is cross-linked with lignin and hemicellulose in the plant cell wall. The cellulosic microfibrils are coated with hemicellulose matrices, and the lignin is found to be projected outside. These structures are tightly packed, forming a rigid frame, and hence it becomes difficult for enzymes to find access to cellulose, thereby helping in the bioconversion process (Mansfield et al., 1999). Cellulose has β-D-anhydro-glucopyranose units linked by β-(1,4)-glycosidic bonds, forming a linear chain polymer. Being tightly linked with lignin and hemicellulose polymers, the hydrolysis of cellulose becomes very difficult. Enzymes, mainly cellulases, help hydrolyze cellulose for further uses. Since having a rigid structure due to tight cross-linking with lignin and hemicellulose, the accessibility for enzymes becomes very difficult. Due to this, many pretreatment processes are being investigated by which enzymes like cellulases may have greater accessibility to cellulose. The various pretreatment techniques are lime, aqueous ammonia, dilute acid, etc. These aspects have been discussed in the earlier chapters and Section 6.3 of this chapter.

Advances in Lignocellulosic Biofuel Production Systems. DOI: https://doi.org/10.1016/B978-0-323-91192-4.00005-5

If the enzymes cannot access cellulose properly, the saccharification process becomes even more complicated, thereby leading to the formation of undesired and toxic products or the efficiency of the product formed by cellulose hydrolysis hindered. Most currently used saccharification processes (i.e., acid and alkali treatment) remain unpreferable due to incomplete degradation of polysaccharides, affecting saccharification (Fang et al., 1999). A single pretreatment might not be sufficient for the complete removal of lignin. Also, the products released upon pretreatment in which hemicellulose might be degraded might inhibit further downstream processes. Different kinds of cellulases exist, each acting on different substrates (Oppert et al., 2010, Fischer et al., 2013). Studies have also shown that cellulose hydrolysis increases with the increase of lignin removal. Among the different forms and types of cellulases, experiments suggested that the allomorphic form of cellulase helps improve the saccharification of cellulose successfully (Cui et al., 2014).

6.2 Cellulase and hydrolysis mechanism

Cellulases are a mixture of glycoside hydrolase, which consists of cellobiohydrolases, endoglucanases, and β-glucosidases (Chatterjee et al., 2015) (Table 6.1). Several fungi and bacterial species, such as *Trichoderma reesei*, *Trichoderma viride*, *Cellulomonas*, and *Clostridium thermocellum*, have been reported to produce extracellular cellulases (Mandels and Weber, 1969). For a long time, it was anticipated that cellulose degradation would occur due to the action of hydrolytic enzymes. However, the anticipation ended with a breakthrough in research when the study suggested the discovery of oxidative polysaccharide degradation via enzymes (Fig. 6.1) (Vaaje-Kolstad et al., 2010).

Cellulose is a condensed linear polymer of D-anhydro glucopyranose joined by glycosidic bonds with adjacent polymeric chains and sheets linked by hydrogen bonds and van der Waal's forces, forming a parallel alignment and crystalline structure with low accessibility (Demain et al., 2005). Cellulose is a stable molecule

Table 6.1 Active enzymes of fungal origin present in cellulase complex.

Enzyme	EC	Mode of action
Cellobiohydrolase	3.2.1.176, 3.2.1.91	Breaking up of cellobiose from reducing and nonreducing ends of cellulosic chain
Endoglucanase	3.2.1.4	Cleaving β-(1→4) linkages from noncrystalline regions of cellulose
β-glucosidase	3.2.1.21	Cleaving D-glucose from nonreducing ends of oligosaccharides
Xyloglucanase	3.2.1.151	Breaking of xyloglucan
β-xylosidase	3.2.1.37	Cleaves D-xylose from nonreducing end of xylo-oligosaccharides
Endo-β-xylanase	3.2.1.8	Cleaves the main xylose chain and aids in release of xylo-oligosaccharides
β-mannosidase	3.2.1.25	Cleaves D-mannose from nonreducing end of glucomanno-oligosaccharides

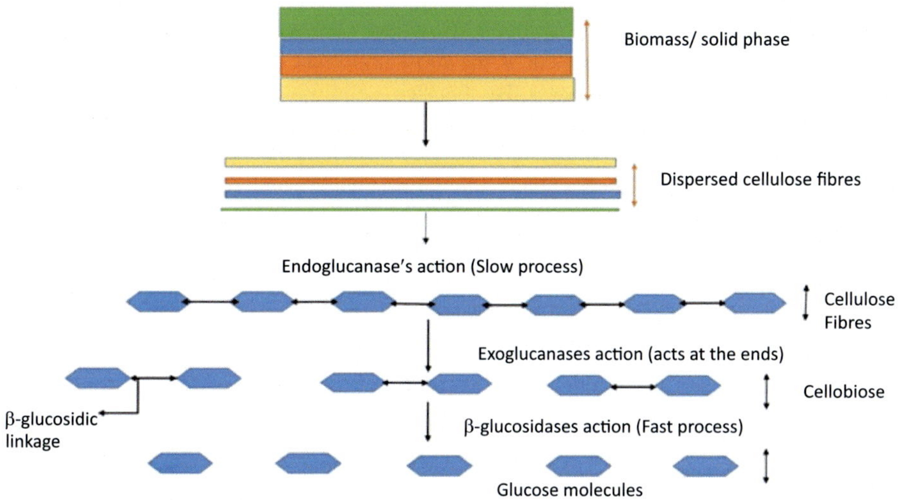

Figure 6.1 Mechanism of cellulose hydrolysis.

with a half-life of approximately 5–8 million years (Wolfenden and Snider, 2001). Cellulose hydrolysis involves the synergistic action of enzymes endoglucanase, exoglucanase, cellobiohydrolase, and β-glucosidase (Henrissat, 1994; Lynd et al., 2002). Endoglucanases hydrolyze the intramolecular β-1,4-glucosidic bonds giving rise to shorter chains, whereas the exoglucanase cleaves the polymeric ends to liberate glucose or soluble cellobiose. *β*-glucosidases, on the other end, hydrolyze the cellobiose to glucose, thereby eliminating the inhibition caused by cellobiose molecules. The primary hydrolysis occurs on the surface of the solid substrates, while the secondary hydrolysis is initiated in the liquid phase. The depolymerization by endoglucanase and exoglucanase is the rate-limiting step in hydrolysis. The surface characteristics of cellulose are seen to be modified over time by endoglucanase and exoglucanase action, thus altering the rate of hydrolysis.

6.3 Pretreatment techniques

Ethanol production from lignocellulosic biomass includes a collection of feedstocks, pretreatment at the time of collection, enzymatic saccharification, fermentation, and product recovery. The viability of biofuel production will depend on feedstock availability, the pretreatment process involved, enzymes used, along with enzymatic efficiency (Yang and Wyman, 2008). Pretreatment techniques widely used to carry out saccharification include wet oxidation, steam explosion, dilute acidic treatment, hydrothermal treatment, and ammonia fiber expansion (AFEX). The feedstock type and desired product decide the type of pretreatment for the process. The study suggests grass-type feedstock requires hydrothermal pretreatment and AFEX, whereas steam explosion and sulfite pulping are utilized for the pretreatment of woody biomass (Bals et al., 2010). While some pretreatment techniques act on biomass fractionation by separating lignin and hemicellulose, others focus on enhancing accessibility in the plant cell wall by reorganizing the cell wall products. The pretreatment is the most important phase in biofuel production, affecting the saccharification process. By this step, cellulose is unlocked and releases lignin and the other components so that the enzymes may have greater access to cellulose. Many techniques are used for this process, and they may be classified into four main types: physical, chemical, physicochemical, and biological pretreatment. Choosing appropriate techniques is vital as energy consumption is always a concern.

The different types of pretreatment strategies are as follows.

6.3.1 Physical/mechanical methods

1. *Milling:* This technique is mainly followed to reduce the particle size, increasing the surface/volume ratio. This technique is not feasible as the energy consumption required for this, particularly from an industrial point of view, is very much.
2. *Ultrasound-based pretreatment:* This technique widely saves energy and produces very small biomass, thereby increasing enzymatic hydrolysis efficiency. Upon proper combination with solvents (i.e., acids), the lignocellulose structure is broken more effectively, thereby providing better fermentation results.

6.3.2 Physiochemical method

AFEX: In this type of pretreatment, the biomass is treated under elevated temperature and pressure in the presence of ammonia, which rapidly depolymerizes the other components attached to cellulose (Krishnan et al., 2010). The significant advantage of this technique is that complete delignification and cellulose hydrolysis occur.

6.3.3 Chemical method

Acid pretreatment: Several acids such as HCl, H_2SO_4, and HNO_3 are used in this process which is conducted under high temperature that leads to hydrolysis of cellulose trapped (Chen et al., 2009). Out of the various acids used, sulfuric acid, phosphoric acid, nitric acid, and hydrochloric acid are the most effective and economical treatments for agricultural residues (Saha et al., 2005).

6.3.4 Biological method

Microbes have also been used in pretreatment and are advantageous over other pretreatment techniques. Reports have been published where fungi are used in the pretreatment techniques in which lignin is degraded selectively. This technique is environmentally efficient and helps boost enzymatic saccharification and fermentation of biomass for ethanol production (Cardona and Sánchez, 2007). Saprophytic fungi such as actinomycetes and basidiomycetes have also been reported to secrete certain ligninolytic enzymes (Ray et al., 2010).

6.3.5 Enzyme testing

Cellulases and their isoforms are enzymes that work on cellulose and cellulose-specific substrates. Among other approaches, two widely employed techniques in enzyme testing are as follows:

1. Quantifying individual enzymatic activities of the various forms of cellulases.
2. Quantification of total enzymatic activity.

The general activity of hydrolase enzymes is observed to occur in two phases: the rate of hydrolysis and the end point of hydrolysis for the complete enzyme mixture (Sheehan and Himmel, 1999; Percival Zhang et al., 2006). The substrates employed for cellulase assays based on water solubility can be further classified into major groups (Dowe, 2009):

1. Soluble substrates with a low degree of polymerization, that is, carboxymethyl cellulose (CMC) (Sharrock, 1988).
2. Insoluble substrates.

6.3.5.1 Testing for endoglucanases

This cellulase type functions randomly, cleaving intramolecular β-1,4-glucosidic linkage. Endoglucanases and exoglucanases differ in their mode of action; the former specifically lowers the CMC viscosity and reduces hydrolysis significantly, while the exoglucanase hydrolyzes the polymeric chains from the ends in a progressive manner (Irwin et al., 1993). Additionally, endoglucanase can be determined by a semiquantitative method involving agar plate staining of the polysaccharide with Congo Red (Kim et al., 2000).

6.3.5.2 Testing for exoglucanases activity

These enzymes separate the accessible end of the cellulose molecule leading to the production of glucose and cellobiose. A compound named Avicel (microcrystalline cellulose) is used in the Testing of exoglucanase activity due to the very high ratio of Filter Paper Unit/mL (FPU/mL) amongst cellulosic substrates (insoluble) (Wood and Bhat, 1988).

There also exist a few selective assay techniques for exoglucanase activity. Deshpande et al. (1984), reported exoglucanase assay in presence of endoglucanases and β-glucosidases. The study stated that exoglucanase hydrolyzes the aglyconic bond of p-nitrophenyl-β-D-cellobioside, producing cellobiose and p-nitrophenol.

β-D-glucosidase activity is quantified using cellobiose, which remains unhydrolyzed by endoglucanases and exoglucanases, whereas in other cases is hydrolyzed by endoglucanases and exoglucanases in the presence of cellodextrin (Gong et al., 1977).

6.3.5.3 Total cellulase assay

Endoglucanase, exoglucanase, and β-glucosidase include total cellulase content for hydrolyzing cellulose. The total enzymatic activity assay involves insoluble substrates, for example, cotton linen, Whatman filter paper, bacterial cellulose, etc. An enzyme dilution solution exists, which helps to bring about the final hydrolysis.

The major advantages of this assay are that widely available substrates can be used, and the process is also very simple.

To detect the sugar inhibition for total cellulase, dyed cellulose can be used as a substrate (Wood and Bhat, 1988). There is also fluorescent-dyed cellulose being utilized for the same purpose, where the higher the emission of these dyed molecules signal, the low the total cellulase activity (Helbert et al., 2003). For commercial purposes, such as cellulase digestibility check, α-cellulose, and pretreated lignocellulose are often used as general substrates.

6.4 Improvement in cellulase strategies

Since most naturally occurring enzymes have low activity and poor stability, their actions and applications in the industrial sector become expensive and restricted. Engineering enzymes have proven to be an effective solution to overcome such issues in this demanding industrial development era (Behera and Ray, 2016). Two effective strategies for engineering enzymes are directed evolution and rational development to obtain a better hydrolysis rate (Zhang and Zhang, 2013; Druzhinina and Kubicek, 2017; Dadwal et al., 2020).

6.5 Directed evolution

Directed evolution can be an amalgamation of two vital technologies: the generation of random genetic diversity and high throughput screening (HTS) for successfully selecting required/desirable enzyme variants (Xiao et al., 2015). One major issue regarding directed evolution is the identification of novel variants from the genomic libraries. The technique of HTS becomes very effective and crucial in this scenario as it enables the inventors to successfully screen these required novel variants in the vast genomic pool generated by random mutagenesis. Two major strategies of HTS for directed evolution are the direct addition of indicators and an indirect indication. The direct addition of indicators is the simplest and easiest way of screening. Usually, this strategy is viable for reactions involved with the release of hydrogen ions or even for reactions in which a cofactor is involved. In this technique, the substrates must have certain characteristic groups which help their detection successfully. For example, pH indicators like bromophenol blue, methyl red, and bromocresol purple are successfully used as an HTS screening technique. On the other hand, when the substrates or a product lacks specific groups which can be used for detection, the indirect indication is the technique that is followed. This technique is quite flexible and can be used to identify a wide range of enzymes.

Over the last few years, another major type of HTS that has been preferred and used instead of the other techniques is virtual-screening-assisted HTS. This computational technique saves time and labor, providing an effective and successful alternative to experimental screening techniques. Furthermore, using these techniques, scientists have successfully altered many such enzymes' catalytic and thermostability properties over the years (Lassila et al., 2005; Korkegian et al., 2005).

6.6 Rational design/rational protein design approach

Like the directed evolution technique, rational design is yet another technique for improving cellulase activity with the help of site-directed mutagenesis (SDM). This technique can be coupled with chemical modifications and improve many enzymatic properties that natural enzymes do not exhibit. Removal of specific amino acid residues in many enzymes has been seen to elevate the thermostability properties of many enzymes, that is, aspargine in α-amylase (Declerck et al., 2000). In addition, removing cystine and methionine amino acid residues in many enzymes enhances oxidation stability. During the (rational design/rational protein design) technique generation of the novel, variants are done based on their existing protein structures, following the SDM technique (Cedrone et al., 2010). While directed evolution uses random mutagenesis, the rational design relies on protein structure, and their SDM is intentional and directed.

One major branch which largely influences new enzymatic properties within the enzyme is chemistry. For example, introducing a catalytic residue through substitution in enzyme papin elevates the nitrile hydratase activity by 104-fold compared to the normal naturally existing one (Dufour et al., 1995). Another way to engineer enzyme property is by assembling essential components on a selected macromolecule template (Fig. 6.2).

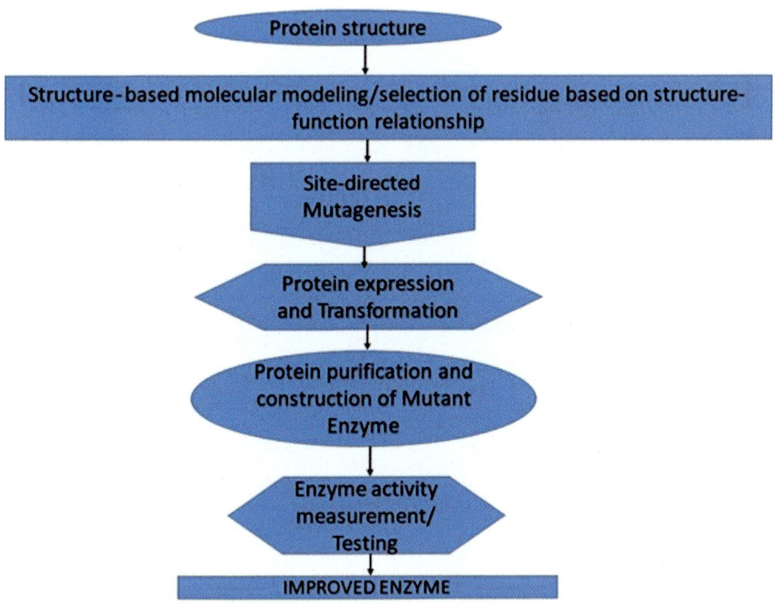

Figure 6.2 Schematic of rational design.

6.7 Synthetic biology in microbial cellulase production

Genetically modified organisms are always considered when it comes to enhanced enzyme activities. Recent research has seen manipulations in several fungal systems, *Neurospora crassa, T. reesei, Fusarium graminearum*, and *Aspergillus* spp. One major transcription activator, xylanase regulator 1 (XYR1), which activates the cellulase genes, is often manipulated, enhancing cellulase and xylanase expression in these systems. This overexpression involves inducers, which play a vital role in the process. One such significant work done by Lv et al. (2015) showed that the gene expression of XYR1 strongly increased when driven by the promoter *tuc1* (encodes for putative copper transporter). Apart from this gene, another gene manipulated to increase cellulase production in *T. reesei* is the *Vel1* (velvet) gene. Conidium development/asexual sporulation is a crucial factor that influences cellulase production in *T. reesei* (Metz et al., 2011). The gene *Vel1* is seen to control both sexual and asexual development in *T. reesei* fungal strains. Karimi Aghcheh et al. (2014) showed that constitutively overexpressing *Vel1* under the promoter *Tef1* leads to improved cellulase production in *T. reesei*. Thus, genetic manipulations in fungal systems can be regarded as a solution to increase cellulase production, thereby reducing the operational and implementation costs in many industries (Druzhinina and Kubicek, 2017) (Fig. 6.3). Another way to enhance cellulase activity is by fusion of nanotechnology and enzyme technology. Enzymes produced by this fusion possess a higher binding efficiency, more stability, and much higher activity than normal free enzymes; it is very helpful from the industrial point of view.

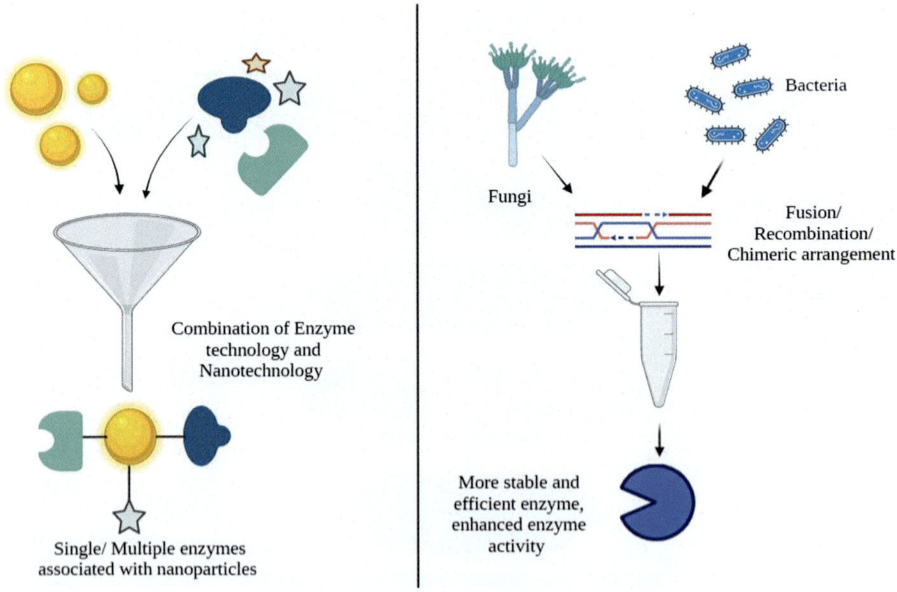

Figure 6.3 Various approaches of synthetic biology for cellulase production.

6.8 Lignocellulosic bioethanol and its economics

The various forms of biofuel existing currently are (1) bioethanol (mostly from sugarcane, wheat, cassava, sorghum, and maize), (2) biodiesel (mostly from soy, oil palm, jatropha), (3) biogas (mostly from industrial and municipal waste), (4) biomethanol (from cellulosic biomass), and (5) biohydrogen (either from biomass or by converting methane through steam reforming). Biofuel generation and utilization rely on various parameters associated with local conditions. Such parameters are (1) pricing and availability of the biomass substrate, which would vary from one country to the other based on available land, labor costs, and agricultural yield; (2) production cost related to biofuel, which would depend on the techniques and equipment used in the production plant along with its size; (3) the price of the biofuel generated in comparison to that of fossil fuels, governed by fluctuating rates of the petroleum production, fossil fuel demand and exchange rates; and (4) the tactical advantage of replacing imported petroleum products with domestic biofuel (Lundgren et al., 2008).

Biofuels, that is, ethanol produced from sugarcane in Brazil and New Zealand with its liquid biofuels research, provide perfect examples of the concerns of the period and the response within the developed and developing world. Many other countries have been extensively involved in the biofuel industry; for example, China is the third largest ethanol producer and produces an average of 10 million liters of ethanol as per data in 2017. The price of the feedstock is a critical factor in the production of biofuel, so the choice of the feedstock is a very critical and significant factor on which the economics of biofuels depends (Lundgren et al., 2008). The Sugarcane Bioethanol (SBE) industry has many advantages, including increased crop production and using bagasse and Bioethanol as fuel. The SBE generation system includes various components such as sugarcane production, product dispensation, sugar-to-ethanol conversion and production of other byproducts, methodological research, distribution, economic support services, and commercialization. Biofuels produced from molasses-based feedstock may cost approximately US$0.97/L, which may be reduced to US$0.40/L when a leaf or lignin-based materials are included. Among the many countries mentioned above, Brazil has the lowest price of bioethanol because of the readily available resources and cheap labor price. Thus the cost of the feedstock plays a vital role in biofuel production (Macrelli et al., 2012).

6.9 Patents and commercial applications

Advancements in the pretreatment processes are one of the major ways to cut down extraction costs and allow faster production of products like biofuels. Therefore patents related to hydrolysis techniques of cellulose using a combination of enzymes have gained much importance over the past few years. For example, the hydrolysis of untreated or pretreated cellulosic biomass is a crucial step in the production of biofuels. A group of inventors has patented a continuous unique process for the enzymatic hydrolysis of such biomass, which upon implementation, can lead to the hydrolysis of larger biomass and thereby lower the investment and operation costs, respectively (Sjoede et al., 2013). Another patent filed by Novozymes in

2016 invented a unique enzyme (cellulase) mixture produced by genetically modified microbes which immensely improved the treatment processes of cellulose-containing goods (Hill et al., 2016).

6.10 Conclusion and future perspectives

Several mutated strains of bacteria and fungi are being developed to improve the cellulases produced by them. The development of enhanced enzyme not only affect the pretreatment techniques but also alter the saccharification process, whereby the production of fuels can get simpler, producing better-quality biofuels. Since lignin is difficult to remove, advanced pretreatment strategies are employed, which again require the involvement of more advanced enzymes. Strains such as *Aspergillus fumigatus* produce better degrading abilities when mutated. SDM, UV radiations in combination with chemicals have been seen to mutate the bacteria and fungi strains, giving rise to better degrading capabilities and thus showing the potential to alter saccharification processes to a greater extent. Presently the rational design approach to improve cellulase strains is widely used. Thus advances in the enzyme have shown a direct impact on the saccharification process, which again alters the production cost of biofuels.

Acknowledgments

The authors would like to thank Lovely Professional University, Punjab, for the infrastructural facilities provided. Vijay Kumar would like to acknowledge the Department of Science and Technology (DST), SERB, Govt. of India, under Startup Research Scheme (File No. SRG/2019/001279).

Abbreviations

AFEX ammonia fiber expansion
CMC carboxymethyl cellulose
FPase filter paperase
FPU/mL Filter Paper Unit/mL
GHG greenhouse gas
SSF solid-state fermentation
SDM site-directed mutagenesis
HTS high throughput screening

References

Bals, B., Rogers, C., Jin, M., Balan, V., Dale, B., 2010. Evaluation of ammonia fibre expansion (AFEX) pretreatment for enzymatic hydrolysis of switchgrass harvested in different seasons and locations. Biotechnol. Biofuels 3, 1.

Behera, S.S., Ray, R.C., 2016. Solid state fermentation for production of microbial cellulases: recent advances and improvement strategies. Int. J. Biol. Macromol. 86, 656–669.

Cardona, C.A., Sánchez, O.J., 2007. Fuel ethanol production: process design trends and integration opportunities. Bioresour. Technol. 98, 2415–2457.

Cedrone, F., Ménez, A., Quéméneur, E., 2010. Tailoring new enzyme functions by rational redesign. Curr. Opin. Str. Biol. 10, 405–410.

Chatterjee, S., Sharma, S., Prasad, R.K., Datta, S., Dubey, D., Meghvansi, M.K., et al., 2015. Cellulase enzyme-based biodegradation of cellulosic materials: an overview. South Asian J. Exp. Biol. 5, 271–282.

Chen, M., Zhao, J., Xia, L., 2009. Comparison of four different chemical pretreatments of corn stover for enhancing enzymatic digestibility. Biomass Bioenerg. 33, 1381–1385.

Cui, T., Li, J., Yan, Z., et al., 2014. The correlation between the enzymatic saccharification and the multidimensional structure of cellulose changed by different pretreatments. Biotechnol. Biofuels 7, 134.

Dadwal, A., Sharma, S., Satyanarayana, T., 2020. Progress in ameliorating beneficial characteristics of microbial cellulases by genetic engineering approaches for cellulose saccharification. Front. Microbiol. 11, 1387. Available from: https://doi.org/10.3389/fmicb.2020.01387.

Declerck, N., Machius, M., Wiegand, G., Huber, R., Gaillardin, C., 2000. Probing structural determinants specifying high thermostability in *Bacillus licheniformis* α-amylase. J. Mol. Biol. 301, 1041–1057.

Demain, A.L., Newcomb, M., Wu, J.H.D., 2005. Cellulase, clostridia, and ethanol. Microbiol. Mol. Biol. Rev. 69, 124–154.

Deshpande, M.V., Eriksson, K.E., Pettersson, L.G., 1984. An assay for selective determination of exo-1,4,-beta-glucanases in a mixture of cellulolytic enzymes. Anal. Biochem. 138, 481–487.

Dowe, N., 2009. Assessing cellulase performance on pretreated lignocellulosic biomass using saccharification and fermentation-based protocols. Meth. Mol. Biol. 581, 233–245.

Druzhinina, I.S., Kubicek, C.P., 2017. Genetic engineering of *Trichoderma reesei* cellulases and their production. Microb. Biotechnol. 10, 1485–1499.

Dufour, E., Storer, A.C., Menard, R., 1995. Engineering nitrile hydratase activity into a cysteine protease by a single mutation. Biochem 34, 16382–16388.

Fang, J.M., Sun, R.C., Salisbury, D., Fowler, P., Tomkinson, J., 1999. Comparative study of hemicelluloses from wheat straw by alkali and hydrogen peroxide extractions. Polym. Degrad. Stab. 66, 423–432.

Fischer, R., Ostafe, R., Twyman, R.M., 2013. Cellulases from insects. Adv. Biochem. Eng. Biotechnol. 136, 51–64.

Gong, C.-S., Ladisch, M.R., Tsao, G., 1977. Cellobiase from *Trichoderma viride*: purification, properties, kinetics, and mechanism. Biotechnol. Bioeng. 19, 959–981.

Hahn-Hägerdal, B., Galbe, M., Gorwa-Grauslund, M.F., Lidén, G., Zacchi, G., 2006. Bio-ethanol—the fuel of tomorrow from the residues of today. Trends Biotechnol. 24, 549–556.

Helbert, W., Chanzy, H., Husum, T.L., Schulein, M., Ernst, S., 2003. Fluorescent cellulose microfibrils as substrate for the detection of cellulase activity. Biomacromolecules 4, 481–487.

Henrissat, B., 1994. Cellulases and their interaction with cellulose. Cellulose 1, 169–196.

Hill, C., Tomashek, J.J., Savich, L.G., et al., 2016. Cellulase enzyme mixtures for depilling and uses thereof (US Patent No. 9.238,806 B2). US Patent and Trademark Office.

Irwin, D., Spezio, M., Walker, L.P., Wilson, D.B., 1993. Activity studies of eight purified cellulases: specificity, synergism, and binding domain effects. Biotechnol. Bioeng. 42, 1002–1013.

Karimi Aghcheh, R., Németh, Z., Atanasova, L., Fekete, E., Paholcsek, M., Sándor, E., et al., 2014. The VELVET A orthologue VEL1 of *Trichoderma reesei* regulates fungal development and is essential for cellulase gene expression. PLoS One 9, e112799.

Kim, Y.S., Jung, H.C., Pan, J.G., 2000. Bacterial cell surface display of an enzyme library for selective screening of improved cellulase variants. Appl. Env. Microbiol. 66, 788–793.

Korkegian, A., Black, M.E., Baker, D., Stoddard, B.L., 2005. Computational thermostabilization of an enzyme. Science 308, 857–860.

Krishnan, C., da Costa Sousa, L., Jin, M., Chang, L., Dale, B.E., Balan, V., 2010. Alkali-based AFEX pretreatment for the conversion of sugarcane bagasse and cane leaf residues to ethanol. Biotechnol. Bioeng. 107, 441–450.

Lassila, J.K., Keeffe, J.R., Oelschlaeger, P., Mayo, S.L., 2005. Computationally designed variants of *Escherichia coli* chorismate mutase show altered catalytic activity. Prot. Eng. Des. Select. 18, 161–163.

Lundgren, T., Marklund, P., Brännlund, R., Kriström, B., 2008. The economics of biofuels. Int Rev. Env. Resour. Econ. 2, 237–280.

Lv, X., Zheng, F., Li, C., Zhang, W., Chen, G., Liu, W., 2015. Characterization of a copper responsive promoter and its mediated overexpression of the xylanase regulator1 results in an induction-independent production of cellulases in *Trichoderma reesei*. Biotech. Biofuels 8, 67.

Lynd, L.R., Weimer, P.J., van Zyl, W.H., Pretorius, I.S., 2002. Microbial cellulose utilization: fundamentals and biotechnology. Microbiol. Mol. Biol. Rev. 66, 506–577.

Macrelli, S., Mogensen, J., Zacchi, G., 2012. Techno-economic evaluation of 2nd generation bioethanol production from sugar cane bagasse and leaves integrated with the sugar-based ethanol process. Biotechnol. Biofuels 5, 22.

Mandels, M., Weber, J., 1969. The production of cellulases. Adv. Chem. Ser. 95, 391–413.

Mansfield, S.D., Mooney, C., Saddler, J.N., 1999. Substrate and enzyme characteristics that limit cellulose hydrolysis. Biotechnol. Prog. 15, 804–816.

Metz, B., Seidl-Seiboth, V., Haarmann, T., Kopchinskiy, A., Lorenz, P., Seiboth, B., et al., 2011. Expression of biomass-degrading enzymes is a major event during conidium development in *Trichoderma reesei*. Eukaryot. Cell 10, 1527–1535.

Oppert, C., Klingeman, W.E., Willis, J.D., Oppert, B., Jurat-Fuentes, J.L., 2010. Prospecting for cellulolytic activity in insect digestive fluids. Comp. Biochem. Physiol. B Biochem Mol. Biol. 155, 145–154.

Percival Zhang, Y.H., Himmel, M.E., Mielenz, J.R., 2006. Outlook for cellulase improvement: screening and selection strategies. Biotechnol. Adv. 24, 452–481.

Ray, M.J., Leak, D.J., Spanu, P.D., Murphy, R.J., 2010. Brown rot fungal early-stage decay mechanism as a biological pretreatment for softwood biomass in biofuel production. Biomass Bioenerg. 34, 1257–1262.

Saha, B.C., Iten, L.B., Cotta, M.A., Wu, Y.V., 2005. Dilute acid pretreatment, enzymatic saccharification and fermentation of wheat straw to ethanol. Process. Biochem. 40, 3693–3700.

Sharrock, K.R., 1988. Cellulase assay methods: a review. J. Biochem. Biophy Meth. 17, 81–105.

Sheehan, J., Himmel, M., 1999. Enzymes, energy, and the environment: a strategic perspective on the U.S. Department of Energy's Research and Development Activities for Bioethanol. Biotechnol. Prog. 15, 817–827.

Sjoede, A., Froelander, A., Lersch, M., et al., 2013. Enzymatic hydrolysis of cellulose (US Patent No. 2013/0217074 A1). US Patent and Trademark Office.

Vaaje-Kolstad, G., Westereng, B., Horn, S.J., Liu, Z., Zhai, H., Sørlie, M., et al., 2010. An oxidative enzyme boosting the enzymatic conversion of recalcitrant polysaccharides. Science 330, 219–222.

Wolfenden, R., Snider, M.J., 2001. The depth of chemical time and the powder of enzyme as catalysts. Acc. Chem. Res. 34, 938–945.

Wood, T.M., Bhat, K.M., 1988. Methods for measuring cellulase activities. Meth. Enzymol. 160, 87–117.

Xiao, H., Bao, Z., Zhao, H., 2015. High throughput screening and selection methods for directed enzyme evolution. Ind. Eng. Chem. Res. 54, 4011–4020.

Yang, B., Wyman, C.E., 2008. Pretreatment: the key to unlocking low-cost cellulosic ethanol. Biofuels. Bioprod. Bioref. 2, 26–40.

Zhang, X.Z., Zhang, Y.H.P., 2013. Cellulases: characteristics, sources, production, and applications. Bioprocessing Technologies in Biorefinery for Sustainable Production of Fuels. Chemicals, and Polymers (Editor(s)Shang-Tian Yang, Hesham A El-Enshasy, Nuttha Thongchul), John Wiley and Sons, pp. 131–146.

Further reading

Bhatt, S.M., Shilpa, 2014. Lignocellulosic feedstock conversion, inhibitor detoxification and cellulosic hydrolysis – a review. Biofuels 5, 633–649.

Lu, Y., Yang, B., Gregg, D., Saddler, J.N., Mansfield, S.D., 2008. Cellulase adsorption and an evaluation of enzyme recycle during hydrolysis of steam-exploded softwood residues. Appl. Biochem. Biotechnol 2002 (98–100), 641–654.

Role of accessory enzymes and proteins in efficient biomass hydrolysis

7

Md Shahadat Hossain[1], Kalyani Ananthakrishnan[1], Ankita Juneja[2], Mukesh Kumar Awasthi[3] and Deepak Kumar[1]
[1]Department of Chemical Engineering, SUNY College of Environmental Science and Forestry, Syracuse, NY, United States, [2]Department of Biology, Syracuse University, Syracuse, NY, United States, [3]College of Natural Resources and Environment, Northwest A&F University, Yangling, Shaanxi Province, P. R. China

7.1 Introduction

Lignocellulosic biomass has a recalcitrant structure primarily comprising cellulose, hemicellulose, and lignin biopolymers, where each of these polymers has a unique structure and characteristics. Cellulose is the most abundant polymer in the lignocellulosic biomass, making up to 40% of the total biomass (Zoghlami and Paës, 2019). Cellulose is a homopolymer with a degree of polymerization of 10,000, where cellobiose dimers (glucose–glucose dimers) are linked together by β-1,4-glycosidic bonds, resulting in microfibrils of cellulose (Tian et al., 2020). Often these microfibrils are cross-linked by hydrogen bonds and Van der Waals forces leading to both the crystalline and amorphous regions of cellulose (Awoyale and Lokhat, 2021). Hemicellulose is an amorphous polymer with a degree of polymerization up to 200, and this polymer contains both hexose (mannose, galactose, and glucose) and pentose sugars (xylose and arabinose) as well as a small amount of acetyl group (Monteiro et al., 2021). Sugars are also interlinked by glycosidic bonds and form a highly branched and short amorphous biopolymer chain. Lignin is a nonsugar phenyl propane-based biopolymer that creates a protective cover for the cellulose and hemicellulose fibers (Dos Santos et al., 2019). Cellulose microfibrils remain dispersed in the hemicellulose fibers and embedded inside the lignin, forming a rigid and complex structure of the lignocellulosic biomass (Woiciechowski et al., 2020).

The most studied and used route of biomass conversion to biofuel/bioproducts involves pretreatment to alter the structure and reduce the recalcitrance of the biomass, followed by enzymatic hydrolysis to release sugar monomers that are available to be fermented to desired products using microbes. Although various natural enzymes are available, cellulase enzyme has gained much popularity because of their versatile use in bioethanol production from different lignocellulosic biomass. Commercial cellulase contains a cocktail of enzymes such as endoglucanase (EG), exoglucanase, hemicellulose, and β-glucosidase, in addition to cellulase (such as Accellerase 1500). A wide range of naturally available microorganisms (fungi and bacteria) can produce

Advances in Lignocellulosic Biofuel Production Systems. DOI: https://doi.org/10.1016/B978-0-323-91192-4.00001-8

these enzymes that can partially or completely hydrolyze the cellulose, hemicellulose, and some of the lignin fraction of biomass (da Silva et al., 2020) into their monomer sugars. Different filamentous fungi species such as *Trichoderma*, *Penicillium*, and *Aspergillus* are used for cellulase enzyme production, which can hydrolyze the β-1,4-glycosidic bonds of lignocellulosic biomass into constituent monomeric sugars (Siqueira et al., 2020). With these commercially available enzymes, >90% cellulose hydrolysis efficiency can be achieved, but overall efficiency (including hemicellulose) is much lower. To make the lignocellulosic biofuel process cost-effective, the hydrolysis efficiency needs to be drastically improved, implying maximum recovery of monomers from hemicellulose too.

The enzyme systems best studied for the depolymerization of cellulose comprise cellobiohydrolase (CBH) (EC 3.2.1.91), endo-β1,4-glucanase (EG) (EC 3.2.1.4), and β-glucosidase (BG). CBH I is the major fraction (up to 60% of total protein) of crude enzyme mixtures. Cellulose inherently possesses both crystalline and amorphous regions. For effective hydrolysis of cellulose, enzymes must act in the solid–liquid interface with a high catalytic concentration. Additionally, enzymes must have a three-dimensional catalytic domain as well as a carbohydrate-binding module (CBM) and linker. CBM remains close to the cellulose and carries out the hydrolysis, while the linker enhances the enzyme mobility and binds the enzyme to the cellulose surface (Sugathan et al., 2017). In the case of processive enzymes (exoglucanase), the enzyme identifies the free chain end of the cellulose, and those chain ends are put into the catalytic domain for threading, resulting in a catalytically active complex (CAC). CBH I acts from the reducing end and CBH II from the nonreducing end. The hydrolysis is then carried out at several events for product formation and final expulsion from the CAC (Fig. 7.1). This processive enzyme cycle is carried out at the new free chain ends after the product expulsion. In this mechanism

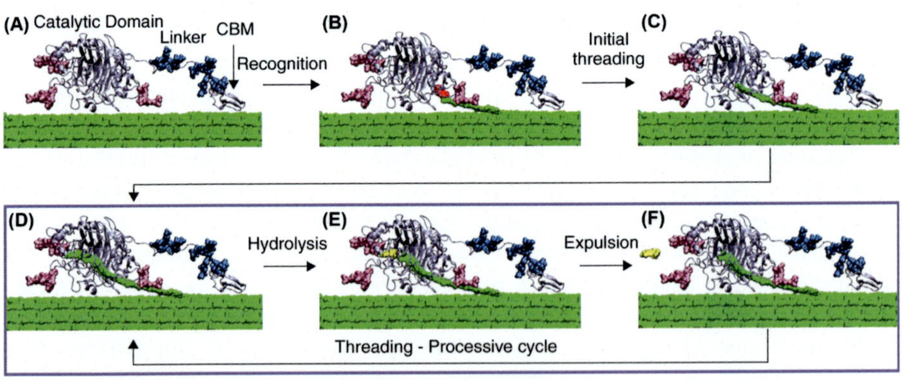

Figure 7.1 Mechanism of cellulose hydrolysis by cellobiohydrolase I: (A) enzyme binding on glucan chain, (B) identification of reducing end, (C) threading of free chin end with the catalytic domain, (D) catalytically active complex formation, (E) product (yellow) formation by hydrolysis, and (F) product expulsion step (Beckham et al., 2011).
Source: Adapted from Beckham, G.T., Bomble, Y.J., Bayer, E.A., Himmel, M.E., and Crowley, M.F. (2011). Applications of computational science for understanding enzymatic deconstruction of cellulose. Curr. Opin. Biotechnol., 22 (2), 231–238.

of cellulosic hydrolysis, enzymes need to be moved to an active site/free chain end for CAC formation, which is considered a rate-limiting step for hydrolysis. That is where enzyme synergism comes into play as the EGs enzyme acts effectively in the amorphous region in a random fashion and produces chain ends for the exocellulase. The cellobiose liberated from the action of CBH is further hydrolyzed by the BGs into glucose.

Our understanding of the enzymology of hemicellulose breakdown is limited, as a breakdown of hemicellulose requires a complex group of enzymes, referred to as hemicellulases. For example, complete xylan hydrolysis would require endo-β-1,4-xylanase, which randomly acts on xylan chains to release xylo-oligosaccharides, β-xylosidase, which releases xylose by hydrolyzing the nonreducing ends of xylose chains, and a variety of accessory enzymes. Those naturally derived enzymes do not provide enough hydrolysis activity and are affected by various other parameters of the reaction environment, such as pH, temperature, inhibitory compounds, etc. (da Silva et al., 2020). Therefore accessory enzyme(s) are recently used to create synergism and overcome the low hydrolysis efficiency for faster lignocellulosic biomass hydrolysis (Ummalyma et al., 2019).

Accessory enzymes can be added as supplementation to other hydrolase enzymes, with an expectation of improving hydrolysis efficiency due to the synergistic effect of all classes of enzymes. Accessory enzymes facilitate biomass hydrolysis in a variety of ways. They can directly act on the crystalline and amorphous structure of the lignocellulosic biomass and disrupt the constituent pentose and hexose sugars for hydrolysis. For example, recently discovered lytic polysaccharide monooxygenases (LPMOs) enzymes can act in the crystalline region of cellulose, which is highly recalcitrant to EG action, and generate new chain ends for CBH enzymes (Fig. 7.2). These auxiliary

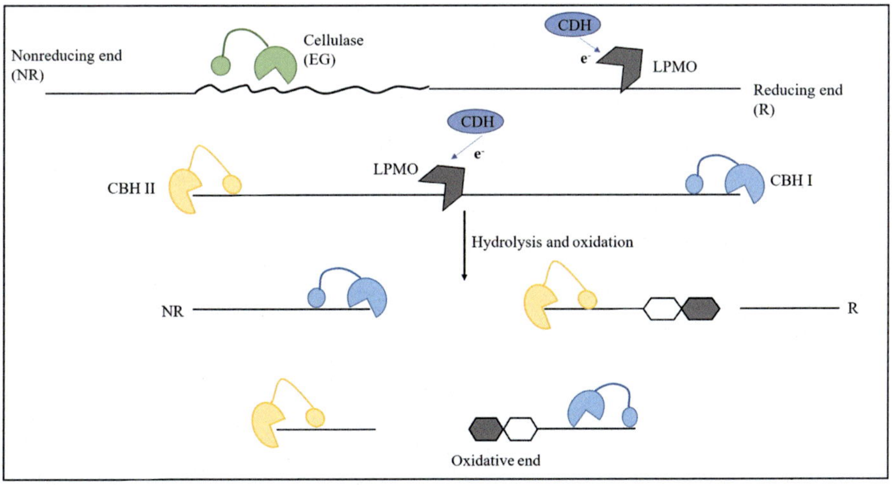

Figure 7.2 Combined cellulosic hydrolysis and oxidative breakdown by cellulases (endoglucanases) and accessory enzyme (lytic polysaccharide monooxygenase). Cellobiose dehydrogenase, cellobiohydrolases I, and cellobiohydrolases II (Sugathan et al., 2017). *Source*: Adapted from Sugathan, S., Pradeep, N.S., Abdulhameed, S., eds., 2017. Bioresources and Bioprocess in Biotechnology: Volume 2: Exploring Potential Biomolecules. Springer Nature, Singapore, 456p.

proteins catalyze the cleavage of glycosidic bonds via an oxidation mechanism and can significantly reduce the cellulase enzyme loading during cellulose hydrolysis. Some of the accessory enzymes can also reduce the inhibition of certain biomolecules so that enzymatic hydrolysis of cellulosic sugars can be accelerated. Another type of accessory enzyme mainly works on the hydrolysis of noncellulolytic fraction of biomass which results in the removal of protective coating around cellulosic sugars (Sun et al., 2015). While an accessory enzyme is coupled with the main cellulase or hemicellulase enzyme, the accessory enzyme works as an essential part of the main enzyme for biomass hydrolysis supplementation. In such a case synergistic effect between the enzymes is considered an important parameter, and an enzyme cocktail is often prepared considering the type of biomass feedstock and employed pretreatment processes (Maitan-Alfenas et al., 2015). This chapter focuses only on accessory enzymes. The details of conventional cellulase enzyme action mechanisms and modeling can be found in our previous publications (Kumar and Murthy, 2016).

7.2 Types of accessory enzyme

Accessory enzymes include enzymes that cleave less abundant chemical linkages in the lignocellulosic structure, such as arabinases, esterases, proteases, pectinases, galactanases, and lyases. Some types of accessory enzymes with their mode of working in synergism with other enzymes are mentioned below.

7.2.1 Auxiliary active enzyme

Auxiliary active enzymes (e.g., LPMOs) are copper-based enzymes identified in several natural and recombinant fungi (Forsberg et al., 2019). These enzymes act on the pretreated lignocellulosic biomass and break down the long crystalline polysaccharide chains of cellulose. Both C1 and C4 carbon of the glucose (from the cellulose) can be oxidized in the presence of an electron donor (Chylenski et al., 2019). Due to the oxidation, more oxidized cellulosic terminals are generated for exocellulases enzyme exposure, resulting in an acceleration of hydrolysis and reduction of crystallinity of cellulosic fraction of biomass. Carbohydrate binding modules' presence in the structure of the auxiliary enzymes enables their effective synergism with the exocellulases and cellobiases enzymes.

LMPOs are better for hydrolysis in aerobic conditions than in anaerobic conditions; however, their employment depends on the pretreatment of the biomass (Valenzuela et al., 2019). For instance, if the pretreated biomass is close to the pure cellulose/glucose, then an additional electron donor is required in the form of redox enzymes or pure chemicals. However, if pretreated biomass contains both hemicellulosic and lignin fractions alongside the cellulose, there is no need to add an external electron source. In such a case, lignin facilitates the oxidation of cellulose/glucose (Hu et al., 2014).

The use of auxiliary enzymes in biomass hydrolysis has its own merits and demerits. Auxiliary enzymes can be advantageous as cellulase enzyme loading can be reduced to 5–6 times with their use (Sun et al., 2015). Similarly, if the auxiliary enzyme is supplemented with cellobiose dehydrogenase (CDH), cellulosic sugar generation increases 2–8 folds (Langston et al., 2011). Additionally, those enzymes help in biomass liquefaction by reducing cellulose viscosity (through oxidation). Finally, as discussed earlier, auxiliary enzymes significantly lower the crystallinity of the cellulosic fraction of biomass compared to its counterpart EG enzyme, which mainly works in the amorphous region of the biomass. Some of the downsides of auxiliary enzymes include a reduction in glucose yields with excessive use, therefore making it important to identify the optimum loading of these auxiliary enzymes. In addition, C1 oxidation of glucose by an auxiliary enzyme has the end product of gluconic acid, which has an inhibitory effect on the β-glucosidase enzyme (Ezeilo et al., 2017). These enzymes also produce gluconic acid that cannot be fermented by the yeast and other sugars, resulting in a lower yield of biofuel (bioethanol) (Hemsworth et al., 2015). The higher cost of auxiliary enzymes and external electron sources mostly offset the benefits of lowering hydrolysis time and decrystallization of biomass. However, it is important to note that a smaller proportion (2%–3%) of the auxiliary enzyme is required in the optimized enzyme cocktail used for biomass hydrolysis to provide positive results.

7.2.2 Hemicellulolytic enzymes

Xylan is the most prevalent polysaccharide present in the hemicellulosic fraction of biomass for pentose sugar xylose hydrolysis, and hemicellulolytic enzyme xylanases mostly act on such type of hydrolysis. Complete hydrolysis of hemicellulose requires two groups of enzymes (1) exo-xylanases, endo-β-1,4-xylanases, and β-xylosidases, which cleave the main chain of xylan, and (2) accessory enzymes such as feruloyl esterases (FEs), acetyl xylan esterases (AXEs), β-mannosidases, α-L-arabinofuranosidase, and α-glucuronidase, which remove the side links and disrupt lignin—hemicellulose interaction within the biomass. The synergistic action of these enzymes leads to increased susceptibility of lignocellulosic biomass to the cellulase enzyme for sugar hydrolysis. Hemicellulolytic enzymes work in two ways for biomass hydrolysis, those enzymes can be used as base enzyme preparation, or they can be supplemented with cellulase enzyme preparation. Both the cellulolytic and accessory hemicellulolytic enzymes are abundant in the fungal and bacterial systems in pure form at various amounts, but their use in pure form is most of the time costly and dependent on the pretreatment and types of feedstock. Therefore, using both types of hemicellulolytic enzymes in enzymatic cocktail preparation is preferable for biomass hydrolysis. This leads to a two-way benefit of reduction in the cost of the enzymes and improvement in the overall yield of sugar hydrolysis. For instance, celluclast (Novozyme[s]) enzyme supplemented with 20% hemicellulolytic enzyme, xylanases, for bagasse hydrolysis led to a 20% reduction in the cellulase enzyme cost in addition to an increase in sugar yield (Li et al., 2014; Adsul et al., 2020; Agrawal et al., 2015).

7.2.3 Carbohydrate esterases

Carbohydrate esterases (Ces) remove the ester substitutions on the O- and N-position of carbohydrates, whereas in hemicellulose, acetyl, glucuronoyl, and feruloyl substitutions present the acid parts and sugars behave as the alcohol of the ester bonds that CE acts on.

7.2.3.1 Feruloyl esterases

Esterases are another family of accessory enzymes derived from a wide range of bacteria and fungi that act synergistically to hydrolyze xylan to its monomers completely, and FE forms a part of this enzyme complex. FE plays a very important role in the hydrolysis of hemicellulose fibers by removing the ferulic acid side chains and cross-links (Wong, 2006). FEs can be classified as type A or type B based on their ability to release diferulic acids from esterified substrates (Crepin et al., 2003). FEs were used in synergism with cellulase and xylanase for enzymatic saccharification of wheat straw for bioethanol production. Synergistic action was observed with low concentrations of enzymes (10 U/g of cellulases, 3 U/g of xylanase, and 10 U/g of FE) (Tabka et al., 2006). FEs are also studied for their ability to hydrolyze quinic esters and ester linkages between phenolic acids and lignin monomer (Benoit et al., 2006).

7.2.3.2 Acetyl xylan esterases

One of the key obstacles to the effective enzymatic hydrolysis of hemicelluloses to fermentable sugars is acetylation. Microbial degradation of hemicellulose in nature involves the action of deacetylating esterases that act synergistically with glycoside hydrolases. In contrast, in industrial processing, pretreatment methods (other than alkaline pretreatment) generate acetylated hemicellulose. Therefore, deacetylating enzymes are important to achieve efficient hydrolysis. AXEs are the enzymes that aid in the breakdown of acetyl xylans, which is the most abundant hemicellulose in nature. AXEs act on acetyl xylan synergistically with endoxylanases, where the former stimulates the rate of hydrolysis of glycosidic linkages, and the latter stimulates the rate of deacetylation (Biely et al., 2014). Ramos-Martinez et al. (2019) investigated engineered *Chlamydomonas reinhardtii* to secrete fungal AXE and prey on pretreated wheat straw. They observed a reduction of biomass acetyl ester contents by 96% (Ramos-Martinez et al., 2019). Another study performed on 1% beechwood xylan, AXE, when it acted synergistically with glycoside hydrolase gene derived from the *Streptomyces* sp. Showed a 2.27-fold increase in the xylan degradation (Marasinghe et al., 2021).

7.2.3.3 Glucuronoyl esterases

Lignin-carbohydrate complexes are formed between lignin alcohols and MeGlcA side groups of the leading polysaccharide chains through ester, ether, and glycosidic bonds (Imamura et al., 1994). GEs are responsible for the cleavage of the covalent ester bonds found between lignin and glucuronoxylan (Fig. 7.3) and selectively remove lignin from hemicellulose, allowing access to other enzymes to the

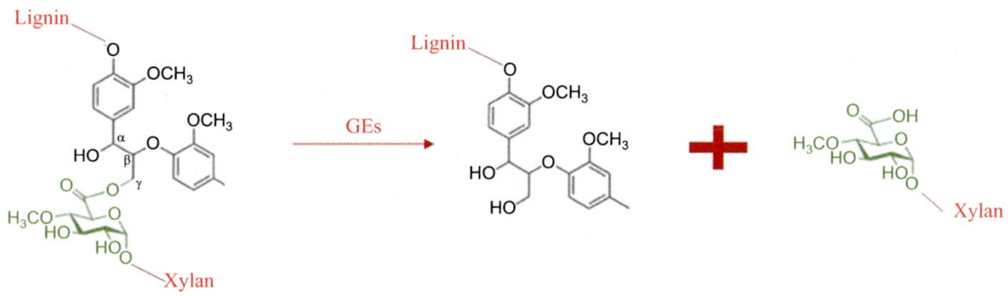

Figure 7.3 Cleavage of covalent ester bonds between lignin and glucuronoxylan by glucuronoyl esterases.

substrate (d'Errico et al., 2016). Preferred substrates for GEs are the glucuronoyl esters of heavier alcohols such as allyl, methyl, and benzyl alcohol (d'Errico et al., 2015). A study investigated the effect of GEs in addition to different commercial cellulase/hemicellulase enzymes for the degradation of heat-pretreated corn fiber and observed improved yields of fermentable sugars due to delignification by GEs (d'Errico et al., 2016). Another study expressed two GEs from *Tribulus terrestris* in *Pichia pastoris* and observed a substantial increase in the release of monomeric sugars and glucuronic acid from autohydrolysis of corn bran when GEs were supplemented with xylanases (Tang et al., 2019).

7.2.3.4 α-l-Arabinofuranosidases

Arabinoxylan (AX) is a major component of lignocellulosic biomass, investigated for biofuel production, and novel AX-degrading enzymes need to be studied and evaluated for degradation of this abundant hemicellulose. The backbone of AX comprises β-1,4-linked xylose residues, where arabinose can be substituted at the C(O)2 and/or C(O)3 positions, and important parameters for AX properties are the degree of substitution (DS) or arabinose to xylose ratio (A/X) (Perlin, 1951; Darvill et al., 1980). AX degrading enzymes, along with other glycosidases, can act either with retention or inversion of the anomeric center, and both mechanisms depend on a proton donor and a base (Sinnott, 1990). EGs act on the backbone of the AX; however, their activity is hindered by arabinose substitution. However, α-l-arabinofuranosidases can act in synergy with endoxylanases by cleaving the arabinose from the backbone. B-xylosidases are also required to complete the AX degradation, as they cleave the xylose residues from the nonreducing end of the xylose chain. The synergistic action of β-xylosidases and endoxylanases can be described as endoxylanases creating more reducing ends for β-xylosidases to act on by hydrolyzing the insoluble xylan backbone into shorter and soluble xylo-oligosaccharides, and β-xylosidases reduces inhibition of oligosaccharides and xylobiose to endoxylanases by hydrolyzing these to xylose (Sunna and Antranikian, 1997). In a recent study, two arabinofuranosidases from the anaerobic cellulolytic bacterium *Acetivibrio mesophilus* were expressed in *Escherichia coli*, and it was observed that for wheat AX, incorporation of these two enzymes increased the yield of arabinose and xylose 13.77-fold and 4.96-fold, respectively (Liu et al., 2021).

7.2.3.5 Pectinase

Pectic substances are the major component of the middle lamella of the cell wall and are composed of partially methyl-esterified galacturonic acid residues linked by α-1,4-glycosidic bonds. Pectinases are the group of enzymes that naturally degrade these substances. The first pectinases identified were pectin methylesterase, which is responsible for the removal of methyl-ester groups, and endo-polygalacturonase (endo-PG), which is responsible for the random hydrolysis reaction of α-1,4 bonds (Reignault et al., 1994; Shen et al., 1996). In synergism with xylanase and cellulase, Pectinases have shown improved saccharification of the lignocellulosic biomass and, therefore, a potential to form an efficient enzyme cocktail for biomass degradation (Thite and Nerurkar, 2018). A study using cellulase in addition to α-l-arabinofuranosidase, pectin esterase, and endo-1,4-β-xylanase reported 214% glucose released with alkali pretreated rice straw compared to cellulase alone (Laothanachareon et al., 2015).

7.2.4 Noncatalytic proteins

Noncatalytic proteins improve the cellulolytic hydrolysis of biomass by reducing the crystallinity of biomass fibrils. Those proteins make microfibrils more dispersed and macrofibrils looser and swollen towards cellulolytic enzymes, leading to reduced cellulose crystallinity. Plant expansins and fungal swollenin are examples of noncatalytic proteins sourced from bacteria. Other types of noncatalytic proteins have a CBM in their structure for substrate carbohydrate binding with the cellulolytic enzymes for biomass hydrolysis improvement. Some noncatalytic proteins improve the thermal property (heat resistance) of the enzymes and subsequently facilitate biomass hydrolysis. Commonly used noncatalytic proteins for biomass hydrolysis and their impacts are tabulated in Table 7.1. Those accessory proteins mostly depend on the pretreatment process and their large-scale production cost.

7.2.5 Accessory enzymes for lignin degradation

Enzyme development for lignin degradation has not gained much popularity for many reasons. Biomass hydrolysis can be carried out without lignin degradation, and lignin content in biomass feedstock varies widely. Lignin degradation is avoided during the biological conversion processes involving the fermentation step (Oliva-Taravilla et al., 2015). This is preferable because of the inhibitory effects of lignin degradation products on sugar fermentation, and as a result, final biofuels and value-added chemicals yield can reduce significantly (Ragauskas et al., 2014). Among the accessory enzymes used for lignin degradation, laccases, lignin peroxidases, and manganese peroxidases are notable (Su et al., 2018). Their synergistic effect with cellulase enzymes remains unclear, and their role in biomass hydrolysis is being investigated (Adsul et al., 2020).

Table 7.1 Nonhydrolytic accessory proteins for lignocellulosic biomass hydrolysis.

Name of accessory enzyme/protein		Source	Impact	References
Plant expansin proteins	BsEXLX1	*Bacillus subtilis*	Fivefold increase in cellulase activity and 240% increase in the synergistic effect	Kim et al. (2009)
	TrSwo1	*Trichoderma reesei*	2–2.5-fold increase in the hydrolysis sugar yield and improved synergism with EG (cellulolytic enzyme)	Wang et al. (2014)
Expansin like protein	Swollenin	*Orpinomyces* strain C1A	7% increment of hydrolysis sugar of corn stover	Morrison et al. (2016)
	CBM		Supplement the enzyme binding with feedstock carbohydrate	Sandhu et al. (2018)
	CbHsp18 (heat shock protein)	*Caldicellulosiruptor bescii*	Enhance thermal stability of glycoside hydrolase and further biomass hydrolysis	Su et al. (2012)

7.3 Enzyme engineering for accessory enzyme development

Enzymes are made of a repeated sequence of various amino acids. Change in this amino acid sequence to improve the efficiency of an existing enzyme or the creation of advanced enzyme activity by this sequence change is termed enzyme engineering. The rational design and molecular evolution, either individually or in amalgamation with the molecular modeling approach, are considered the general approach for enzyme engineering. Various other approaches are also used widely for enzyme engineering, such as enzyme immobilization, peptidomimetics, flow cytometry, etc. Recently, recombinant DNA technology has become very popular for enzyme engineering. In this technology, amino acid sequences are altered for enzyme activity enhancement based on gene alterations. Essentially, this technology transfers the desired genes from less active microorganisms into highly active microorganisms. Genes can be transferred stepwise, or all the chosen genes can be transferred altogether to bring the wanted activity in the new enzyme or biocatalyst. This technology has given a promising potential for utilizing enzymology to produce desired traits in a microorganism. For instance, in a study, GE characterized by bacterial species could use substrates lacking 4-OH methyl substitutions and efficiently cleave esters of galacturonate, both of the properties not present in fungal GEs (Arnling Bååth et al., 2018).

The first step in the process of enzyme engineering is the identification and isolation of the desired enzyme, followed by the determination of its structure and property. The structure of the isolated enzyme (primary, secondary, or tertiary) is determined by X-ray diffraction, or nuclear magnetic resonance is used to establish the new (native) enzyme's structural-property relationship. This information is then compared with the existing database of amino acids to identify the amino acid sequence change that will alter/add properties to the new enzyme. Molecular modeling aids in this step in finding the exact amino acid sequence for the desired structural–property relationship. The next step is site-directed mutagenesis, where a gene capable of encoding the desired amino acid sequence is constructed, generally with the help of a natural gene. Once the recombinant gene is created, it is placed inside a suitable host for recombinant enzyme production. Then produced enzyme is isolated and purified, and a further structural–property relationship is identified. Finally, the structural–property relationship of both native and recombinant enzymes is compared to see whether enzyme engineering has brought the desired property in the recombinant enzyme. If the desired property is achieved, the exact sequence of amino acids is stored in the database for future comparison during new enzyme engineering (Basheer and Chellappan, 2017). Some examples of enzyme engineering are mentioned below.

7.3.1 *Xylanase*

Xylanase is the most widely used enzyme for hydrolyzing hemicellulosic pentose sugar complex xylan into xylose. Natural xylanase was isolated from the plaques of

50,000 samples of fresh bovine manure. To improve the xylanase activity, molecular modeling identified the nine interesting amino acid sequences/mutations. Site-directed mutagenesis prepared 70,000 genes for encoding the above amino acid mutations (Palackal et al., 2004). While nine mutations were combinedly applied to the native xylanase enzyme, its melting temperature was increased by 34.2°C (Basheer and Chellappan, 2017). This is one of the best examples of molecular evolution of enzyme engineering, which was reported by Palackal et al. (2004)

7.3.2 β-Glucosidase

β-glucosidases enzymes are used to hydrolyze β-glucosidic bonds of β-D-glucosides or oligosaccharides into glucose. However, those enzymes later gained popularity to increase the aroma of the final tea products (Romo-Sánchez et al., 2014). Fresh leaves of tea contain monosaccharide and polysaccharide precursors, β-D-glucopyranosides. These precursors are hydrolyzed into the floral aroma ingredients such as moterpene alcohol and aromatic alcohol by the action of the β-D-glucosidase enzyme (Wang et al., 2001). But the activity of this β-glucosidase enzyme is very limited under natural conditions, and it can be degraded highly during the whole tea manufacturing process. Hence, the enzyme engineering technique, enzyme immobilization, is used to trap the β-glucosidase in a cross-linked structure for the enzyme to reach the final tea manufacturing process in intact condition (Romo-Sánchez et al., 2014). Then this engineered β-glucosidase enzyme works on the polysaccharide precursors to release the floral aroma ingredients. In addition, this engineered enzyme can be used up to 500 times repeatedly without substantial loss of activity (around 7% compared to its initial activity) (Su et al., 2010). However, thermal activity enhanced β-glucosidase enzymes can have a potential application in the lignocellulosic biomass hydrolysis into glucose.

7.4 Industrial applications of accessory enzymes

With the capability of hydrolyzing the bonds that cannot be cleaved by main enzymes, accessory enzymes have many industrial applications. As explained above, one of the major applications of accessory enzymes is biomass processing. Many accessory enzymes are essential to complete the hydrolysis of lignocellulosic biomass for bioethanol and other biorefineries. The activity of accessory enzymes in combination with other hydrolases and oxidases significantly increases the degradation of biomass and availability of fermentable sugars and, in addition, aids in the release of toxic esters that can inhibit fermentation. Some enzymes are modified (mutagenesis) to improve the efficiency of biomass conversion; for instance, strategies to create thermostable FE mutants were applied for application in biomass processing (Yin et al., 2015; Zhang et al., 2012; Cao et al., 2015). FEs are also used for manipulating the structure of oligosaccharides to produce xylooligosaccharides, which are important functional food additives with prebiotic properties (Aachary and Prapulla, 2011).

FEs are used in combination with other hydrolases to produce ferulic acid, which has wide applications in the cosmetic and food industry. For example, to obtain products with altered chemical and biological properties, FEs can be used to synthesize ester-linked hydroxycinnamic acids via a transesterification reaction, exchanging the organic acid of the ester with organic acid of alcohol (Dilokpimol et al., 2016). Another application of accessory enzymes is in the paper and pulp industry, where not only the enzymatic process leads to lignin degradation, but it also results in reduced energy consumption and a significant lowering of chemical oxidation demand of the pulping wastewater (Xu et al., 2013). Fungi produce oxidases that generate H_2O_2 required by their peroxidases for lignin degradation, in addition to producing dehydrogenases that degrades lignin and reduce lignin-derived compounds such as aryl-alcohol dehydrogenases and quinone reductases (Guillén et al., 1997).

Proteases and FEs are also used with cellulase enzymes in corn wet milling industries for starch production to increase the production yield (Dilokpimol et al., 2016). Proteases in wet milling are also applied to reduce the steep time and sulfur dioxide requirement in the steeping process of corn wet milling process (Johnston and Singh, 2001). FE has found its application in many other industries, such as the feed industry (Cao et al., 2016), the beverage industry (Kanauchi et al., 2008; Kanauchi, 2012; Uno et al., 2009), the seasoning industry as well as in medical applications (Neu et al., 2016).

7.5 Summary and conclusion

Enzymatic hydrolysis of cellulose is a critical but cost-intensive step in the lignocellulosic biomass conversion to biofuel and biomaterials. Due to its complex crystalline structure, cellulose is difficult to degrade and requires a large amount of enzymes. As biofuel and bioproduct yield is directly proportional to sugars produced during hydrolysis, achieving high yields and hydrolysis rates is necessary for the economic feasibility of the process. The addition of accessory enzymes to the naturally derived cellulase enzymes can significantly improve cellulose hydrolysis efficiency in various ways, including their synergistic interaction with cellulases, reducing inhibition, disrupting the crystalline lattices and enhancing accessibility, and helping in creating active sites and release of enzymes. However, the benefits of accessory enzyme addition are highly dependent on the type of substrate. Also, there is limited information on the trade-off between higher efficiency and additional cost associated with accessory enzymes, which needs to be investigated.

Abbreviations

AX arabinoxylan
AXEs acetyl xylan esterases
BG β-glucosidase

CAC catalytically active complex
CBH cellobiohydrolase
CBH I cellobiohydrolases I
CBH II cellobiohydrolases II
CBM carbohydrate binding module
CDH cellobiose dehydrogenase
CE carbohydrate estérases
DNA deoxyribonucleic Acid
DS degree of substitution
EG endo-β1,4-glucanase
endo-PG endo-polygalacturonase
FE feruloyl esterases
GEs glucuronoyl esterases
LPMOs lytic polysaccharide monooxygenases
NR nonreducing ends
R reducing ends

References

Aachary, A., Prapulla, S., 2011. Xylooligosaccharides (XOS) as an emerging prebiotic: microbial synthesis, utilization, structural characterization, bioactive properties, and applications. Compr. Rev. Food Sci. Food Saf. 10 (1), 2–16.

Adsul, M., Sandhu, S.K., Singhania, R.R., Gupta, R., Puri, S.K., Mathur, A., 2020. Designing a cellulolytic enzyme cocktail for the efficient and economical conversion of lignocellulosic biomass to biofuels. Enzyme Microb. Technol. 133, 109442.

Agrawal, R., Gaur, R., Mathur, A., Kumar, R., Gupta, R.P., Tuli, D.K., et al., 2015. Improved saccharification of pilot-scale acid pretreated wheat straw by exploiting the synergistic behavior of lignocellulose degrading enzymes. RSC Adv. 5 (87), 71462–71471.

Arnling Bååth, J., Mazurkewich, S., Knudsen, R., Poulsen, J., Olsson, L., Lo Leggio, L., et al., 2018. Biochemical and structural features of diverse bacterial glucuronoyl esterases facilitating recalcitrant biomass conversion. Biotechnol. Biofuels 11 (1), 1–14.

Awoyale, A.A., Lokhat, D., 2021. Experimental determination of the effects of retreatment on selected Nigerian lignocellulosic biomass in bioethanol production. Sci. Rep. 11 (1), 1–16.

Basheer, S., Chellappan, S., 2017. Enzyme engineering. Bioresour. Bioprocess. Biotechnol. 151–168.

Beckham, G.T., Bomble, Y.J., Bayer, E.A., Himmel, M.E., Crowley, M.F., 2011. Applications of computational science for understanding enzymatic deconstruction of cellulose. Curr. Opin. Biotechnol. 22 (2), 231–238.

Benoit, I., Navarro, D., Marnet, N., Rakotomanomana, N., Lesage-Meessen, L., Sigoillot, J.C., et al., 2006. Feruloyl esterases as a tool for the release of phenolic compounds from agro-industrial by-products. Carbohydr. Res. 341 (11), 1820–1827.

Biely, P., Westereng, B., Puchart, V., De Maayer, P., Cowan, D.A., 2014. Recent progress in understanding mode of action of acetylxylan esterases. J. Appl. Glycosci. 61 (2), 35–44.

Cao, B., Jin, X., Li, S., Li, S., Jiang, L., 2016. Microbial release of ferulic and p-coumaric acids from forages and their digestibility in lactating cows fed total mixed rations with different forage combinations. J. Sci. Food Agricult. 96 (2), 650–655.

Cao, L., Chen, R., Xie, W., Liu, Y., 2015. Enhancing the thermostability of feruloyl esterase EstF27 by directed evolution and the underlying structural basis. J. Agric. Food Chem. 63 (37), 8225–8233.

Chylenski, P., Bissaro, B., Sørlie, M., Røhr, A.K., Varnai, A., Horn, S.J., et al., 2019. Lytic polysaccharide monooxygenases in enzymatic processing of lignocellulosic biomass. ACS Catal. 9 (6), 4970–4991.

Crepin, V.F., Faulds, C.B., Connerton, I.F., 2003. A non-modular type B feruloyl esterase from Neurospora crassa exhibits concentration-dependent substrate inhibition. Biochem. J. 370 (2), 417–427.

da Silva, A.A., Espinheira, R.P., Teixeira, R.S., de Souza, M.F., Ferreira-Leitão, V., Bon, E.P., 2020. Constraints and advances in high-solids enzymatic hydrolysis of lignocellulosic biomass: a critical review. Biotechnol. Biofuels 13 (1), 1–28.

Darvill, A., McNeil, M., Albersheim, P., Delmer, D., 1980. The primary cell walls of flowering plants. Biochem. Plants 1, 91–162.

d'Errico, C., Börjesson, J., Ding, H., Krogh, K., Spodsberg, N., Madsen, R., et al., 2016. Improved biomass degradation using fungal glucuronoyl—esterases—hydrolysis of natural corn fiber substrate. J. Biotechnol. 219, 117–123.

d'Errico, C., Jørgensen, J., Krogh, K., Spodsberg, N., Madsen, R., Monrad, R., 2015. Enzymatic degradation of lignin-carbohydrate complexes (LCCs): Model studies using a fungal glucuronoyl esterase from *Cerrena unicolor*. Biotechnol. Bioeng. 112 (5), 914–922.

Dilokpimol, A., Mäkelä, M., Aguilar-Pontes, M., Benoit-Gelber, I., Hildén, K., de Vries, R., 2016. Diversity of fungal feruloyl esterases: Updated phylogenetic classification, properties, and industrial applications. Biotechnol. Biofuels 9 (1), 1–18.

Dos Santos, A.C., Ximenes, E., Kim, Y., Ladisch, M.R., 2019. Lignin–enzyme interactions in the hydrolysis of lignocellulosic biomass. Trends Biotechnol. 37 (5), 518–531.

Ezeilo, U.R., Zakaria, I.I., Huyop, F., Wahab, R.A., 2017. Enzymatic breakdown of lignocellulosic biomass: the role of glycosyl hydrolases and lytic polysaccharide monooxygenases. Biotechnol. Biotechnol. Equip. 31 (4), 647–662.

Forsberg, Z., Sørlie, M., Petrović, D., Courtade, G., Aachmann, F.L., Vaaje-Kolstad, G., et al., 2019. Polysaccharide degradation by lytic polysaccharide monooxygenases. Curr. Opin. Struct. Biol. 59, 54–64.

Guillén, F., Martínez, M., Muñoz, C., Martínez, A., 1997. Quinone redox cycling in the ligninolytic funguspleurotus eryngiileading to extracellular production of superoxide anion radical. Arch. Biochem. Biophys. 339 (1), 190–199.

Hemsworth, G.R., Johnston, E.M., Davies, G.J., Walton, P.H., 2015. Lytic polysaccharide monooxygenases in biomass conversion. Trends Biotechnol. 33 (12), 747–761.

Hu, J., Arantes, V., Pribowo, A., Gourlay, K., Saddler, J.N., 2014. Substrate factors that influence the synergistic interaction of AA9 and cellulases during the enzymatic hydrolysis of biomass. Energy Environ. Sci. 7 (7), 2308–2315.

Imamura, T., Watanabe, T., Kuwahara, M., Koshijima, T., 1994. Ester linkages between lignin and glucuronic acid in lignin-carbohydrate complexes from *Fagus crenata*. Phytochemistry 37 (4), 1165–1173.

Johnston, D., Singh, V., 2001. Use of proteases to reduce steep time and SO_2 requirements in a corn wet-milling process. Cereal Chem. 78 (4), 405–411.

Kanauchi, M., 2012. Characteristics and role of feruloyl esterase from *Aspergillus awamori* in Japanese spirits, '*Awamori*' production. Scientific, Health and Social Aspects of the Food Industry. InTech, Rijeka, pp. 145–162.

Kanauchi, M., Watanabe, S., Tsukada, T., Atta, K., Kakuta, T., Koizumi, T., 2008. Purification and characteristics of feruloyl esterase from *Aspergillus awamori* G-2 strain. J. Food Sci. 73 (6), C458–C463.

Kim, E., Lee, H., Bang, W., Choi, I., Kim, K., 2009. Functional characterization of a bacterial expansin from *Bacillus subtilis* for enhanced enzymatic hydrolysis of cellulose. Biotechnol. Bioeng. 102 (5), 1342–1353.

Kumar, D., Murthy, G.S., 2016. Enzymatic hydrolysis of cellulose for ethanol production: fundamentals, optimal enzyme ratio, and hydrolysis modeling. In: Gupta, V. (Ed.), New and Future Developments in Microbial Biotechnology and Bioengineering, Ed. Elsevier, pp. 65–78.

Langston, J.A., Shaghasi, T., Abbate, E., Xu, F., Vlasenko, E., Sweeney, M.D., 2011. Oxidoreductive cellulose depolymerization by the enzymes cellobiose dehydrogenase and glycoside hydrolase 61. Appl. Environ. Microbiol. 77 (19), 7007–7015.

Laothanachareon, T., Bunterngsook, B., Suwannarangsee, S., Eurwilaichitr, L., Champreda, V., 2015. Synergistic action of recombinant accessory hemicellulolytic and pectinolytic enzymes to *Trichoderma reesei* cellulase on rice straw degradation. Bioresour. Technol. 198, 682–690.

Li, J., Zhou, P., Liu, H., Xiong, C., Lin, J., Xiao, W., et al., 2014. Synergism of cellulase, xylanase, and pectinase on hydrolyzing sugarcane bagasse resulting from different pretreatment technologies. Bioresour. Technol. 155, 258–265.

Liu, Y., Vanderhaeghen, S., Feiler, W., Angelov, A., Baudrexl, M., Zverlov, V., et al., 2021. Characterization of two α-l-arabinofuranosidases from *Acetivibrio mesophilus* and their synergistic effect in degradation of arabinose-containing substrates. Microorganisms 9 (7), 1467.

Maitan-Alfenas, G.P., Visser, E.M., Guimarães, V.M., 2015. Enzymatic hydrolysis of lignocellulosic biomass: converting food waste in valuable products. Curr. Opin. Food Sci. 1, 44–49.

Marasinghe, S.D., Jo, E., Hettiarachchi, S.A., Lee, Y., Eom, T.Y., Gang, Y., et al., 2021. Characterization of glycoside hydrolase family 11 xylanase from *Streptomyces* sp. strain J103; its synergetic effect with acetyl xylan esterase and enhancement of enzymatic hydrolysis of lignocellulosic biomass. Microb. Cell Factories 20, 129. Available from: https://doi.org/10.1186/s12934-021-01619-x.

Monteiro, C.R., Avila, P.F., Pereira, M.F., Pereira, G.N., Bordignon, S.E., Zanella, E., et al., 2021. Hydrothermal treatment on depolymerization of hemicellulose of mango seed shell for the production of xylooligosaccharides. Carbohydr. Polym. 253, 117274.

Morrison, J., Elshahed, M., Youssef, N., 2016. Defined enzyme cocktail from the anaerobic fungus *Orpinomyces* sp. strain C1A effectively releases sugars from pretreated corn stover and switchgrass. Sci. Rep. 6 (1), 1–12.

Neu, J., Lorca, G., Triplett, E., Atkinson, M., Schatz, D., 2016. Washington, DC: U.S. Patent and Trademark Office Patent No. U.S. Patent No. 9,474,773.

Oliva-Taravilla, A., Tomás-Pejó, E., Demuez, M., González-Fernández, C., Ballesteros, M., 2015. Inhibition of cellulose enzymatic hydrolysis by laccase-derived compounds from phenols. Biotechnol. Prog. 31 (3), 700–706.

Palackal, N., Brennan, Y., Callen, W., Dupree, P., Frey, G., Goubet, F., et al., 2004. An evolutionary route to xylanase process fitness. Protein Sci. 13 (2), 494–503.

Perlin, A., 1951. Isolation and composition of the soluble pentosans of wheat flours. Cereal Chem. 28, 370–381.

Ragauskas, A., Beckham, G., Biddy, M., Chandra, R., Chen, F., Davis, M., et al., 2014. Lignin valorization: improving lignin processing in the biorefinery. Science 344 (6185), 1246843.

Ramos-Martinez, E.M., Fimognari, L., Rasmussen, M.K., Sakuragi, Y., 2019. Secretion of acetylxylan esterase from *Chlamydomonas reinhardtii* enables utilization of lignocellulosic biomass as a carbon source. Front. Bioeng. Biotechnol. 7, 35.

Reignault, P., Mercier, M., Bompeix, G., Boccara, M., 1994. Pectin methylesterase from Botrytis cinerea: physiological, biochemical and immunochemical studies. Microbiology 140 (12), 3249–3255.

Romo-Sánchez, S., Arévalo-Villena, M., García Romero, E., Ramirez, H., Briones Pérez, A., 2014. Immobilization of β-glucosidase and its application for enhancement of aroma precursors in muscat wine. Food Bioprocess. Technol. 7 (5), 1381–1392.

Sandhu, S.K., Mathur, A., Gupta, R., Puri, S.K., Adsul, M., 2018. Cellulosic biomass-hydrolyzing enzymes. In: Singhania, R.R., Agarwal, R.A., Kumar, R.P., Sukumaran, R.K. (Eds.), Waste to Wealth. Springer, Singapore, pp. 441–456. Available from: https://doi.org/10.1007/978-981-10-7431-8_1.

Shen, Z., Reese, J., Reeck, G., 1996. Purification and characterization of polygalacturonase from the rice weevil, *Sitophilus oryzae* (Coleoptera: Curculionidae). Insect Biochem. Mol. Biol. 26 (5), 427–433.

Sinnott, M., 1990. Catalytic mechanism of enzymic glycosyl transfer. Chem. Rev. 90 (7), 1171–1202.

Siqueira, J.W., Rodrigues, C., de Souza Vandenberghe, L.P., Woiciechowski, A.L., Soccol, C.R., 2020. Current advances in on-site cellulase production and application on lignocellulosic biomass conversion to biofuels: a review. Biomass Bioenergy 132, 105419.

Su, E., Xia, T., Gao, L., Dai, Q., Zhang, Z., 2010. Immobilization of β-glucosidase and its aroma-increasing effect on tea beverage. Food Bioprod. Process. 88 (2–3), 83–89.

Su, X., Zhang, J., Mackie, R., Cann, I., 2012. Supplementing with non-glycoside hydrolase proteins enhances enzymatic deconstruction of plant biomass. PLoS One 7 (8), e43828. Available from: https://doi.org/10.1371/journal.pone.0043828. Epub 2012 Aug 27.

Su, Y., Yu, X., Sun, Y., Wang, G., Chen, H., Chen, G., 2018. Evaluation of screened lignin-degrading fungi for the biological pretreatment of corn stover. Sci. Rep. 8 (1), 1–11.

Sugathan, S., Pradeep, N.S., Abdulhameed, S., eds., 2017. Bioresources and Bioprocess in Biotechnology: Volume 2: Exploring Potential Biomolecules. Springer Nature, Singapore, 456p.

Sun, F.F., Hong, J., Hu, J., Saddler, J.N., Fang, X., Zhang, Z., et al., 2015. Accessory enzymes influence cellulase hydrolysis of the model substrate and the realistic lignocellulosic biomass. Enzyme Microb. Technol. 79, 42–48.

Sunna, A., Antranikian, G., 1997. Xylanolytic enzymes from fungi and bacteria. Crit. Rev. Biotechnol. 17 (1), 39–67.

Tabka, M.G., Herpoël-Gimbert, I., Monod, F., Asther, M., Sigoillot, J.C., 2006. Enzymatic saccharification of wheat straw for bioethanol production by a combined cellulase xylanase and feruloyl esterase treatment. Enzyme Microb. Technol. 39 (4), 897–902.

Tang, J., Long, L., Cao, Y., Ding, S., 2019. Expression and characterization of two glucuronoyl esterases from *Thielavia terrestris* and their application in enzymatic hydrolysis of corn bran. Appl. Microbiol. Niotechnol. 103 (7), 3037–3048.

Thite, V., Nerurkar, A., 2018. Physicochemical characterization of pectinase activity from *Bacillus* spp. and their accessory role in synergism with crude xylanase and commercial cellulase in enzyme cocktail mediated saccharification of agrowaste biomass. J. Appl. Microbiol. 124 (5), 1147–1163.

Tian, D., Guo, Y., Hu, J., Yang, G., Zhang, J., Luo, L., et al., 2020. Acidic deep eutectic solvents pretreatment for selective lignocellulosic biomass fractionation with enhanced cellulose reactivity. Int. J. Biol. Macromol. 142, 288–297.

Ummalyma, S.B., Supriya, R.D., Sindhu, R., Binod, P., Nair, R.B., Pandey, A., et al., 2019. Biological pretreatment of lignocellulosic biomass—current trends and future perspectives. In: Basile, A., Dalena, F. (Eds.), Second and Third Generation of Feedstocks.The Evolution of Biofuels. Elsevier, pp. 197–212.

Uno, T., Itoh, A., Miyamoto, T., Kubo, M., Kanamaru, K., Yamagata, H., et al., 2009. Ferulic acid production in the brewing of rice wine (*Sake*). J. Inst. Brew. 115 (2), 116–121.

Valenzuela, S.V., Valls, C., Schink, V., Sanchez, D., Roncero, M.B., Diaz, P., et al., 2019. Differential activity of lytic polysaccharide monooxygenases on celluloses of different crystallinity. Effectiveness in the sustainable production of the nanofibrils. Carbohydr. Polym. 207, 59–67.

Wang, D., Kubota, K., Kobayashi, A., Juan, I., 2001. Analysis of glycosidically bound aroma precursors in tea leaves. 3. Change in the glycoside content of tea leaves during the oolong tea manufacturing process. J. Agric. Food Chem. 49 (11), 5391–5396.

Wang, W., Liu, C., Ma, Y., Liu, X., Zhang, K., Zhang, M., 2014. Improved production of two expansin-like proteins in *Pichia pastoris* and investigation of their functional properties. Biochem. Eng. J. 84, 16–27.

Woiciechowski, A.L., Neto, C.D., de Souza Vandenberghe, L.P., de Carvalho Neto, D.P., Sydney, A.N., Letti, L.J., et al., 2020. Lignocellulosic biomass: acid and alkaline pretreatments and their effects on biomass recalcitrance–conventional processing and recent advances. Bioresour. Technol. 304, 122848.

Wong, D.W., 2006. Feruloyl esterase. Appl. Biochem. Biotechnol. 133 (2), 87–112.

Xu, Z., M. J., Men, Z., Chen, Z., Wang, X., Tan, F., et al., 2013. Complex enzyme preparation and technique for preparing dissolving pulp by using same, in C. Patent (ed.), Yellow River Delta, Jingbo Chemical Research Institute Co Ltd, China.

Yin, X., Li, J.F., Wang, C.J., Hu, D., Wu, Q., Gu, Y., et al., 2015. Improvement in the thermostability of a type A feruloyl esterase, AuFaeA, from *Aspergillus usamii* by iterative saturation mutagenesis. Appl. Microbiology Biotechnol. 99 (23), 10047–10056.

Zhang, S., Pei, X., Wu, Z., 2012. Multiple amino acid substitutions significantly improve the thermostability of feruloyl esterase A from *Aspergillus niger*. Bioresour. Technol. 117, 140–147.

Zoghlami, A., Paës, G., 2019. Lignocellulosic biomass: understanding recalcitrance and predicting hydrolysis. Front. Chem. 7, 874. Available from: https://www.frontiersin.org/research-topics/9182#articles.

Fermentable sugars as bioprocessing feedstocks from lignocellulosic biomass pretreated with acid mine drainage

8

Edzai Magowo[1], *Craig Sheridan*[1,2] and *Karl Rumbold*[3]
[1]School of Chemical and Metallurgical Engineering, University of the Witwatersrand, Johannesburg, South Africa, [2]School Geography, Archaeology and Environmental Studies, University of the Witwatersrand, Johannesburg, South Africa, [3]School of Molecular and Cell Biology, University of the Witwatersrand, Johannesburg, South Africa

8.1 Introduction

The depletion of fossil fuels and the demand for cleaner energy has led to searches for alternative fuel sources such as biofuel. Biofuels are produced from biological material. Currently, biofuels are produced mainly from the edible parts of crops, which strains the provision of food and causes a rise in food prices. The focus is now shifting to the production of biofuels from lignocellulosic material. The utilization of lignocellulosic biomass in the production of biofuels has large potential due to its abundance and the fact that it is usually treated as waste material. The conversion of lignocellulose to fermentable sugars is the main challenge to the implementation of lignocellulosic processes. Lignocellulose material must be pretreated to unlock the energy potential hidden in lignocellulosic biomass (Zuliani et al., 2021).

Dilute acid pretreatments using sulfuric acid, or hydrochloric acid have been used for the pretreatment of lignocellulose before enzymatic hydrolysis and fermentation (Swiatek et al., 2020; Chen et al., 2014). The dilute acid treatment process has some drawbacks, including harsh operating conditions, difficulties in acid recovery, and chemical wastes from the neutralization of acid (Chen et al., 2014). Recently acid mine drainage (AMD) has been recognized as a source of acid for the pretreatment of lignocellulosic biomass in the production of fermentable sugars to feed the biorefinery industry; concurrently, the AMD is remediated through dissimilatory sulfate reduction (DSR) (Burman et al., 2019).

8.2 Acid mine drainage: an overview

AMD refers to the outflow of acidic water from (usually abandoned) metal and coal mines. It is characterized by low pH, high sulfate concentrations, and heavy metals

Advances in Lignocellulosic Biofuel Production Systems. DOI: https://doi.org/10.1016/B978-0-323-91192-4.00006-7

(Magowo et al., 2020a,b). It is considered the most critical mining-related pollution problem (Rob and Robson, 1995; Johnson and Hallberg, 2005; Kuyacak, 2011). AMD occurs when sulfide ores are exposed to the atmosphere, which can be magnified through mining and milling activities, where oxidation reactions are initiated (Jennings et al., 2008; Zipper et al., 2011; McCarthy, 2011; Johnson and Hallberg, 2005; Saharan et al., 1995). Upon exposure to oxygen and water, pyritic minerals oxidize to form acidic, iron, and sulfate-rich drainage. The oxidation of the pyritic minerals is the primary mechanism by which acid is released into AMD (Ackil and Kodas, 2006). The reactions involved in forming AMD are a multistep process involving oxygen-independent and oxygen-dependent reactions. A general equation for the process is [Eq. (8.1)]:

$$FeS_2(s) + 14Fe^{3+}(aq) + 8H_2O \rightarrow 15Fe^{2+}(aq) + 2SO4^{2-} + 16H^{+}(aq) \tag{8.1}$$

The net effect of these reactions is to release hydrogen ions (H^+), which lowers pH and maintains the solubility of ferric ions (Blodau, 2006). Because of the low pH of AMD, metal ions, including iron, manganese, zinc, aluminum, cobalt copper, radium, and aluminum, are mobilized (Neto et al., 2018). These dissolved metals have varying degrees of toxicity and radioactivity (Kousi et al., 2015). As a result, surface and groundwater contamination with metal-rich effluents like AMD reduces water availability for domestic and industrial applications. Furthermore, the impact of metal contaminants persists over long periods because of bioaccumulation (Kosinka and Miskiewisc, 2012), affecting microscopic life cycles and food chains and leading to a lack of biodiversity (Sangita et al., 2010).

AMD treatment can be achieved through active, semipassive, or passive methods (Greben et al., 2009). A broad range of passive and active treatment approaches are available for dealing with AMD. Neutralization (pH control) is the AMD treatment mechanism for both passive and active systems. By increasing pH to create alkaline conditions, the solubility of most metals is significantly decreased by precipitation (Taylor, 2005).

The treatment technology can be classified as physicochemical and biological (Magowo et al., 2020a,b; Sheridan et al., 2018). Physicochemical processes include chemical neutralization, reverse osmosis, filtration, and ion exchange, among others (Fahruddin et al., 2020). These treatment technologies are effective in treating AMD; however, their application may be limited due to high capital, operational, and maintenance costs. Biological treatment technologies use sulfate-reducing bacteria (SRB) to reduce sulfate to sulfide under anaerobic conditions and in the presence of an organic source (Mukwevho et al., 2019; Magowo et al., 2020a,b). The sulfide reacts with metals to form insoluble metal sulfides (Mukwevho et al., 2019). In addition, biological sulfate reduction generates alkalinity, which neutralizes the acidity of AMD, thereby improving the pH (Van den Berg et al., 2016). Precipitation of metals such as sulfides is more desirable as it offers advantages to traditional hydroxide precipitation. This is because the volume of metal sulfide precipitates is much lower than that of the corresponding metal hydroxide. In addition, metal sulfide precipitates have better thickening and dewatering characteristics and are stable over a broad range of pH, making their disposal easier (Macinova and Luptakova, 2012).

Furthermore, the precipitation of metals as sulfides lowers their toxicity; for example, metal sulfides of arsenic, copper, lead, and mercury is less toxic to anaerobic biomass (Cirne et al., 2015). In addition, the precipitation of metal sulfides allows for the selective recovery of valuable minerals due to differences in solubility at different pH values (Kousi et al., 2015). Finally, SRB can also reduce the dissolved metal ion concentration through extracellular binding, cellular uptake, accumulation, and surface-mediated mineral precipitation (Siite et al., 2010).

The treatment of AMD using biological sulfate reduction requires electron donors and carbon sources. AMD does not contain sufficient organic substances to sustain biological sulfate reduction. Therefore it is necessary to add enough electron donors to sustain complete sulfate reduction (Neto et al., 2018; Magowo et al., 2020a,b). The carbon and electron sources can be low-cost solid materials like sawdust, grass, and spent mushrooms (Sahinkaya et al., 2017). These lignocellulose-based substrates have been successfully used to provide electrons for biological sulfate reduction in the treatment of sulfate-rich wastewater (Ramla and Sheridan, 2014). The organic matter stimulates the activity of hydrolytic, fermentative, and sulfate-reducing microorganisms and acts as support media for microbial growth. With the help of sulfuric acid in AMD, the hydrolytic enzymes break down the macromolecules in the lignocellulosic material into small molecules. The action of the hydrolytic enzymes provides substrates for fermentative microorganisms, which convert these products to organic acids and alcohols. SRB can utilize the acids and alcohols while reducing sulfate to sulfide. The sulfide reacts with dissolved metals to form insoluble metal sulfide precipitates. In addition, biological sulfate reduction produces alkalinity, which neutralizes the acidity in AMD. Lignocellulose contains numerous reactive groups, such as phenolic, hydroxyl, and carbonyl groups, which can bind to metals and remove them from contaminated water (Magowo, 2014).

The abundance of these lignocellulosic wastes in regions where AMD is generated determines their suitability for an application. Podja et al. (2017) discussed the potential of using cassava peel to treat AMD in Mozambique. They concluded that this was a viable option due to the abundance of cassava in Mozambique. In South Africa, AMD is prevalent in the Witwatersrand region due to the legacy of gold mining activities. In addition, AMD is produced in Mpumalanga and the KwaZulu Natal region due to coal mining activities. These regions are situated in the Grassland and Savannah biomes that are rich in long-stem grasses, which are a good source of lignocellulose biomass (Burman et al., 2019). In addition, The Mpumalanga and KwaZulu Natal regions are producers and processors of agricultural crops like sugarcane bagasse (Magowo et al., 2020a,b). The lignocellulose waste obtained after harvesting and processing the crops can be pretreated with AMD in preparation for enzyme hydrolysis and fermentation to produce biofuels and other valuable products (Magowo, 2014).

8.3 Lignocellulose and biofuels

Lignocellulosic biomass refers to organic-based plant material sourced from forest and forest residues, agricultural residues, and dedicated energy crops (Hisham and Megeed, 2008).

Plant-based biomass has been used to generate clean power and is already the fourth largest energy source in the world after oil, coal, and gas. The fuels generated from biomass are called biofuels (Lee and Lavole, 2013). The two most common biofuels in use today are ethanol and biodiesel. Ethanol has been produced from starches and sugars through the fermentation of C6 sugars (mostly glucose) using yeast strains such as *Saccharomyces cerevisiae* (Lee and Lavole, 2013). The most common feedstock used to produce bioethanol is sugarcane and corn, while whey, barley, sugar beet, potatoes, and rice can also be fermented to ethanol. Fuels produced from plant material dedicated to food are termed first-generation fuels. However, first-generation fuels are unsustainable because of the potential stress that their production places on food commodities, notably on food price increases (Burman et al., 2019).

Research has been ongoing in developing new technologies to allow biofuels to be obtained from lignocellulose, the fibrous material that makes most plants' bulk. Biofuels produced from nonfood plant material are termed second-generation biofuels (Havlik et al., 2011; Burman et al., 2019). The production of cellulosic biofuels is attractive and sustainable because it does not compete with food crops, is less water and land intensive, and can be produced quickly at a significantly lower cost (Carriquirry et al., 2010; Ray and Ramachandran, 2018). Lignocellulose can also be obtained from agricultural waste as it constitutes the nonedible portion of the plant, which is currently underutilized (Huber, 2008) and is usually disposed of through burning, contributing to environmental pollution. Westernsee et al. (2018) estimate that for every ton of maize, wheat, sugarcane, or sunflower seeds, an estimated ton of dry residue is produced, leading to the production of 12.2 Mt/year of agricultural residue and 9.14 Mt/year of lignocellulosic residue in South Africa alone.

Lignocellulose material is composed of cellulosic fibrils surrounded by a matrix of hemicelluloses and lignin (Burman et al., 2019). Cellulose is a major component of cell walls and provides mechanical strength and chemical stability. Plants absorb solar energy through photosynthesis and store it in the form of cellulose. Cellulose is a linear polymer of glucose molecules linked together through B-1−4 bonds. Hemicellulose is a copolymer of different C5 and C6 sugars that coexist in the plant cell wall (Magowo, 2014; Zuliani et al., 2021). Lignin is a polymer of aromatic compounds, which form a protective layer for the plant cell walls.

Furthermore, minor amounts of proteins, minerals, and other components can be found in lignocellulose composition. Every plant consists of different ratios of cell wall compositions in different parts of the plant (Magowo, 2014; Zuliani et al., 2021). The main compositions of typical grass biomass are cellulose, xylan, galactan, arabinan, lignin, and mannan.

The conversion of lignocellulose to valuable products is significantly hindered by structural and chemical complexity. For example, lignin acts as a physical barrier to restrict access of chemicals and cellulolytic enzymes to cellulose. It may inhibit the activity of cellulase by irreversibly adsorbing to cellulase enzymes (Zuliani et al., 2021). Hemicellulose also acts as a physical barrier to limit the accessibility of cellulase (Zhao et al., 2012). Cellulose and hemicelluloses, when hydrolyzed into their component sugars, can be converted to ethanol or other valuable products through well-established fermentation technologies. However, the sugars in cellulose are challenging to recover because they are trapped inside the cross-linking structure of the lignocellulose. Enzymatic hydrolysis can, therefore, not be used successfully alone. A pretreatment procedure is necessary to modify and/or remove the surrounding matrix of lignin and hemicelluloses prior to enzymatic hydrolysis. Pretreatment refers to a process that converts lignocellulose from its native form, in which it is recalcitrant to cellulase enzyme systems, into a form, which can be easily hydrolyzed by cellulase. The pretreatment reduces cellulose crystallinity and increases the accessible surface area to enzymes. Thus, improving the formation of sugars by creating a reactive cellulose fiber for enzymes to attack (Sun and Cheng, 2002). The lignocellulose degrading enzymes work under mild conditions and therefore have lower utility costs, the enzymes themselves, however, are costly. This limits their application on a commercial scale (Duff and Murray, 1996).

The pretreatment of lignocellulosic biomass is an important step in the recovery of monomeric sugars through the enzymatic digestion of the lignocellulosic residue. Several pretreatment methods have been applied to lignocellulosic biomass before enzymic digestion to recover fermentable sugars for the production of biofuels. These treatment methods include physical (milling), chemical, and biological (Kristiani et al., 2013).

After the pretreatment step, the lignocellulosic residue is subjected to further enzymatic hydrolysis. In this process, cellulose is converted to glucose using cellulase. This process is termed enzymatic saccharification. Hemicellulose is a branched polymer composed of pentose [5-carbon (5C)] and hexose [6-carbon (6C)] sugars. Hemicellulases or acids can be used to hydrolyze the hemicelluloses to release their component sugars, including xylose, arabinose, galactose, glucose, and/or mannose. Hexoses such as glucose, galactose, and mannose are readily fermented to ethanol by naturally occurring microorganisms. Still, the pentose, including xylose and arabinose, is fermented to ethanol by native strains and usually produces low ethanol yields (Petrova and Ivanova, 2010, cited in Magowo, 2014).

The mixed 5C and 6C platforms produced by the hydrolysis of hemicelluloses can be used to produce a variety of organic acids. Microbial production of organic acids may be used to produce building blocks for industrial processes. Most acids are natural products of microorganisms or at least natural intermediates in major metabolic pathways (Magowo, 2014). Of particular interest in biorefinery are those platform intermediates from fermentation that can be converted into numerous consumer and industrial products, including succinic acid, butanol, itaconic acid, 1,3-propanediol, polyhydroxypropionic acid, lactic acid, citric acid, gluconic acid, and malic acid. Because of their functional groups, these organic acids are beneficial as starting materials for the chemical industry (Magowo, 2014). Lactic acid, for example, is used as a starting material for biodegradable and biocompatible polylactic polymers, an environmentally friendly alternative to nonbiodegradable plastics derived from petrochemicals. Another promising substance within organic acids is itaconic acid. Itaconic acid can be regarded as a substituted acrylic or metacyclic acid and is isomeric with cis-itaconic and mesaconic acids (Blatt, 1943). Itaconic acid is industrially produced by the fermentation of starch hydrolysates or molasses by *Aspergillus niger* and *A. itaconic* (Kubicek and Miskiewisc, 2012). Its polymerized esters (methyl, ethyl, and vinyl) are used in adhesives and coatings. Itaconic acid is also used in emulsion paints to aid in polymer adhesion and as a hardening agent for organo-siloxanes, which are used as contact lenses (Okabe et al., 2009). Itaconic acid has two carbonyl groups, which make it suitable to be incorporated into polymers. The economics of producing these chemical precursors depend on many factors, of which the cost of raw materials is the most significant. It is expensive when sugars such as commercial-grade glucose, sucrose, and starch are used as feedstock. Lignocellulosic biomass is a promising feedstock in the production of these chemicals.

In previous work, lignocellulosic biomass, including hay, alfalfa, grass, sawdust, and woodchips, has successfully treated AMD. The results indicated that grasses hold promise for remediating AMD, as a maximum of 99% iron removal, 80% sulfate removal, and a final pH of 8.5 was achieved from initial conditions of 2000 mg/L of dissolved iron 6000 mg/L sulfate, and a pH of 3. The pH was tested using universal indicator paper. Iron and sulfate were analyzed using Merck test kits (No. 114761 for iron and 114791 for sulfate) and the Merck Spectroquant (Sheridan et al., 2013). The residual lignocellulosic biomass left after AMD treatment still contains a lot of cellulose, a potential source of reducing sugars for the production of biofuels. The residue is currently treated as waste and needs to be disposed of. This waste can be further processed to recover sugars.

8.4 Case study: the utilization of acid mine drainage for the pretreatment of lignocellulosic biomass and the subsequent recovery of fermentable sugars

This study hypothesized that the lignin fraction of grasses and other cellulosic feedstocks contain numerous reactive groups which bind to and remove metals from AMD. Simultaneously the sulfuric acid in the AMD hydrolyzes hemicellulose and exposes the cellular structure of cellulose, whereby it can be subjected to a subsequent enzymatic hydrolysis step. In this study, we assess the potential use of the acidity present in AMD to (at least) partially hydrolyze cellulose. This pretreatment of lignocellulosic material is analogous to dilute acid pretreatment which uses commercial sulfuric acid (Kristiani et al., 2013) and would result in partially remediated AMD and a fermentable feedstock that could be utilized to produce, for example, bioethanol.

This study tested two different cellulosic feedstocks: sugarcane bagasse and switch grass.

8.4.1 Lignocellulose biomass

Sugarcane bagasse was obtained initially from Illovo Sugar Company based in Kwazulu Natal, South Africa. The switch grass, *Panicum virgitum* (an indigenous perennial grass), was milled at the School of Chemical and Metallurgical Engineering laboratory at the University of the Witwatersrand. The second batch of switch grass was collected from the African Leadership Academy Garden. It was obtained already dry.

8.4.2 Enzyme

The enzyme (cellulase) was supplied by Yakult Pharmaceutical Ind. Co., Ltd from Tokyo, Japan. The enzyme was obtained in powdered form. The product name of the enzyme is Cellulase "Onozuka" FA. The enzyme was composed of cellulase at 12% and lactose at 88%.

8.4.3 Methodology

8.4.3.1 Pretreatment of lignocellulose

A Synthetic AMD feed with 500 and 3000 mg/L of dissolved iron and sulfate, respectively, and a pH of 2.1 was used for this investigation. The pH, iron, and sulfate content of the AMD was modeled on the findings of Bell et al. (2001). Dried sugarcane bagasse (particle size <5 cm) and switch grass (particle size <2 mm and 2–5 cm) were separately used as the organic source. The AMD was mixed with the organic material at a liquid/solid ratio of 13:1. The reactors were inoculated with SRB obtained from passive bioreactors on the roof of the Richard Ward building at the University of The Witwatersrand. The reactors were sealed and allowed to incubate anaerobically at ambient temperature for 14 weeks. The pH of the AMD exposed to the biomass was measured using a benchtop pH meter. The iron concentration was analyzed using the Merck test kit 114761 and the Merck spectroquant Pharo 300 according to the manufacturer's manual. Reducing sugars were measured using a 1200 series Agilent high-performance liquid chromatography (HPLC). Scanning electron microscopy (SEM) was done using an FEI Quanta 400 E-SEM.

8.4.3.2 Determination of glucose

Samples were taken at weekly intervals to determine the release of glucose during pretreatment. The samples were incubated at 95°C for 3 h in a benchtop water bath to complete the hydrolysis. After cooling, the samples were neutralized with sodium hydroxide to a pH between 5.5–6.5. The samples were centrifuged for 20 min in a desktop centrifuge (model HKSC-220 globe). Centrifugation was done to remove the precipitated iron sulfide and other solids (lignin) from the liquid containing glucose. This was followed by sterile filtration using a 0.45 microfilter. The amount of glucose in the filtrate was determined using HPLC.

8.4.3.3 Enzyme hydrolysis

The milled switch grass (<2 mm) was chosen for enzymatic hydrolysis as it affected the most significant remediation of the simulated AMD. The lignocellulose residue was washed to remove adsorbed metals and divided into 8 g portions. The 8 g lignocellulose portions were autoclaved at 121°C for 20 min. The portions were subjected to different enzyme concentrations ranging from 0% to 10% using a buffer at pH 6. These were incubated at 50°C with shaking for 24 h. Clear filtrates containing the digest were obtained by filtration using 0.45 microfilters. Filtrates were analyzed using HPLC.

8.4.4 Results

8.4.4.1 Dissolved iron and pH changes

Figs. 8.1 and 8.2 show the increase in pH and the decrease in the dissolved iron concentration during the treatment of AMD using different lignocellulosic biomass over the 14-week experimental period.

The data presented in Figs. 8.2 and 8.3 indicate that the three sources of cellulosic biomass could affect some AMD remediation. In all three cases, dissolved iron concentration was reduced, and pH was increased. The milled switch grass (≤2 mm) was the most effective in the remediation of AMD. The milled switch grass removed 65.2% of dissolved iron and increased the pH from 2.11 to 5.30. The maximum pH was obtained in a relatively shorter period (5 weeks). The switch grass (5 cm) removed 42% dissolved iron and increased the pH from 2.11 to 3.86. Sugarcane bagasse had the least remedial effect on AMD. Sugarcane bagasse removed 28% of iron from AMD and increased the pH from 2.11 to 2.36; this was achieved in 14 weeks. The pH and the iron concentration of the control, which contained only AMD, did not change much during the experimental period, indicating

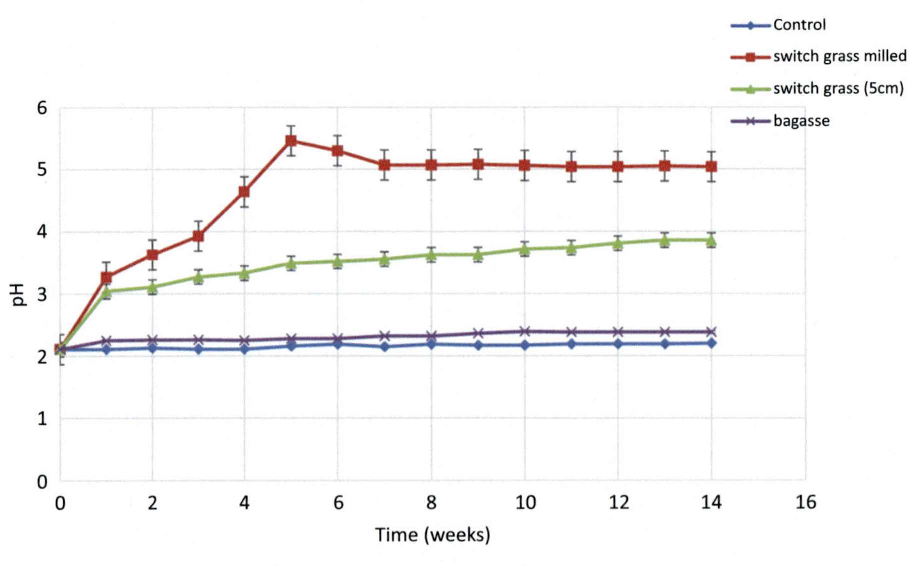

Figure 8.1 Changes in pH as affected by different types of lignocellulosic biomass.

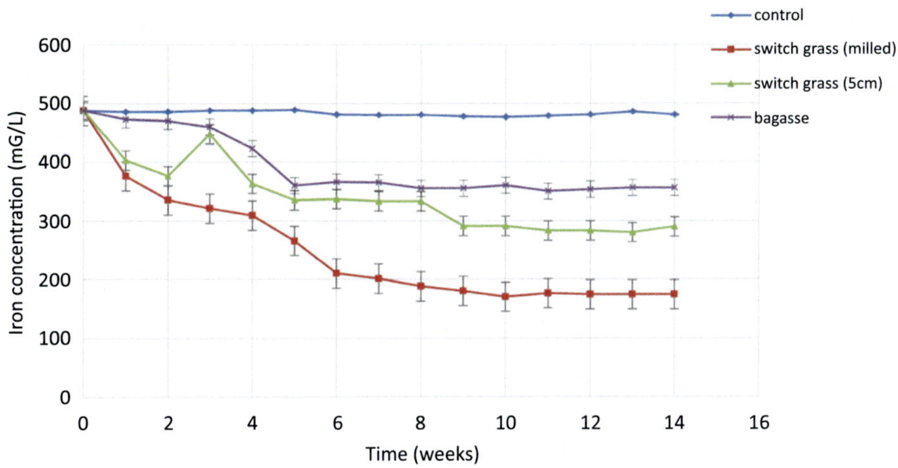

Figure 8.2 Changes in iron concentration as affected by different types of lignocellulosic material.

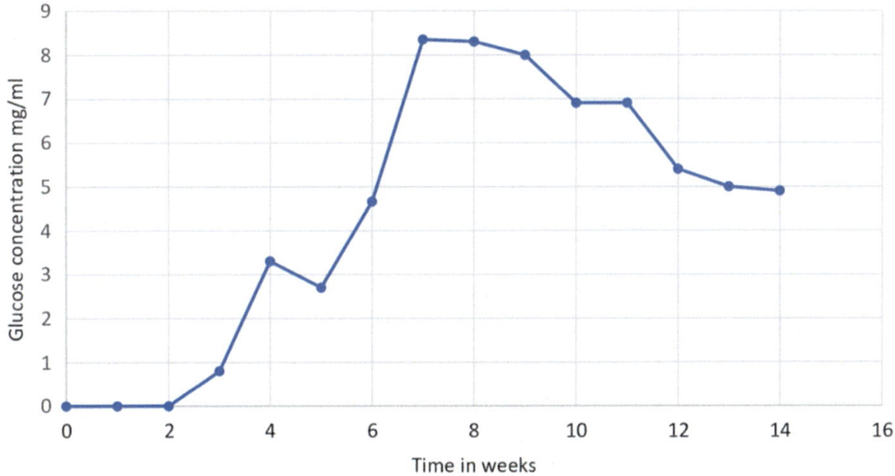

Figure 8.3 The release of glucose during treatment of acid mine drainage using milled switchgrass.

that the biomass (switch grass and bagasse) caused the increase in alkalinity and decreased dissolved iron concentration. Similar results were obtained by Ramla (2012). Ramla's investigation used two indigenous grass species, *Hyparrheia hirta*, and *Setaria spacenta*. Maximum pH values of 8.48% and 99% percentage removal of iron from the AMD solution was reported in that study.

These values are comparable with the ones reported in the literature for the biosorption and precipitation of metals on switch grass and other biomasses. According to the literature, the main mechanisms of metal ion removal in bioreactors are precipitation in sulfides, hydroxides, and carbonates and sorption mechanisms such as adsorption (Neculita et al., 2007). The most significant effect on remediating the AMD was observed for the milled switchgrass. The decrease in dissolved iron in the solution can be explained by the presence of many active sites on the milled switch grass due to the increased surface area caused by milling. Functional groups capable of metal sorption in biomass include phenolic, hydroxyl, and carboxylic groups. Milling exposes these functional groups to dissolved ions. These functional groups in the biomass are protonated at higher pH and presumably available for binding dissolved metals (Neculita et al., 2007). This explains that the increase in pH is accompanied by a marked decrease in dissolved iron concentration. Increases in the pH of acidic water effectively remove some metals due to the precipitation of metals in the form of hydroxides (Goatham, 2013).

Microbial activity also plays a role in the remediation of AMD. Microbial activity was observed in the two switch grass series, with the milled switch grass seeming to support more growth. It is most probable that this microbial population is comprised of SRB. Ramla (2012) explains that the increase in pH can be directly related to SRB population growth. The substrate may have been providing nutrients to SRB, which contributed to the development of the microbial population. The SRB activity reduces sulfate to sulfide in the presents of an organic source and produces hydrogen carbonate [Eq. (8.2)]. The carbonate generated by biological sulfate reduction helps in precipitating metals as carbonates, as well as improving pH by removing acidity from the AMD [Eq. (8.4)], while the hydrogen sulfide reacts with dissolved metals in the AMD, resulting in the precipitation of metals as metal sulfides [Eq. (8.3)] (Taylor, 2005).

$$SO_4^{2-} + 2CH_2O + 2H^+ \rightarrow H_2S + 2H_2CO_3 \tag{8.2}$$

$$H_2S + M^{2+} \rightarrow MS + 2H^+ \tag{8.3}$$

$$HCO_3^- + H^+ \rightarrow CO_2 + H_2O \tag{8.4}$$

where CH_2O is the sugars, M^{2+} and MS represent dissolved metal ions and metal sulfide precipitates, respectively.

8.4.4.2 The release of glucose during the treatment of acid mine drainage using milled switch grass (≤2 mm)

The release of glucose was observed only in the milled switch grass, suggesting that milling was a significant factor in the generation of glucose. Milling reduces cellulose crystallinity and disrupts the lignin-carbohydrate complexes. This aids the acid treatment process and the subsequent enzyme hydrolysis (Maise et al., 2002: Canam et al., 2013). The change in concentration of glucose is shown in Fig. 8.3. Glucose started to appear in the third week of treatment, where a rapid increase was observed. The maximum amount of glucose obtained was 8.4 mg/mL in the seventh week. After that, the concentration decreased to 4.9 mg/mL after 14 weeks. The consumption of glucose by SRB may explain the decrease in glucose concentration.

The generation of glucose and other reducing sugars during AMD treatment is attributed to acid catalysis, where the sulfuric acid in the AMD catalyzes the breakdown of cellulose and hemicellulose to release glucose, xylose, and other degradation products. Hemicellulose, comprised mainly of mannan and xylan, accounts for up to a third of the total carbohydrate in many lignocellulosic materials (Pingali et al., 2010). According to the literature, the cellulose content of switch grass is estimated between 36% and 38%. In comparison, hemicellulose content is estimated at 28% of the total biomass, making switch grass suitable for producing fermentable sugars for bioprocessing.

In a study by Roman (2004), high levels of reducing sugars (306 g/L) were produced. In Roman's study, production was also followed by consumption. The consumption was accompanied by an increase in pH from 5.8 to 7.0. The increase in pH was related to sulfate reduction and alkalinity production within the bioreactor. Burman et al. (2019) achieved ethanol concentration between 4.11 and 14.82 g/L by fermenting the hydroxylate from AMD-pretreated biomass residue after the subsequent enzyme hydrolysis step. Mohapatra et al. (2017) reported ethanol yields of between 4.72 and 16 g/L from grass pretreated with AMD.

8.4.4.3 Scanning electron microscope analysis of switch grass treated by water and acid mine drainage before and after enzymatic hydrolysis

SEM was used to study the morphological features and surface characteristics of the switch grass treated with water and AMD before and after enzyme hydrolysis (Fig. 8.4).

The SEM images reveal that pretreatment induced significant morphological changes in the switch grass. Water-treated switch grass exhibited rigid and highly ordered fibrils (Fig. 8.4A). The AMD-treated switch grass fibers appear distorted, with some holes on the surface (Fig. 8.4B). This renders the switch grass more accessible to cellulase, resulting in higher enzymatic hydrolysis rates and cellulose digestibility (Moxley et al., 2008). The molecular structure of the pretreated and enzyme-treated switch grass remained intact, showing no digestion occurred during the hydrolysis of the water-treated switch grass (Fig. 8.4C). The switch grass's molecular structure, which was pretreated by AMD followed by enzyme digestion, was destroyed (Fig. 8.4D). The improved/altered morphological properties caused by AMD pretreatment of switch grass appeared to be the primary source for enhancing enzymatic hydrolysis (Goshadrou et al., 2011). These micrographs agree with the HPLC results in Table 8.1, which offer high digestibility of AMD-treated switch grass and little digestibility in the water-treated switch grass.

8.4.5 Enzymatic hydrolysis of acid mine drainage treated and untreated switch grass

Table 8.1 shows the concentrations of xylose and glucose after the enzymatic hydrolysis of the water-treated and AMD-treated switch grass as analyzed using HPLC.

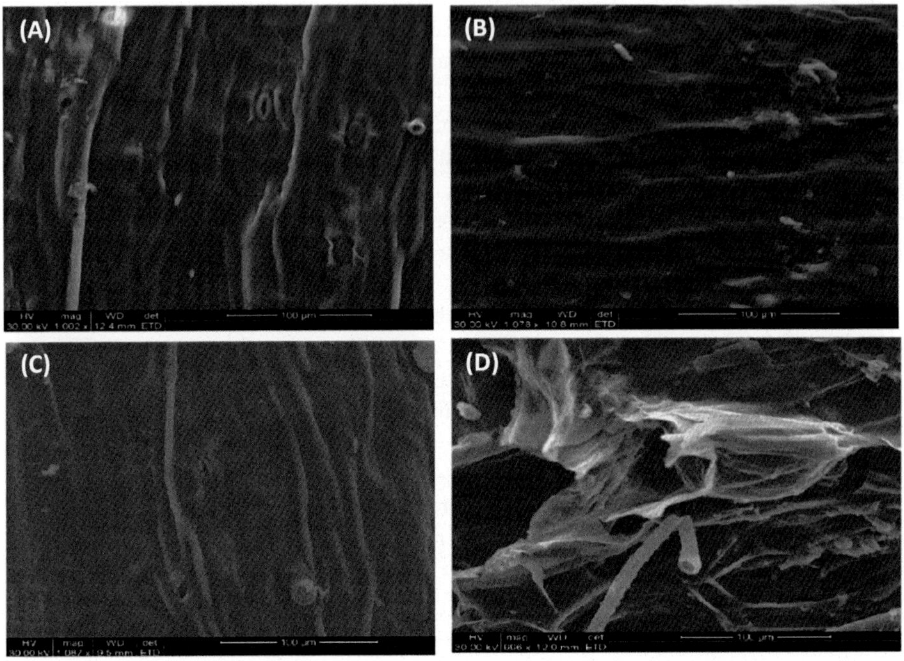

Figure 8.4 Scanning electron microscope images of switch grass (A) water treated, (B) acid mine drainage (AMD) treated, (C) water treated followed by enzyme treatment and (D) AMD treated followed by enzyme treatment.

Table 8.1 Enzymatic hydrolysis of water-treated and acid mine drainage-treated switch grass.

Enzyme (%)	**Water treated**		**Acid mine drainage treated**	
	Xylose (mg/mL)	**Glucose (mg/mL)**	**Xylose (mg/mL)**	**Glucose (mg/mL)**
0.5	0	0	1.9	0
2.5	0	0	3.3	3.8
5	0	0	0	4.8
10	1.4	0	3.4	3.7

The water-treated sample was the control for this experiment. It is noted from Table 8.1 that the glucose yield for the water-treated switch grass was zero for all the enzyme concentrations. The xylose yield was zero except for the 10% enzyme concentration, which had a xylose concentration of 1.4 mg/mL. The SEM images (Fig. 8.4) showed that the water-treated switch grass displayed no obvious mechanical disruption of the general surface morphology. This rigid and compact structure of untreated lignocellulosic biomass hinders the effectivity of cellulase and prevents it from digesting the grass. The enzymatic digestion of the AMD-treated switch grass shows the concentrations of glucose and xylose in mg/mL. The maximum sugar concentration achieved with the milled switch grass was 4.8 mg/mL glucose at 5% enzyme concentration and 3.4 mg/mL xylose at 10% enzyme concentration. The enzyme concentration had a significant effect on the amounts of glucose and xylose obtained. There is a general increase in the glucose and xylose concentration with the increase in enzyme concentration. Small quantities of enzymes were ineffective in digesting the treated switch grass. The AMD treatment of the switch grass was, therefore able to significantly improve the accessibility of the cell walls to enzymatic digestion, leading to greater sugar yields than the water-treated switch grass.

8.5 Discussion and conclusion

Switchgrass has been shown in this study to have the capacity to remove iron and acidity from AMD effectively. The results indicate that milled switch grass (2 mm) was more effective in remediating AMD than sugar cane bagasse or the large particle size switch grass (<5 cm). The potential of switchgrass, sugarcane bagasse, and other lignocellulosic material to remove metals by adsorption is expected since materials contain functional groups such as phenolic, hydroxyl, carboxyl, and carboxylic (Harman et al., 2007). This theory is supported by Munawar and Riwandi (2010). They attributed the decrease in dissolved iron in the first two weeks of AMD treatment to iron retention by organic matter rather than iron sulfide precipitation. In this study, the milled switch grass, which had a higher surface area, retained more iron resulting in lower dissolved iron concentration. Zagury et al. (2006) (cited in Munawar and Riwandi, 2010) explained that metal removal tends to occur sooner than sulfate reduction through the adsorption of metals to organic matter. Once reducing conditions are established, metal precipitation rather than adsorption becomes the predominant removal mechanism. Munawar and Riwandi (2010) reported a decrease in sulfate concentration in the third week of their study, using chicken manure mixed with AMD. This was attributed to sulfate reduction and the precipitation of metal sulfides. In our study, microbial activity seemed to dominate in the third week of AMD treatment, supporting Munawar and Riwandis' (2010) findings that sulfate reduction occurs at a later stage of AMD treatment. SRB activity produces hydrogen carbonate, which increases the alkalinity of the AMD.

The milled switch grass was able to increase the pH of the AMD from 2.11 to 5.30 and remove dissolved iron from the AMD by 65.2% by the fifth week. Both the pH and the dissolved iron content started to stabilize thereafter. To further increase the alkalinity and reduce the dissolved iron concentration, periodic substrate loading may be considered in the future for the long-term remediation of AMD (Ramla, 2012). Santos et al. (2004) worked with different biomass concentrations. They showed that an increase of the biomass (grape stalks) concentration up to 30 g/L leads to a 20% reduction of the total content of dissolved iron in the

AMD. In a similar study, Greben et al. (2008) used to cut grass serving as a carbon and energy source for continuous sulfate removal and obtained between 80% and 90% removal efficiency by adding cut grass loadings every two weeks. In addition, the increase in grass loadings provides a new surface area for the adsorption of metal ions, further reducing their concentration.

In our study, the residue switch grass obtained from the AMD treatment process was prepared for a further enzymatic hydrolysis step. The preparation included washing to remove metal ions adsorbed to the surface. SEM was used to study the effects of AMD on the switch grass. The results showed that AMD induced morphological changes in the switch grass (Fig. 8.4). These changes are due to the hydrolysis of hemicellulose and, to a lesser extent, cellulose by the sulfuric acid in the AMD. The hemicellulose fraction is converted to monomeric sugars, including pentose and glucose. This treatment is analogous to the pretreatment of lignocellulosic biomass using acid (Burman et al., 2019). In literature, dilute acid pretreatments of lignocellulosic material are done at higher temperatures and pressure and can recover relatively higher amounts of fermentable sugars from the hemicellulose fraction of lignocellulose. Barrier et al. (1985) reported the conversion and recovery of sugars from the hemicellulose fraction treated with dilute acid to be more than 90% efficient. Dilute sulfuric acid in the range of 0.5%–1.5% and temperatures above 160°C have been the most favored in industrial applications. Under these conditions, high sugar yields from hemicellulose are obtained (Hisham and Megeed, 2008). In our study, the conditions are mild (room temperature and pressure), so we expected reduced rates.

Glucose was detected during the treatment of AMD during this study (Fig. 8.3). A maximum glucose concentration of 8.4 mg/mL was obtained in the seventh week of the study. After AMD treatment, some of the sugars from the AMD/sugar liquor must be recovered. The sugars in the solution in the AMD are highly contaminated with metal ions and would need purification if they are recovered and used for further bioprocessing. Recovering the hemicellulose fraction of sugars from the AMD/sugar solution is complex. The AMD/sugar liquor can be fermented to produce ethanol (Burman et al., 2019) which can be recovered by distillation. Alternatively, the AMD/sugar liquor could be used as a carbon and energy source for SRB in subsequent treatments. Instead of being discarded, the remaining cellulose residue could be further processed by additional enzymatic hydrolysis as the cellulose residue contains the bulk of the glucose. Cellulase enzymes were used in our study to recover some of the sugars in the lignocellulose, with maximum values of glucose and xylose at 4.8 and 3.4 mg/mL, respectively. The xylose produced in the processing may also be fermented or used to produce other value-added products. One possible use of the unconverted xylose is to produce fodder yeast (single cell protein) for animal feed (Barrier et al., 1985).

The yields obtained in this study compare well to the yields obtained by Burman et al. (2019), even though Burman et al. (2019) used higher temperatures in the pretreatment process. This study has evaluated the potential of AMD-treated switch grass in the production of fermentable sugars. From this study, it can be concluded that:

Lignocellulosic material can treat AMD:

1. The AMD-treated switch grass has the potential to produce fermentable quantities of glucose upon enzymatic hydrolysis.
2. AMD is not only remediated but assists in a bioprocess to prepare second-generation feedstock for industrial biotechnology.

The research is, however, at a very early stage of development. Substantial additional work is required to optimize the production of fermentable sugars and understand the rates of AMD treatment. Much work is still needed to develop a process that could be used commercially. This is the subject of ongoing research.

References

Ackil, A., Kodas, S., 2006. Acid mine drainage: causes, treatment and case studies. J. Clean. Prod. 14 (12–13), 1139–1145.

Barrier, J.W., Moore, M.R., Farina, G.E., Broder, J.D., Forsythe, M.L., 1985. Experimental production of ethanol from cellulosic material at low temperature acid hydrolysis. Biomass Congresses 1985, 587–600.

Bell, F.G., Bullock, S.E.T., Halbich, T.F.J., 2001. Environmental impact associated with an abandoned mine in the Witbank coalfields. South Africa. Int. J. Coal Geol. 11–25.

Blatt, A.H., 1943. Org. Synth. 11, 328.

Blodau, C., 2006. A review of acidity generation and consumption in acid coal mine lakes and their watershed. Sci. Total Environ. 369 (1–3), 307–332.

Burman, N.W., Sheridan, C.M., Harding, K.G., 2019. Lignocellulosic bioethanol production from grasses pretreated with acid mine drainage. Modelling and comparison of SHF and SSF. Bioresour. Technol. Rep. 7 (2019). Available from: https://doi.org/10.1016/J.biteb.2019.100299.

Canam, T., Town, J., Iroba, K., Tabil, C., Dumachdeaux, T., 2013. Pretreatment of lignocellulosic biomass using microorganisms: approaches, advantages and limitations. pp. 181–189. Available from: https://doi.org/10.5772/55088.

Carriquirry, M.A., Du, X., Timilsina, G.T., 2010. Second generation biofuels economics and policies. Policy Res. 5406.

Chen, R., Zhu, S., Cheng, B., Chen, J., Wu, Y., 2014. Reviving the acidic hydrolysis process of lignocellulosic material in biorefinery. BioResources. 9 (2), 1824–1827.

Cirne, D.G., Van Der Zee, F.P., Fernandez-Polanco, M., Fernandez-Polanco, F., 2015. Control of sulfide during anaerobic treatment of s-containing wastewater by adding limited amounts of oxygen and nitrate. Rev. Environ. Sci. Biotechnol. 7, 93–105.

Duff, S.J.B., Murray, W.D., 1996. Bioconversion of forest industry waste cellulosics to fuels: a review. Bioresour. Technol. 55 (1), 1–33.

Fahruddin, K., Abdulla, A., Nurhaedar, L.A., Nafie, N., 2020. Estuary sediment treatment for reducing sulfate in acid mine wáter. Environ. Nat. Resour. J. 18 (2), 191–199.

Goatham, V., 2013. Treatment methods of acid mine drainage and case study on selective recovery of metals. 13M160R09. Min. Eng. 1–31.

Goshadrou, A., Karimi, K., Taherdazeh, M.S., 2011. Improved of sweet sorghum bagasse hydrolysis by alkali and acidic pretreatments. World Renewable Energy Congress. Linkoping, Sweeden.

Greben, H.A., Baloyi, J., Sigama, J., Venter, N.S., 2008. Bioremediation of sulfate rich effluents using grass cuttings and rumen fluid microorganisms. J. Geochem. Exploration. Available from: https://doi.org/10.1016/j.gexpo.2008.01.004.

Greben, H., Sigama, J., Burke, L., Venter, S.N., 2009. Cellulose fermentation products as energy source for biological sulfate reduction of acid mine drainage type wastewater. Rep. Water Res. Comm. 5–7.

Harman, G., Patrick, R., Spittler, T., 2007. Removal of heavy metals from polluted waters using lignocellulosic agriculture waste products. Ind. Biotechnol. Winter. 366–374.

Havlik, P., Schneider, A.W., Schmid, E., Bottcher, H., Fritz, S., Skasky, R., et al., 2011. Global land use implications of first and second-generation biofuels targets. Energy Policy 39, 5690. -5602.

Hisham, F.M., Megeed, M.E., 2008. Lignocellulosic biomass conversión technologies to biofuels potentials in Egypt. A report submitted to UNIDO-Cairo, Egypt IMC. 6–20.

Huber, G.W. (ed.), 2008. Breaking the chemical and engineering barriers to lignocellulosic biofuels. University of Massachusetts Amherst. National Science Foundation, Chemical Bioengineering, and Transport System División, Washington D.C., p. 180.

Jennings, S.R., Neuman, D.R., Blicker, P.S., 2008. Acid Mine Drainage and Effects on Fish Health and Ecology: A Review. Reclamation Research Group Publication, Bozeman, MT, pp. 1–29.

Johnson, D.B., Hallberg, K.B., 2005. Acid mine drainage remediation options: a review. Sci. Total. Environ. 338, 3–14.

Kosinka, K., Miskiewisc, Z., 2012. Precipitation of heavy metals from industrial wastewater by *Desulfovibrio desulfuricans*. Environ. Prot. Eng. 38 (2), 52–62.

Kousi, P., Remoundaki, E., Hatzikioxylan, A., Tsezos, M., 2015. Sulfate reducing bioreactors: current practices and perspectives. IWA Balk. Young Wáter Professionals 409–413.

Kubicek, C.D., and Miskiewisc, Z., 2012. Fungi Cellulosic Biomass. pp. 223–224.

Kristiani, A., Abimanyu, H., Setiawan, A.H., Sudiyarmanto, Aulia, F., 2013. Effect of pretreatment process by using dilute acid characteristics of oil palm's frond. Energy Proc. 32 (2013), 183–189.

Kuyacak, N., 2011. Minewater Treatment: Application Policy. pp. 360–390.

Lee, A.R., Lavole, J.N., 2013. First to third generation biofuels: challenges of producing a commodity from biomass of increasing complexity. Anim. Front. 3 (2), 6–11.

Macinova, E., Luptakova, A., 2012. Recovery of metals from AMD. Chem. Eng. Trans. 28, 109–114.

Magowo, W.E., 2014. The utilization of cellulosic biomass in the treatment of AMD and the subsequent production of sugars for bioprocessing. Dissertation.

Magowo, W.E., Sheridan, C., Rumbold, K., 2020a. Global co-occurrence of acid mine drainage and organic rich industrial and domestic effluent: biological sulfate reduction as a co-treatment option. J. Water Process. Eng. 38 (2020). Available from: https://doi.org/10.1016/j.jwpe.2020.101650.

Magowo, W.E., Sheridan, C., Rumbold, K., 2020b. Bioremediation of acid mine drainage using Fischer Tropsch wastewater as a feedstock for dissimilatory sulfate reduction. J. Water Process. Eng. 35 (2020). Available from: https://doi.org/10.1016/j.jwpe.2020.10.1219.

Maise, U., Esteghhlalian, A., Saddler, J., Mansfield, S., 2002. Enhancing enzymatic hydrolysis of cellulosic material using simultaneous ball milling. Appl. Biochem. Biotechnol. 98, 815–839.

McCarthy, T.S., 2011. The impact of acid mine drainage in South Africa. South. Afr. J. Sci. 107 (5/6), 1–7.

Mohapatra, S., Mishra, C., Behera, S.S., Thatol, H., 2017. Application of treatment, fermentation, and molecular techniques for enhancing bioethanol production from biomass – a review. Renew. Sust. Energ. Rev. 78, 1007–1032.

Moxley, A., Zhu, Z., Zhang, P.Y.-H., 2008. Efficient sugar reléase by the cellulose solvent-based lignocellulose fractionation technology and enzymatic cellulose hydrolysis. J. Agric. Foodchem. 57, 7885–7890.

Mukwevho, M.J., Chirwa, E.M.N., Maharaj, D., 2019. Effect of pH and temperature on biological sulfate reduction. Chem. Eng. Trans. 74, 517–522.

Munawar, A., Riwandi, 2010. Chemical characteristics of organic wastes and their potential use for acid mine drainage. J. Nat. Indonesia 12 (2), 167–172.

Neculita, M.C., Zagury, G.J., Bussire, B., 2007. Effectiveness of sulfate reducing passive bioreactors for treating highly contaminated acid mine drainage: metal removal mechanisms and potential mobility. Appl. Geochem. 23, 3545–3560.

Neto, E.S.C., Aquiar, A.B.S., Rodriguez, R.P., Saneti, G.P., 2018. Acid mine drainage treatment and metal removal based biological sulfate reduction process. Braz. J. Chem. Eng. 35 (2), 543–552.

Okabe, M., Lies, D., Kanamasa, S., Park, E., 2009. Biotechnological production of itaconic acid and its biosynthesis in *Aspergillus terreus*. Appl. Microbiol. Biotechnol. 84 (4), 597–606.

Petrova, P., Ivanova, V., 2010. Perspectives for the production of bioetanol from lignocellulosic material. Biotechnol. Biotechnol. Equip. 24 (01), 529–546.

Pingali, S.V., Urban, V.S., Heller, W.T., 2010. Breakdown of cell wall nanostructure in dilute acid pretreated biomass. Bio Macromol. 11 (9), 2329–2335.

Podja, A.E., Persson, K.M., Masinhe, D.M., 2017. The potential use of cassava peel for treatment of AMD in Mozambique. J. Environ. Prot. 8, 277–289.

Ramla, B., Sheridan, C., 2014. The potential of indigenous South African grasses for AMD remediation. In Conference Paper: Water Institute of South Africa (WISA) Biennial Conference, Mbombela, 25–29 May 2014.

Ramla, B., 2012. Potential utilization of indigenous grass for AMD remediation. Master's thesis, University of the Witwatersrand, p. 122.

Ray, R.C., Ramachandran, S., 2018. Bioethanol production from food crops: sustainable sources. Interventions, and Challenges. Academic Press.

Rob, G.A., Robson, J.D.F., 1995. Acid drainage from mines. Geograph. J. 107 (5/6), 38–39.

Roman, 2004. Degrad. Lignocellulose A Biologically-Generated Sulfidic Environ. pp. 1–133.

Saharan, M.R., Gupta, K.K., Jamal, A., Sheoran, A.S., 1995. Management of acidic effluents from tailings dams in metalliferous mines. Mine Water Environ. 14, 85–93.

Sahinkaya, E., Ucar, D., and Kaksonen, A.H., 2017. Bioprecipitation of metals and metalloids. In Sustainable Heavy Metal Remediation. pp. 199–231.

Sangita, G., Prasad, U., Prasad, B., 2010. Studies on the environmental impact of acid mine drainage generation and treatment: an appraisal. Indian. J. Environ. Prot. 30 (11), 953–967.

Santos, S., Machado, R., Correira, N.J.M., and Carvalho, J.R., 2004. Treat. Acid. Mine Waters. 17, 225–232.

Sheridan, C., Ackil, U., Kappelmeyer, U., Moodley, I., 2018. In Stefanakis, M. (Ed.), A Review on the Use of Constructed Wetlands for Industrial Wastewater Treatment, first ed. Willy and Sons Ltd., pp. 249–262.

Sheridan, C., Harding, K., Koller, E., de Preto, 2013. A comparison of charcoal and slag based constructed wetlands for acid mine remediation. Water SA 39 (3), 369–372.

Siite, J., Akob, D.M., Kauffman, C., Finster, K., Banerjee, D., Burkhardt, E.M., et al., 2010. Microbial link between sulfate reduction and metal retention in uranium and heavy metal contaminated soil. Appl. Environ. Microbiol. 76 (10), 3143–3152.

Sun, Y., Cheng, J., 2002. Hydrolysis of lignocellulosic material for ethanol production: a review. Bioresour. Technol. 83, 1–11.

Swiatek, K., Gaag, S., Klier, A., Kruse, A., Sauer, J., Steinbach, D., 2020. Acid hydrolysis of lignocellulosic biomass: sugars and furfurals formation. Catalysis. 2020 (10), 437. Available from: https://doi.org/10.3390/catal10040437.

Taylor, G., 2005. Biofuels and biorefinery concept. Energy Policy 36 (12), 4406–4409.

Van den Berg, M., Botes, M., Cloete, T.E., 2016. The formulation of synthetic domestic wastewater sludge médium to study anaerobic treatment of AMD in the laboratory. Water SA 42 (2), 350–354.

Westernsee, D.K., Rumbold, K., Harding, G.H., Sheridan, C.M., Van Dyke, L.D., Simate, G. S., et al., 2018. The availability of second generation feedstocks for treatment of acid mine drainage and to improve South Africa's bio-based economy. Sci. Total Environ. 637–638, 132–136.

Zagury, G.J., Neculita, C., Bussiere, B., 2006. Passive treatment of acid mine drainage in bioreactors: short review, application, and research needs. Ott. Geo 14439–14446. 2007.

Zhao, X., Zhang, L., Liu, D., 2012. Biomass recalcitrance. Part 1: the chemical and physical structures affecting enzymatic hydrolysis of cellulose. Bio-products Bio-refining 6 (4), 465–482.

Zipper, C., Skousen, J., Jade, C., 2011. Passive Treatment of Acid Mine Drainage. Virginia Cooperative Extension Publication, pp. 1–14, 460-133.

Zuliani, L., Serpico, A., DeSimone, M., Frison, N., Fusco, S., 2021. Biorefinery gets hot: thermophilic enzymes and microorganisms for second-generation bioethanol production. Processes 2021 (9), 1583. Available from: https://doi/10.3390/pr9091583.

Section III

Recent trends in bioprocessing

Metabolic engineering of microorganisms in advancing biofuel production

9

Abhishek Joshi[1], *Krishan K. Verma*[2], *Vishnu D. Rajput*[3] *and Jaya Arora*[1]
[1]Laboratory of Biomolecular Technology, Department of Botany, Mohanlal Sukhadia University, Udaipur, Rajasthan, India, [2]Key Laboratory of Sugarcane Biotechnology and Genetic Improvement (Guangxi), Ministry of Agriculture and Rural Affairs/Guangxi Key Laboratory of Sugarcane Genetic improvement/Sugarcane Research Institute, Guangxi Academy of Agricultural Sciences, Nanning, Guangxi, P. R. China, [3]Academy of Biology and Biotechnology, Southern Federal University, Rostov-on-Don, Russia

9.1 Introduction

Globally, petroleum-based fossil fuels are at high risk of depletion due to their massive utilization, mainly in the transportation and industrial sectors. Further, its widespread use has brought many environmental, economic, and health concerns, such as global warming, the greenhouse effect, the emergence of pandemics, rising prices, and more (Fulton et al., 2015). There is a growing demand for eco-friendly, cost-effective, and sustainable energy sources to drastically minimize these demerits. The United Nations (UN) also has presented affordable and clean energy [Sustainable Development Goals 7 (SDG7)], responsible production and consumption (SDG12), and global climate action (SDG13) as prime concerns among SDG and set the path for these goals to be achieved by 2030 (Joshi et al., 2019).

Biofuels derived from bio-based materials, primarily lignocellulosic biomass, can be considered environmentally friendly and cost-effective alternatives to fossil fuels (Mat Aron et al., 2020). As far as bio-based materials are concerned, lignocellulose is the cheapest and most abundant. It is well documented that lignocellulosic feedstock can be converted into bioethanol, biodiesel, biohydrogen, and other petrochemicals like butanol, iso-pentanol, and terpenes (Sharma et al., 2017; Joshi et al., 2020). However, lignocellulosic biofuels have become a key option to minimize the demerits of fossil fuels, but their sustainable production is not free of challenges. Several difficulties have been associated with the cost-competitive bioconversion of biomass into desired biofuels (Joshi et al., 2021; Okolie et al., 2021).

Considering bioconversion processes, microorganisms are a central aspect of lignocellulose conversion, functioning as biocatalysts that convert substrates into desired products through a series of metabolic reactions (Lakshmi et al., 2021). It is essential that the microbe or strain used for scale-up and biorefinery use can utilize a variety of substrates, produce better yields of end products, and withstand inhibitors. The application of metabolic and bioprocess engineering approaches will improve the

Advances in Lignocellulosic Biofuel Production Systems. DOI: https://doi.org/10.1016/B978-0-323-91192-4.00016-X

promise of robust industrial phenotypes by redirecting metabolic fluxes toward biofuel pathways (Fig. 9.1). There has been substantial progress in this regard over the last couple of years. A number of robust phenotypes have been successfully constructed, mostly from yeast and bacteria domains that have improved substrate spectrum, product yield, and resistance to hydrolysate toxins (Lauer et al., 2022; Lee et al., 2021; Joshi et al., 2022). Moreover, microbial cells have improved their biocatalytic capabilities through precise modifications of surface display enzymes.

Thus, the current study summarizes the fundamental aspects of microorganisms and their metabolic pathways to highlight significant accomplishments in bioconversion processes discussed in the literature. This work also highlights microbial cells and biomolecules that have potential in biorefineries and would contribute significantly to a more sustainable bioeconomy.

9.2 Overview of metabolic pathways of microorganisms for biofuels

Microorganisms convert xylose and glucose into ethanol via the Entner–Doudoroff and Embden– Mayer Hoff–Parnas pathways. Linear-chain alcohols and branched-chain

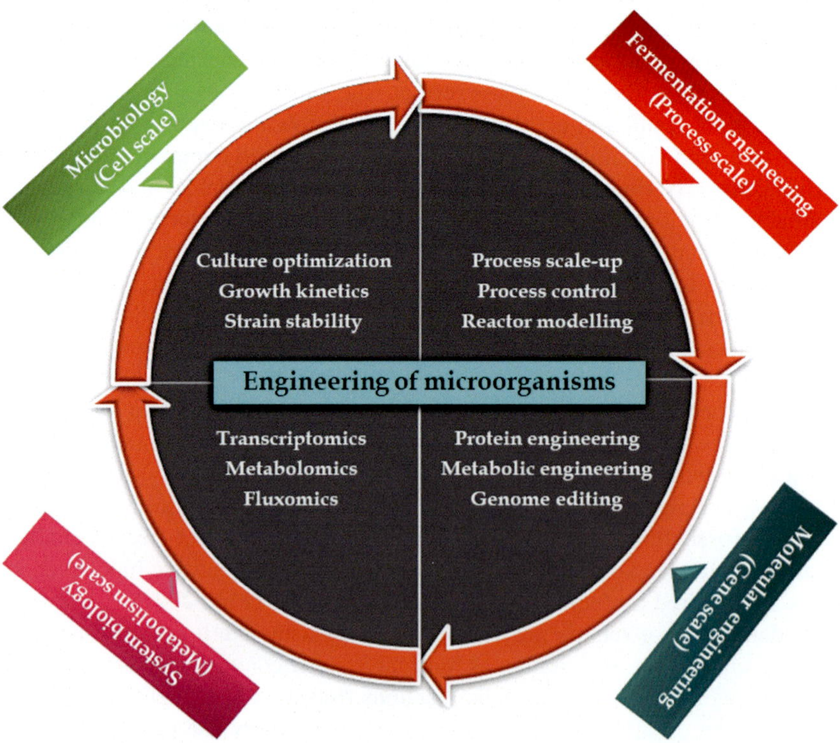

Figure 9.1 Different metabolic engineering approaches for biofuel production.

alcohols can be synthesized either by fatty acid or amino acid pathways. Moreover, linear-chain alcohols are generated by fatty acid metabolism, whereas branched-chain alcohols are fashioned by amino acid metabolism. Recent reports have highlighted that advanced biofuels can be synthesized mainly through the keto-acid pathway, the isoprenoid pathway, the fatty acid synthesis pathway, and the CoA dependent-oxidation pathway or reverse B-oxidation pathway (Jiang et al., 2020; Keasling et al., 2021). Pyruvate is a precursor molecule for initiating keto-acid and isoprenoid pathways, which are responsible for producing isobutanol, 1-propanol, 1-isopropyl-4, 2-methyl-1-butanol, and bisabolene. The CoA-dependent β-oxidation pathway and fatty acid synthesis pathway initiate with acetyl-CoA. The CoA-dependent β-oxidation pathway is advantageous for producing isopropanol and 1-butanol, while the fatty acid synthesis pathway is advantageous for producing alkanes, alkenes, and fatty alcohols. Recently, alkanes and alkenes have also been produced by the polyketide biosynthetic pathway mediated by polyketide synthases and cognate thioesterase. Fig. 9.2 illustrates the major metabolic pathways of microorganisms for biofuel production.

9.3 Metabolic engineering of microorganisms for biofuel production

The benefits of microorganisms extend to various fields, including agriculture, food and beverage, medicine, pharmaceuticals, and environmental clean-up. Recently, its application in generating climate-neutral energy has received tremendous research interest. Many native and nonindigenous microorganisms have been discovered that can convert renewable carbon sources into various biofuels. It has been reported that microorganisms for biofuels must be capable of consuming a broad substrate

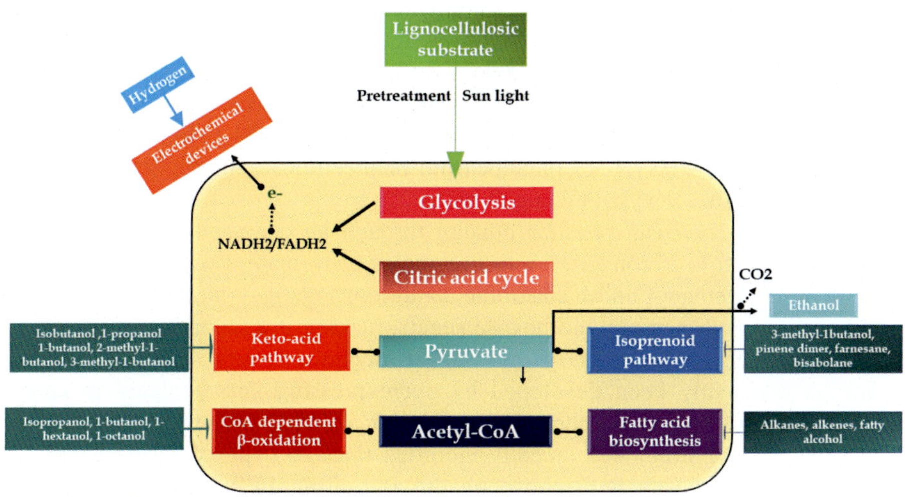

Figure 9.2 Major metabolic pathways of microorganisms for biofuels.

spectrum, be strong against inhibitors, and have a high product yield. For this reason, metabolically engineered microorganisms that redirect metabolic fluxes to fuel products at a faster rate and a lower cost could be the best choice for producing biofuels at an industrial scale. In recent years, several efforts have been made to engineer the metabolic pathways of bacteria, cyanobacteria, filamentous fungi, and yeast. A summary of the significant achievements in the metabolic engineering of such microorganisms for advancing biofuel production is presented in Table 9.1, with more details below.

9.3.1 Metabolic engineering of bacteria for biofuel production

Bacteria are promising hosts for biofuel production. In addition, several features make bacterial metabolism an attractive target for metabolic engineering, such as substrate suitability, growth rate, and tractable physiology and genetics. For this purpose, several bacterial hosts are available. Still, only *Clostridium* sp., *Escherichia coli*, and *Zymomonas mobilis* have been used in recent studies to develop a model microbial chassis for sustainable biorefinery practices.

The *Clostridium* genus is characterized by rod-shaped, anaerobic species with Gram-positive staining. A nutrient-depleting environment activates alternative metabolic pathways in *Clostridium* species. In addition to affecting end products related to energy metabolism, sporulation initiates at the beginning of the stationary phase, involving several metabolic pathways related to biochemical, morphological, and physiological alterations (Liberato et al., 2019). A variety of substrates (carbon monoxide, cellulose, glycerol, and more) can be metabolized by *Clostridium* sp. to generate valuable fuel-like compounds such as ethanol, butanol, isopropanol, and hydroxybutyrate (Xin et al., 2017; Shanmugam et al., 2018). Recently, a *C. ljungdahlii* strain was engineered by introducing heterologous gene clusters derived from the acetoacetyl-CoA pathway. The recombinant strain produced 109 mg/L of butanol and 393 mg/L of hexanol from carbon monoxide and CO_2 (Lauer et al., 2022). In another investigation, a *C. ljungdahlii* strain that overexpressed dehydrogenase produced 13.4 g/L of isopropanol, 28.4 g/L of ethanol, and 3.0 g/L of hydroxybutyrate from one carbon (C1) source (Jia et al., 2021). *C. acetobutylicum* strain has recently been engineered by overexpressing xylanase (*XynB*) genes. The recombinant strain yielded 4.03 g/L of butanol from hemicellulose, the highest yield from either hemicellulose or xylan (Wang et al., 2020a,b).

The native *E. coli* can produce ethanol through an endogenous process where one mole of glucose is converted into two moles of formate, two moles of acetate, and one mole of ethanol under anaerobic conditions (Wang et al., 2017). Moreover, *E. coli* utilizes both pentose and hexose sugars, making it a favorable candidate for producing ethanol, hydrogen (H_2), and higher alcohols. Several engineered *E. coli* strains have recently been developed by overexpressing genes related to central metabolism (Förster and Gescher, 2014; Saini et al., 2017). The mevalonate pathway of the *E. coli* strain was recently modified through overexpressing *NudB* (dihydroneopterin triphosphate di-phosphohydrolase), *IspA* (FPP synthase), and *AphA* (endogenous phosphatase). The recombinant strain produced up to 2027 mg/L of C_5

Table 9.1 Summary of recent progress in metabolic engineering of microorganisms for advancing biofuel production.

Microorganism types	Name of species	Targeted substrate/ metabolites	Metabolic engineering approach	Outcomes	Reference
Bacteria	*Clostridium acetobutylicum*	Hemicellulose	Overexpression of indigenous xylanase (XynB)	Butanol yield- 4.03 g/L	Wang et al. (2020a,b)
	Clostridium carboxidivorans	Glucose	Overexpression of aldehyde: ferredoxin oxidoreductase (AOR), aldehyde dehydrogenase (AdhE2), and ferredoxin oxidoreductase (FNR)	Ethanol yield- 2.38–3 g/L; butanol yield- 0.20–0.35 g/L	Cheng et al. (2019)
	Clostridium cellulovorans	Cellulose	Overexpression of AdhE2 and butanol dehydrogenase (Bdh)	n-Butanol—4/l with a high yield of 0.22 g/g	Bao et al. (2019)
	C. cellulolyticum	Cellulose	Introduction of CoA-dependent pathway	n-Butanol yield—120 mg/L	Gaida et al. (2016)
	Clostridium ljungdahlii	CO and CO_2	Integration of the heterologous gene clusters of acetoacetyl-CoA-derived pathway	Butanol yield— 109 mg/L; hexanol yield- 393 mg/L	Lauer et al. (2022)
	C. ljungdahlii	One-carbon (C_1)	Introduction and expression of heterologous functional genes of dehydrogenase	Isopropanol yield—13.4 g/L; ethanol yield—28.4 g/L; 3 hydroxybutyrate yield- 3.0 g/L	Jia et al. (2021)
	Clostridium thermocellum	Cellulose	Disruption in hydrogenase (*hfsB*) and mutation in *adhE*	Isobutanol titer—5.1 g/L; ethanol yield—29.9 g/L	Holwerda et al. (2020)
	C. thermocellum	Cellulose	Overexpression of thiolase–hydroxybutyryl-CoA dehydrogenase–crotonase (Thl-Hbd-Crt) and trans-enoyl-CoA reductase (Ter) and butyraldehyde dehydrogenase and alcohol dehydrogenase (Bad-Bdh) module	n-Butanol titer—2.2 folds higher; yield—357 mg/L	Tian et al. (2019)
	Corynebacterium glutamicum	Glucose and xylose	Downregulation of glyceraldehyde 3-phosphate dehydrogenase and phosphoenolpyruvate-dependent glucose uptake system (PTS) by inositol permeases (IolT1) and glucokinase (GLK)	3-Hydroxypropionic acid titer—62.6 g/L; yield—0.51 g/g glucose	Chen et al. (2017)
	Enterobacter aerogenes	Glucose and fructose	Knocking out NADH dehydrogenase and lactate dehydrogenase (LDH)	2,3-Butanediol titer—up to 71.2%	Wu et al. (2021)
	Enterobacter cloacae	Glucose and xylose	Inactivation of PtsG (glucose transporter); overexpression *galP* (galactose permease encoding gene)	2,3-Butanediol productivity—2.3 g/L/h with yield of 95.0%	Li et al. (2015)
	Escherichia coli	Levoglucosan	Overexpressing levoglucosan kinase, adh and pyruvate decarboxylase (*pdc*)	Ethanol yield—0.43 g/g levoglucosan	Chang et al. (2021)

(*Continued*)

Table 9.1 (Continued)

Microorganism types	Name of species	Targeted substrate/ metabolites	Metabolic engineering approach	Outcomes	Reference
	E. coli	Hemicellulose	Deletion of D-lactate dehydrogenase (*ldhA*) and fumarate reductase (*frdD*)genes	Hydrogen (H_2) yield—342.7 mL H_2/g total reducing sugar (TRS); ethanol yield—0.48 g ethanol/g TRS	Lopez-Hidalgo et al. (2021)
	E. coli	Glucose and xylose	Elimination of glucose permease-enzyme IIB (*ptsG*), d-lactate dehydrogenase (*ldhA*) and/or fumarate reductase (*frdD*) genes	H_2 yield—0.27 mol C/mol	Balderas-Hernandez et al. (2020)
	E. coli	Cellulose	Overexpressing NudB (dihydroneopterin triphosphate diphosphohydrolase), IspA (FPP synthase) along with a heterologous mevalonate (MVA) pathway	Mixture of C_5 and C_{15} alcohols and derivatives—1652 mg/L	Zada et al. (2018)
	Klebsiella pneumoniae	Glucose	Introduction of diol dehydratase and AdhE2	2-Butanol titer—1030 mg/L	Chen et al. (2015)
	Pseudomonas putida	Ethanol	Acetylating aldehyde dehydrogenase (ADA)	Fatty acid ethyl esters (FAEEs) yield—152.09 mg FAEEs/g ethanol	Sarwar et al. (2022)
	Rhodobacter capsulatus	Glucose	Promoter screening, improving the NADPH availability and precursor supply, introducing a heterologous MVA pathway	Bisabolene yield—9.8 g/L (78% of theoretical maximum)	Zhang et al. (2021a,b)
	Thermoanaerobacterium aotearoense	Cellulose	Overexpression of pyruvate formate lyase-activating protein (PflA)	Ethanol production—45.01 g/L	Fu et al. (2021)
	Zymomonas mobilis	Cellulose	Overexpression of the *phaCAB* operon, and genes associated with cofactor balance, and ethanol utilization pathway (EUP)	Bioethanol and poly-3-hydroxybutyrate yield—50.30% (dry cell weight, DCW)	Li et al. (2022)
	Z. moblis	Lignocellulose	Heterologous expression of synthetic ethylene-forming enzyme (EFE)	Ethylene production—5.8 nmol/OD600/ml	He et al. (2021)
	Z. mobilis	Cellulose	Overexpression of 2-ketoisovalerate decarboxylase (*kdcA*), *als*, *ilvC* and *ilvD*	Isobutanol titer—4.0 g/L	Qiu et al. (2020)

Cyanobacteria	*Anabaena* sp. PCC 7120	CO_2	Overexpression of *pdc*, *adh* and *invB* gene	Ethanol—1681.9 mg/L	Ehira et al. (2018)
	Synechocystis PCC 6803	CO_2	Overexpression of fructose-1,6/sedoheptulose-1,7-bisphosphatase (FBP/SBPase) and transketolase (TK)	Ethanol—two to ninefold higher	Roussou et al. (2021)
	Synechococcus PCC7002	CO_2	Blocking glycogen synthesis genes (*glgA*)	Ethanol yield—0.8 g/L in 7 days with an average of 100 mg/L/day	Wang et al. (2020a,b)
	Synechococcus PCC7002	CO2	Introduction of NADH-dependent nitrite reductase	2-Methyl-1-butanol titer—177 mg/L; isobutanol titer—50 mg/L	Purdy et al. (2022)
	Synechocystis PCC 6803	CO_2	Overexpression of 1-deoxy-D-xylulose-5-phosphate synthase (DXS) and IPP/DMAPP isomerase (IDI)	Bisabolene titer—180 mg/L	Rodrigues and Lindberg (2021)
	Synechocystis PCC 6803	CO_2	Overexpression of α-ketoisocaproate dioxygenase (KICD)	Isobutene yield—91 ng OD_{750}/h	Mustila et al. (2021)
	Synechococcus elongatus PCC 7942	CO_2	Overexpression of phosphate acetyltransferase (*pta*)	Isopropanol titer—33.1 mg/L	Hirokawa et al. (2017)
	Synechocystis PCC 6803	CO_2	Deletions of glycogen and polyhydroxybutyrate (*phb*) genes	Ethanol production rate—2.01 g/L/day	Namakoshi et al. (2016)
Filamentous fungi	*Aspergillus niger*	Glucose	Overexpression of trehalase	Ethanol conversion rate—1698.83 U/mL	Dong et al. (2020a,b)
	Fusarium oxysporum	Glucose	Overexpression of translaldolase (TAL) and phosphoglucomutase (PGM)	Ethanol productivity—0.146 g/L/h	Anasontzis et al. (2016)
	Myceliophthora thermophila	Glucose and cellobiose	Over expression of cellodextrin transport system (CDT-1/CDT-2) and pyc gene	Ethanol yield—11.3 g/L	Li et al. (2020)
	M. thermophila	Glucose	Deletion of the fumarate reductase	Fumaric acid yield—17 g/L	Gu et al. (2018)
Yeast	*Rhodotorula toruloides*	Glycerol	Deletion of DGA1 and LRO1 (acyltransferases) with overexpression of ACC1 (cytosolic acetyl-CoA) and ACL1(malonyl-CoA production)	Fatty alcohol titer—3.7 g/L	Schultz et al. (2022)
	R. toruloides	Glucose	Overexpression of bifunctional wax ester synthase (WSs)/acyl-CoA-diacyglycerol acyltransferase (DGAT)	FAEEs titer—9.97 g/L	Zhang et al. (2021a,b)
	R. toruloides	Wheat straw hydrolysates	Overexpression of DGAT1 and SCD1	Lipid yield—0.179 g/g	Díaz et al. (2018)
	Saccharomyces cerevisiae	Lignocellulose	Overexpression of XYL1, XYL2 and XKS1 (xylose metabolism encoding genes)	Ethanol yield—2.08 g/L	He et al. (2022)
	S. cerevisiae	Xylose	Overexpression of SIP4, ADR1 and HAP4 (transcription factors for xylose catabolism)	Ethanol yield—0.41 g/g	Dzanaeva et al. (2021)

(*Continued*)

Table 9.1 (Continued)

Microorganism types	Name of species	Targeted substrate/metabolites	Metabolic engineering approach	Outcomes	Reference
	S. cerevisiae	Lignocellulose	Overexpressing xylose reductase (XR) and xylose dehydrogenase (XDH)	Ethanol yield—0.48 g/g sugars	Liu et al. (2021)
	S. cerevisiae	Xylose	Expression of hexokinase 2 (Hxk2p)	Ethanol yield—42.6%	Zheng et al. (2020)
	S. cerevisiae	Glucose and xylose	Overexpressing SFA1	Ethanol yield—0.492 g/g total sugars (TS) within 48 h	Zhu et al. (2020)
	Schizosaccharomyces pombe	Glucose and cellobiose	Overexpression of malonyl-CoA reductase (MCR)	3-Hydroxypropionic acid yield—11.4 g/L	Takayama et al. (2018)
	Yarrowia lipolytica	Waste cooking oil	Combining mitochondrial and cytoplasmic engineering	α-bisabolene titer—1058.1 mg/L	Zhu et al. (2022)
	Y. lipolytica	Glycerol	Overexpression of glycerol kinase (GK), and transketolase (TKL)	Erythritol yield—58.8 g/L	Jagtap et al. (2021)
	Y. lipolytica	Waste cooking oil	Overexpression of a lipase (Lip2)	FAEEs titer—12.5 g/L	Wei et al. (2021)
	Y. lipolytica	Acetate	Overexpression of acetyl-CoA carboxylase (ACC1) and fatty acid synthase (FAS)	Lipid content—95%	Chen et al. (2021)
	Y. lipolytica	Glycerol	Overexpression of thioesterase (TesA), fatty acyl-CoA synthetases (FAA), and peroxisomal acyl-CoA transporter (PXA1)	Fatty acids titer—2.3 g/L	Ghogare et al. (2020)
	Y. lipolytica	Glycerol	Overexpression of WSs and fatty acyl reductases (FARs)	FAEEs titer—2.0 g/L	Gao et al. (2020)
	Y. lipolytica	Glycerol	Overexpression of WSs	FAEEs titer—1.18 g/L	Gao et al. (2018)

and C_{15} alcohols and their derivatives (Zada et al., 2018). Recently, a recombinant *E. coli* was developed by deleting D-lactate dehydrogenase (*LDHA*) and fumarate reductase (*FRDD*) genes. The engineered strain generated 342.7 mL H_2/g total reducing sugar (TRS) and 0.48 g ethanol/g TRS (Lopez-Hidalgo et al., 2021). In another study, Chang et al. (2021) developed an ethanologenic *E. coli* strain by overexpressing genes of levoglucosan kinase, alcohol dehydrogenase (*ADH*), and pyruvate decarboxylase (*PDC*). The recombinant strain was able to produce 0.43 g ethanol/g levoglucosan. Cao et al. (2020) also constructed an engineered *E. coli* strain through overexpressing alcohol dehydrogenase (ADH) and aldehyde dehydrogenase, which could utilize ethanol as the sole carbon source. The engineered strain showed better growth when ethanol was added after glucose starvation, and 2.24 g/L of ethanol was consumed during shake-flask fermentation.

The naturally occurring strain of *Z. mobilis* exhibits excellent industrial properties like high substrate uptake rates, high ethanol tolerance, and low aeration costs due to anaerobic exposure fermentation (Behera et al., 2010, 2011; Kalnenieks et al., 2014). A wild-type *Z. mobilis*, however, is incompetent at utilizing lignocellulosic hydrolysates containing pentoses like arabinose and xylose. Due to this, numerous metabolic engineering efforts have been conducted to broaden its substrate spectrum via heterologous expression of pentoses-assimilating genes (Wang et al., 2018; Zhang et al., 2019). Several recombinant *Z. mobilis* strains have been developed by employing expression and cassette optimization of genes related to xylose transporter (XylE and ABC type transporter) and fermentation inhibitors (HFQ and SIGE). These engineered strains produced higher ethanol yield from lignocellulosic hydrolysates (Yang et al., 2018; Sarkar et al., 2021; Lou et al., 2021). Similarly, He et al. (2021) developed a recombinant *Z. mobilis* strain by heterologous expression of the synthetic ethylene-forming enzyme (EFE). The recombinant strain, ZM532-EFE, produced 5.8 nmol ethylene/OD600/mL from the corn straw hydrolysates. In another study, Li et al. (2022) constructed a recombinant *Z. mobilis* strain by overexpressing the phaCAB operon, genes coupled with cofactor balance, and the ethanol utilization pathway (EUP). As a result, the recombinant strain $ZMPt^{N2\text{-}EUP}$ coproduced high-titer ethanol and poly-3-hydroxybutyrate (PHB) without affecting ethanol productivity.

9.3.2 Metabolic engineering of cyanobacteria for biofuel production

Cyanobacteria are gram-negative photosynthetic microorganisms that play an influential role in carbon sequestration, nitrogen fixation, and oxygen evolution. An endogenous ethanologenic pathway makes cyanobacteria an excellent choice for solar-powered ethanol production. They can also convert CO_2 and water into carbon-rich lipid substrates for biodiesel production. Furthermore, they have well-established genetic backgrounds, are flexible in genetic manipulation, and are tolerant to foreign genes, which makes them an attractive metabolic engineering target (Angermayr et al., 2015). It has been demonstrated that metabolic engineering applied to these photosynthetic hosts can result in the conception of cell factories that can produce biofuel molecules in a single step from CO_2, light, and water (Lai and Lan, 2015; Gomaa et al., 2016; Choi and Woo, 2020). In recent research, *Synechocystis* PCC 6803 strain was developed by overexpressing 1-deoxy-D-xylulose-5-phosphate synthase (DXS) and IPP/DMAPP isomerase (IDI). The recombinant strain produced 180 mg/L titer of bisabolene (Rodrigues and Lindberg, 2021). Similarly, the strain overexpressing fructose-1, 6/sedoheptulose-1, 7-bisphosphatase (FBP/SBPase), and transketolase (TK) was reported to yield 2-to-ninefold higher ethanol compared to the control strain (Roussou et al., 2021). In another study, a recombinant *Synechococcus* PCC7002 strain was constructed by blocking glycogen synthesis genes (*glgA*). The engineered strain produced 0.8 g/L of ethanol in 7 days with an average of 100 mg/L/day (Wang et al., 2020a,b). Similarly, a recombinant *Synechococcus* PCC7002 strain was developed by introducing NADH-dependent nitrite reductase. The recombinant *Synechococcus* strain was reported to produce 177 mg/L titer of 2-methyl-1-butanol and 50 mg/L titer of isobutanol (Purdy et al., 2022).

9.3.3 Metabolic engineering fungi for biofuel production

In the past decades, filamentous fungi have been considered challenging and painstaking to metabolically engineer due to their slow growth rate, the lack of efficient genetic transformation systems, the absence of auxotrophic selection markers, and poor transformation rates compared to bacteria and yeast (Workman et al., 2013; Brandl and Andersen, 2015). Nowadays, filamentous fungal cell factories are available with several efficient transformation techniques, a wide range of selection markers, and numerous constitutive or inducible expression systems, making them an ideal target for metabolic engineering to advance biofuel production (Meyer, 2021). Recent metabolic engineering strategies have been adopted to increase lipid accumulation in oleaginous *Aspergillus*, *Cunninghamella*, *Mortierella*, and *Mucor*, resulting in 20%–70% of oil yield from a single cell (Yang et al., 2020; Mhlongo et al., 2021). Hussain et al. (2019) also developed recombinant *Mucor circinelloides* M65 stains by introducing four different acyl-acyl carrier proteins (ACP) and thioesterase (TE). Among all the recombinant strains, M65-TE-03 and M65-TE-04 displayed more nonnative C8–C10 and C12 fatty acids, respectively, compared to the control strain. Furthermore, these strains synthesized medium-chain fatty acids within the range from 1.14 to 2.77 g/L of total cell lipids. In recent research, recombinant *Aspergillus niger* strains were constructed through overexpression of trehalase. The engineered strain exhibited a 1698.83 U/mL ethanol conversion rate (Dong et al., 2020a,b). Recombinant *Myceliophthora thermophila* was developed by introducing the CDT-1 and CDT-2 genes into the cellodextrin transport system. The engineered strain produced 11.3 g/L of ethanol from glucose and cellobiose (Li et al., 2020).

9.3.4 Metabolic engineering of yeast for biofuel production

Among the yeasts, *Saccharomyces cerevisiae* has been extensively studied for several reasons, including its safety (generally regarded as safe, GRAS), high fermentation, ethanol production, tolerance, and well-studied genome (Behera et al., 2010). Since native *S. cerevisiae* cannot ferment pentose (xylose or cellobiose) for ethanol production, metabolic engineers must develop pentose-fermenting yeast for sustainable biofuel production. Over the past few years, several metabolic engineering strategies have been undertaken and implemented in yeast to address this issue. It has been shown that engineering the pentose phosphate pathway and xylose assimilating pathway maximizes xylose utilization and alleviates glucose repression (Lee et al., 2014; Jagtap et al., 2019). Recently, Dzanaeva et al. (2021) constructed a recombinant strain of *S. cerevisiae* by overexpressing SIP4, ADR1, and HAP4 transcription factors. The engineered strain produced 0.41 g/g of ethanol from xylose. Similarly, Liu et al. (2021) developed a recombinant strain of *S. cerevisiae* through overexpressing xylose reductase (XR) and xylose dehydrogenase (XDH). The engineered strain produced 0.48 g/g ethanol from lignocellulose hydrolysates. In another study, overexpression of xylose metabolism encoding genes *XYL1, XYL2*, and *XKS1* in *S. cerevisiae* resulted in 2.08 g/L of ethanol yield from lignocellulose hydrolysates (He et al., 2022). In addition to *S. cerevisiae*, nonconventional yeasts *Spathaspora passalidarum* and *Pichia stipites* have also been engineered for natural xylose metabolism to improve ethanol production (Rodrussamee et al., 2018; Selim et al., 2020).

Oleaginous yeasts such as *Rhodotorula toruloides* and *Yarrowia lipolytica* have also been explored for natural and engineered lipid metabolism to improve biodiesel production. Recently, Zhang et al. (2021a,b) developed a recombinant strain of *R. toruloides* through overexpressing bifunctional wax ester synthase (WSs)/acyl-CoA-diacyglycerol acyltransferase (DGAT). The engineered strain was reported to produce 9.97 g/L titer of fatty acid ethyl esters (FAEEs) from glucose. Similarly, the deletion of genes *DGA1* and *LRO1* (acyltransferases) in *R. toruloides* resulted in 3.7 g/L titer of fatty alcohol from glycerol (Schultz et al., 2022). In another study, overexpression of thioesterase (*tesA*), fatty acyl-CoA synthetases (*faa*), and peroxisomal acyl-CoA transporter (*pxa1*) in *Y. lipolytica* resulted in 2.3 g/L titer of fatty acids from glycerol (Ghogare et al., 2020). More recently, Zhu et al. (2022) developed recombinant strains of *Y. lipolytica* by combining mitochondrial and cytoplasmic engineering. These recombinant strains were reported to produce 1058.1 mg/L titer of α-bisabolene from waste cooking oil.

9.4 Cell surface display engineering of microorganisms for biofuel production

Microorganism-based cell surface display engineering is an innovative approach for building whole-cell biocatalysts and engineering proteins through high-throughput screening of protein libraries (Tanaka and Kondo, 2015; Han et al., 2018). In this approach, functional heterologous proteins or enzymes are immobilized on the cell surface of microbial hosts by fusion with cell wall or plasma membrane-anchoring domains (Fig. 9.3). This process has been well established in a wide variety of

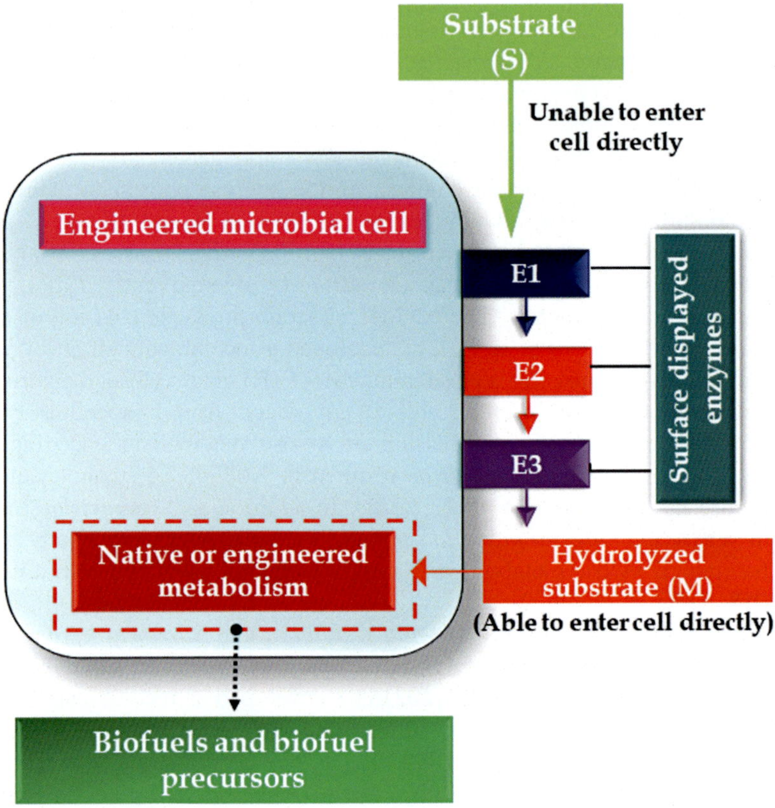

Figure 9.3 A model microbial cell with surface displayed enzymes E1–E3.

microorganisms, including bacteria, fungi, phage, and yeast. Particularly, yeasts are the most appropriate hosts for displaying functional proteins with high molecular mass since they possess a good quality control system involving protein folding and modification systems common to eukaryotic pathways (Smith et al., 2015). Over the last couple of years, several studies have demonstrated the importance of microbial cell surface display engineering in the field of medicine, environmental remediation, and bioenergy production (Jia et al., 2017; Wang et al., 2021). It has become increasingly clear that microbial cell surface display can be used to design and optimize novel biocatalysts for sustainable biofuel processes (Fan et al., 2020; Madavi et al., 2022).

The use of yeast surface display to express cellulolytic enzymes such as cellulase, endoglucanase, and β-glucosidase displayed better adhesion ability associated with increased surface area and roughness of cellulose fibers, resulting in higher hydrolysis efficiency (Liu et al., 2016a,b). Various cellulase-displaying yeast strains demonstrated clear cell-to-cellulose adhesion and tearing cellulose degradation patterns, resulting in significantly higher ethanol yields and titers from lignocellulosic substrates with a 50% reduction in enzyme dosage needed for high-density fermentation (Liu et al., 2016a,b; Dong et al., 2020a,b; Ye et al., 2021). Similarly, Chetty et al. (2022) overexpressed genes previously shown to enhance cellulase secretion in *S. cerevisiae*. The recombinant strain exhibited approximately 101%, 99%, and 231% higher cellobiohydrolase, β-glucosidase, and endoglucanase activities, respectively, compared to parental strains. This strain also displayed improved hydrolysis of crystalline cellulose up to 186% and improved ethanol yields up to 40%–50%. In another study, Raoufi and Gargari (2018) successfully displayed the lipase A (Lip A) genes from *Pseudomonas aeruginosa* on *Pichia pastoris* X33 surface. The recombinant *Pichia* strain X33 yielded high biodiesel yields from microalgae oil (87.6%) up to 10 cycles of batch production.

In addition to yeasts, certain bacterial Clostridia can produce cellulosomes or minicellulosomes, an assembly of scaffolding proteins requiring more than one enzyme to hydrolyze lignocellulosic substrates (Bhattacharya et al., 2015; Tang et al., 2018). For example, in a study by Chin et al. (2017), outer-membrane-targeted tilapia metallothionein (OmpC-TMT) and the clostridial n-butanol pathway were coexpressed in *E. coli*. The modified strain, BUT3-DE, was reported to improve a 3.3-fold higher n-butanol yield than the control strain (95.7 mg/L). Similarly, Xu et al. (2019) developed a novel multi-enzyme complex for lipase, carboxylic acid reductase, and aldehyde reductase through a specific interaction between cohesion-dockerin and functional assembly of *E. coli*-ice nucleation protein. An optimized *E. coli* cell associated with such enzyme complexes has been reported to convert 73% of triacylglycerols into fatty alcohols. In another study, Lee et al. (2022) developed a novel whole-cell biocatalyst displaying the enzyme complex consisting of endoglucanase E (cCelE) and *N*-acetylglucosaminidase (cNAG) on *Corynebacterium glutamicum*. The optimized *C. glutamicum* cells showed 2.4- and 4.43-fold higher cCelE and cNAG activity than single enzyme-secreting *C. glutamicum* cells. Furthermore, these cells produced a maximum titer of 218.9 mg/L of isopropanol and a 90% efficiency rate after four recycling cycles. Similarly, cyanobacterium *Synechocystis* PCC 6803 was selected to overexpress aldolase, ribulose-1, 5-bisphosphate carboxylase/oxygenase, transketolase, and fructose-1, 6/sedoheptulose-1, 7-bisphosphatase, together with pyruvate decarboxylase and alcohol dehydrogenase and alcohol dehydrogenase enzyme complex. The optimized *Synechocystis* PCC 6803 cells produced 55%, 67%, 37%, and 69% more ethanol and 7.7%, 15.1%, 8.8%, and 10.1% more biomass than the strain expressing only the ethanol biosynthesis pathway, respectively (Liang et al., 2018). A summary of the recent achievements in cell surface display engineering of microorganisms for advancing biofuel production is presented in Table 9.2.

9.5 Conclusion and future prospects

Numerous challenges associated with biofuel production have been addressed by applying microbial metabolic engineering strategies in recent years. Many biofuels

Table 9.2 Summary of recent advances in microorganism cell surface display engineering for biofuel production.

Microorganism	Used approach	Targeted substrates	Outcomes	Reference
Corynebacterium glutamicum	Expression of N-acetylglucosaminidase (cNAG) and endoglucanse E (cCelE)	Cellulose	Isopropanol titer—218.9 mg/L	Lee et al. (2022)
Escherichia coli	Expression of an α-amylase in WDHFP strain containing *hycA*, *frdD* and *pta* genes	Starch	H_2 yield—1689.68 cm^3/dm^3; Ethanol yield—0.28 g/g	Gutiérrez-García et al. (2020)
E. coli	Coexpression of lipase (LIP), carboxylic acid reductase (CAR) and aldehyde reductase (AHR) ice nucleation protein (INP)	Triacylglycerols	Fatty alcohols yield—73%	Xu et al. (2019)
E. coli	Coexpression of OmpC (osmoregulatory membrane protein) and TMT (fused tilapia metallothioneins)	Lignocelluosic substrate	n-butanol titer—320 mg/L	Chin et al. (2017)
Pichia pastoris	Establishing minicellulosomes containing endoglucanase (EG), an exoglucanase (CBH) and a β-glucosidase (BGL), and carbohydrate-binding module (CBM)	Cellulose	Ethanol titer—5.1 g/L	Dong et al. (2020a,b)
P. pastoris	Coexpression of Lip A and glycosylphosphatidylinositol (GPI)-anchored protein (Gcw61)	Microalgae oil	Biodiesel yield—87.6%	Raoufi and Gargari (2018)
Saccharomyces cerevisiae	Expression of codisplaying CWP2, YGP1 and BGL	Cellobiose	Ethanol production—59%	Arnthong et al. (2022)
S. cerevisiae	Expression of cell-adhered cellobiohydrolase and GPI-anchors	Cellulose	Ethanol yields—40%–50%	Chetty et al. (2022)
S. cerevisiae	Expression of codisplaying cellulase and BGL	Corn cob	Ethanol titer—50 g/L	Cunha et al. (2021)
S. cerevisiae	Expressing bifunctional minicellulosomes by galactose induction and cellodextrin	Sucrose	Ethanol yield—0.43 g/g	Li et al. (2017)

have been produced by engineering the metabolic pathways of bacteria, yeasts, and cyanobacteria with a high yield close to its theoretical maximum. Thermophilic and halophilic bacteria could also drive the production of commercially viable biofuels. It appears that cyanobacteria and other photosynthetic bacteria may also produce biofuels in the future that are commercially viable. Using yeast strains engineered to display hydrolytic enzymes on their surfaces may provide a more significant potential for commercial-scale bioethanol production directly from lignocellulosic biomass in the near future. To produce more sustainable biofuels, it is necessary to discover novel enzymes that catalyze unknown reactions, engineer more efficient microorganisms, and develop synthetic pathways that increase metabolic flux toward biofuels and other value-added products.

References

Anasontzis, G.E., Kourtoglou, E., Villas-Boas, S.G., Hatzinikolaou, D.G., Christakopoulos, P., 2016. Metabolic engineering of *Fusarium oxysporum* to improve its ethanol-producing capability. Front. Microbiol. 7, 632.

Angermayr, S.A., Rovira, A.G., Hellingwerf, K.J., 2015. Metabolic engineering of cyanobacteria for the synthesis of commodity products. Trends Biotechnol. 33, 352–361.

Arnthong, J., Ponjarat, J., Bussadee, P., Deenarn, P., Prommana, P., Phienluphon, A., et al., 2022. Enhanced surface display efficiency of β-glucosidase in *Saccharomyces cerevisiae* by disruption of cell wall protein-encoding genes YGP1 and CWP2. Biochem. Eng. J. 179, 108305.

Balderas-Hernandez, V.E., Maldonado, K.P.L., Sánchez, A., Smoliński, A., Rodriguez, A.D. L., 2020. Improvement of hydrogen production by metabolic engineering of *Escherichia coli*: modification on both the PTS system and central carbon metabolism. Int. J. Hydro. Energy. 45, 5687–5696.

Bao, T., Zhao, J., Li, J., Liu, X., Yang, S.T., 2019. n-Butanol and ethanol production from cellulose by *Clostridium cellulovorans* overexpressing heterologous aldehyde/alcohol dehydrogenases. Bioresour. Technol. 285, 121316.

Behera, S., Mohanty, R.C., Ray, R.C., 2010. Comparative study of bio-ethanol production from mahula (*Madhuca latifolia* L.) flowers by *Saccharomyces cerevisiae* and *Zymomonas mobilis*. Appl. Energy. 87, 2352–2355.

Behera, S., Mohanty, R.C., Ray, R.C., 2011. Ethanol production from mahula (*Madhuca latifolia* L.) flowers using free and immobilized (in *Luffa cylindrica* L. sponge discs) cells of *Zymomonas mobilis* MTCC 92. Ann. Microbiol. 61, 469–474.

Bhattacharya, A.S., Bhattacharya, A., Pletschke, B.I., 2015. Synergism of fungal and bacterial cellulases and hemicellulases: a novel perspective for enhanced bio-ethanol production. Biotechnol. Lett. 37, 1117–1129.

Brandl, J., Andersen, M.R., 2015. Current state of genome-scale modeling in filamentous fungi. Biotechnol. Lett. 7, 1131–1139.

Cao, Y., Mu, H., Guo, J., Liu, H., Zhang, R., Liu, W., et al., 2020. Metabolic engineering of *Escherichia coli* for the utilization of ethanol. J. Biol. Res.-Thessaloniki. 27, 1–10.

Chang, D., Islam, Z.U., Yang, Z., Thompson, I.P., Yu, Z., 2021. Conversion efficiency of bioethanol from levoglucosan was improved by the newly engineered *Escherichia coli*. Environ. Prog. Sustain. Energy. 40, 13687.

Chen, Z., Wu, Y., Huang, J., Liu, D., 2015. Metabolic engineering of *Klebsiella pneumoniae* for the de novo production of 2-butanol as a potential biofuel. Bioresour. Technol. 197, 260–265.

Chen, Z., Huang, J., Wu, Y., Wu, W., Zhang, Y., Liu, D., 2017. Metabolic engineering of *Corynebacterium glutamicum* for the production of 3-hydroxypropionic acid from glucose and xylose. Metabol. Eng. 39, 151–158.

Chen, L., Yan, W., Qian, X., Chen, M., Zhang, X., Xin, F., et al., 2021. Increased lipid production in *Yarrowia lipolytica* from acetate through metabolic engineering and co-substrate fermentation. ACS Synth. Biol. 10, 3129–3138.

Cheng, C., Li, W., Lin, M., Yang, S.T., 2019. Metabolic engineering of *Clostridium carboxidivorans* for enhanced ethanol and butanol production from syngas and glucose. Bioresour. Technol. 284, 415–423.

Chetty, B.J., Inokuma, K., Hasunuma, T., van Zyl, W.H., den Haan, R., 2022. Improvement of cell-tethered cellulase activity in recombinant strains of *Saccharomyces cerevisiae*. Appl. Microbiol. Biotechnol. 1–15.

Chin, W.C., Lin, K.H., Liu, C.C., Tsuge, K., Huang, C.C., 2017. Improved n-butanol production via co-expression of membrane-targeted tilapia metallothionein and the clostridial metabolic pathway in *Escherichia coli*. BMC Biotechnol. 17, 1–14.

Choi, S.Y., Woo, H.M., 2020. CRISPRi-dCas12a: a dCas12a-mediated CRISPR interference for repression of multiple genes and metabolic engineering in cyanobacteria. ACS Synth. Biol. 9, 2351–2361.

Cunha, J.T., Gomes, D.G., Romaní, A., Inokuma, K., Hasunuma, T., Kondo, A., et al., 2021. Cell surface engineering of *Saccharomyces cerevisiae* for the simultaneous valorization of corn cob and cheese whey *via* ethanol production. Energy Convers. Manage. 243, 114359.

Díaz, T., Fillet, S., Campoy, S., Vázquez, R., Viña, J., Murillo, J., et al., 2018. Combining evolutionary and metabolic engineering in *Rhodosporidium toruloides* for lipid production with non-detoxified wheat straw hydrolysates. Appl. Microbiol. Biotechnol. 102, 3287–3300.

Dong, C., Qiao, J., Wang, X., Sun, W., Chen, L., Li, S., et al., 2020a. Engineering *Pichia pastoris* with surface-display minicellulosomes for carboxymethyl cellulose hydrolysis and ethanol production. Biotechnol. Biofuels 13, 1–9.

Dong, L., Lin, X., Yu, D., Huang, L., Wang, B., Pan, L., 2020b. High-level expression of highly active and thermostable trehalase from *Myceliophthora thermophila* in *Aspergillus niger* by using the CRISPR/Cas9 tool and its application in ethanol fermentation. J. Indus. Microbiol. Biotechnol. 47, 133–144.

Dzanaeva, L., Kruk, B., Ruchala, J., Sibirny, A., Dmytruk, K., 2021. The impact of transcription factors Znf1, Sip4, Adr1, Tup1, and Hap4 on xylose alcoholic fermentation in the engineered yeast *Saccharomyces cerevisiae*. Antonie Van Leeuwenhoek 114, 1373–1385.

Ehira, S., Takeuchi, T., Higo, A., 2018. Spatial separation of photosynthesis and ethanol production by cell type-specific metabolic engineering of filamentous cyanobacteria. Appl. Microbiol. Biotechnol. 102, 1523–1531.

Fan, S., Liang, B., Xiao, X., Bai, L., Tang, X., Lojou, E., et al., 2020. Controllable display of sequential enzymes on yeast surface with enhanced biocatalytic activity toward efficient enzymatic biofuel cells. J. Am. Chem. Soc. 142, 3222–3230.

Förster, A.H., Gescher, J., 2014. Metabolic engineering of *Escherichia coli* for production of mixed-acid fermentation end products. Front. Bioeng.. Biotechnol. 2, 16.

Fu, H., Luo, S., Dai, K., Qu, C., Wang, J., 2021. Engineering *Thermoanaerobacterium aotearoense* SCUT27/Δldh with pyruvate formate lyase-activating protein (PflA) knockout for enhanced ethanol tolerance and production. Process Biochem. 106, 184–190.

Fulton, L.M., Lynd, L.R., Körner, A., Greene, N., Tonachel, L.R., 2015. The need for biofuels as part of a low carbon energy future. Biofuels, Bioprod. Bioref. 9, 476–483.

Gaida, S.M., Liedtke, A., Jentges, A.H.W., Engels, B., Jennewein, S., 2016. Metabolic engineering of *Clostridium cellulolyticum* for the production of n-butanol from crystalline cellulose. Microb. Cell Fact. 15, 1–11.

Gao, Q., Cao, X., Huang, Y.Y., Yang, J.L., Chen, J., Wei, L.J., et al., 2018. Overproduction of fatty acid ethyl esters by the oleaginous yeast *Yarrowia lipolytica* through metabolic engineering and process optimization. ACS Synth. Biol. 7, 1371–1380.

Gao, Q., Yang, J.L., Zhao, X.R., Liu, S.C., Liu, Z.J., Wei, L.J., et al., 2020. *Yarrowia lipolytica* as a metabolic engineering platform for the production of very-long-chain wax esters. J. Agricult. Food Chem. 68, 10730–10740.

Ghogare, R., Chen, S., Xiong, X., 2020. Metabolic engineering of oleaginous yeast *Yarrowia lipolytica* for overproduction of fatty acids. Front. Microbiol. 11, 1717.

Gomaa, M.A., Al-Haj, L., Abed, R.M.M., 2016. Metabolic engineering of cyanobacteria and microalgae for enhanced production of biofuels and high-value products. J. Appl. Microbiol. 121, 919–931.

Gu, S., Li, J., Chen, B., Sun, T., Liu, Q., Xiao, D., et al., 2018. Metabolic engineering of the thermophilic filamentous fungus *Myceliophthora thermophila* to produce fumaric acid. Biotechnol. Biofuels 11, 1–10.

Gutiérrez-García, A.K., Alvarez-Guzmán, C.L., De Leon-Rodriguez, A., 2020. Autodisplay of alpha amylase from *Bacillus megaterium* in *E. coli* for the bioconversion of starch into hydrogen, ethanol and succinic acid. Enzyme Microbial. Technol. 134, 109477.

Han, L., Zhao, Y., Cui, S., Liang, B., 2018. Redesigning of microbial cell surface and its application to whole-cell biocatalysis and biosensors. Appl. Biochem. Biotechnol. 185, 396–418.

He, Y., Wu, B., Xia, W., Zhao, K.Y., Qin, Y., Tan, Q., et al., 2021. Metabolic engineering of *Zymomonas moblis* for ethylene production from straw hydrolysate. Appl. Microbiol. Biotechnol. 105, 1709–1720.

He, B., Hao, B., Yu, H., Tu, F., Wei, X., Xiong, K., et al., 2022. Double integrating XYL2 into engineered *Saccharomyces cerevisiae* strains for consistently enhanced bioethanol production by effective xylose and hexose co-consumption of steam-exploded lignocellulose in bioenergy crops. Renewable Energy 186, 341–349.

Hirokawa, Y., Dempo, Y., Fukusaki, E., Hanai, T., 2017. Metabolic engineering for isopropanol production by an engineered cyanobacterium, *Synechococcus elongatus* PCC 7942, under photosynthetic conditions. J. Biosci. Bioeng. 123, 39–45.

Holwerda, E.K., Olson, D.G., Ruppertsberger, N.M., Stevenson, D.M., Murphy, S.J., Maloney, M.I., et al., 2020. Metabolic and evolutionary responses of *Clostridium thermocellum* to genetic interventions aimed at improving ethanol production. Biotechnol. Biofuels 13, 1–20.

Hussain, S.A., Hameed, A., Khan, M.A.K., Zhang, Y., Zhang, H., Garre, V., et al., 2019. Engineering of fatty acid synthases (FASs) to boost the production of medium-chain fatty acids (MCFAs) in *Mucor circinelloides*. Int. J. Molecular Sci. 20, 786.

Jagtap, R.S., Mahajan, D.M., Mistry, S.R., Bilaiya, M., Singh, R.K., Jain, R., 2019. Improving ethanol yields in sugarcane molasses fermentation by engineering the high osmolarity glycerol pathway while maintaining osmotolerance in *Saccharomyces cerevisiae*. Appl. Microbiol. Biotechnol. 103, 1031–1042.

Jagtap, S.S., Bedekar, A.A., Singh, V., Jin, Y.S., Rao, C.V., 2021. Metabolic engineering of the oleaginous yeast *Yarrowia lipolytica* PO1f for production of erythritol from glycerol. Biotechnol. Biofuels 14, 1–17.

Jia, H.R., Zhu, Y.X., Chen, Z., Wu, F.G., 2017. Cholesterol-assisted bacterial cell surface engineering for photodynamic inactivation of Gram-positive and Gram-negative bacteria. ACS Appl. Mater. Interfaces 9, 15943–15951.

Jia, D., He, M., Tian, Y., Shen, S., Zhu, X., Wang, Y., et al., 2021. Metabolic engineering of gas-fermenting *Clostridium ljungdahlii* for efficient co-production of isopropanol, 3-hydroxybutyrate, and ethanol. ACS Synth. Biol. 10, 2628–2638.

Jiang, Y., Dong, W., Xin, F., Jiang, M., 2020. Designing synthetic microbial consortia for biofuel production. Trends Biotechnol. 38, 828–831.

Joshi, A., Kanthaliya, B., Arora, J., 2019. Current scenario of potential renewable energy sources for sustainable development in India. The J. Plant Sci. Res. 35, 205–214.

Joshi, A., Kanthaliya, B., Arora, J., 2020. Halophytes: the nonconventional crops as source of biofuel production. In: Grigore, M.N. (Ed.), Handbook of Halophytes: From Molecules to Ecosystems towards Biosaline Agriculture. Springer Nature, Switzerland AG, pp. 1–28.

Joshi, A., Kanthaliya, B., Meena, S., Khan, F., Arora, J., 2021. Process consolidation approaches for cellulosic ethanol production. In: Ray, R.C. (Ed.), Sustainable Biofuels. Academic Press, pp. 43–72.

Joshi, A., Verma, K.K., Rajput, V.D., Minkina, T., Arora, J., 2022. Recent advances in metabolic engineering of microorganisms for advancing lignocellulose-derived biofuels, Bioengineered., 13, pp. 8135–8163.

Kalnenieks, U., Pentjuss, A., Rutkis, R., Stalidzans, E., Fell, D.A., 2014. Modeling of *Zymomonas mobilis* central metabolism for novel metabolic engineering strategies. Front. Microbiol. 5, 42.

Keasling, J., Martin, H., Lee, T.S., Mukhopadhyay, A., Singer, S.W., Sundstrom, E., 2021. Microbial production of advanced biofuels. Nat. Rev. Microbiol. 1–15.

Lai, M.C., Lan, E.I., 2015. Advances in metabolic engineering of cyanobacteria for photosynthetic biochemical production. Metabolites. 5, 636–658.

Lakshmi, N.M., Binod, P., Sindhu, R., Awasthi, M.K., Pandey, A., 2021. Microbial engineering for the production of isobutanol: current status and future directions. Bioengineered. 12, 12308–12321.

Lauer, I., Philipps, G., Jennewein, S., 2022. Metabolic engineering of *Clostridium ljungdahlii* for the production of hexanol and butanol from CO_2 and H_2. Microb. Cell Fact. 21, 1–18.

Lee, S.M., Jellison, T., Alper, H.S., 2014. Systematic and evolutionary engineering of a xylose isomerase-based pathway in *Saccharomyces cerevisiae* for efficient conversion yields. Biotechnol. Biofuels 7, 1–8.

Lee, S., Akeprathumchai, S., Bundidamorn, D., Salaipeth, L., Poomputsa, K., Ratanakhanokchai, K., et al., 2021. Interplays of enzyme, substrate, and surfactant on hydrolysis of native lignocellulosic biomass. Bioengineered 12, 5110–5124.

Lee, M.E., Ko, Y.J., Hwang, D.H., Cho, B.H., Jeong, W.Y., Bhardwaj, N., et al., 2022. Surface display of enzyme complex on *Corynebacterium glutamicum* as a whole cell biocatalyst and its consolidated bioprocessing using fungal-pretreated lignocellulosic biomass. Bioresour. Technol. 127758.

Li, L., Li, K., Wang, Y., Chen, C., Xu, Y., Zhang, L., et al., 2015. Metabolic engineering of *Enterobacter cloacae* for high-yield production of enantiopure (2R, 3R)-2, 3-butanediol from lignocellulose-derived sugars. Metabol. Eng. 28, 19–27.

Li, Y.J., Lu, Y.Y., Zhang, Z.J., Mei, S., Tan, T.W., Fan, L.H., 2017. Co-fermentation of cellulose and sucrose/xylose by engineered yeasts for bioethanol production. Energy Fuels 31, 4061–4067.

Li, J., Zhang, Y., Li, J., Sun, T., Tian, C., 2020. Metabolic engineering of the cellulolytic thermophilic fungus *Myceliophthora thermophila* to produce ethanol from cellobiose. Biotechnol. Biofuels 13, 1–15.

Li, Y., Wang, Y., Wang, R., Yan, X., Wang, J., Wang, X., et al., 2022. Metabolic engineering of *Zymomonas mobilis* for continuous co-production of bioethanol and poly-3-hydroxybutyrate (PHB). Green Chem. 24, 2588–2601.

Liang, F., Englund, E., Lindberg, P., Lindblad, P., 2018. Engineered cyanobacteria with enhanced growth show increased ethanol production and higher biofuel to biomass ratio. Metabol. Eng. 46, 51–59.

Liberato, V., Benevenuti, C., Coelho, F., Botelho, A., Amaral, P., Pereira Jr, N., et al., 2019. *Clostridium* sp. as bio-catalyst for fuels and chemicals production in a biorefinery context. Catalysts. 9, 962.

Liu, Z., Ho, S.H., Hasunuma, T., Chang, J.S., Ren, N.Q., Kondo, A., 2016a. Recent advances in yeast cell-surface display technologies for waste biorefineries. Biores. Technol 215, 324–333.

Liu, Z., Ho, S.H., Sasaki, K., Den Haan, R., Inokuma, K., Ogino, C., et al., 2016b. Engineering of a novel cellulose-adherent cellulolytic *Saccharomyces cerevisiae* for cellulosic biofuel production. Sci. Rep. 6, 1–10.

Liu, L., Jin, M., Huang, M., Zhu, Y., Yuan, W., Kang, Y., et al., 2021. Engineered Polyploid Yeast Strains Enable Efficient Xylose Utilization and Ethanol Production in Corn Hydrolysates. Front. Bioeng. Biotechnol. 9, 655272.

Lopez-Hidalgo, A.M., Hernandez, V.E.B., De Leon-Rodriguez, A., 2021. Scale-up of hydrogen and ethanol co-production by an engineered *Escherichia coli*. Fuel. 300, 121002.

Lou, J., Wang, J., Yang, Y., Yang, Q., Li, R., Hu, M., et al., 2021. Development and characterization of efficient xylose utilization strains of *Zymomonas mobilis*. Biotechnol. Biofuels 14, 1–18.

Madavi, T.B., Chauhan, S., Keshri, A., Alavilli, H., Choi, K.Y., Pamidimarri, S.D., 2022. Whole-cell biocatalysis: Advancements toward the biosynthesis of fuels. Biofuels. Bioproducts Biorefining 16, 859–876.

Mat Aron, N.S., Khoo, K.S., Chew, K.W., Show, P.L., Chen, W.H., Nguyen, T.H.P., 2020. Sustainability of the four generations of biofuels—a review. Int. J. Energy Res 44, 9266–9282.

Meyer, V., 2021. Metabolic engineering of filamentous fungi. In: Nielsen, J., Stephanopoulos, G., Lee, S.Y. (Eds.), Metabolic Engineering: Concepts and Applications, 13, pp. 765–801.

Mhlongo, S.I., Ezeokoli, O.T., Roopnarain, A., Ndaba, B., Sekoai, P.T., Habimana, O., et al., 2021. The potential of single-cell oils derived from filamentous fungi as alternative feedstock sources for biodiesel production. Front. Microbiol. 12, 57.

Mustila, H., Kugler, A., Stensjö, K., 2021. Isobutene production in *Synechocystis* sp. PCC 6803 by introducing α-ketoisocaproate dioxygenase from *Rattus norvegicus*. Metabol. Eng. Commun. 12, 00163.

Namakoshi, K., Nakajima, T., Yoshikawa, K., Toya, Y., Shimizu, H., 2016. Combinatorial deletions of *glgC* and *phaCE* enhance ethanol production in *Synechocystis sp.* PCC 6803. J. Biotechnol. 239, 13–19.

Okolie, J.A., Mukherjee, A., Nanda, S., Dalai, A.K., Kozinski, J.A., 2021. Next-generation biofuels and platform biochemicals from lignocellulosic biomass. Int. J. Energy Res. 45, 14145–14169.

Purdy, H.M., Pfleger, B.F., Reed, J.L., 2022. Introduction of NADH-dependent nitrate assimilation in *Synechococcus* sp. PCC 7002 improves photosynthetic production of 2-methyl-1-butanol and isobutanol. Metabol. Eng. 69, 87–97.

Qiu, M., Shen, W., Yan, X., He, Q., Cai, D., Chen, S., et al., 2020. Metabolic engineering of *Zymomonas mobilis* for anaerobic isobutanol production. Biotechnol. Biofuels 13, 1–14.

Raoufi, Z., Gargari, S.L.M., 2018. Biodiesel production from microalgae oil by lipase from *Pseudomonas aeruginosa* displayed on yeast cell surface. Biochem. Eng. J. 140, 1–8.

Rodrigues, J.S., Lindberg, P., 2021. Metabolic engineering of *Synechocystis* sp. PCC 6803 for improved bisabolene production. Metabol. Eng. Comm. 12, e00159.

Rodrussamee, N., Sattayawat, P., Yamada, M., 2018. Highly efficient conversion of xylose to ethanol without glucose repression by newly isolated thermotolerant *Spathaspora passalidarum* CMUWF1–2. BMC Microbial. 18, 1–11.

Roussou, S., Albergati, A., Liang, F., Lindblad, P., 2021. Engineered cyanobacteria with additional overexpression of selected Calvin-Benson-Bassham enzymes show further increased ethanol production. Metabol. Eng. Comm. 12, e00161.

Saini, M., Wang, Z.W., Chiang, C.J., Chao, Y.P., 2017. Metabolic engineering of *Escherichia coli* for production of n-butanol from crude glycerol. Biotechnol. Biofuel. 10, 1–8.

Sarkar, P., Goswami, G., Mukherjee, M., Das, D., 2021. Heterologous expression of xylose specific transporter improves xylose utilization by recombinant *Zymomonas mobilis* strain in presence of glucose. Process Biochem. 102, 190–198.

Sarwar, A., Nguyen, L.T., Lee, E.Y., 2022. Bio-upgrading of ethanol to fatty acid ethyl esters by metabolic engineering of *Pseudomonas putida* KT2440. Bioresour. Technol. 350, 126899.

Schultz, J.C., Mishra, S., Gaither, E., Mejia, A., Dinh, H., Maranas, C., et al., 2022. Metabolic engineering of *Rhodotorula toruloides* IFO0880 improves C16 and C18 fatty alcohol production from synthetic media. Microb. Cell Fact. 21, 1–14.

Selim, K.A., Easa, S.M., El-Diwany, A.I., 2020. The xylose metabolizing yeast *Spathaspora passalidarum* is a promising genetic treasure for improving bioethanol production. Fermentation 6, 33.

Shanmugam, S., Sun, C., Zeng, X., Wu, Y.R., 2018. High-efficient production of biobutanol by a novel *Clostridium* sp. strain WST with uncontrolled pH strategy. Bioresour. Technol. 256, 543–547.

Sharma, V., Joshi, A., Ramawat, K.G., Arora, J., 2017. Bioethanol production from halophytes of Thar desert: a Green Gold. In: Basu, S.K., Zandi, P., Chalaras, S.K. (Eds.), Environment at Crossroads: Challenges, Dynamics and Solutions. Haghshenass Publishing, Iran, pp. 219–235.

Smith, M.R., Khera, E., Wen, F., 2015. Engineering novel and improved biocatalysts by cell surface display. Indus. Eng. Chem Res. 54, 4021–4032.

Takayama, S., Ozaki, A., Konishi, R., Otomo, C., Kishida, M., Hirata, Y., et al., 2018. Enhancing 3-hydroxypropionic acid production in combination with sugar supply engineering by cell surface-display and metabolic engineering of *Schizosaccharomyces pombe*. Microb. Cell Fact. 17, 1–11.

Tanaka, T., Kondo, A., 2015. Cell surface engineering of industrial microorganisms for biorefining applications. Biotechnol. Adv. 33, 1403–1411.

Tang, H., Wang, J., Wang, S., Shen, Y., Petranovic, D., Hou, J., et al., 2018. Efficient yeast surface-display of novel complex synthetic cellulosomes. Microb. Cell Fact. 17, 1–13.

Tian, L., Conway, P.M., Cervenka, N.D., Cui, J., Maloney, M., Olson, D.G., et al., 2019. Metabolic engineering of *Clostridium thermocellum* for n-butanol production from cellulose. Biotechnol. Biofuels 12, 1–13.

Wang, C., Pfleger, B.F., Kim, S.W., 2017. Reassessing *Escherichia coli* as a cell factory for biofuel production. Curr. Opin. Biotechnol. 45, 92–103.

Wang, X., He, Q., Yang, Y., Wang, J., Haning, K., Hu, Y., et al., 2018. Advances and prospects in metabolic engineering of *Zymomonas mobilis*. Metabol. Eng. 50, 57–73.

Wang, M., Luan, G., Lu, X., 2020a. Engineering ethanol production in a marine cyanobacterium *Synechococcus* sp. PCC7002 through simultaneously removing glycogen synthesis genes and introducing ethanolgenic cassettes. J. Biotechnol. 317, 1–4.

Wang, Z., Cao, X., Li, N., Yang, Z., Lei, M., Zhao, Y., et al., 2020b. Production of butanol directly from hemicellulose through secretory expression of a xylanase in *Clostridium acetobutylicum*. Energy Fuels 34, 3376–3382.

Wang, Y., Selvamani, V., Yoo, I.K., Kim, T.W., Hong, S.H., 2021. A novel strategy for the microbial removal of heavy metals: Cell-surface display of peptides. Biotechnol. Biopr. Eng. 26, 1–9.

Wei, L.J., Ma, Y.Y., Cheng, B.Q., Gao, Q., Hua, Q., 2021. Metabolic engineering *Yarrowia lipolytica* for a dual biocatalytic system to produce fatty acid ethyl esters from renewable feedstock in situ and in one pot. Appl. Microbiol. Biotechnol. 105, 8561–8573.

Workman, M., Andersen, M.R., Thykaer, J., 2013. Integrated approaches for assessment of cellular performance in industrially relevant filamentous fungi. Industr. Biotechnol. 9, 337–344.

Wu, Y., Chu, W., Yang, J., Xu, Y., Shen, Q., Yang, H., et al., 2021. Metabolic engineering of *Enterobacter aerogenes* for improved 2,3-butanediol production by manipulating NADH levels and overexpressing the small RNA RyhB. Front. Microbiol. 12, 754306.

Xin, F., Chen, T., Jiang, Y., Lu, J., Dong, W., Zhang, W., et al., 2017. Enhanced biobutanol production with high yield from crude glycerol by acetone uncoupled *Clostridium sp.* strain CT7. Biores. Technol. 244, 575–581.

Xu, Y., Li, F., Yang, K., Qiao, Y., Yan, Y., Yan, J., 2019. A facile and robust non-natural three enzyme biocatalytic cascade based on *Escherichia coli* surface assembly for fatty alcohol production. Energy Conver. Management. 181, 501–506.

Yang, Y., Hu, M., Tang, Y., Geng, B., Qiu, M., He, Q., et al., 2018. Progress and perspective on lignocellulosic hydrolysate inhibitor tolerance improvement in *Zymomonas mobilis*. Biores. Bioprocess. 5, 1–12.

Yang, J., Khan, M.A.K., López-García, S., Nosheen, S., Nazir, Y., Zhang, H., et al., 2020. Improved SDA production in high lipid accumulating strain of *Mucor circinelloides* WJ11 by genetic modification. Am. J. Biochem. Biotechnol. 16, 147.

Ye, M., Ye, Y., Du, Z., Chen, G., 2021. Cell-surface engineering of yeasts for whole-cell biocatalysts. Bioprocess Biosyst. Eng. 44, 1003–1019.

Zada, B., Wang, C., Park, J.B., Jeong, S.H., Park, J.E., Singh, H.B., et al., 2018. Metabolic engineering of *Escherichia coli* for production of mixed isoprenoid alcohols and their derivatives. Biotechnol. Biofuels 11, 1–12.

Zhang, K., Lu, X., Li, Y., Jiang, X., Liu, L., Wang, H., 2019. New technologies provide more metabolic engineering strategies for bioethanol production in *Zymomonas mobilis*. Appl. Microbiol. Biotechnol. 103, 2087–2099.

Zhang, Y., Peng, J., Zhao, H., Shi, S., 2021a. Engineering oleaginous yeast *Rhodotorula toruloides* for overproduction of fatty acid ethyl esters. Biotechnol. Biofuels 14, 1–11.

Zhang, Y., Song, X., Lai, Y., Mo, Q., Yuan, J., 2021b. High-yielding terpene-based biofuel production in *Rhodobacter capsulatus*. ACS Synth. Biol. 10, 1545–1552.

Zheng, L., Wei, S., Wu, M., Zhu, X., Bao, X., Hou, J., et al., 2020. Improving xylose fermentation in *Saccharomyces cerevisiae* by expressing nuclear-localized hexokinase 2. Microorganisms. 8, 856.

Zhu, L., Li, P., Sun, T., Kong, M., Li, X., Ali, S., et al., 2020. Overexpression of SFA1 in engineered *Saccharomyces cerevisiae* to increase xylose utilization and ethanol production from different lignocellulose hydrolysates. Bioresour. Technol. 313, 123724.

Zhu, K., Zhao, B., Zhang, Y., Kong, J., Rong, L., Liu, S., et al., 2022. Mitochondrial engineering of *Yarrowia lipolytica* for sustainable production of α-bisabolene from waste cooking oil. ACS Sustain. Chem. Eng. 10, 9644–9653.

Lignocellulosic biofuel production: Insight into microbial factories

10

Yeshona Sewsynker-Sukai[1], Anthea N. David[2], Milesh Laltha[2], Edson L. Meyer[1] and Evariste B. Gueguim Kana[2]
[1]Fort Hare Institute of Technology, University of Fort Hare, Alice, South Africa, [2]School of Life Sciences, University of KwaZulu-Natal, Pietermaritzburg, South Africa

10.1 Introduction

Microbial factories produce an array of bioproducts that include biofuels and biochemicals with the potential to revolutionize the global bioeconomy and address key challenges (Banu et al., 2021). Currently, one major challenge is the dependence on nonrenewable petroleum-based energy sources with their subsequent depletion due to the ever-changing growth in the human population coupled with social development and industrialization (Patel and Shah, 2021). Therefore the development of alternative energy sources and efficiently harnessing energy from them is imperative (Sewsynker-Sukai et al., 2020). In part, biofuels have emerged as potential alternative energy sources to replace conventional fossil fuels. They have shown to be promising since they are renewable, sustainable, and minimize the carbon footprint (Sewsynker-Sukai et al., 2018). However, biofuels have been traditionally produced using food crops such as sugarcane, corn, and wheat, which has questioned their application due to their negative impact on global food security (Mohapatra et al., 2018; Ray and Ramachandran, 2018; Sarawan et al., 2019).

Over the past decade, lignocellulosic biofuel production has increasingly gained widespread attention to curb the use of food crops since these substrates are considered wastes, abundant, and renewable (Moodley et al., 2020; Moodley, 2021). However, the presence of resistant lignin structures within the lignocellulosic biomass (LCB) prevents its direct utilization by microorganisms. Therefore, the use of LCB for microbial bioprocesses requires pretreatment to release fermentable sugars that microorganisms can harness for their growth, replication, and metabolic functioning (Jugwanth et al., 2020). Examples of biofuels that can be generated using LCB include bioethanol, biobutanol, biohydrogen, and biogas. These fuels represent a green energy perspective, each with its merits and limitations. Biofuels have previously been produced using wastes from sugarcane, corn, wheat, or rice, including the bagasse, leaves, stalks, and husks. However, various key technological issues have challenged economic lignocellulosic biofuel production (Sewsynker-Sukai et al., 2018). Identifying and addressing the challenges of lignocellulosic biofuel processes pave the way toward industrial-scale application. The major bottleneck is the high cost that renders these processes unfeasible for large-scale production.

Advances in Lignocellulosic Biofuel Production Systems. DOI: https://doi.org/10.1016/B978-0-323-91192-4.00013-4

The lignocellulosic biofuel production process entails pretreatment, enzymatic hydrolysis, and microbial fermentation. The lignocellulosic biofuel production system may be carried out using different bioprocess types that include:

1. Separate hydrolysis and fermentation (SHF).
2. Simultaneous saccharification and fermentation (SSF).
3. SSF with a prehydrolysis step (PSSF) (Moodley et al., 2020; David et al., 2020).

Several strategies are currently being undertaken globally to enhance the lignocellulosic biofuel production process toward reducing the cost and achieving an industrial-scale system that can be commercialized. These approaches are tailored in line with developing and enhancing: (1) cost-effective pretreatment methods that produce high sugar yields from LCB, (2) microbial fermentation process types that result in high biofuel yields and shorter fermentation times, and (3) kinetic knowledge of bioprocesses with regards to microbial growth and product formation for a translatable outcome when replicated at industrial scale (David et al., 2020).

Kinetic modeling of cell growth and product formation has recently been tagged as a practical approach to determining the efficiency of the fermentation process with regard to specific parameters (David et al., 2020). Some examples of kinetic parameters include the maximum specific growth of the cells, lag time, and yield productivity since these parameters can be used as translation indices during scale-up toward industrialization (Sewsynker-Sukai and Gueguim Kana, 2018). Despite the abundance of knowledge that is conferred from these kinetic parameters, the majority of reports have been concentrated on bioethanol (David et al., 2020; Sewsynker-Sukai and Gueguim Kana, 2018; Moodley and Gueguim Kana, 2019; Jugwanth et al., 2020; Sarawan et al., 2019) and to a lesser extent biobutanol (Chen et al., 2013; Rochón et al., 2018, 2020). Nonetheless, biofuel production and the microbial metabolic pathways involved are influenced by factors such as temperature, pH, and nutrient concentration, revealing key information for maximizing the process yield.

This chapter provides insight into the advancement of microbial biofuel production from LCB. Biofuels such as bioethanol, biobutanol, biohydrogen, and biogas are highlighted with the current progress on these processes pertaining to the kinetics, microbial metabolic pathways, and key process parameters briefly discussed. Furthermore, the current challenges and progress in lignocellulosic biofuel production are outlined.

10.2 Lignocellulosic biomass and pretreatment

LCB substrates are plant waste materials presently being sought as potential feedstocks for biofuel production without negative consequences on food security (Jugwanth et al., 2020). The annual global production of LCB is approximately 120 billion tons (Abraham et al., 2020). LCB substrates include corn stover, cobs, sugarcane bagasse, banana pseudostem, switchgrass, *Miscanthus*, and wheat straw.

Nevertheless, LCB substrates consist of recalcitrant structures such as cellulose, hemicellulose, and lignin that cannot be directly utilized by microorganisms (Muthuvelu et al., 2019). The composition of LCB ranges between 30% and 50% cellulose, 10% and 45% hemicellulose, and 2% and 40% lignin, depending on the type of substrate (McKendry, 2002). Cellulose, hemicellulose, and lignin structures are connected by hydrogen and covalent bonds that influence the recalcitrance of lignocellulosic substrates (Rebello et al., 2020). Consequently, a pretreatment step is crucial for the degradation of the recalcitrant structures within LCB (Kumar et al., 2020).

Pretreatment of LCB is a prerequisite step for its conversion to biofuels and other biobased products (Kim, 2013). A major obstacle that hinders the use of LCB as an energy source in biorefinery systems is the presence of recalcitrant structures within these materials (Himmel et al., 2007). Pretreatment methods are crucial for disrupting resistant lignin structures and enhancing subsequent cellulose conversion to glucose during enzymatic hydrolysis reactions. Lignocellulosic pretreatment consists of physical (mechanical, microwave, and ultrasound), chemical (acids, bases, inorganic salts, ionic liquids, and organosolvents), physicochemical (steam explosion, ammonia fiber explosion, CO_2 explosion, and liquid hot water), and biological methods (enzymes and whole cells) (Fig. 10.1) (Jędrzejczyk et al., 2019). Due to the higher efficiency, hybrid strategies that combine some of the abovementioned methods are typically used. However, present pretreatment methods are expensive, release high concentrations of enzymatic and fermentation inhibitory compounds, produce too low sugar yields, and are resource-demanding and energy exhaustive, thus posing a significant hindrance to the use of lignocellulosic substrates. The development of high-performance pretreatments is ongoing and remains a crucial aspect for use of lignocellulosic bioprocesses. Criteria for an ideal pretreatment method include using cost-effective chemicals or wastes with low energy that release high sugar yields and low inhibitor concentrations and do not negatively impact the environment.

10.3 Microbial fermentation and process types

Fermentation is a metabolic process by which various microorganisms metabolize sugars for the production of valuable products such as biofuels and biochemicals (Kang et al., 2014). The fermentation of lignocellulosic wastes into useful bio-based products or renewable biofuels can be achieved through various strategies, including SHF, SSF, and PSSF (Fig. 10.2) (Morales-Martínez et al., 2017, Dahnum et al., 2015). The SHF process is a conventional method performed in two separate biochemical steps with different temperatures. It includes an enzymatic hydrolysis stage followed by the fermentation process. On the other hand, the SSF process allows hydrolysis and fermentation to be carried out in a single bioreactor at a median temperature that can accommodate the cellulase enzyme and the microbial culture, respectively (David et al., 2020). A variation of SSF and SHF processes is the PSSF process. In the PSSF process type, pretreated biomass is prehydrolyzed at the optimal temperature for cellulolytic enzymes for a short period of time (6–24 h), which is followed by a change in temperature for

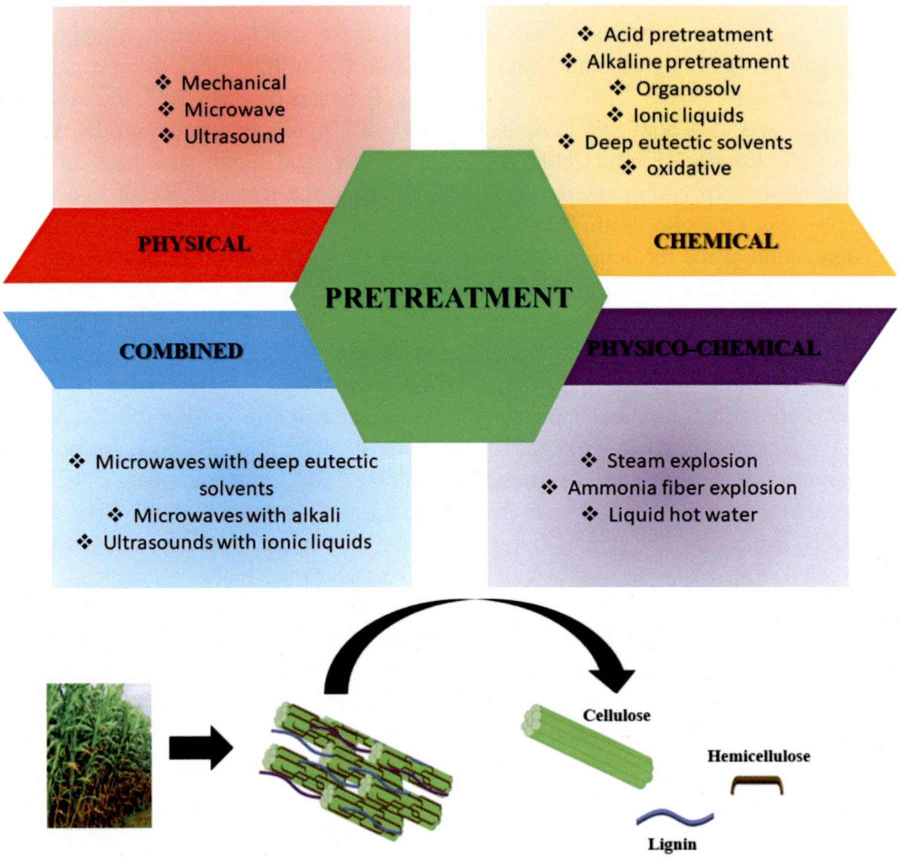

Figure 10.1 Different methods of lignocellulosic biomass pretreatment.

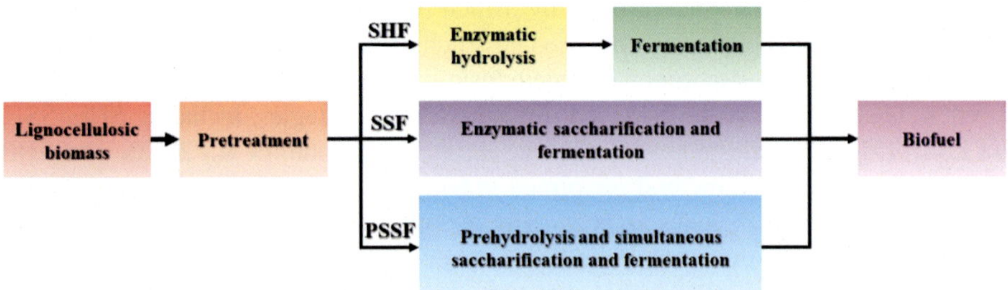

Figure 10.2 Process flow of separate hydrolysis and fermentation; simultaneous saccharification and fermentation, and prehydrolysis and simultaneous saccharification and fermentation.

inoculation of the microbial culture during fermentation (Morales-Martínez et al., 2017, Guerrero et al., 2018). The selection of these process types at an industrial scale remains contentious due to their advantages and limitations. For instance, the SHF system ensures that enzymatic saccharification and fermentation occur in different vessels at their optimal conditions. During these steps, the pretreated substrate is enzymatically hydrolyzed to fermentable sugars, followed by the fermentation of available sugars to the product of interest (Guerrero et al., 2018). The main limitations of the SHF process include the inhibitory effect on cellulase enzymes due to the accumulation of fermentable sugars (Portero Barahona et al., 2020). In addition, there is a risk of contamination when transferring the enzymatic hydrolysate to the fermentation reactor. The SHF process exhibits high operational costs due to long processing times in separate vessels (Giang et al., 2019).

Major drawbacks of the SHF process can be overcome by the implementation of the SSF approach since hydrolysis and fermentation occur simultaneously in a single vessel and eliminates long process times (Da Silva et al., 2018). Additionally, the SSF process does not cause cellulase inhibition due to the simultaneous conversion of simple sugars (glucose) to bioproducts while reducing the risk of contamination (Carrillo-Nieves et al., 2017). When compared to SHF, SSF necessitates a shorter process time, leading to higher productivity yields (Giang et al., 2019). Nevertheless, the SSF process type also presents limitations attributable to ineffective hydrolysis stages carried out under suboptimal conditions (temperature and pH) that may yield significantly lower sugar and have a negative impact on the enzyme. The PSSF system, therefore bridges the gap between the SHF and SSF process types since it includes a short prehydrolysis step followed by fermentation. The benefits of enzymatic saccharification at its optimum setpoint promote the release of fermentable sugars before fermentation (Carrillo-Nieves et al., 2017). However, the prehydrolysis stages incur higher costs than the SSF process due to energy consumption and longer processing times, reducing its feasibility on a large scale. Compared to the SHF and PSSF process types, the SSF system offers several advantages for large-scale applications since it reduces process time and energy required by completely omitting separate hydrolysis steps (David et al., 2020).

For this reason, the SSF process type has been widely investigated for bioethanol production from various substrates, including banana waste (Guerrero et al., 2018), mango stem bark (Carrillo-Nieves et al., 2017), corn cobs (David et al., 2020), sugarcane wastes (Jugwanth et al., 2020; Moodley and Gueguim Kana, 2019) and *Agave lechuguilla* (Morales-Martínez et al., 2017), among several others. To date, all three different process types (SSF, SHF, and PSSF) have been extensively explored for bioethanol production. However, the literature on other biofuels such as biobutanol, biogas, and biohydrogen remains scanty.

10.4 Kinetic modeling for bioprocess development

Kinetic models provide significant insight into understanding the rational design of microbial metabolic processes and explore the phenomena in complex biological systems. This information leads to accelerated cell factory engineering and defines the parameters for bioprocess control when considering industrial-scale applications (Oliveira et al., 2016). Bioprocess modeling represents the biological, physical, and chemical aspects of a bioreactor and aims to select and optimize the process parameters affecting microbial growth and product formation (Fedailaine et al., 2015). The kinetics models provide insight into the information gathered from the empirical observations through the experimental output. Thus predicting the optimal conditions within a fermentation process assists in (1) increasing the product yield and productivity, (2) minimizing undesired byproduct formation, (3) managing resource utilization, (4) cost and time efficiency, and (5) transforming lab-scale reactions to industrial-scale reactions, while maintaining high product purity (Kucharska et al., 2018a).

The kinetic growth modeling approaches are required to account for the biological system's regulatory effects and dynamics at a metabolite level (Costa et al., 2016). These models are usually based on a set of mathematical expressions that quantitatively describe the systematic occurrences and predict the response to various input results. There are many types of kinetic models used to describe specific growth rates of microorganisms and product formation, namely, the Monod model, logistic model and Haldane model, modified Gompertz model, and modified logistic model, amongst others. These models interpret microorganisms as independent components that interact with the environment, and their growth rate is determined by the biomass and substrate concentrations (Dong et al., 2015; Panikov, 1999). The Monod model describes the relationship between the specific growth rate and substrate utilization within the bioreactor (Monod, 1949). This model type can interpret data based on the substrate and biomass concentration as well as the substrate concentration only (Muloiwa et al., 2020). Although the model is most widely used, it presents many limitations, such as assuming that the bioreactor only consists of one growth-limiting substrate and its inability to be used when substrate inhibition occurs. It does not consider the lag and death phase of the growth cycle of the microbes resulting in overestimating the specific growth rate (Kong, 2017).

On the other hand, the logistic model, also known as the Verhulst model, is a substrate-independent kinetic model that observes changes in biomass growth and relates it to the growth rate and initial and maximum concentrations over time (Muloiwa et al., 2020). The model also accommodates the toxicity of the substrate, which results in substrate inhibition (Ali et al., 2017). The Haldane model is an extension of the Monod model that addresses the discrepancies of the latter model. The model includes an inhibition constant that allows it to handle toxic and nontoxic substrates and specific growth rate inhibition at low and high substrate concentrations (Muloiwa et al., 2020). Furthermore, the model can account for all growth phases, including the lag, exponential, stationary, and death phases.

Regarding product formation, the modified Gompertz model has been widely used for nonlinear modeling of cumulative biofuel production. The resultant data that fits the modified Gompertz equation assumes that the biofuel production in a batch mode is a function of the specific growth rate of the microorganisms within the bioreactor. The model can determine the production lag time, maximum production rate, and maximum production concentration on a given substrate (Dodić et al., 2012). The modified logistic model is similar to that of the modified Gompertz model, which contains many constants of biological meaning that facilitates a better understanding of the reaction.

The abovementioned kinetic models, amongst others, consider the reliability of the results, sampling methods, analysis, and degree of accuracy when conducting experiments. The resultant outcome of these procedures enables the model to predict the behavioral trends of biochemical reactions. This knowledge is a critical aspect that contributes significantly to the technoeconomic analysis and interprets the phenomena within a process more precisely, thus enhancing productivity (González-Figueredo et al., 2018).

10.5 Lignocellulosic biofuel production

The global energy demand has escalated in light of population spikes, modern living, and urbanization, with its consumption, predicted to reach over 106 quadrillions British thermal units (Btu) by the year 2050 (Tabatabaei et al., 2019; Energy Information Administration, 2019). However, the finite supply of fossil fuels has jeopardized energy security since it accounts for more than 80% of the total energy markets (Zabed et al., 2016). Additionally, the excessive release of anthropogenic greenhouse gases has triggered unexpected environmental fluctuations (Soltanian et al., 2020). These drawbacks have strengthened the need to switch from fossil-based fuels to clean, renewable, and cheaper alternative sources such as biofuels produced from lignocellulosic waste (Kumari and Singh, 2018). Kinetic modeling plays a key role during the development, optimization, and scale-up of bioprocesses and has been applied in the lignocellulosic production of different biofuels. The different biofuels and processes are briefly discussed below.

10.5.1 Bioethanol

Bioethanol is known for its excellent fuel properties, such as higher flame speeds, broader flammability limits, and a high-octane number (108) that increases the heat of vaporization and compression ratio. These characteristics enable bioethanol to be blended with gasoline, which results in the reduced emission of CO_2, unburnt hydrocarbons, and carcinogens. Bioethanol is a product of microbial metabolism that has shown promise as an effective and eco-friendly alternative to fossil fuels. During bioethanol production, hexoses (glucose, fructose, mannose, and galactose) undergo oxidation in the absence of oxygen to generate ethanol, CO_2, and adenosine triphosphate (ATP). When considering commercial bioethanol production, several factors must be considered. For instance, microorganisms must have a high fermentative activity for monosaccharides, provide high bioethanol yields, maintain genetic stability and tolerate environmental stressors (inhibitors, osmotic changes, and high alcohol concentrations, amongst others) (Robak and Balcerek, 2018).

Traditionally, *Saccharomyces cerevisiae* and *Zymomonas mobilis* have been most commonly involved in bioethanol production. *S. cerevisiae* is a robust and well-established microorganism used frequently in industrial processes (alcohol and baking). This microbe can metabolize hexoses using the Embden–Meyerhof–Parnas (EMP) pathway in which pyruvate is generated and converted into ethanol and CO_2 by specific enzymes (Marin et al., 2017). On the other hand, *Z. mobilis*, a Gram-negative bacterium, produces bioethanol at an efficient rate from glucose, fructose, or sucrose as a carbon source. This microorganism follows the Entner–Doudoroff (ED) pathway that converts the fermentable sugars into ethanol, making the ethanologenic process 50% more efficient than the EMP pathway (Majidian et al., 2018). Even though both these microorganisms possess favorable characteristics for bioethanol production, their ineffective pentose phosphate pathway does not accommodate the utilization of pentose sugars for bioethanol (Robak and Balcerek, 2018). This presents a major drawback since pentose sugars are abundantly available in LCB and can drastically reduce expenses within the biorefinery system. As a result, metabolic and genetic engineering aims to address this issue by manipulating these pathways within microbial species for enhanced biofuel production. In addition to bioengineering, kinetic modeling has provided insight into the parameters that will enhance microbial growth and product formation toward industrialization.

Numerous studies have evaluated the kinetics of bioethanol production from different lignocellulosic wastes, as noted in Table 10.1. For instance, Srimachai et al. (2015) varied the age of oil palm and evaluated the kinetics of the oil palm frond juice (OPFJ) utilization for bioethanol production. The OPFJ at 3–4 years gave a maximum specific growth rate (μ_{max}) and Monod constant (K_s) of 0.29/h and 47.05 g/L, respectively, for the Monod model while the maximum ethanol concentration (P_m) of 11.50 g/L and maximum ethanol production rate ($r_{p.m}$) of 0.24 g/L/h was observed for the modified Gompertz model (Srimachai et al., 2015). These results showed that as the oil palm age increased, the μ_{max}, K_s, P_m, and $r_{p.m}$ decreased (Srimachai et al., 2015). Phukoetphim et al. (2017) reported on the production of bioethanol from sweet sorghum juice under high gravity conditions with and without the supplementation of yeast extract. Under high gravity conditions with the supplementation of yeast extract, the u_{max} (0.185/h) and X_{max} (10.6 g/L) were higher than those with no supplementation, which indicates that yeast extract promoted cell growth (Phukoetphim et al., 2017). Furthermore, an increase in the P_m (112.06 g/L) and $r_{p.m}$ (5.25 g/L/h) show that the presence of yeast extract in media enhances cell growth and, in turn, leads to higher sugar consumption and ethanol production (Phukoetphim et al., 2017). Additionally, a comparison of the SSF and PSSF bioprocesses using pretreated corn cob waste was channeled toward bioethanol production under microaerophilic conditions (Sewsynker-Sukai and Gueguim Kana, 2018). It was observed that the SSF process produced a higher μ_{max} of 0.274/h while the PSSF bioprocess gave a higher P_m of 42.24 g/L (Sewsynker-Sukai and Gueguim Kana, 2018). In the same vein, Moodley and Gueguim Kana (2019) assessed the kinetics of two SHF process modes using either filtered enzymatic hydrolysate (SHF-filtered) and unfiltered enzymatic hydrolysate (SHF-unfiltered) for the production of bioethanol by *S. cerevisiae* BY4743. The

Table 10.1 Bioethanol production from various lignocellulosic waste.

Microorganism	Substrate	Process type	Product formation kinetics	Microbial growth kinetics	Reference
Saccharomyces cerevisiae	Oil palm frond juice (3–4 y)	ND	11.5 g/L[a], 0.24 g/L/h[b]	0.29/h[c], 47.05 g/L[d]	Srimachai et al. (2015)
S. cerevisiae	Oil palm frond juice (20–25 y)	ND	2.34 g/L[a], 0.05 g/L/h[b]	0.11/h[c], 1.82 g/L[d]	Srimachai et al. (2015)
S. cerevisiae NP 01	Sweet sorghum juice	SSF	88.48 g/L[a], 2.17 g/L/h[b]	0.164 h^{-1}[c], 6.31 g/L[e]	Phukoetphim et al. (2017)
S. cerevisiae NP 01	Sweet sorghum juice + yeast extract	SSF	112.06 g/L[a], 5.25 g/L/h[b]	0.185/h[c], 10.6 g/L[e]	Phukoetphim et al. (2017)
S. cerevisiae ATCC 24858	Sweet sorghum bagasse	SSF	16.64 g/L[a], 0.51 g/L/h[b]	ND	Wang et al. (2013)
S. cerevisiae BY4743	Sugarcane bagasse	SSF	3.12 g/L[a], 0.29 g/L/h[b]	0.15/h[c], 2.58 g/L[e]	Jugwanth et al. (2020)
S. cerevisiae BY4743	Corn cobs	PSSF	42.24 g/L[a], 2.39 g/L/h[b]	0.216/h[c], 3.65 g/L[e]	Sewsynker-Sukai and Gueguim Kana (2018)
S. cerevisiae BY4743	Corn cobs	SSF	37.87 g/L[a], 2.14 g/L/h[b]	0.274/h[c], 3.52 g/L[e]	Sewsynker-Sukai and Gueguim Kana (2018)
S. cerevisiae BY4743	Sugarcane leaf waste	SHF-unfiltered	31.06 g/L[a], 2.44 g/L/h[b]	0.153/h[c], 4.19 g/L[d]	Moodley and Gueguim Kana (2019)
S. cerevisiae BY4743	Sugarcane leaf waste	SHF-filtered	30.49 g/L[a], 2.81 g/L/h[b]	0.153/h[c], 4.19 g/L[d]	Moodley and Gueguim Kana (2019)
S. cerevisiae BY4743	Corn cobs	SHF	26.82 g/L[a], 1.48 g/L/h[b]	0.28/h[c], 0.70 g/L[e]	David et al. (2020)
S. cerevisiae BY4743	Corn cobs	SSF	16.49 g/L[a], 2.06 g/L/h[b]	0.23/h[c], 1.63 g/L[e]	David et al. (2020)
S. cerevisiae BY4743	Corn cobs	PSSF	18.28 g/L[a], 3.08 g/L/h[b]	0.61/h[c], 0.87 g/L[e]	David et al. (2020)

ND: not determined; *PSSF*: simultaneous saccharification and fermentation with prehydrolysis; *SHF*: separate hydrolysis and fermentation; *SSF*: simultaneous saccharification and fermentation.

[a] P_m—potential maximum bioethanol concentration (g/L),

[b] r_{pm}—maximum bioethanol production rate (g/L/h),

[c] μ_{max}—maximum specific growth rate (per h),

[d] K_s—Monod constant (g/L),

[e] X_{max}—maximum cell biomass concentration (g/L).

SHF-unfiltered process observed a higher P_m of 31.06 g/L compared to the SHF-filtered process (30.49 g/L) (Moodley and Gueguim Kana, 2019). However, the SHF-filtered and SHF-unfiltered processes showed a μ_{max} of 0.153/h. These results indicate that these systems had an insignificant effect on the μ_{max}. Therefore carrying out the SHF-unfiltered process for fermentation can reduce the costs attached to the biomass separation stage usually required in the conventional SHF system (Moodley and Gueguim Kana, 2019). In a recent study, David et al. (2020) reported on the kinetics of microbial growth and ethanol production of all three bioprocess types (SHF, PSSF, and SFF) using the logistic and modified Gompertz models, respectively. The PSSF system obtained the highest μ_{max} (0.61/h), while the SSF process showed the highest X_{max} (1.63 g/L) (David et al., 2020). The high μ_{max} was attributed to the initial glucose produced in the prehydrolysis stage and was readily available to the yeast cells for its growth (David et al., 2020). Furthermore, the high X_{max} for the SSF type results from low initial glucose concentrations that drive the microbial metabolism toward cell growth instead of ethanol production (David et al., 2020). On the other hand, the SHF system produced the highest P_m (26.82 g/L) due to the fermentable sugars readily available to drive metabolism to product formation (David et al., 2020). Even though the SHF process is known to be the most effective for producing bioethanol, its prolonged enzymatic hydrolysis, biomass separation, and risk of contamination due to reactor transfer may incur more expenses, and increase time and energy (David et al., 2020).

10.5.2 Biobutanol

Biobutanol has gained significant attention as a biofuel due to its similarities with gasoline, a well-established petroleum-derived liquid used in spark-ignited internal combustion engines (Majidian et al., 2018). The four-carbon primary straight-chain fuel is produced using the acetone–butanol–ethanol (ABE) fermentation process. The ABE process involves two distinct fermentation phases: acidogenesis and solventogenesis (Ibrahim et al., 2018). During the acidogenic phase, cell growth occurs and converts the substrate to acetic and butyric acids. The acids generated causes the pH of the fermentation media to decrease below 5.0, inhibiting cell growth (Majidian et al., 2018). The low extracellular pH signals and intracellular acyl phosphates signals shift the ABE fermentation from the acidogenesis phase to the solventogenesis phase (Jiang et al., 2019). This phase causes 70%–80% of the viable bacterial cells to sporulate (Al-Hinai et al., 2014). The organic acids are transformed into acetone, butanol, and ethanol in a 3:6:1 ratio (Nithyanandan et al., 2016). Biobutanol is typically produced by *Clostridium* species, including *C. acetobutylicum, C. beijerinckii, C. saccharoperbutylacetonicum*, and *C. saccahrobutylicum*, amongst others. However, the low production profile of *Clostridium* species poses many challenges in the commercialization of the ABE fermentation process. To improve biobutanol production during ABE fermentation, the focus has shifted towards genetically engineering the Clostridia fermentative pathway in microorganisms such as *Escherichia coli* and *Bacillus subtilis*. Despite these attempts, the lack of information regarding the metabolic shift from acidogenesis to solventogenesis and its regulatory interactions hinders the functional butanol biosynthesis pathway. Biobutanol has several benefits compared to traditional biofuels, such as higher energy content, lower volatility, reduced ignition issues, and a higher heating value (Ibrahim et al., 2018). Furthermore, butanol presents similar chemical and physical characteristics to gasoline, allowing it to be easily blended in a pure or mixed form. It may be used in automobile engines without modifications (Ibrahim et al., 2018). When mixed with diesel, it could also decrease soot emission and improve thermal efficiency (Rakopoulos et al., 2010). However, butanol is not void of drawbacks such as higher fuel consumption and lower engine performance. Additionally, the industrial-scale application has been hampered due to the high expenses and low production of the ABE fermentation system. For this reason, researchers are searching for a system that will reduce the cost of production and increase the biobutanol yields for large-scale implementation. Recently, Rochón et al. (2020) carried out an isopropanol–butanol–ethanol (IBE) batch fermentation using in situ gas stripping-pervaporation to retrieve the biobutanol (Table 10.2). The modified Monod kinetic model was used to estimate the consumption of a sugarcane and sorghum juice mixture by *C. beijerinckii* DSM 6423 for the IBE solvent production. The IBE solvent fermentation gave a $\mu_{\max}$, K_s, product yield ($Y_{p/s}$), and $r_{p.m}$ of 0.23/h, 2 g/L, 0.22 g/g, 0.21 g/L/h, respectively, for the modified Monod model (Rochón et al., 2020). A previous study by Rochón et al. (2018) carried out an ABE batch fermentation using in situ gas stripping-pervaporation to recover biobutanol. The metabolism of a sugarcane and sorghum juice mixture by *C. beijerinckii* DSM 6423 was evaluated for ABE solvent production. This type of solvent fermentation gave a lower $\mu_{\max}$ (0.22/h), K_s (0.2 g/L), and $Y_{p/s}$ (0.20 g/g) (Rochón et al., 2018). The results showed significant improvements in the microbial growth and productivity of the IBE fermentation process compared to the ABE process. In an earlier study, Chen et al. (2013) compared the product formation kinetics of the SHF process under sterile and nonsterile conditions. The SHF-sterile process gave a higher P_m (6.23 g/L) and $r_{p.m}$ (4.82 g/L/d) in comparison to the SHF-nonsterile process, which obtained a P_m and $r_{p.m}$ of 6.51 g/L and 3.21 g/L/d, respectively, for the modified Gompertz model (Chen et al., 2013). Regardless of the higher butanol concentration and the production rate for the SHF under sterile conditions, the study points out that fermentation under nonsterile conditions did not present a significant difference (Chen et al., 2013). Therefore the presterilization step can be omitted within the biobutanol production system, resulting in lower operation costs (Chen et al., 2013). Several studies have been reported on the production of biobutanol. However, there is a shortage of knowledge on the kinetic modeling of butanol bioprocesses. For this reason, bioprocess commercialization may be severely hindered since substantial information regarding microbial growth and product formation has not been established.

10.5.3 Biohydrogen

The depletion of fossil fuel reserves, combined with rising environmental concerns resulting from fossil fuel consumption, has prompted research into biohydrogen

Table 10.2 Biobutanol production from various lignocellulosic waste.

Microorganism	Substrate	Process type	Product formation kinetics	Microbial growth kinetics	Reference
Clostridium saccharoper butylacetonicum N1−4	Rice straw	SHF-sterile	6.23 g/L[a], 4.82 g/L/d[b]	ND	Chen et al. (2013)
C. saccharoper butylacetonicum N1−4	Rice straw	SHF-nonsterile	6.51 g/L[a], 3.21 g/L/d[b]	ND	Chen et al. (2013)
Clostridium beijerinckii DSM 6423	Sugarcane and sorghum juice	SHF	0.20 g/g[c]	0.22/h[d], 0.2 g/L[e]	Rochón et al. (2018)
C. beijerinckii DSM 6423	Sugarcane and sorghum juice	SHF	0.21 g/L/h[b], 0.22 g/g[c]	0.23/h[d], 2 g/L[e]	Rochón et al. (2020)

ND: not determined; *PSSF*: simultaneous saccharification and fermentation with prehydrolysis; *SHF*: separate hydrolysis and fermentation; *SSF*: simultaneous saccharification and fermentation.
[a]P_m—potential maximum biobutanol concentration (g/L),
[b]$r_{p,m}$—maximum bioethanol production rate (g/L/h or g/L/d),
[c]$Y_{p/s}$—biobutanol yield coefficient (g/g),
[d]μ_{max}—maximum specific growth rate (per h),
[e]K_s—Monod constant (g/L).

synthesis using a variety of biological processes (Giang et al., 2019). These include (1) dark fermentation with anaerobic bacteria, (2) photofermentation by photosynthetic bacteria, (3) direct/indirect biophotolysis using cyanobacteria and green algae, and (4) microbial electrolysis cell (MEC) (Hay et al., 2013). Dark fermentation has emerged as a promising technology for biological hydrogen production and offers several advantages over other processes. The transformation of LCB to biohydrogen is a multistep process that includes: (1) physical, chemical, and physicochemical pretreatment, (2) enzymatic hydrolysis of LCB to simple or complex sugars, (3) conversion of these sugars into organic acids, carbon dioxide (CO_2) and biohydrogen (H_2) through dark fermentation (Patil et al., 2021). During dark fermentation, LCB is converted to volatile fatty acids (acetic acid and butyric acid), H_2, and CO_2 by fermentative bacteria in the absence of light (Soares et al., 2020). Through the glycolytic pathway, bacteria convert the simple and complex saccharides obtained from the hydrolysis of cellulose to pyruvic acid by simultaneously producing ATP from ADP and NADH. Pyruvic acid is further converted to CO_2 and H_2 through the assistance of pyruvate ferredoxin oxidoreductase and hydrogenase enzymes. There are three types of hydrogenase enzymes primarily expressed by microorganisms (e–Fe, Fe–Fe, and Ni–Fe hydrogenase). The Fe–Fe hydrogenase can accelerate both the reduction of H^+ and the oxidation of H_2. Whereas Ni–Fe hydrogenase only increases the oxidation of hydrogen. Hence, Fe–Fe hydrogenase plays a primary role in fermentative H_2 production (Patil et al., 2021). Pyruvate can be further converted to acetyl-CoA, acetate, butyrate, and ethanol. In theory, Eqs. (10.1) and (10.2) exemplify that 4 mol of H_2/mol glucose can be produced when acetic acid is the byproduct, whereas 2 mol H_2/mol glucose is produced when butyric acid is the byproduct. However, not all glucose is converted to biohydrogen since some glucose is required to support and maintain microbial growth (Hay et al., 2013). In addition to LCB, pure sugars (glucose, xylose, lactose, galactose, sucrose), food waste, municipal solid waste, industrial wastewater, or glycerol can be utilized as feedstocks during dark fermentation (Soares et al., 2020). Furthermore, mixed cultures can be used, such as anaerobic sludge, bovine manure, and organic compost (Soares et al., 2020). Most biohydrogen-generating microorganisms are facultative anaerobes and obligate anaerobes (Sarangi et al., 2020). Examples of these anaerobic microorganisms include mesophilic (*Escherichia coli*, *Bacillus*, *Enterobacteria*) and strict anaerobic (*Clostridia*) and thermophilic bacteria (*Thermoanaerobacterium*).

$$C_6H_{12}O_6 + 2H_2O \rightarrow 4H_2 + 2CH_3COOH + 2CO_2 \quad (10.1)$$

$$C_6H_{12}O_6 + 2H_2O \rightarrow 2H_2 + CH_3CH_2CH_2COOH + 2CO_2 \quad (10.2)$$

Several benefits and limitations exist for biohydrogen as gas and its microbial production. For example, biohydrogen is a renewable and green energy resource that does not emit greenhouse gases (GHG) when combusted. It is also environmentally friendly since the combustion of biohydrogen produces only water vapor and heat as a byproduct (Giang et al., 2019). In addition, biohydrogen has a high calorific value of (141 kJ/g) (Sarangi and Nanda, 2020); therefore it can be used to generate a vast amount of energy per unit mass (Hay et al., 2013). Furthermore, biohydrogen can be converted to hydrocarbon fuels or electricity by incorporating fuel cell technologies (Sarangi and Nanda, 2020). The microbial dark fermentation process can be carried out at ambient temperatures and pressures; therefore, it is less energy-intensive (Hay et al., 2013). The dark fermentation process also does not require a light source, making it economically viable and sustainable (Giang et al., 2019). However, major drawbacks arise from several aspects pertaining to the fermentation process. The fermentative bacteria are sensitive to pH and temperature as both fluctuations may negatively impact the metabolic pathway and growth rate of microorganisms (Hay et al., 2013). For biohydrogen production, an optimal pH of 5.5 during dark fermentation is required (Sarangi and Nanda, 2020). Dark fermentation can be operated at a temperature for mesophilic (25°C–40°C), thermophilic (40°C–65°C), and hyperthermophilic ($>$80°C) microorganisms (Sarangi and Nanda, 2020). Compared to mixed cultures, pure cultures require a sterile environment to prevent contamination. However, this is problematic and expensive to achieve at an industrial scale (Soares et al., 2020). For this reason, mixed cultures have been given first preference for scale-up operations. In addition, mixed cultures are desirable because they do not require a sterile process and create synergies between microorganisms (Tapia-Venegas et al., 2015). Another underlying issue is the heterogeneity of waste materials, usually composed of various substrates. Furthermore, microorganisms present in the waste may consume these monosaccharides, thus leading to a lower biohydrogen yield than expected (Hay et al., 2013).

Several studies have focused on producing biohydrogen from various lignocellulosic substrates (Table 10.3). Cheng et al. (2015) pretreated water hyacinth with microwave-assisted dilute acid (1% H_2SO_4) and AC (activated charcoal detoxification) under dark fermentation and obtained a maximum hydrogen yield and production rate of 134.9 mL/g and 89.2 mL/L/h, respectively when the hydrogen-producing bacteria (HPB) were utilized for hydrogen fermentation. The findings indicated that AC could effectively remove the fermentative inhibitors produced in the acid pretreatment process. A slightly higher maximum biohydrogen yield of 150 mL H_2/g was obtained by Morales-Martínez et al. (2020) from Agave biomass using a synthetic *Clostridium* medium. Similarly, Wei et al. (2021) achieved a maximum biohydrogen yield and rate of 153.96 mL/g and 3.55 mL/h from sugarcane bagasse by employing ammonium-tolerant *Rhodobacter capsulatus*. In a study by Nasr et al. (2014), corn cobs were anaerobically digested, resulting in a maximum biohydrogen potential of 141 mL/g and biohydrogen yield of 265 mL/g. The highest biohydrogen production rate was 8.8 mL/h. A positive correlation between H_2 production rates and yields was observed. Moodley and Gueguim Kana (2018) assessed the pretreatment effectiveness of three acids (HCl, H_2SO_4, and HNO_3) on sugarcane leaf waste for the successive generation of biohydrogen. A potential maximum biohydrogen yield of 191.53 mL/g was obtained after optimized HCl (4.9%) pretreated corn cobs. A biohydrogen yield of 18.6 mL/g was attained. Previous studies have reported on the production of biohydrogen from various lignocellulosic

Table 10.3 Biohydrogen production from various lignocellulosic substrates.

Inoculum	Substrate	Process type	Parameter conditions	Biohydrogen production	Reference
Anaerobic sludge (HPB)	Water hyacinth	SHF	35°C, pH 6, 40 h	89.2 mL/L/h[b], 134.9 mL/g[c]	Cheng et al. (2015)
Synthetic medium (*Clostridium*)	Agave biomass	SHF	35°C, pH 5.5, 264 h	150 mL/g[c]	Morales-Martínez et al. (2020)
Anaerobic sludge (HPB)	Corn cobs	SHF	70°C, pH 5.5, 30 min	141 mL/g[a], 8.8 mL/h[b], 265 mL/g[c]	Nasr et al. (2014)
MedA medium (*Rhodobacter capsulatus* SB1003)	Sugarcane bagasse	SHF	30°C, pH 7, 108 h	3.55 mL/h[b], 153.96 mL/g[c]	Wei et al. (2021)
Anaerobic sludge	Sugarcane leaf waste	ND	37°C, pH 6.5, 180 rpm, 72 h	191.53 mL/g[a], 18.6 mL/g[c]	Moodley and Gueguim Kana (2018)

HPB: hydrogen producing bacteria; *ND*: not determined; *SHF*: separate hydrolysis and fermentation.
[a]P_m—potential maximum biohydrogen concentration (mL/g),
[b]$r_{p,m}$—maximum biohydrogen production rate (mL/L/h or mL/h),
[c]$Y_{p/s}$—biohydrogen yield (mL/g).

substrates; however, there has been a lack of microbial growth kinetics data due to the use of mixed microbial consortia.

10.5.4 Biogas

Biogas production has been encouraged globally because of its versatility and applicability for common purposes such as household uses (Meyer et al., 2021). In a recent study, Meyer et al. (2021) indicated that biodigesters can supply the energy required for cooking within households in remote areas and could result in monetary savings of approximately \$2/d or \$55/mon. Biogas consists of methane (55%–75%), carbon dioxide (CO_2) (25%–45%) and hydrogen (5%–10%) (de Mes et al., 2003; Kamali et al., 2016). This fuel can be upgraded to a higher quality (biomethane) or used for electricity and heat generation (de Mes et al., 2003). Biogas is a renewable energy resource derived from organic wastes (such as LCB) or industrial wastewater, mainly using anaerobic digestion (AD) (Moghaddam et al., 2019, Li et al., 2017). Prior to AD, a pretreatment process is required to enhance biogas production from LCB. Biogas production through AD (nonoxidative metabolism) is a naturally occurring biological process that can be divided into four steps: hydrolysis, acidogenesis, acetogenesis, and methanogenesis (Xu et al., 2019). During hydrolysis, various microbes (*Bacteriodes, Clostridium*, and *Acetivibrio*) secrete hydrolytic enzymes that cleave the complex organic macromolecules (proteins, lipids, and carbohydrates) to their equivalent component units (amino acids, fatty acids, and sugars). Following hydrolysis, the monomers are converted by

Enterobacterium, Acetobacterium, and *Eubacterium* into various organic acids (acetic, propionic acid, butyric acid, succinic acid, and lactic acid), alcohols, and ammonia. The acetogenic bacteria (*Syntrophomonas, Syntrophus, Clostridium,* and *Syntrobacter*) further convert these compounds to acetate, carbon dioxide, and hydrogen, which are natural substrates for methanogenesis (Xu et al., 2019, Goswami et al., 2016). Methanogenesis involves the fermentation of various organic compounds (carbon dioxide and hydrogen gas) to generate biogas as the end product (Goswami et al., 2016). Common inoculum sources for AD and biomass gasification include agricultural waste, municipal solid waste, and waste-activated sludge (Li et al., 2017). In addition, animal manure and food wastes contain a significant amount of biodegradable organic carbon, which can be used for biogas production.

Biogas production has several benefits and drawbacks. An advantage of the AD process is the low requirements for heat and electricity. Furthermore, it is a sustainable process that generates a residual fertilizer (digestate rich in nitrogen and phosphorus), which can be used for soil conditioning (Ersahin et al., 2011). Moreover, mixing several wastes (codigestion) increases biogas yields and cost efficiency. Ammonia reduction, carbon to nitrogen ratio, and essential trace elements contribute to a controlled reaction environment for microorganisms, subsequently influencing the yield of biogas produced (Xu et al., 2019). Nevertheless, several disadvantages persist, whereby biogas produced from AD comprises a large amount of carbon dioxide (25%–45%). For biogas to be used as fuel for vehicles, it must be upgraded by removing CO_2 to achieve a high purity ($>95\%$) of biomethane. The cost of upgrading is relatively high and represents a significant bottleneck for increasing biogas yields (Li et al., 2017). Factors that affect the efficiency of the generation of biogas involve pH, temperature, alkalinity, hydraulic retention time and organic loading rates, presence of toxic and inhibitory compounds, type of substrate and total solids (TS) and volatile solids (VS) content (Ersahin et al., 2011).

Numerous studies have obtained biogas from various lignocellulosic feedstocks (Table 10.4). For example, Ugwu and Enweremadu (2019) mechanically pretreated Orka pod waste and fermented it with anaerobic digest for 25 days generating a biogas yield of 270.98 mL/g. Similarly, Mustafa et al. (2016) subjected rice straw to fungal pretreatment with *Pleurotus ostreatus* prior to AD to enhance the biodegradability of rice straw and biogas production yield. A biogas yield of 263 mL/g was obtained at a moisture content of 75% after 20 days. A 120% higher biogas yield was observed from pretreated rice straw than from untreated rice straw.

Similarly, Fu et al. (2015) pretreated corn straw using a thermophilic microaerobic pretreatment. Structural analysis showed disruption of cellulosic structures and pores of corn straw during pretreatment. A maximum biogas yield (325.7 mL/g) was obtained at an oxygen load of 5 mL/g substrates. The highest biogas value received was 16.24% higher than untreated corn straw. In another study, Oleszek et al. (2014) compared biogas yields from two types of reed canary grass (cultivated and wild). A higher biogas yield of 406 mL/g (cultivated reed grass) was obtained compared to 120 mL/g (wild reed grass). It was determined that the chemical composition and indigestible crude fiber and ash within the reed grass contributed to the

Table 10.4 Biogas production from various lignocellulosic substrates.

Inoculum	Substrate	Process type	Parameter conditions	Biomethane production	References
Anaerobic digest	Okra pod waste	SHF	37°C, pH 7.68, 25 d	268.4 mL/g[a], 270.98 mL/g[c]	Ugwu and Enweremadu (2019)
Anaerobic sludge (*P. ostreatus*)	Rice straw	SSF	37°C, 20 d, 75% moisture content	263 mL/g[c]	Mustafa et al. (2016)
Biogas slurry	Corn straw	SHF	37°C, 130 rpm, oxygen 5 mL/g substrate	325.7 mL/g[c]	Fu et al. (2015)
Sewage sludge	Reed canary grass (cultivated)	ND	35°C, pH 7.0, 20 to 40 d	406 mL/g[c]	Oleszek et al. (2014)
Sewage sludge	Reed canary grass (wild-type)	ND	35°C, pH 7.0, 20 to 40 d	120 mL/g[c]	Oleszek et al. (2014)
MPB (methane producing bacteria)	Water hyacinth hydrolysate	SHF	35°C, pH 7.0, 5 d	123 mL/g[c], 189.4 mL/L/d[b]	Cheng et al. (2015)

ND: not determined; *SHF*: separate hydrolysis and fermentation; *SSF*: simultaneous saccharification and fermentation.
[a]P_m—potential maximum biomethane concentration (mL/g).
[b]$r_{p.m}$—maximum biomethane production rate (mL/L/d).
[c]$Y_{p/s}$—biomethane yield (mL/g).

variations observed. Furthermore, these components could reduce biogas quantity and quality. The study by Cheng et al. (2015) subjected residual hydrolysates of water hyacinth to biogas production and produced a yield and production rate of 123 mL/g and 189.4 mL/L/d, respectively.

10.6 Current challenges of lignocellulosic biofuel production

Even though microbially produced biofuels from LCB sources have proven advantageous, their production presents many technological issues. These drawbacks occur at different biofuel processing stages and significantly hinder its large-scale implementation. Pretreatment is necessary to make the cellulose moieties within LCB accessible to enzymatic attack for enhanced conversion of polysaccharides to monosaccharides. Numerous pretreatments have been established with chemical methods displaying high efficacy since they can recover up to 80% of fermentable sugars compared to pretreatments without chemicals. However, these pretreatment strategies incur high costs due to the chemicals required, equipment, and reactor compatibility. Additionally, water is a finite resource required in large quantities during pretreatment, further placing strain on resources (Kumar and Sharma, 2017). Heating processes in pretreatment are also associated with high energy and cost (Moodley et al., 2020). Furthermore, the generation of inhibitors due to chemical methods reduces the product yields during enzymatic saccharification and fermentation and reduces the product's purity during recovery. More specifically, regarding the biofuel production process, one major challenge is the significant disparity in the number of studies, with some fuels receiving higher priority than others. For instance, numerous studies have assessed the SHF, SSF, and PSSF systems for bioethanol production, while a significant knowledge gap exists for other biofuels (biobutanol, biohydrogen, and biogas). Likewise, kinetic modeling plays a fundamental role in understanding the metabolic fluxes occurring within the cell and product formation (David et al., 2020), but most studies have focused on bioethanol production with a lack of data for fuels such as biohydrogen and biogas. Moreover, the production of biofuels requires a significant amount of energy input for heating the bioreactors and thermal activation of biochemical reactions (Baruah et al., 2018). On the research front, most studies on biofuel production have been carried out at a laboratory scale; therefore there is a dearth of translatable data for scale-up. Apart from the drawbacks mentioned above, safety limitations of biofuels such as biohydrogen include its explosiveness when reacted with oxygen. Therefore it has to be stored in specialized vessels under pressure (Sarangi and Nanda, 2020).

10.7 Advancements in lignocellulosic biofuel production

The commercialization of biofuel production is still at an early and dynamic stage in the current markets due to the high costs attached. Many technologies have been developed to mitigate the limitations of lignocellulosic waste-based biofuel production with minimal capital investments and increased product outputs. To date, studies have aimed to improve the systematic approaches to the four stages of biofuel production: pretreatment, enzymatic saccharification, fermentation, and downstream processes. Pretreatment is imperative when considering LCB and accounts for 40% of the total production expenses. The pretreatment step is a key process that gets the ball rolling toward high product yields, improved downstream applications, and reduced capital costs (Kucharska et al., 2018b; Limayem and Ricke, 2012). Researchers have been investigating the suitability of new lignocellulosic waste sources, assessing new pretreatment strategies, and optimizing the existing methods to enhance this process (David et al., 2020; Rorke et al., 2021). For example, the development of alkali and inorganic salt pretreatments has emerged as a cheaper and more environmentally friendly alternative to more expensive chemicals such as NaOH (Moodley and Gueguim Kana, 2017a; Qing et al., 2016; Sewsynker-Sukai and Guegium Kana, 2018; Bhardwaj et al., 2020). Particularly, inorganic salts such as sodium chloride (NaCl) are emerging as efficient pretreatments (Moodley and Gueguim Kana, 2017b) that offer impurified equivalents in the form of table salt, that is widely accessible, noncorrosive, and do not negatively impact on the environment. However, recent studies have focused on using chemicals generated from industrial chemical lines to pretreat LCB. For instance, green and black liquor from the Kraft paper and pulp industry has previously been investigated for their pretreatment abilities (Zhou et al., 2017; Goshadrou, 2019). However, these residues are usually recycled into the Kraft paper and pulp industry to be used again. On the other hand, green liquor dregs have been brought to the forefront as a potential pretreatment catalyst for lignocellulosic pretreatments since these residues possess alkaline characteristics that have been shown to be beneficial during pretreatment (David et al., 2020; Rorke et al., 2021). Moreover, these residues are discarded after the Kraft paper and pulping process since they cannot remain in the filter lines due to blockages and hydraulic conductivity issues. Therefore diversion of these waste residues to the pretreatment aspect of biofuel production will drastically reduce the expenses attached to the cost of pretreatment chemicals. Besides the high chemical cost, water usage also significantly impacts operation costs. To solve this problem, the use of industrial wastewater has been proposed to be coupled with pretreatment chemicals to omit the use of freshwater streams (Sewsynker-Sukai et al., 2020). Implementing these strategies may become a suitable method that can be assessed for scale-up of biofuel production. In conjunction with pretreatment strategies, important aspects such as process development, optimization, and mathematical kinetic modeling during fermentation are imperative for industrial-scale biofuel production. For this reason, researchers need to understand the kinetics of each process type (SHF, SSF, and PSSF) to enhance biofuel production's feasibility and productivity. Recent biofuel production studies have proposed the application of nanotechnology and surfactants as catalysts that could improve the biological dynamics such as enzymatic saccharification and microbial cellular processes of the system (Sanusi et al., 2021; Rorke et al., 2021). In addition to catalyst inclusion and kinetic studies, genetically modified fermentative microorganisms capable of resisting adverse conditions such as tolerance to inhibitors, high temperatures, and the ability to uptake co-ferment hexoses and pentoses should be explored (Kumar and Sharma, 2017). Furthermore, employing a single microorganism that can consume all available sugars (broaden substrate utilization) simultaneously at high rates in a medium could shorten fermentation time, consequently improving volumetric productivity and increasing productivity yields (Hay et al., 2013). With the abovementioned advancements in mind, it is interesting to note that the production of high-value industrial commodities such as itaconic acid, lactic acid, and citric acid, amongst others, may present another avenue to alleviate high costs incurred during biofuel production processes. Following the fermentation of value-added products, the effluent generated still contains residual components of vitamins, peptides, and carbon sources. This fermentation effluent can be channeled toward the production of biofuels such as bioethanol, biobutanol, biohydrogen, biogas, or other desired biofuels. This proposed concept of using the effluent from the production of other commodities intends to manipulate every aspect of waste to achieve a circular bioeconomy with complete waste valorization (Pinales-Márquez et al., 2021; Patel and Shah, 2021).

10.8 Conclusion and future perspectives

Recently, second generation (2G) biofuels from lignocellulosic feedstocks have gained widespread attention regarding their sustainability and nonpolluting nature. An abundance of literature on lignocellulosic biofuel production systems pertaining to pretreatment, enzymatic hydrolysis, and microbial fermentation toward biofuels and other bioproducts. Biorefinery systems incorporate various thermal, chemical, and biological processes to produce multiple products (biofuels and bioproducts) and utilize the feedstocks involved completely. Nevertheless, the interconnection between research and applied biotechnology requires swift action to identify the knowledge gaps to overcome these systems' technological, economical and environmental barriers. Multidisciplinary approaches that bring experts from various backgrounds seem to be the solution to achieving successful biorefineries, with a long path ahead toward industrialization.

Abbreviations

2G	second generation
ABE	acetone-butanol-ethanol
AC	activated charcoal detoxification
AD	anaerobic digestion
ADP	adenosine diphosphate
ATP	adenosine triphosphate
Btu	British thermal units
ED	Entner–Doudoroff
EMP	Embden–Meyerhof–Parnas
GHG	greenhouse gases
HPB	hydrogen-producing bacteria
IBE	isopropanol-butanol-ethanol fermentation
$\mathbf{K_s}$	monod constant
LCB	lignocellulosic biomass
MEC	microbial electrolysis cell
NADH	nicotinamide adenine dinucleotide hydride
ND	not determined
OPFJ	oil palm frond juice
P_m	maximum product concentration
PSSF	SSF with a prehydrolysis step
$r_{p.m}$	maximum product production rate
SHF	separate hydrolysis and fermentation
SSF	simultaneous saccharification and fermentation
TS	total solids
VS	volatile solids
X_{max}	maximum cell biomass concentration
$Y_{p/s}$	product yield
μ_{max}	maximum specific growth rate

References

Abraham, A., Mathew, A.K., Park, H., Choi, O., Sindhu, R., Parameswaran, B., et al., 2020. Pretreatment strategies for enhanced biogas production from lignocellulosic biomass. Bioresour. Technol. 301, 122725.

Al-Hinai, M.A., Jones, S.W., Papoutsakis, E.T., 2014. σK of *Clostridium acetobutylicum* is the first known sporulation-specific sigma factor with two developmentally separated roles, one early and one late in sporulation. J. Bacteriol. 196 (2), 287−299.

Ali, M.K., Hiligsmann, S., Outili, N., Cherfia, R., Chaouche, N.K., 2017. Kinetic models and parameters estimation study of biomass and ethanol production from inulin by *Pichia caribbica* (KC977491). Afr. J. Biotechnol. 16 (3), 124−131.

Annual Energy Outlook. U.S. Energy Information Administration (EIA), 2019. <https://www.eia.gov/>.

Banu, J.R., Kavitha, P.S., Tyagi, V.K., Gunasekaran, M., Karthikeyan, O.P., Kumar, G., 2021. Lignocellulosic biomass based biorefinery: a successful platform towards circular bioeconomy. Fuel 302, 121086.

Baruah, J., Nath, B.K., Sharma, R., Kumar, S., Deka, R.C., Baruah, D.C., et al., 2018. Recent trends in the pretreatment of lignocellulosic biomass for value-added products. Front. Energy Res. 6, 141.

Bhardwaj, N., Kumar, B., Verma, P., 2020. Microwave-assisted pretreatment using alkali metal salt in combination with orthophosphoric acid for generation of enhanced sugar and bioethanol. Biomass Conv. Biorefin. 12, 1069−1076.

Carrillo-Nieves, D., Ruiz, H.A., Aguilar, C.N., Ilyina, A., Parra-Saldivar, R., Torres, J.A., et al., 2017. Process alternatives for bioethanol production from mango stem bark residues. Bioresour. Technol. 239, 430−436.

Chen, W.H., Chen, Y.C., Lin, J.G., 2013. Evaluation of biobutanol production from non-pretreated rice straw hydrolysate under non-sterile environmental conditions. Bioresour. Technol. 135, 262−268.

Cheng, J., Lin, R., Song, W., Xia, A., Zhou, J., Cen, K.J.I., 2015. Enhancement of fermentative hydrogen production from hydrolyzed water hyacinth with activated carbon detoxification and bacteria domestication. Int. J. Hydrogen Energy 40, 2545−2551.

Costa, R.S., Hartmann, A., Vinga, S., 2016. Kinetic modeling of cell metabolism for microbial production. J. Biotechnol. 219, 126−141.

Da Silva, F.L., De Oliveira Campos, A., Dos Santos, D.A., Batista Magalhães, E.R., De Macedo, G.R., Dos Santos, E.S., 2018. Valorization of an agroextractive residue—Carnauba straw—for the production of bioethanol by simultaneous saccharification and fermentation (SSF). Renew. Energy 127, 661−669.

de Mes, T.Z.D., Stams, A.J.M., Reith, J.H., Zeeman, G., 2003. Methane production by anaerobic digestion of wastewater and solid wastes. In: Reith, J.H., Wijffels, R.H., Barten, H. (Eds.), Biomethane & Biohydrogen: Status and Perspectives of Biological Methane and Hydrogen Production. Dutch Biological Hydrogen Foundation, Petten, Netherlands, pp. 58−102.

Dahnum, D., Tasum, S.O., Triwahyuni, E., Nurdin, M., Abimanyu, H., 2015. Comparison of SHF and SSF processes using enzyme and dry yeast for optimization of bioethanol production from empty fruit bunch. Energy 68, 107−116.

David, A.N., Sewsynker-Sukai, Y., Sithole, B., Gueguim Kana, E.B., 2020. Development of a green liquor dregs pretreatment for enhanced glucose recovery from corn cobs and kinetic assessment on various bioethanol fermentation types. Fuel 274, 117797.

Dodić, J.M., Vučurović, D.G., Dodić, S.N., Grahovac, J.A., Popov, S.D., Nedeljković, N.M., 2012. Kinetic modelling of batch ethanol production from sugar beet raw juice. Appl. Energy 99, 192–197.

Dong, Y., Yan, X., Lu, F., Guo, M., Zhuang, Y., 2015. Development and optimization of an unstructured kinetic model for sodium gluconate fermentation process. CMC-Comput. Mater. Continua 48 (1), 43–55.

Ersahin, M.E., Yangin Gomec, C., Dereli, R.K., Arikan, O., Ozturk, I.J., 2010. Biomethane production as an alternative bioenergy source from codigesters treating municipal sludge and organic fraction of municipal solid wastes. Biomed. Biotechnol. 2011, 953065.

Fedailaine, M., Moussi, K., Khitous, M., Abada, S., Saber, M., Tirichine, N., 2015. Modeling of the anaerobic digestion of organic waste for biogas production. Procedia Comput. Sci. 52, 730–737.

Fu, S.F., Wang, F., Yuan, X.Z., Yang, Z.M., Luo, S.J., Wang, C.S., et al., 2015. The thermophilic (55 C) microaerobic pretreatment of corn straw for anaerobic digestion. Bioresour. Technol. 175, 203–208.

Giang, T.T., Lunprom, S., Liao, Q., Reungsang, A., Salakkam, A., 2019. Enhancing hydrogen production from *Chlorella* sp. biomass by prehydrolysis with simultaneous saccharification and fermentation (PSSF). Energy 12, 908.

González-Figueredo, C., Flores-Estrella, R.A., Rojas-Rejón, O.A., 2018. Fermentation: metabolism, kinetic models, and bioprocessing. In: Shiomi, S. (Ed.), Current Topics in Biochemical Engineering. TechOpen limited, London, United Kingdom, pp. 11–28.

Goshadrou, A., 2019. Bioethanol production from Cogongrass by sequential recycling of black liquor and wastewater in a mild-alkali pretreatment. Fuel 258, 116141.

Goswami, R., Chattopadhyay, P., Shome, A., Banerjee, S.N., Chakraborty, A.K., Mathew, A. K., et al., 2016. An overview of physico-chemical mechanisms of biogas production by microbial communities: a step towards sustainable waste management. Biotechnology 6, 72.

Guerrero, A.B., Ballesteros, I., Ballesteros, M., 2018. The potential of agricultural banana waste for bioethanol production. Fuel 213, 176–185.

Hay, J.X.W., Wu, T.Y., Juan, J.C., Jahim, J.Md, 2013. Biohydrogen production through photo fermentation or dark fermentation using waste as a substrate: overview, economics, and future prospects of hydrogen usage. Bioproducts Biorefin. 7, 334–352.

Himmel, M.E., Ding, S.Y., Johnson, D.K., Adney, W.S., Nimlos, M.R., Brady, J.W., et al., 2007. Biomass recalcitrance: engineering plants and enzymes for biofuels production. Science 315 (5813), 804–807.

Ibrahim, M.F., Kim, S.W., Abd-Aziz, S., 2018. Advanced bioprocessing strategies for biobutanol production from biomass. Renew. Sustain. Energy Rev. 91, 1192–1204.

Jędrzejczyk, M., Soszka, E., Czapnik, M., Ruppert, A.M., Grams, J., 2019. Physical and chemical pretreatment of lignocellulosic biomass. In: Basile, A., Dalena, F. (Eds.), Second and Third Generation of Feedstocks: The Evolution of Biofuels. Elsevier Incorporated, Amsterdam, Netherlands, pp. 143–196.

Jiang, Y., Lv, Y., Wu, R., Sui, Y., Chen, C., Xin, F., et al., 2019. Current status and perspectives on biobutanol production using lignocellulosic feedstocks. Biores. Technol. Rep. 7, 100245.

Jugwanth, Y., Sewsynker-Sukai, Y., Gueguim Kana, E.B., 2020. Valorization of sugarcane bagasse for bioethanol production through simultaneous saccharification and fermentation: optimization and kinetic studies. Fuel 262, 116552.

Kamali, M., Gameiro, T., Costa, M.E.V., Capela, I., 2016. Anaerobic digestion of pulp and paper mill wastes – an overview of the developments and improvement opportunities. Chem. Eng. J. 298, 162–182.

Kang, Q., Appels, L., Tan, T., Dewil, R., 2014. Bioethanol from lignocellulosic biomass: current findings determine research priorities. The Sci. World J. 2014, 1–13.

Kim, T.H., 2013. Pretreatment of lignocellulosic biomass. In: Yang, S.T., El-Enshasy, H.A., Thongchul, N. (Eds.), Bioprocessing Technologies in Biorefinery for Sustainable Production of Fuels. Chemicals, and Polymers, Wiley, New York, United States of America, pp. 91–109.

Kong, J.D., 2017. Modeling microbial dynamics: effects on environmental and human health. Doctoral dissertation, University of Alberta, Canada. <https://doi.org/10.7939/R3FT8DZ73>.

Kucharska, K., Hołowacz, I., Konopacka-Łyskawa, D., Rybarczyk, P., Kamiński, M., 2018a. Key issues in modeling and optimization of lignocellulosic biomass fermentative conversion to gaseous biofuels. Renew. Energy 129, 384–408.

Kucharska, K., Rybarczyk, P., Hołowacz, I., Łukajtis, R., Glinka, M., Kamiński, M., 2018b. Pretreatment of lignocellulosic materials as substrates for fermentation processes. Molecules 23, 2937.

Kumar, A.K., Sharma, S., 2017. Recent updates on different methods of pretreatment of lignocellulosic feedstocks: a review. Biores. Bioprocess. 4, 1–19.

Kumar, B., Bhardwaj, N., Agrawal, K., Chaturvedi, V., Verma, P., 2020. Current perspective on pretreatment technologies using lignocellulosic biomass: an emerging biorefinery concept. Fuel Process. Technol. 199, 106244.

Kumari, D., Singh, R., 2018. Pretreatment of lignocellulosic wastes for biofuel production: a critical review. Renew. Sustain. Energy Rev. 90, 877–891.

Li, H., Mehmood, D., Thorin, E., Yu, Z., 2017. Biomethane production via anaerobic digestion and biomass gasification. Energy Proc. 105, 1172–1177.

Limayem, A., Ricke, S.C., 2012. Lignocellulosic biomass for bioethanol production: current perspectives, potential issues and future prospects. Prog. Energy Combust. Sci. 38 (4), 449–467.

Majidian, P., Tabatabaei, M., Zeinolabedini, M., Naghshbandi, M.P., Chisti, Y., 2018. Metabolic engineering of microorganisms for biofuel production. Renew. Sustain. Energy Rev. 82, 3863–3885.

Marin, A.R., Lopez, Y.C., Garcia, A.N., Heredia, J.C.R., Loria, J.D.C.Z., 2017. Potential production of ethanol by *Saccharomyces cerevisiae* immobilized and coimmobilized with *Zymomonas mobilis*: alternative for the reuse of a waste organic. In: Morata, A., Loira, L. (Eds.), Yeast: Industrial Applications. InTechOpen limited, Croatia, United Kingdom, pp. 249–266.

McKendry, P., 2002. Energy production from biomass (part 1): overview of biomass. Bioresour. Technol. 83, 37–46.

Meyer, E.L., Overen, O.K., Obileke, K., Botha, J.J., Anderson, J.J., Koatla, T.A.B., et al., 2021. Financial and economic feasibility of bio-digesters for rural residential demand-side management and sustainable development. Energy Rep. 7, 1728–1741.

Moghaddam, E.A., Ericsson, N., Hansson, P.A., Nordberg, Å., 2019. Exploring the potential for biomethane production by willow pyrolysis using life cycle assessment methodology. Energy Sustain. Soc. 9, 1–18.

Monod, J., 1949. The growth of bacterial cultures. Annu. Rev. Microbiol. 3 (1), 371–394.

Moodley, P., Gueguim Kana, E.B., 2017a. Development of a steam or microwave-assisted sequential salt-alkali pretreatment for lignocellulosic waste: effect on delignification and enzymatic hydrolysis. Energy Conver. Manage. 148, 801–808.

Moodley, P., Gueguim Kana, E.B., 2017b. Microwave-assisted inorganic salt pretreatment of sugarcane leaf waste: effect on physiochemical structure and enzymatic saccharification. Bioresour. Technol. 235, 35–42.

Moodley, P., Gueguim Kana, E.B., 2018. Comparative study of three optimized acid-based pretreatments for sugar recovery from sugarcane leaf waste: a sustainable feedstock for biohydrogen production. Eng. Sci. Technol. 21, 107–116.

Moodley, P., Gueguim Kana, E.B., 2019. Bioethanol production from sugarcane leaf waste: effect of various optimized pretreatments and fermentation conditions on process kinetics. Biotechnol. Rep. 22, 00329.

Moodley, P., Sewsynker-Sukai, Y., Gueguim Kana, E.B., 2020. Progress in the development of alkali and metal salt catalysed lignocellulosic pretreatment regimes: potential for bioethanol production. Bioresour. Technol. 310, 123372.

Morales-Martínez, T.K., Díaz-Blanco, D.I., Rodríguez-De La Garza, J.A., Morlett-Chávez, J., Castro-Montoya, A.J., Quintero, J., et al., 2017. Assessment of different saccharification and fermentation configurations for ethanol production from *Agave lechuguilla*. Bioresour. Technol. 12, 8093–8105.

Morales-Martínez, T.K., Medina-Morales, M.A., Ortíz-Cruz, A.L., Rodríguez-De La Garza, J.A., Moreno-Dávila, M., López-Badillo, C.M., et al., 2020. Consolidated bioprocessing of hydrogen production from agave biomass by *Clostridium acetobutylicum* and bovine ruminal fluid. Int. J. Hydrogen Energy 45, 13707–13716.

Muloiwa, M., Nyende-Byakika, S., Dinka, M., 2020. Comparison of unstructured kinetic bacterial growth models. South African J. Chem. Eng. 33, 141–150.

Mustafa, A.M., Poulsen, T.G., Sheng, K., 2016. Fungal pretreatment of rice straw with *Pleurotus ostreatus* and *Trichoderma reesei* to enhance methane production under solid-state anaerobic digestion. Appl. Energy Environ. Biol. 180, 661–671.

Muthuvelu, K.S., Rajarathinam, R., Kanagaraj, L.P., Ranganathan, R.V., Dhanasekaran, K., Manickam, N.K., 2019. Evaluation and characterization of novel sources of sustainable lignocellulosic residues for bioethanol production using ultrasound-assisted alkaline pretreatment. Waste Manage. 87, 368–374.

Nasr, N., Gupta, M., Elbeshbishy, E., Hafez, H., El Naggar, M.H., Nakhla, G., 2014. Biohydrogen production from pretreated corn cobs. Int. J. Hydrogen Energy 39, 19921–19927.

Nithyanandan, K., Zhang, J., Li, Y., Wu, H., Lee, T.H., Lin, Y., et al., 2016. Improved SI engine efficiency using acetone–butanol–ethanol (ABE). Fuel 174, 333–343.

Oleszek, M., Król, A., Tys, J., Matyka, M., Kulik, M., 2014. Comparison of biogas production from wild and cultivated varieties of reed canary grass. Bioresour. Technol. 156, 303–306.

Oliveira, S.C., Oliveira, R.C., Tacin, M.V., Gattás, E.A.L., 2016. Kinetic modeling and optimization of a batch ethanol fermentation process. Bioprocess Biotechnol. 6 (266), 2.

Panikov, N.S. (1999). Kinetics, microbial growth. Encyclopedia of Industrial Biotechnology: Bioprocess, Bioseparation, and Cell Technology. pp. 1513–1543.

Patel, A., Shah, A.R., 2021. Integrated lignocellulosic biorefinery: gateway for production of second generation ethanol and value added products. J. Biores. Bioproducts 6, 108–128.

Patil, P.D., Tiwari, M.S., Bhange, V.P., Marghade, D.T., Kumaran, S., 2021. Bioconversion of lignocellulosic residues into hydrogen. In: Inamuddin, Khan, A. (Eds.), Sustainable Bioconversion of Waste to Value Added Products. Springer International Publishing., Cham, Switzerland, pp. 59–80.

Phukoetphim, N., Salakkam, A., Laopaiboon, P., Laopaiboon, L., 2017. Kinetic models for batch ethanol production from sweet sorghum juice under normal and high gravity fermentations: logistic and modified Gompertz models. J. Biotechnol. 243, 69–75. Available from: https://doi.org/10.1016/j.jbiotec.2016.12.012.

Pinales-Márquez, C.D., Rodríguez-Jasso, R.M., Araújo, R.G., Loredo-Treviño, A., Nabarlatz, D., Gullón, B., et al., 2021. Circular bioeconomy and integrated biorefinery in the production of xylooligosaccharides from lignocellulosic biomass: a review. Industr. Crops Products 162, 113274.

Portero Barahona, P., Bastidas Mayorga, B., Martín-Gil, J., Martín-Ramos, P., Carvajal Barriga, E., 2020. Cellulosic ethanol: improving cost efficiency by coupling semi-continuous fermentation and simultaneous saccharification strategies. Processes 8, 1459.

Qing, Q., Zhou, L., Huang, M., Guo, Q., He, Y., Wang, L., et al., 2016. Improving enzymatic saccharification of bamboo shoot shell by alkalic salt pretreatment with H_2O_2. Bioresour. Technol. 201, 230–236.

Rakopoulos, D.C., Rakopoulos, C.D., Giakoumis, E.G., Dimaratos, A.M., Kyritsis, D.C., 2010. Effects of butanol–diesel fuel blends on the performance and emissions of a high-speed DI diesel engine. Energy Conver. Manage. 51 (10), 1989–1997.

Moodley, P., 2021. Sustainable biofuels: opportunities and challenges. In: Ray, R.C. (Ed.), Sustainable Biofuels. Academic Press, Elsevier, London, United Kingdom, pp. 1–20.

Mohapatra, S., Ray, R.C., Ramachandran, S., 2018. General perspectives of bioethanol production technologies. In: Ray, R.C., Ramachandran, S. (Eds.), Bioethanol Production from Food Crops Sustainable Sources, Interventions, and Challenges. Academic Press, Elsevier, London, United KingdomCambridge, pp. 3–27.

Ray, R.C., Ramachandran, S. (Eds.), 2018. Bioethanol Production from Food Crops: Sustainable Sources, Interventions, and Challenges. Academic Press, p. 460.

Rebello, S., Anoopkumar, A.N., Aneesh, E.M., Sindhu, R., Binod, P., Pandey, A., 2020. Sustainability and life cycle assessments of lignocellulosic and algal pretreatments. Bioresour. Technol. 301, 122678.

Robak, K., Balcerek, M., 2018. Review of second generation bioethanol production from residual biomass. Food Technol. Biotechnol. 56 (2), 174–187.

Rochón, E., Ferrari, M.D., Lareo, C., 2018. Process energy evaluation of fuel butanol production from sugar cane–sweet sorghum juices by Acetone–Butanol–Ethanol fermentation associated with a gas stripping system. Energy Fuels 32 (9), 9470–9477.

Rochón, E., Cortizo, G., Cabot, M.I., Cubero, M.T.G., Coca, M., Ferrari, M.D., et al., 2020. Bioprocess intensification for isopropanol, butanol and ethanol (IBE) production by fermentation from sugarcane and sweet sorghum juices through a gas stripping-pervaporation recovery process. Fuel 281, 118593.

Rorke, D.C., Lekha, P., Gueguim Kana, E.B., Sithole, B.B., 2021. Surfactant-assisted green liquor dregs pretreatment to enhance the digestibility of paper mill sludge. Int. J. Hydrogen Energy 46 (41), 21359–21371.

Sanusi, I.A., Suinyuy, T.N., Gueguim Kana, E.B., 2021. Impact of nanoparticle inclusion on bioethanol production process kinetic and inhibitor profile. Biotechnol. Rep. 29, e00585.

Sarangi, P.K., Nanda, S., 2020. Biohydrogen production through dark fermentation. Chem. Eng. Technol. 43, 601–612.

Sarawan, C., Suinyuy, T.N., Sewsynker-Sukai, Y., Gueguim Kana, E.B., 2019. Optimized activated charcoal detoxification of acid-pretreated lignocellulosic substrate and assessment for bioethanol production. Bioresour. Technol. 286, 121403.

Sewsynker-Sukai, Y., Gueguim Kana, E.B., 2018. Simultaneous saccharification and bioethanol production from corn cobs: process optimization and kinetic studies. Bioresour. Technol. 262, 32–41.

Sewsynker-Sukai, Y., Suinyuy, T.N., Gueguim Kana, E.B., 2018. Development of a sequential alkalic salt and dilute acid pretreatment for enhanced sugar recovery from corn cobs. Energy Conver. Manage. 160, 22–30.

Sewsynker-Sukai, Y., David, A.N., Gueguim Kana, E.B., 2020. Recent developments in the application of kraft pulping alkaline chemicals for lignocellulosic pretreatment: potential beneficiation of green liquor dregs waste. Bioresour. Technol. 306, 123225.

Soares, J.F., Confortin, T.C., Todero, I., Mayer, F.D., Mazutti, M., 2020. Dark fermentative biohydrogen production from lignocellulosic biomass: technological challenges and future prospects. Renew. Sustain. Energy Rev. 117, 109484.

Soltanian, S., Aghbashlo, M., Almasi, F., Hosseinzadeh-Bandbafha, H., Nizami, A.S., Ok, Y. S., et al., 2020. A critical review of the effects of pretreatment methods on the exergetic aspects of lignocellulosic biofuels. Energy Conver. Manage. 212, 112792.

Srimachai, T., Nuithitikul, K., O-thong, S., Kongjan, P., Panpong, K., 2015. Optimization and kinetic modeling of ethanol production from oil palm frond juice in batch fermentation. Energy Proc. 79, 111–118. Available from: https://doi.org/10.1016/j.egypro.2015.11.490.

Tabatabaei, M., Soltanian, S., Aghbashlo, M., Nizami, A.S., 2019. Fast pyrolysis of biomass: advances in science and technology: a book review. J. Clean Product 213, 1411–1413.

Tapia-Venegas, E., Ramirez-Morales, J.E., Silva-Illanes, F., Toledo-Alarcón, J., Paillet, F., Escudie, R., et al., 2015. Biohydrogen production by dark fermentation: scaling-up and technologies integration for a sustainable system. Rev. Environ. Sci. Biotechnol. 14, 761–785.

Ugwu, S.N., Enweremadu, C.C., 2019. Biodegradability and kinetic studies on biomethane production from okra (*Abelmoschus esculentus*) waste. S. Afr. J. Sci. 115, 1–5.

Wang, L., Luo, Z., Shahbazi, A., 2013. Optimization of simultaneous saccharification and fermentation for the production of ethanol from sweet sorghum (*Sorghum bicolor*) bagasse using response surface methodology. Industr. Crops Products 42, 280–291.

Wei, X., Feng, J., Cao, W., Guo, L., 2021. Enhanced biohydrogen production by an ammonium-tolerant *Rhodobacter capsulatus* from sugarcane bagasse. Fuel 300, 121009.

Xu, N., Liu, S., Xin, F., Zhou, J., Jia, H., Xu, J., et al., 2019. Biomethane production from lignocellulose: biomass recalcitrance and its impacts on anaerobic digestion. Front. Bioeng. Biotechnol. 7, 191.

Zabed, H., Sahu, J.N., Boyce, A.N., Faruq, G., 2016. Fuel ethanol production from lignocellulosic biomass: an overview on feedstocks and technological approaches. Renew. Sustain. Energy Rev. 66, 751–774.

Zhou, Z., You, Y., Lei, F., Li, P., Jiang, J., Zhu, L., 2017. Enhancement of enzymatic hydrolysis of sugarcane bagasse by pretreatment combined green liquor and sulfite. Fuel 203, 707–714.

Cell immobilization strategies to enhance yield of liquid biofuels

11

Sudhanshu S. Behera[1] *and Ramesh C. Ray*[2]
[1]Department of Biotechnology, National Institute of Technology Raipur, Raipur, Chhattisgarh, India, [2]Centre for Food Biology and Environment Studies, Bhubaneswar, Odisha, India

11.1 Introduction

Rising population, surging globalization, soaring energy needs, the rapid rise in oil costs, and dwindling fossil fuels (coal, oil, and natural gas) have urged researchers to explore/investigate alternate fuels (Razack and Duraiarasan, 2016; Adelabu et al., 2019). Biofuels have been the most likely replacement for petrofuels due to the later unrelenting demand and limited availability (Razack and Duraiarasan, 2016). Biofuels can compete with fossil fuels and augment them because they are more environmentally friendly and have higher energy efficiency (Shanmugam et al., 2020). Based on the source of feedstocks, biofuels are classified as first-generation (1G), second-generation (2G), and third-generation (3G) biofuels. Biofuels from lignocellulosic biomass (LCB) come under 2G biofuels.

Several constraints, such as low process yields and high production costs, have limited the industrial production of lignocellulosic liquid biofuels (i.e., bioethanol, biodiesel, biohydrogen, and biobutanol). This necessitates researching/developing new strategies/solutions to address the issue (Malhotra and Basir, 2020). For example, biofuel cells involving enzymes and whole cells of microorganisms are used to produce liquid biofuels (De Poulpiquet et al., 2014). On the other hand, free enzyme/whole cells are relatively unstable, expensive, and challenging to recover after an industrial operation in the biofuel production process (Rehman et al., 2014). Immobilization of enzyme/whole cells is a widely used empiric approach that improves stability and also aids in separation from the product, reducing the risk of contamination. Furthermore, enzyme/whole cell immobilization facilitates the easy recovery of enzyme/cells and allows for the reuse, being expensive for continuous processes (Rehman et al., 2014; Santos et al., 2018; Malhotra and Basir, 2020; Chen et al., 2021; Jigajinni and Meti, 2021).

This chapter discusses a few immobilization techniques such as entrapment, adsorption, encapsulation, and cross-linking. An overview of immobilized enzyme/cells and stability, support/substrate used in immobilization, and different immobilized bioreactor systems for an enhanced yield of liquid biofuel production are focused on. In addition, the sustainable biofuels, life cycle analysis (LCA), and research gaps associated with biofuel production are summarized.

Advances in Lignocellulosic Biofuel Production Systems. DOI: https://doi.org/10.1016/B978-0-323-91192-4.00014-6

11.2 Biofuels from lignocellulosic biomass

LCB is an important source for producing different categories of liquid transport fuels (Singh et al., 2021). Multistep processes are generally required to convert LCB components (cellulose, hemicelluloses, and lignin) into hydrocarbons and biofuels (Singh et al., 2021). The multistep processes include the selection of suitable biomass, effective pretreatment methods, appropriate enzyme cocktails for saccharification, fermentation of mixed sugars, and biofuel purification (Zhu et al., 2020; Singh et al., 2021). Immobilization of whole cells/enzymes is an appropriate technology for saccharification and fermentation of LCB to produce various biofuels (Fig. 11.1). Immobilization of cell/enzyme offers several advantages:

1. Easy recovery of cells or enzymes.
2. Pure product isolation (biofuels) from the reaction mixture.
3. Reusing cells and enzymes
4. Operation of enzymatic/fermentation reactions continuously, and
5. Improvements in the efficiency of cells and enzymes (Malhotra and Basir, 2020).

11.3 Immobilization methods/techniques

Selection of the most appropriate immobilization technique is essential in determining the activity of immobilized enzymes or cells (Mathew et al., 2013; Santos et al., 2018). Several immobilization techniques based on various interactions between the support and the enzyme or cells are as follows:

1. Entrapment/encapsulation,
2. Adsorption,
3. Covalent bonding, and
4. Cross-linking (Fig. 11.2) (Santos et al., 2018).

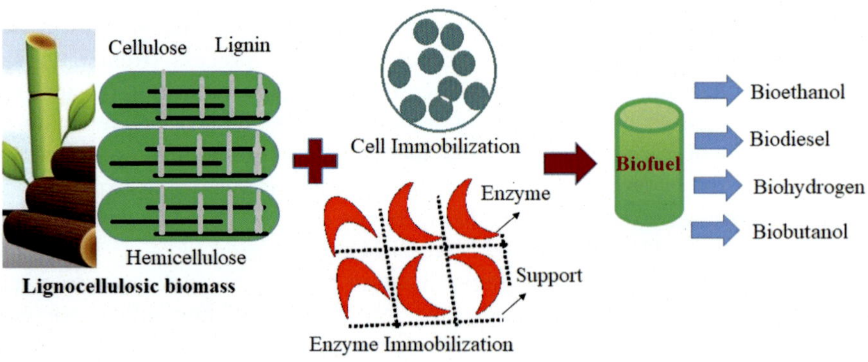

Figure 11.1 A consolidated figure using lignocellulosic biomass and whole cell/enzyme immobilization into liquid biofuels.

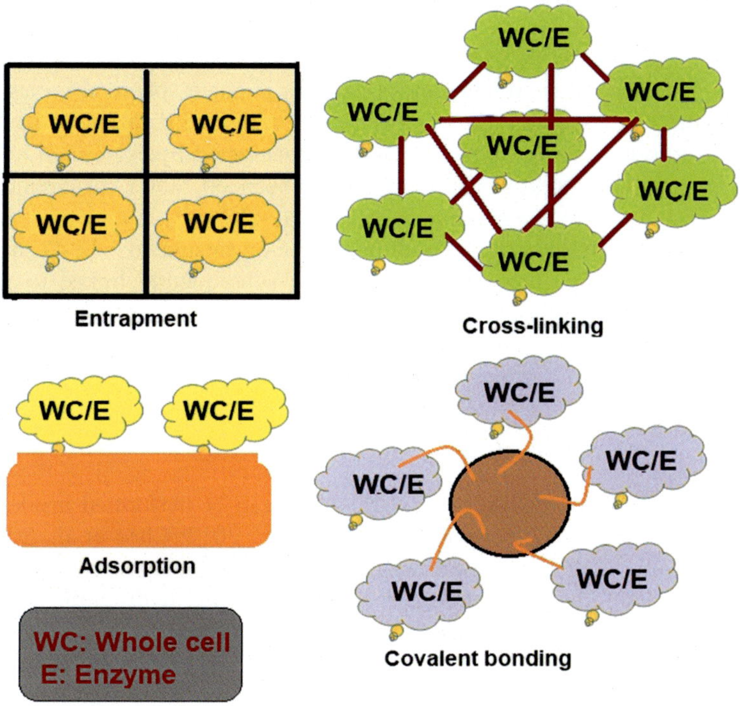

Figure 11.2 Different types of immobilization methods.

11.3.1 Entrapment/encapsulation

Among the various immobilization techniques, cellular entrapment in polymeric supports (i.e., agar agar, calcium alginate, gelatin, chitosan, k-carrageenan, and polyvinyl alcohol) has been thoroughly investigated for the generation of bioethanol, because of cost-efficiency and simplicity (Mathew et al., 2013; Zhou et al., 2018). Mathew et al. (2013) researched the suitability of entrapping *Saccharomyces cerevisiae* cells using alginate beads. The maximum bioethanol production rate by cells immobilized in alginate beads was 169.26 g bioethanol/kg oilseed rape straw after 3 h of fermentation. In a study, Jigajinni and Meti (2021) prepared the calcium alginate gel beads as a support structure for lipase (from *Lysinibacillus macroides* FS1) immobilization *via* the entrapment method in biodiesel production. The highest immobilization efficiency of 69% and lipase activity of 46% were recorded.

The enzyme invertase (β-D-fructofuranosidase) entrapped in the Ca-alginate-kappa-carrageenan matrix had the best reusability with high retention in activity, highest pH, and highest thermal and storage stability (Malhotra and Basir, 2020). Further, the researchers reported that when invertase-entrapped in Ca-alginate-kappa-carrageenan was used as a catalyst, the production increased by 9%.

Razack and Duraiarasan (2016) studied the transesterification of used cooking oil to produce biodiesel using encapsulated lipase from mixed microorganisms.

This process was less expensive because the immobilized lipase beads continued to function and remain stable after 20 cycles, providing a maximum of 93.61% biodiesel.

Applying the encapsulated silk fibroin-calcium alginate spherical mixes, the use of lipase (from *Pseudomonas fluorescens*) in the transesterification of soybean oil with ethanol to produce ethylic biodiesel was investigated by Ferreira et al. (2017). Under these reaction conditions (150 mg of soybean oil, 450 L of ethanol, 32°C, and a 48 h reaction period), the biodiesel yield was 42%.

Xie et al. (2017) synthesized the hydroxyapatite-encapsulated γ-Fe_2O_3 nanoparticles (HAp-γ-Fe_2O_3), and organic biguanide was subsequently linked to the magnetic carrier by covalent attachments, which was used to make biodiesel fuel. The inorganic—organic hybrid material was found to be environmentally friendly, offering attractive approaches for biodiesel production.

11.3.2 Physical adsorption

Adsorption is an easy immobilization method that can be performed inside reactors without using chemicals (Vučurović and Razmovski, 2012; Chen et al., 2021). The ionic composition of the material primarily influences cell adsorption on the support material, the surface charge of the microorganisms, ionic, hydrophilic, and hydrophobic interactions, and electrostatic forces like the Van der Waals force (Zhou et al., 2018). Vučurović and Razmovski (2012) described the use of sugar beet pulp (thick juice) as a biomaterial for *S. cerevisiae* adsorption by natural attachment for ethanol production. Without the addition of extra nutrients, efficient bioethanol production was carried out in recurrent batch modes. The seventh batch of fermentation with thick beet pulp juice produced a maximum ethanol concentration of 52.26 g/L and a yield of 0.446 g/g (equivalent to 87.30% of its theoretical value).

Chen et al. (2021) reported that the *S. cerevisiae* cells adsorbed on sorghum stalks exhibited healthy growth. At 16 h of fermentation, *S. cerevisiae* stably adhered to the loose and porous surfaces of sorghum stalk cores and increased bioethanol productivity.

Godoy-Salinas et al. (2021) evaluated using unprocessed sugarcane bagasse to immobilize *S. cerevisiae* ITV-01 RD cells by adsorption to increase ethanol production. Employing glucose medium at a dilution rate of 0.11/h, 48 g/L of ethanol was generated in a packed bed reactor (PBR), and the ethanol productivity was 12.1 g/L/h. Using sugarcane juice diluted to 100 g/L of total sugars, ethanol could be produced at a rate of 5.03 g/L/h. Further, the adsorption process could improve the fermentation process because it provided stable biomass, greater mass transfer efficiency, better biomass retention capability, and efficient substrate use under a short hydraulic retention period (Zhou et al., 2018).

11.3.3 Covalent binding

In contrast to other noncovalent immobilization approaches, covalent coupling (co-immobilization) of enzymes usually results in the development of a stable enzyme electrode (Shim et al., 2011). Shim et al. (2011) described the coimmobilization of glucose oxidase (GOx) and Fc electron transfer mediator (ferrocene derivative) on a graphite carbon felt (felts are insulating materials for application temp up to 1000°C) electrode. The findings showed that coimmobilization in an enzymatic biofuel cell was a highly effective means of mediating the electron transfer between the electrode and enzyme.

In the transesterification of sunflower oil, ethanol, and other low molecular weight alcohols, new second-generation biodiesel was produced using 1,3-specific pig pancreatic lipase, covalently immobilized on $AlPO_4$/sepiolite substrate (Luna et al., 2012).

11.3.4 Cross-linking

Cell immobilization has been extensively explored in terms of adsorption, trapping, and encapsulation. However, although they are quite effective at immobilizing cells, these techniques have a high rate of cell leakage, diffusion restrictions, high immobilization cost, and poor cell loading (Adelabu et al., 2019). On the other hand, the cross-linking method is practical, inexpensive, and durable for cells and/or enzymes (Adelabu et al., 2019).

Modenez et al. (2018) studied the impact of glutaraldehyde cross-linking mechanisms on the recyclability of lipase B from immobilized *Candida antarctica* (CAL-B) for converting soybean oil into biodiesel through transesterification. The potential of the immobilized CAL-B enzyme in biodiesel production by microorganisms was proven by the transesterification of lipids extracted from oleaginous yeasts, which produced an 89% yield. In a wide range of conditions, the strategy to prevent enzyme desorption via enzymatic intermolecular cross-linking on hydrophobic support is useful (Fernandez-Lopez et al., 2017).

11.4 Immobilized bioprocess components

The choice of immobilization technique depends on the nature of the substrate, enzyme types, and reaction conditions (Pikula et al., 2020).

11.4.1 Immobilization of whole cells

Immobilized whole cells (bacteria or yeasts) can increase cell mass concentration in bioreactors, improve substrate utilization effectiveness, and increase the productivity of fermentation processes while incurring the minimum possible production costs (Vučurović and Razmovski, 2012). Immobilized cells have commonly been used for batch or continuous ethanol fermentation because of their feasibility for repeated usage and high biomass retention rate (Wirawan et al., 2012).

Several reports describe using immobilized cells (particularly for immobilized *S. cerevisiae*) to produce bioethanol (Behera et al., 2010a,b; Wirawan et al., 2012).

However, less research has been published using immobilized *Zymomonas mobilis* for ethanol fermentation (Behera et al., 2010c,d; Wirawan et al., 2012).

Compared to free cells, the cell immobilization technique increases fermentation efficiency by improving the tolerance of liquid biofuels (biobutanol) and providing a more significant cell density that can enhance biological activity and process stability (Zhou et al., 2018).

11.4.2 Immobilization of enzymes

Enzyme immobilization has several benefits, including ease of recovery, repeated applications, increased stability under adverse conditions, and decreased total process economy (Shanmugam et al., 2020). Different enzyme systems (e.g., lipase and cellulase) are involved in liquid biofuel production.

Lipase and cellulase, two main types of enzymes that can produce fatty acid alkyl esters and glucose from lipids and cellulose, respectively, have been used extensively in biodiesel production (De Poulpiquet et al., 2014). The most effective and environmentally friendly biocatalysts for the efficient breakdown of LCB into simple sugars are enzymes (cellulases), which have high specificity, selectivity, and potential catalytic activity (Rai et al., 2019).

Because of their outstanding intrinsic features, such as selectivity toward the substrate and high catalytic activity under optimum operating conditions, enzymes are anticipated to be better choices for biocatalysts than microorganisms (De Poulpiquet et al., 2014).

11.4.3 Substrates for immobilization

To produce liquid biofuels, several natural polymers (i.e., agar agar, alginate, and chitosan) and synthetic polymers (i.e., polyacrylamide or polyvinyl alcohol) have been used for the immobilization of cells and enzymes (Wirawan et al., 2012). Of these polymers, alginate, in the form of Ca-alginate, is the most popular and effective one (Behera et al., 2010a) because it is relatively easy to prepare. However, Ca-alginate is less suited for continuous biofuel production systems due to its high cost and limited mechanical strength (Wirawan et al., 2012; Mathew et al., 2013). Agar is a polysaccharide composed of agarose and is also commonly used in immobilizing agents. It has a robust gelling ability, is acid stable, reduces the risk of contamination, and enhances liquid biofuel productivity compared to free cells (Rehman et al., 2014). Conversely, synthetic polymers, such as polyvinyl alcohol, have exceptional mechanical strength but often cause a loss in cell viability due to high toxicity, especially when synthesized by the boric acid method (Takei et al., 2011).

11.4.4 Immobilized bioreactor system

Immobilization of the whole cell or enzyme would raise the cost of batch production of liquid biofuels. However, because the cells would be retained in the bioreactor and the amount of downstream processing would be minimized, it might save time, energy, and cost in a continuous system (Mathew et al., 2013). Usually, cells/enzymes are immobilized on carriers and used continuously in PBRs or continuous-stirred tank reactors (CSTRs) to produce biofuels (Tran et al., 2016).

Tran et al. (2016) reported the methanolysis of sunflower oil catalyzed by the immobilized lipase to produce biodiesel using a PBR system. The accumulation of glycerol has been the major reason preventing biodiesel conversion due to pressure drop build-up and increased mass transfer resistance. Using serial PBRs (three columns in a sequence) with subsequent removal of glycerol from the effluent of each column could overcome this issue because biodiesel conversion was increased by up to 85%. The PBR series system could continuously produce biodiesel at the yield of 4413 kg/kg lipase without deactivating the biocatalyst. Dias et al. (2017) scaled up the process for the ethyl-esterification of fatty acids using fermented solids while maintaining process performance. The packed bed containing 120 g of dry fermented solids and 1000 g of fatty acids/olein (molar ratio of ethanol to fatty acids of 1.5:1) and 245 g of ethanol gave 88% conversion in 24 h. Delignified sugarcane bagasse was used by Sowatad and Todhanakasem (2020) to immobilize *S. cerevisiae* SC90 and produced ethanol in a PBR. The maximum ethanol yield ($Y_{P/S}$) of 0.42 g/g (82.35% theoretical yield) was reported.

The jacketed PBR system using immobilized lipase was found to be suitable for the production of H_2 with high stability (Kumar et al., 2020). The researchers investigated biodiesel production from nonedible oil in a jacketed PBR using immobilized lipase (from *Pseudomonas cepacia*). The experimental data were evaluated by the response surface methodology method. The results showed that the 78.58% biodiesel yield was obtained under optimized conditions.

Applying a fluidized bed reactor (FBR) system, Cubides-Roman et al. (2017) studied the feasibility of enzymatically synthesizing ethyl esters (biodiesel) in a magnetically stabilized FBR. The enzyme lipase (from *P. fluorescens*) was immobilized on chitosan with magnetic properties as a biocatalyst. With a molar ratio of 1:11.25, a reaction temperature of 47.5°C, and a magnetic induction of 9.7 mT, the ideal reaction parameters were maximized at a differential mode reactor for 12 h. The conversion achieved under these circumstances was 9.6%, validating the model.

11.5 Production of sustainable biofuels

The immobilization-based approach has been employed as an economically viable and safe alternative for the environmentally friendly manufacture of liquid biofuels (Ingle et al., 2017) (Table 11.1).

11.5.1 Bioethanol

Higher ethanol tolerance and productivity and lower substrate inhibition have been shown to indicate that the immobilized cell system is preferable to the free cell system (Lee et al., 2012). Raw materials (i.e., sugarcane molasses, corn, and cassava bagasse)

Table 11.1 Comparison of the performance of liquid biofuel production from using various immobilized enzyme/cell systems.

Liquid biofuels	Immobilization of whole cells/enzymes	Immobilized substrate	Performance/yield	Reference
Bioethanol	*Saccharomyces cerevisiae* ITV-01 RD	Sugarcane bagasse	5.03 g/L/h	Godoy-Salinas et al. (2021)
	Aspergillus oryzae and *Monascus purpureus*	Ca-alginate	0.41[a]	Lee et al. (2012)
	Saccharomyces diaststicus	*Mucuna urens*	55.27 g/L	Adelabu et al. (2019)
	S. cerevisiae	Maize stem Ground tissue	83.20 g/L 132.39 g/L	Razmovski and Vučurović (2012)
	S. cerevisiae TISTR 5020	Ca-alginate	65.43 g/L	Khammee et al. (2021)
	Cellulase	MNPs	–	Ingle et al. (2017)
	Laccase, cellulase and β-glucosidase	Lignocellulosic biomasses	63.43[b]	Sankar et al. (2018)
	S. cerevisiae	Calcium alginate films	120 g/L	Santos et al. (2018)
	S. cerevisiae	Ca-alginate	189 g/kg	Behera and Ray (2015)
	S. cerevisiae	*Luffa cylindrica* L.	223.2 g/kg	Behera et al. (2011)
	Zymomonas mobilis MTCC 92	Ca-alginate	58.7 g/L	Behera et al. (2012)
	S. cerevisiae (strain CTCRI)	Ca-alginate	152–205 g/kg	Swain et al. (2007)
	S. cerevisiae	Agar agar Calcium alginate	150.6 g/kg 152.8 g/kg	Behera et al. (2010a,b,c,d)
	S. cerevisiae var. *ellipsoideus*	Ca-alginate	176 g/L	Nikolić et al. (2010)
	S. cerevisiae	Sugar beet pulp	0.446 g/g	Vučurović and Razmovski (2012)
	Scheffersomyces stipitis	Ca-alginate	0.33 g/g	Milessi et al. (2015)
	S. cerevisiae SC90	Sugarcane bagasse	0.42[a]	Sowatad and Todhanakasem (2020)
Biodiesel	Lipase (*Pseudomonas fluorescens*)	Porous kaolinite particle	–	Iso et al. (2001)

	Lipase (*Burkholderia* sp. C20)	Fe_3O_4 core with silica shell	97[b]	Tran et al. (2012)
	Lipase (Candida rugosa)	MGO	92.8[b]	Xie and Huang, (2018)
Biohydrogen/ Bio-H_2	*Enterobacter cancerogeous* HG6 2 A and *Enterobacter homaechei* 83	Activated carbon	1.203 mol/mol-glucose	Zhang et al. (2017)
	Trichoderma asperellum	Chitosan coated Fe_3O_4@SiO_2 NPs	2.8 mol H_2/mol reducing sugar	Shanmugam et al. (2020)
	Chlorella sp.	Agar	238 mL/h/L	Song et al. (2011)
	Enterobacter aerogenes	MWCNT-COOH	2.2 mol/mol glucose	Boshagh et al. (2019)
	Clostridium beijerinckii NCIMB8052	Magnetite nanoparticles	2.1 mol H_2/mol glucose	Seelert et al. (2015)
	C. butyricum EB6	PEG gel	5.35 L H_2/L POME	Singh et al. (2013)
	Clostridium sp. T2	Mycelia pellets	2.76 mmol H_2/L/h	Zhao et al. (2012)
	Clostridium pasteurianum	NP-Ni and NP-Fe	–	Wannapokin et al. (2021)
	Tetraspora sp. CU2551	Calcium alginate matrix	12.8 mL	Maswanna et al. (2020)
	C. butyricum TISTR 1032	Sugarcane bagasse	1.52 mol H_2/mol hexose	Plangklang et al. (2012)
	C. butyricum and *Rhodopseudomonas palustris*	Alginate gel	–	Kao et al. (2014)
	Clostridium sp.	Calcium alginate beads	294.83 mL H_2/g TVS	Sekoai et al. (2018)
Biobutanol	*Clostridium acetobutylicum* ATCC 824	Lignocellulosic biomass (rice straw)	0.58 mol/mol glucose	Tsai et al. (2020)
	C. beijerinckii TISTR 1461	Vetiver grass	11.6 g/L	Srisuk et al. (2021)
	C. acetobutylicum	Alginate	21.64 g/L	Ye et al. (2018)
	C. acetobutylicum CICC 8012	Biofilm	5.8 g/L	Zhou et al. (2018)

Fe_3O_4@SiO_2 NPs, Fe_3O_4@SiO_2- nanocomposite; *MGO*, magnetic graphene oxide; *MNPs*, Magnetic nanoparticles; *MWCNT-COOH*, multiwalled carbon nanotube-carboxylic acid; *NP*, nanoparticle; *PEG*, polyethylene glycol; *POME*, palm oil mill effluent; *TVS*, total volatile solids.

[a]$Y_{E/s}$, Yield of ethanol production/starch consumption.

[b]Percentage yield.

are widely used for industrial bioethanol production using an immobilized system (Ariyajaroenwong et al., 2012). In a batch system, Nikolić et al. (2010) investigated the ethanol fermentation of enzymatically saccharified maize meal hydrolysates using free and immobilized *S. cerevisiae* var. *ellipsoideus* yeast. Optimal initial glucose concentrations of 150 and 176 g/L for the free and immobilized systems were achieved, respectively.

Mahula flowers (*Madhuca latifolia* L.), a tree commonly found in tropical rain forests, a renewable, nonfood-grade cheap carbohydrate substrate from a nonagricultural environment and can serve as an alternative to food-grade sugar/starchy crops such as maize, sugarcane for bioethanol production (Behera et al., 2010a,b,c; Behera and Ray, 2019). In a study, Behera et al. (2010a) reported the batch fermentation of mahula flowers for ethanol production using immobilized cells (in the agar agar and calcium alginate) and free cells of *S. cerevisiae*. The ethanol yields were 151.2, 154.5, and 149.1 g/kg flowers using immobilized (in the agar agar and calcium alginate) and free cells, respectively. In another study, Behera et al. (2011) studied the ethanol production from mahula flowers in submerged fermentation using whole cells of *S. cerevisiae* immobilized in luffa (*Luffa cylindrica* L.), a sponge disks. The ethanol production by immobilized cells was 8.96% higher than the free cells.

Behera et al. (2012) studied the bioethanol production from cane molasses (diluted to 15% sugar w/v) using the bacterium *Z. mobilis* MTCC 92 entrapped in luffa sponge disks and Ca-alginate gel beads as the immobilizing matrices. The final ethanol concentrations (after 96 h fermentation) were 58.7 ± 0.09 and 59.1 ± 0.08 g/L molasses with luffa and Ca-alginate entrapped *Z. mobilis* cells, respectively, exhibiting 83.25 ± 0.03 and 84.6 ± 0.02% sugar conversion.

Milessi et al. (2015) used the yeast *Scheffersomyces stipitis* immobilized in calcium alginate matrix to examine the impact of two crucial variables (such as cell concentration and stirring) on the synthesis of ethanol from sugarcane bagasse hydrolysate. Maximum ethanol production (8.90 g/L) with ethanol yield ($Y_{p/s}$) of 0.33 g/g and ethanol productivity (Q_p) of 0.185 g/L/h were obtained under the optimized process conditions (10.0 g/L of cells and 100 rpm).

Aspergillus oryzae and *Monascus purpureus*, two saccharolytic molds co-immobilized with *S. cerevisiae* were studied for bioethanol production from sweet potatoes (Lee et al., 2012). The vitality of yeast cells was 95.7% at a final ethanol concentration of 6% because immobilization increased the ethanol tolerance of yeast cells. After nine days of fermentation, co-immobilization of *S. cerevisiae* with a mixed culture of *A. oryzae* and *M. purpureus* produced bioethanol at a rate of 4.08% (v/v) with a $Y_E/_s$ of 0.41.

Compared to molasses, the thick juice from sugar beets provides a more convenient industrial substrate for repeated batch ethanol fermentation and does not require any nutrient supplementation(Vučurović and Razmovski, 2012). The researchers investigated the possibility of sugar beet pulp as a low-cost, renewable support for the immobilization of *S. cerevisiae* cells. The seventh fermentation batch of thick juice (with an initial sugar of 120 g/L) produced an ethanol output of 0.446 g/g and a maximum ethanol concentration of 52.26 g/L, with nearly complete sugar utilization

(97.52%). In another study, Razmovski and Vučurović (2012) reported the biocatalyst prepared by immobilizing *S. cerevisiae* on maize stem ground tissue for batch fermentation of sugar beet molasses and thick juice under normal and very high gravity conditions. The maximum ethanol concentration of 83.20 g/L for molasses and 132.39 g/L for thick juice were achieved in very high gravity fermentation by immobilized cells. In contrast, the free yeast cells were unsuccessful in the same media. Sugar beet thick juice was found to be economically favorable, compared to molasses, particularly in the very high gravity fermentation process.

Behera and Ray (2015) investigated the saccharification of cassava flour (by acid-enzyme and enzyme-enzyme hydrolysis) and subsequent fermentation into ethanol using *S. cerevisiae* cells immobilized in Ca-alginate beads. The saccharification resulted in a higher total sugar recovery of 72.88% (acid-enzyme hydrolysis) than enzyme-enzyme hydrolysis (58.1%). The yeast cells began to grow in the log phage, and during the stationary phase, maximum ethanol production (189 g ethanol/kg flour) was attained with 94.74% sugar conversion.

In a study, Sankar et al. (2018) developed a novel one-pot pretreatment technology by co-immobilizing a tri-enzyme biocatalyst (laccase, cellulase, and β-glucosidase) to evaluate the bioethanol production potential of four sustainable LCBs (i.e., *Typha angustifolia*, *Arundo donax*, *Saccharum arundinaceum*, and *Ipomoea carnea*). During enzymatic saccharification, *S. arundinaceum* yielded a higher total reducing sugar of 205 mg/g compared to other biomass. However, the highest percentage of bioethanol yield of $63.43 \pm 9.35\%$ was obtained with *I. carnea*.

The performance of the alginate film with immobilized cells and the customized hybrid membrane suggests an integrated reaction-separation device for the production and purification of bioethanol (Santos et al., 2018). Santos et al. (2018) studied sugar fermentation for ethanol production using *S. cerevisiae* cells immobilized in calcium alginate films. Yeast immobilized on alginate films produced a higher ethanol yield, with a flux of 140.6 $g/m^2/h$ (separation factor of 37.52) than free yeast cells under the same fermentation conditions.

Aquatic biomass is now receiving increased attention for the production of biofuels due to its abundant availability, renewable, and environmentally friendly qualities. Macroalgae are regarded as a third-generation feedstock for the manufacture of biofuels and are one of the rich LCB materials (Khammee et al., 2021). Khammee et al. (2021) studied the potential of bioethanol production from macroalgae (*Rhizoclonium* sp.) biomass, and the fermentation process was carried out in the incubator (30°C–35°C). Immobilized yeast (*S. cerevisiae*), which could be used several times, and the highest concentration of ethanol (65.43 18.13 g/L) was achieved after 96 h of fermentation, helped to speed up the fermentation process.

11.5.2 Biodiesel

Biodiesel, referred to as fatty acid methyl esters, is renewable, nontoxic, sulfur and carbon-dioxide-free, and clean fuel (Razack and Duraiarasan, 2016). Vegetable seed oils as a potential raw material for the manufacturing of biodiesel have garnered a lot of attention in several countries (Ferreira et al., 2017). Many immobilized enzymes are commercially available to meet the large-scale industrialization of biodiesel, as opposed to the traditional chemical/biochemical catalysis approach for producing biodiesel (Zhang et al., 2012).

In a study, Nagaraj et al. (2010) reported a low-cost lab scale set up for biodiesel production using immobilization of fungus (*Rhizopus oryzae*) whole cells on fibrous support. The efficiency has improved with the use of whole-cell transesterification for biodiesel production.

Palm oil mill effluent has been used as a sustainable feedstock for biodiesel. Rachmadona et al. (2021) investigated the potential of immobilized *A. oryzae* whole cells expressing *Candida antartica* lipase B (r-CALB) using a palm oil mill effluent-based medium as a low-cost carbon source within biomass support particles. This immobilized r-CALB was used as a catalyst to esterify palm oil mill effluent into biodiesel. This synthesis produced an ester content of 97.52% (w/w) after 96 h. Immobilized lipase r-CALB demonstrated excellent potential as a biocatalyst for converting palm oil mill effluent to biodiesel with a better conversion rate than the commercial Novozyme 435 lipase.

Tosylated cloisite as a new heterofunctional carrier for covalent immobilization of lipase and its utilization for biodiesel production from waste frying oil was studied. Nezhad and Aghaei (2021) immobilized lipase (from *Candida rugosa*) covalently on the tosylated cloisite and used it to produce biodiesel from waste frying oil. The maximum yield of biodiesel (97.1%) was obtained. However, the relative yield of biodiesel for tosylated cloisite and free lipase was 70.6% and 33.4%, respectively, after 30 days of fermentation.

Magnetic nanoparticles are an efficient and significant carrier in lipase immobilization (Thangaraj et al., 2019). Xie and Huang (2018) encapsulated Fe_3O_4 nanoparticles in graphene oxides and then employed them for the immobilization of lipase (from *C. rugosa*) *via* the interfacial activation of enzymes on hydrophobic surfaces. Using this immobilization strategy, the enzyme immobilization efficiency of 85.5% and biodiesel yield of 92.8% were achieved. In a report, Thangaraj et al. (2019) studied the organosilane-modified Fe_3O_4@SiO_2 core magnetic nanocomposites with immobilized lipase and explored their potential application in the biodiesel field. Maximum activity recovery of 84% and biodiesel yield greater than 90% were obtained by lipase immobilized on Fe_3O_4@SiO_2 support.

11.5.3 Biohydrogen

Biological hydrogen production processes are environmentally friendly and hardly energy intensive. Different biohydrogen production methods have been followed; among them, the two-stage system (dark fermentation and photofermentation) is one of the most effective hydrogen production methods (Gundogdu et al., 2013). There are various benefits to producing hydrogen by photofermentation via immobilized reactor systems, such as preventing cell washout at low hydraulic retention times and maintaining a high biomass inventory in the bioreactor (Gundogdu et al., 2013). In a study, Gundogdu et al. (2013) entrapped two bacterial strains (*Rubrivax gelatinosus* and *Stenotrophomas maltophilia*) in LentiKat and the LentiKat disks in an up-flow bioreactor on photofermentative hydrogen production. The LentiKat technique demonstrated that cell immobilization was a reliable and efficient way to maintain a higher biomass inventory in a bioreactor that produced hydrogen. Additionally, its transparent structure allowed efficient light penetration, which was advantageous for photofermentative hydrogen production.

The development of improved photobioreactor systems is one route for raising hydrogen productivity (Ross and Pott, 2021). In a reactor that was either running as a packed bed or a fluidized bed, Ross and Pott (2021) examined the hydrogen productivity of *Rhodopseudomonas palustris* under planktonic and immobilized cell conditions. The fluidized bed photobioreactor outperformed the traditional planktonic culture and the packed bed photobioreactor, achieving a maximum specific hydrogen generation rate and substrate conversion efficiency of 15.74 mL/g/h and 43%, respectively (Ross and Pott, 2021).

The simplest way to produce hydrogen is by the dark fermentation process, which is more economically feasible due to its low energy requirements, wide range of substrates, and diversified range of microorganisms (Sekoai et al., 2018). Zhang et al. (2017) studied dark fermentation performance to improve biohydrogen production using co-cultures (*Enterobacter cancerogeous* and *E. homaechei*) and the immobilization system with activated carbon. The surface attachment and the synergistic effect explained the higher performance (biohydrogen yield of 1.203 mol/mol glucose) of the cell immobilization system. Using suspended and immobilized cells of anaerobic mixed sludge, Sekoai et al. (2018) evaluated the impact of nitrogen gas sparging on dark fermentative biohydrogen generation performance. A dark fermentation method using N_2-sparged cells immobilized on calcium alginate beads resulted in a maximum biohydrogen fraction (56.98%) equated to a biohydrogen yield of 294.83 mL H_2/g total volatile solids.

Using ceramic balls as immobilization materials for mixed culture is an effective way of biomass hold-up and high hydrogen yields. Keskin et al. (2012) investigated the performance of a new cell immobilization medium, a ceramic ball, for continuous biohydrogen production compared to the suspended cell culture system. Significant H_2 generation activity was observed in the cells adsorbed strongly on ceramic balls. An ideal hydrogen generation rate of 2.7 L H_2/L/day was also used to produce H_2 from 10 g/L sucrose successfully.

The nanoparticles with active polymers endow functional groups that serve as better anchorages between enzymes and supports which improves the overall economy of the process (Fang et al., 2016). Particles of Fe_3O_4@SiO_2-chitosan show a significant potential for usage in the synthesis of biohydrogen. Shanmugam et al. (2020) explored the potential of immobilized laccase (from *Trichoderma asperellum*) onto Fe_3O_4@SiO_2-chitosan nanosupport in the delignification of biomass for biohydrogen production. In a CSTR system, biohydrogen yield and production rates were reported at 2.8 mol H_2/mol reducing sugar and 25 L H_2/L/day, respectively. Similarly, Wannapokin et al. (2021) reported the biohydrogen production by *C. pasteurianum* co-immobilized with selected nano-metal (NP–Ni and NP–Fe) particles. The added 400 mL/L of NP-Ni and 200 mL/L of NP-Fe enhanced the composition percentage of hydrogen gas to 46.90% and 42.03%, respectively.

11.5.4 Biobutanol

Biobutanol can be synthesized via the petrochemical route or by the fermentation route using anaerobic bacteria, predominantly of the genera Clostridia, by acetone–biobutanol–ethanol (ABE) fermentation (Van Hecke et al., 2018; Srisuk et al., 2021). The performance of ABE fermentation is affected by several factors, including the microorganism used (Zhou et al., 2018). *C. acetobutylicum* and *C. beijerinckii* are known to produce biobutanol (Van Hecke et al., 2018; Zhou et al., 2018). The characteristics of *C. acetobutylicum* to thrive on a variety of carbohydrates, including glucose, sucrose, and starch, is one of its traits (Zhou et al., 2018).

In the ABE fermentation process, Tsai et al. (2020) evaluated the use of renewable feedstock (rice straw, sugarcane bagasse, and microalgal hydrolysate) for biobutanol production. Immobilization of microbial cells has been considered a potential alternative for enhancing biobutanol productivity through ABE fermentation. *C. acetobutylicum* was immobilized in polyvinyl alcohol to improve the cell loading in fermentation and protect the cells from biobutanol toxicity. The results revealed that rice straw hydrolysate proved to be a better substrate/feedstock, and in the absence of yeast extract, a biobutanol titer, yield, and productivity of 13.80 g/L, 0.90 g/L/h, and 0.58 mol biobutanol/mol glucose (0.23 g biobutanol/g glucose), respectively were obtained.

To enable the enzyme's access to its substrates and the subsequent conversion of carbohydrates into fermentable sugars, a pretreatment step is necessary to convert LCB to ABE synthesis (Zhou et al., 2018). According to a study by Srisuk et al. (2021), vetiver grass (*Vetiveria zizanoides*) pretreated with sodium hydroxide or sulfuric acid as a carrier for immobilized *C. beijerinckii* TISTR 1461 was used in ABE fermentation to produce biobutanol. During the pretreatment, the lignin and hemicellulose components of the vetiver grass were removed. This led to structural modifications, particularly an increase in crystallinity and surface area. The vetiver grass pretreated with 6% (w/v) NaOH at 120°C for 6 h had the best potential as a carrier in ABE fermentation, giving butanol and a total ABE concentration of 11.6 and 19.8 g/L, respectively, approximately 1.3-fold greater than that of the free-cell system.

Ye et al. (2018) effectively established the adsorbent (alginate) immobilization method to support in situ extraction utilizing octanol, which is efficient in extracting butanol but presents a significant harmful effect on the bacteria. In adsorbent-assisted ABE fermentation, there was 21.64 g/L butanol in the removing layer compared to butanol produced without the adsorbent, which can reduce damaging exposure of *C. acetobutylicum* to octanol. The choice of carrier medium is crucial for cell adsorption and biofilm formation. According to Zhou et al. (2018), *C. acetobutylicum* was immobilized in a biofilm on active carbon, bagasse, and bricks for the synthesis of ABE. With bagasse as the carrier, the best fermentation performance was attained, resulting in 10.21 g/L of total solvent concentration (5.8 g/L of butanol), 0.42 g/g of yield, and 0.32 g/L/h of productivity.

11.6 Life cycle analysis of liquid biofuels using immobilization techniques

LCA is a useful method for evaluating the environmental impacts of a product (i.e., liquid biofuels) over the entire process lifetime, from raw material extraction to production, use, recycling, and waste disposal (Chamkalani et al., 2020). This approach considers the entire life cycle of a product or the so-called "from the cradle to the grave" principle (Pikula et al., 2020). In a "cradle to grave approach study," Fernandez et al. (2017) reported the LCA studies comparing alkaline and immobilized enzyme catalyst processes for biodiesel production under Brazilian conditions. The method provided a sustainable audit and claimed that the immobilized enzyme-catalyzed process decreased the environmental impact and was beneficial for biofuel production compared with the alkali-catalyzed transesterification process.

The LCA study adopted a "cradle to gate" approach to estimate the environmental impact of different catalytic processes such as immobilized, soluble biocatalysts, and alkali catalysts (Raman et al., 2011). The results revealed that biodiesel production using immobilized biocatalyst had less environmental impact than alkali and soluble biocatalyst. Although biodiesel production has tremendous environmental benefits, a positive energy balance must also be acquired to ensure its economic feasibility (Wang et al., 2017; Chamkalani et al., 2020). Wang et al. (2017) conducted life cycle energy analysis and biodiesel production for the coupling system based on the LCA framework. The results showed immobilization efficiently reduced the energy consumption in the dewatering stage, and it was necessary to cultivate microalgae with secondary effluents.

11.7 Patents, commercial applications, and research gaps

In this section, some information is given on the available patents on immobilized enzymes for liquid biofuel production, commercial applications of immobilization technology, and research gaps.

11.7.1 Patents

A recent patent (WO2020239983) reports on the lipolytic polymer particles for esterification and interesterification of biofuel production (Andric et al., 2020). The invention provides that the enzyme particles (immobilized lipolytic enzyme), hydrophobic polymer, organic filter aid, and water-soluble polyol are selected from carbohydrates and sugar alcohols. The particles are found suitable for enzymatic interesterification of triglycerides and subsequent separation of the enzyme and triglycerides by filtration (WO2020239983) (Andric et al., 2020).

A US patent (No. 7,144,716) claims the recovery method for immobilized biocatalysts. The process includes the enzymatic treatment of a soluble or particulate substrate, immobilized enzyme, and a bioreactor to produce a slurry which provides for the reclamation and reuse of the immobilized enzyme even when a further particulate solid is present in the effluent/product stream (Saville, 2006).

The patent (No. 9,695,426) describes the metabolically-modified microorganisms as useful for producing biofuels. More specifically, producing high alcohols including isobutanol, 1-butanol, 1-propanol, 2-methyl-1-butanol, 3-methyl-1-butanol, and 2-phenyl ethanol from a suitable substrate are methodized (Liao et al., 2017).

11.7.2 Commercial applications

Commercial-scale enzyme catalysis has been implemented in several industries for biofuel production (Agrawal et al., 2018). In a pilot scale (250 kg/day capacity), Agrawal et al. (2018) pretreated (162°C for 10 min) the rice straw with dilute acid (0.35%) for biofuel production. Further, enzymatic hydrolysis was carried out by batch and fed-batch approaches through substrate and enzyme splitting to improve the hydrolysis yield.

Transbiodiesel entered the biofuel arena with its new IP (intellectual property) protected technology related to the preparation of immobilized enzymes, used in a continuous, stirred tank reaction process for biodiesel production (*Biodiesel Magazine*, 2014). The company offers an advanced enzymatic technology platform to assist biodiesel-producing companies in fueling the world by producing American Society for Testing and Materials and European standards biodiesel from any grade of feedstocks with free fatty acid (up to 100%), including used cooking oil, palm fatty acid distillates, crude plant oils, and brown grease (Tafesh, 2014).

11.7.3 Research gaps

Many researchers are interested in immobilization technology for manufacturing liquid biofuels owing to its merits, such as increased process yields, decreased microbial contamination, and greater uniformity (Tran et al., 2016; Zhou et al., 2018; Kumar et al., 2020). However, the following research shortcomings should be noted:

1. A significant obstacle to biofuel generation is the recalcitrant lignin in the LCB. Therefore it must be removed from the LCB before being converted into biofuels. Enzyme-mediated degradation and immobilization onto the support matrix are regarded as "green approaches" since they boost the production of biofuel while generating no toxic intermediates (Shanmugam et al., 2020).
2. Despite studies describing enzyme immobilization, little is known about how the structure of the enzyme is affected when it is immobilized on various support types and sizes. Even less is known about how the cross-linking agents affect the binding between enzyme and support, changing enzymatic catalysis (Modenez et al., 2018).
3. It is crucial to comprehend the primary, secondary, and tertiary structure of the enzyme to adapt it to carry out simple immobilization while keeping activity, which can produce outstanding results (Nigam et al., 2014).

11.8 Conclusion and future perspectives

Various low-cost carriers from plant biomass are suitable as support materials for enzyme/cell immobilization. Specific biofuel production rate using immobilized enzymes/cells is higher than free cells or enzymes and an attractive option to maximize yield in industrial scale studies. Coimmobilizing the appropriate substrate and hydrogen-producing microorganisms increased hydrogen production compared with the traditional suspended system. In a nutshell, to carry out a successful immobilization process for the production of liquid biofuels, intelligent judgment and manipulation of approaches are necessary on an industrial/large scale.

References

Adelabu, B.A., Kareem, S.O., Oluwafemi, F., Adeogun, I.A., 2019. Bioconversion of corn straw to ethanol by cellulolytic yeasts immobilized *in Mucuna urens* matrix. J. King Saud. University-Science 31 (1), 136–141.

Agrawal, R., Bhadana, B., Mathur, A.S., Kumar, R., Gupta, R.P., Satlewal, A., 2018. Improved enzymatic hydrolysis of pilot scale pretreated rice straw at high total solids loading. Front. Energy Res. 6, 115.

Andric, P., Nielsen, M., Holm Per, C., 2020. Lipolytic polymer particles for esterification and interesterification, WO/2020/239983.

Ariyajaroenwong, P., Laopaiboon, P., Jaisil, P., Laopaiboon, L., 2012. Repeated-batch ethanol production from sweet sorghum juice by *Saccharomyces cerevisiae* immobilized on sweet sorghum stalks. Energies 5 (4), 1215–1228.

Behera, S., Kar, S., Mohanty, R.C., Ray, R.C., 2010a. Comparative study of bio-ethanol from mahula (*Madhuca latifolia* L.) flowers by *Saccharomyces cerevisiae* cells immobilized in agar-agar and Ca-alginate matrices. Appl. Energy. 87, 96–100.

Behera, S., Mohanty, R.C., Ray, R.C., 2010b. Comparative study of bio-ethanol production from mahula (*Madhuca latifolia* L.) flowers by the immobilized cells of *Saccharomyces cerevisiae* and *Zymomonas mobilis* in calcium alginate beads. J. Sci. Ind. Res. India. 69, 472–475.

Behera, S., Mohanty, R.C., Ray, R.C., 2010c. Comparative study of bio-ethanol production from mahula (*Madhuca latifolia* L.) flowers by *Saccharomyces cerevisiae* and *Zymomonas mobilis*. Appl. Energy. 87, 2352–2355.

Behera, S., Mohanty, R.C., Ray, R.C., 2010d. Ethanol fermentation of mahula (*Madhuca latifolia* L.) flowers using free and immobilized bacteria *Zymomonas mobilis* MTCC 92. Biologia. Sect. Cell. Mol. Biol. 65, 416–421.

Behera, S.S., Ray, R.C., 2019. Forest bioresources for bioethanol and biodiesel production with emphasis on mohua (*Madhuca latifolia* L.) flowers and seeds. In: Ray, R.C., Ramachandran, S. (Eds.), Bioethanol Production from Food Crops. Academic Press, pp. 233–247.

Behera, S., Ray, R.C., 2015. Batch ethanol production from cassava (*Manihot esculenta* Crantz.) flour using *Saccharomyces cerevisiae* cells immobilized in calcium alginate. Ann. Microbiol. 65 (2), 779–783.

Behera, S., Mohanty, R.C., Ray, R.C., 2011. Ethanol production from mahula (*Madhuca latifolia* L.) flowers with immobilized cells of *Saccharomyces cerevisiae* in *Luffa cylindrica* L. sponge discs. Appl. Energy 88 (1), 212–215.

Behera, S., Mohanty, R.C., Ray, R.C., 2012. Ethanol fermentation of sugarcane molasses by *Zymomonas mobilis* MTCC 92 immobilized in *Luffa cylindrica* L. sponge discs and Ca-alginate matrices. Braz. J. Microbiol. 43 (4), 1499–1507.

Boshagh, F., Rostami, K., Moazami, N., 2019. Biohydrogen production by immobilized *Enterobacter aerogenes* on functionalized multi-walled carbon nanotube. Int. J. Hydrog. Energy 44 (28), 14395–14405.

Chamkalani, A., Zendehboudi, S., Rezaei, N., Hawboldt, K., 2020. A critical review on life cycle analysis of algae biodiesel: current challenges and future prospects. Renew. Sustain. Energy Rev. 134, 110143.

Chen, Z., Niu, Y., Chen, C., Li, H., 2021. Optimization of bioethanol fermentation productivity in *Saccharomyces cerevisiae* by regulation of social behavior. Chem. Eng. Sci. 246, 116980.

Cubides-Roman, D.C., Pérez, V.H., de Castro, H.F., Orrego, C.E., Giraldo, O.H., Silveira, E. G., et al., 2017. Ethyl esters (biodiesel) production by *Pseudomonas fluorescens* lipase immobilized on chitosan with magnetic properties in a bioreactor assisted by electromagnetic field. Fuel 196, 481–487.

De Poulpiquet, A., Ciaccafava, A., Lojou, E., 2014. New trends in enzyme immobilization at nanostructured interfaces for efficient electrocatalysis in biofuel cells. Electrochim. Acta 126, 104–114.

Dias, G.S., de Lima Luz Jr, L.F., Mitchell, D.A., Krieger, N., 2017. Scale-up of biodiesel synthesis in a closed-loop packed-bed bioreactor system using the fermented solid produced by Burkholderia lata LTEB11. Chem. Eng. J. 316, 341–349.

Fang, G., Chen, H., Zhang, Y., Chen, A., 2016. Immobilization of pectinase onto Fe3O4@ SiO2–NH2 and its activity and stability. Int. J. Biol. Macromol. 88, 189–195.

Fernandez, I.A.P., Liu, D.H., Zhao, J., 2017. LCA studies comparing alkaline and immobilized enzyme catalyst processes for biodiesel production under Brazilian conditions. Resources, Conserv. Recycling 119, 117–127.

Fernandez-Lopez, L., Pedrero, S.G., Lopez-Carrobles, N., Virgen-Ortíz, J.J., Gorines, B.C., Otero, C., et al., 2017. Physical cross-linking of lipase from *Rhizomucor miehei* immobilized on octyl agarose via coating with ionic polymers: avoiding enzyme release from the support. Process. Biochem. 54, 81–88.

Ferreira, I.M., Ganzeli, L.D.S., Rosset, I.G., Yoshioka, S.A., Porto, A.L., 2017. Ethylic biodiesel production using lipase immobilized in silk fibroin-alginate spheres by encapsulation. Catal. Lett. 147 (1), 269–280.

Godoy-Salinas, A.J., Ortiz-Muñiz, B., Rodríguez, J.G., Gutiérrez-Rivera, B., Aguilar-Uscanga, M.G., 2021. Bioethanol production by *S. cerevisiae* ITV-01 RD immobilized on pre-treated sugarcane bagasse. Biomass Convers. Biorefinery 1–8. Available from: https://doi.org/10.1007/s13399-021-01602-x.

Gundogdu, T.K., Akboncuk, M.B., Azbar, N., 2013. Biohydrogen production via a novel immobilized cell bioreactor. Biofuels 4 (6), 595–603.

Ingle, A.P., Rathod, J., Pandit, R., da Silva, S.S., Rai, M., 2017. Comparative evaluation of free and immobilized cellulase for enzymatic hydrolysis of lignocellulosic biomass for sustainable bioethanol production. Cellulose 24 (12), 5529–5540.

Iso, M., Chen, B., Eguchi, M., Kudo, T., Shrestha, S., 2001. Production of biodiesel fuel from triglycerides and alcohol using immobilized lipase. J. Mol. Catal. B: Enzymatic 16 (1), 53–58.

Jigajinni, S.K., Meti, B.S., 2021. Immobilization optimization and characterization of immobilized lipase from Lysinibacillus macroides FS1 for biodiesel production. Int. J. Curr. Microbiol. App. Sci. 10 (04), 232–245.

Kao, P.M., Hsu, B.M., Huang, K.H., Tao, C.W., Chang, C.M., Ji, W.T., 2014. Biohydrogen production by immobilized co-culture of *Clostridium butyricum* and *Rhodopseudomonas palustris*. Energy Proc. 61, 834–837.

Keskin, T., Giusti, L., Azbar, N., 2012. Continuous biohydrogen production in immobilized biofilm system versus suspended cell culture. Int. J. Hydrog. Energy 37 (2), 1418–1424.

Khammee, P., Ramaraj, R., Whangchai, N., Bhuyar, P., Unpaprom, Y., 2021. The immobilization of yeast for fermentation of macroalgae *Rhizoclonium* sp. for efficient conversion into bioethanol. Biomass Convers. Biorefinery 11, 827–835.

Kumar, D., Das, T., Giri, B.S., Verma, B., 2020. Optimization of biodiesel synthesis from nonedible oil using immobilized bio-support catalysts in jacketed packed bed bioreactor by response surface methodology. J. Clean. Prod. 244, 118700.

Lee, W.S., Chen, I.C., Chang, C.H., Yang, S.S., 2012. Bioethanol production from sweet potato by co-immobilization of saccharolytic molds and *Saccharomyces cerevisiae*. Renew. Energy 39 (1), 216–222.

Liao, J.C., Atsumi, S., Cann, A.F., 2017. Washington, DC: U.S. Patent and Trademark Offic., US Patent No. 9,695,426.

Luna, D., Posadillo, A., Caballero, V., Verdugo, C., Bautista, F.M., Romero, A.A., et al., 2012. New biofuel integrating glycerol into its composition through the use of covalent immobilized pig pancreatic lipase. Int. J. Mol. Sci. 13 (8), 10091–10112.

Malhotra, I., Basir, S.F., 2020. Immobilization of invertase in calcium alginate and calcium alginate-kappa-carrageenan beads and its application in bioethanol production. Prep. Biochem. Biotechnol. 50 (5), 494–503.

Maswanna, T., Lindblad, P., Maneeruttanarungroj, C., 2020. Improved biohydrogen production by immobilized cells of the green alga Tetraspora sp. CU2551 incubated under aerobic condition. J. Appl. Phycol. 32 (5), 2937–2945.

Mathew, A.K., Crook, M., Chaney, K., Humphries, A.C., 2013. Comparison of entrapment and biofilm mode of immobilisation for bioethanol production from oilseed rape straw using Saccharomyces cerevisiae cells. Biomass Bioenergy 52, 1–7.

Milessi, T.S., Antunes, F.A., Chandel, A.K., da Silva, S.S., 2015. Hemicellulosic ethanol production by immobilized cells of *Scheffersomyces stipitis*: effect of cell concentration and stirring. Bioengineered 6 (1), 26–32.

Modenez, I.A., Sastre, D.E., C Moraes, F., Marques Netto, C.G., 2018. Influence of glutaraldehyde cross-linking modes on the recyclability of immobilized lipase B from *Candida antarctica* for transesterification of soy bean oil. Molecules 23 (9), 2230.

Nagaraj, B., Liny, P., Sreedhar Reddy, P., Krishnamurthy, N.B., Jhansi Rani, D., Mazumdar, V.B., 2010. Biodiesel production catalyzed by fungus cell immobilization in fibrous support. Biomed. Pharmacol. J. 3 (2), 391–396.

Nezhad, M.K., Aghaei, H., 2021. Tosylated cloisite as a new heterofunctional carrier for covalent immobilization of lipase and its utilization for production of biodiesel from waste frying oil. Renew. Energy 164, 876–888.

Nigam, S., Mehrotra, S., Vani, B., Mehrotra, R., 2014. Lipase immobilization techniques for biodiesel production: an overview. Int. J. Renew. Energy Biofuels 2014, 664708.

Nikolić, S., Mojović, L., Pejin, D., Rakin, M., Vukašinović, M., 2010. Production of bioethanol from corn meal hydrolyzates by free and immobilized cells of *Saccharomyces cerevisiae* var. ellipsoideus. Biomass Bioenergy 34 (10), 1449–1456.

Pikula, K., Zakharenko, A., Stratidakis, A., Razgonova, M., Nosyrev, A., Mezhuev, Y., et al., 2020. The advances and limitations in biodiesel production: feedstocks, oil extraction methods, production, and environmental life cycle assessment. Green. Chem. Lett. Rev. 13 (4), 275–294.

Plangklang, P., Reungsang, A., Pattra, S., 2012. Enhanced bio-hydrogen production from sugarcane juice by immobilized *Clostridium butyricum* on sugarcane bagasse. Int. J. Hydrog. Energy 37 (20), 15525–15532.

Rachmadona, N., Quayson, E., Amoah, J., Alfaro-Sayes, D.A., Hama, S., Aznury, M., et al., 2021. Utilizing palm oil mill effluent (POME) for the immobilization of *Aspergillus oryzae* whole-cell lipase strains for biodiesel synthesis. Biofuels Bioprod. Bioref. 15 (3), 804–814.

Rai, M., Ingle, A.P., Pandit, R., Paralikar, P., Biswas, J.K., da Silva, S.S., 2019. Emerging role of nanobiocatalysts in hydrolysis of lignocellulosic biomass leading to sustainable bioethanol production. Catal. Rev. 61 (1), 1–26.

Raman, J.K., Ting, V.F.W., Pogaku, R., 2011. Life cycle assessment of biodiesel production using alkali, soluble and immobilized enzyme catalyst processes. Biomass Bioenergy 35 (10), 4221–4229.

Razack, S.A., Duraiarasan, S., 2016. Response surface methodology assisted biodiesel production from waste cooking oil using encapsulated mixed enzyme. Waste Manag. 47, 98–104.

Razmovski, R., Vučurović, V., 2012. Bioethanol production from sugar beet molasses and thick juice using *Saccharomyces cerevisiae* immobilized on maize stem ground tissue. Fuel 92 (1), 1–8.

Rehman, H.U., Aman, A., Zohra, R.R., Qader, S.A.U., 2014. Immobilization of pectin degrading enzyme from *Bacillus licheniformis* KIBGE IB-21 using agar-agar as a support. Carbohydr. Polym. 102, 622–626.

Ross, B.S., Pott, R.W., 2021. Hydrogen production by immobilized *Rhodopseudomonas palustris* in packed or fluidized bed photobioreactor systems. Int. J. Hydrog. Energy 46 (2), 1715–1727.

Sankar, M.K., Ravikumar, R., Kumar, M.N., Sivakumar, U., 2018. Development of co-immobilized tri-enzyme biocatalytic system for one-pot pretreatment of four different perennial lignocellulosic biomass and evaluation of their bioethanol production potential. Bioresour. Technol. 269, 227–236.

Santos, E.L.I., Rostro-Alanís, M., Parra-Saldívar, R., Alvarez, A.J., 2018. A novel method for bioethanol production using immobilized yeast cells in calcium-alginate films and hybrid composite pervaporation membrane. Bioresour. Technol. 247, 165–173.

Saville, B.A., 2006. Washington, DC: U.S. Patent and Trademark Office. US Patent No. 7,144,716.

Seelert, T., Ghosh, D., Yargeau, V., 2015. Improving biohydrogen production using *Clostridium beijerinckii* immobilized with magnetite nanoparticles. Appl. Microbiol. Biotechnol. 99 (9), 4107–4116.

Sekoai, P.T., Yoro, K.O., Daramola, M.O., 2018. Effect of nitrogen gas sparging on dark fermentative biohydrogen production using suspended and immobilized cells of anaerobic mixed bacteria from potato waste. Biofuels 9 (5), 595–604.

Shanmugam, S., Krishnaswamy, S., Chandrababu, R., Veerabagu, U., Pugazhendhi, A., Mathimani, T., 2020. Optimal immobilization of *Trichoderma asperellum* laccase on polymer coated Fe_3O_4@ SiO_2 nanoparticles for enhanced biohydrogen production from delignified lignocellulosic biomass. Fuel 273, 117777.

Shim, J., Kim, G.Y., Moon, S.H., 2011. Covalent co-immobilization of glucose oxidase and ferrocenedicarboxylic acid for an enzymatic biofuel cell. J. Electroanal. Chem. 653 (1–2), 14–20.

Singh, L., Wahid, Z.A., Siddiqui, M.F., Ahmad, A., Rahim, M.H.A., Sakinah, M., 2013. Biohydrogen production from palm oil mill effluent using immobilized *Clostridium butyricum* EB6 in polyethylene glycol. Process. Biochem. 48 (2), 294–298.

Singh, M.K., Sahni, S., Narang, A., 2021. Production of liquid biofuels from lignocellulosic biomass. Energy: Crises, Chall. Solut. 208–230. Available from: https://doi.org/10.1002/9781119741503.ch12.

Song, W., Rashid, N., Choi, W., Lee, K., 2011. Biohydrogen production by immobilized Chlorella sp. using cycles of oxygenic photosynthesis and anaerobiosis. Bioresour. Technol. 102 (18), 8676–8681.

Sowatad, A., Todhanakasem, T., 2020. Bioethanol production by repeated batch using immobilized yeast cells on sugarcane bagasse. Waste Biomass Valoriz. 11 (5), 2009–2016.

Srisuk, F., Chinwatpaiboon, P., Atjayutpokin, T., Boonsombuti, A., Savarajara, A., Luengnaruemitchai, A., 2021. Comparison of different Vetiver grass pretreatment techniques and their impact on immobilized butanol production by *Clostridium beijerinckii* TISTR 1461. Cellulose 1–18.

Swain, M.R., Kar, S., Sahoo, A.K., Ray, R.C., 2007. Ethanol fermentation of mahula (*Madhuca latifolia* L.) flowers using free and immobilized yeast *Saccharomyces cerevisiae*. Microbiol. Res. 162 (2), 93–98.

Tafesh, A., 2014. Immobilized enzyme technology enters commercial stage. Biodiesel Mag.

Takei, T., Ikeda, K., Ijima, H., Kawakami, K., 2011. Fabrication of poly (vinyl alcohol) hydrogel beads cross-linked using sodium sulfate for microorganism immobilization. Process. Biochem. 46 (2), 566–571.

Thangaraj, B., Jia, Z., Dai, L., Liu, D., Du, W., 2019. Effect of silica coating on Fe_3O_4 magnetic nanoparticles for lipase immobilization and their application for biodiesel production. Arab. J. Chem. 12 (8), 4694–4706.

Tran, D.T., Chen, C.L., Chang, J.S., 2012. Immobilization of Burkholderia sp. lipase on a ferric silica nanocomposite for biodiesel production. J. Biotechnol. 158 (3), 112–119.

Tran, D.T., Chen, C.L., Chang, J.S., 2016. Continuous biodiesel conversion via enzymatic transesterification catalyzed by immobilized Burkholderia lipase in a packed-bed bioreactor. Appl. Energy 168, 340–350.

Tsai, T.Y., Lo, Y.C., Dong, C.D., Nagarajan, D., Chang, J.S., Lee, D.J., 2020. Biobutanol production from lignocellulosic biomass using immobilized *Clostridium acetobutylicum*. Appl. Energy 277, 115531.

Van Hecke, W., Joossen-Meyvis, E., Beckers, H., De Wever, H., 2018. Prospects and potential of biobutanol production integrated with organophilic pervaporation—a techno-economic assessment. Appl. Energy 228, 437–449.

Vučurović, V.M., Razmovski, R.N., 2012. Sugar beet pulp as support for *Saccharomyces cerivisiae* immobilization in bioethanol production. Ind. Crop. Products 39, 128–134.

Wang, P.T., Zhang, J.J., Li, Z.Y., Bai, J., 2017. Life cycle energy analysis and biodiesel production potential evaluation for immobilized microalgae cultivated with secondary effluent in China. In New Energy and Sustainable Development: Proceedings of 2016 International Conference on New Energy and Sustainable Development (NESD 2016) (pp. 568–576).

Wannapokin, A., Cheng, Y.T., Wu, S.Z., Hsieh, P.H., Hung, C.H., 2021. Potential of biohydrogen production by *C. pasteurianum* co-immobilized with selected nano-metal particles. Int. J. Hydrog. Energy 46 (20), 11337–11344.

Wirawan, F., Cheng, C.L., Kao, W.C., Lee, D.J., Chang, J.S., 2012. Cellulosic ethanol production performance with SSF and SHF processes using immobilized *Zymomonas mobilis*. Appl. Energy 100, 19–26.

Xie, W., Huang, M., 2018. Immobilization of *Candida rugosa* lipase onto graphene oxide Fe_3O_4 nanocomposite: characterization and application for biodiesel production. Energy Convers. Manag. 159, 42–53.

Xie, W., Han, Y., Tai, S., 2017. Biodiesel production using biguanide-functionalized hydroxyapatite-encapsulated-γ-Fe_2O_3 nanoparticles. Fuel 210, 83–90.

Ye, Z., Song, J., Zhu, E., Song, X., Chen, X., Hong, X., 2018. Alginate adsorbent immobilization technique promotes biobutanol production by *Clostridium acetobutylicum* under extreme condition of high concentration of organic solvent. Front. Microbiol. 9, 1071. Available from: https://doi.org/10.3389/fmicb.2018.01071.

Zhang, B., Weng, Y., Xu, H., Mao, Z., 2012. Enzyme immobilization for biodiesel production. Appl. Microbiol. Biotechnol. 93 (1), 61–70.

Zhang, C., Kang, X., Liang, N., Abdullah, A., 2017. Improvement of biohydrogen production from dark fermentation by cocultures and activated carbon immobilization. Energy Fuels 31 (11), 12217–12222.

Zhao, L., Cao, G.L., Wang, A.J., Guo, W.Q., Liu, B.F., Ren, H.Y., et al., 2012. Enhanced bio-hydrogen production by immobilized *Clostridium* sp. T2 on a new biological carrier. Int. J. Hydrog. Energy 37 (1), 162–166.

Zhou, W., Liu, J., Fan, S., Xiao, Z., Qiu, B., Wang, Y., et al., 2018. Biofilm immobilization of *Clostridium acetobutylicum* on particulate carriers for acetone-butanol-ethanol (ABE) production. Bioresour. Technol. Rep. 3, 211–217.

Zhu, P., Abdelaziz, O.Y., Hulteberg, C.P., Riisager, A., 2020. New synthetic approaches to biofuels from lignocellulosic biomass. Curr. Opin. Green. Sustain. Chem. 21, 16–21.

Section IV

Advances in modeling and development

Artificial intelligence as a tool for yield prediction in biofuel production systems

12

Preshanthan Moodley and Cristina Trois
SARCHI Chair Waste and Climate Change, School of Engineering, University of KwaZulu-Natal, Durban, South Africa

12.1 Introduction

The fundamental conversion processes that makeup bioenergy systems are complex, high-dimensional, and multiscale; they occur across time, and geographical variation and are the consequence of the interplay of several elements. Anaerobic digestion (AD), for instance, has a higher success rate due to the importance of the raw materials, operational parameters, and inoculum (Feng et al., 2021). To construct theory-driven (also known as model-driven or hypothesis-driven) models, repetitive trial-and-error is used in conventional research. Several experience-based hypotheses or conjectures are produced by empirical judgment. These models enable job rationale and viability to be anticipated beforehand, significantly lowering the burden. To model the AD process, the Anaerobic Digestion Model No. 1 (ADM1) was created (Batstone et al., 2002; Weinrich and Nelles, 2021). However, due to the intricacy of the AD system and a dearth of thorough substrate characterization, ADM1 has few commercial applications while performing well in laboratory-scale AD (Derbal et al., 2009; Weinrich et al., 2021). Consequently, creating an effective theory-driven model for a bioenergy system is difficult.

On the other hand, analyzing large data sets from bioenergy systems makes it challenging to build complete knowledge based on physical principles regarding data gathered from complicated occurrences. Only advanced data analysis techniques can fully realize the value of such data. Artificial intelligence (AI) has been described as the fourth industrial revolution of the 21st century (Diez-Olivan et al., 2019). Machine learning (ML) has seen tremendous development as a crucial data processing tool in the last 10 years. This has substantially expanded AI's problem-solving capabilities. ML is a group of sophisticated data analysis techniques that find patterns between characteristics in a data set and make predictions when faced with fresh inputs. These techniques are founded on the concepts of probability and statistics, and mathematics. Business, medicine (Stokes et al., 2020), biology (Kim et al., 2018), physical sciences (Mehta et al., 2019), chemistry (Dral, 2020), and geoscience (Reichstein et al., 2019) are just a few of the fields where ML has been effectively used. Multimodal learning from several perspectives may be used to

Advances in Lignocellulosic Biofuel Production Systems. DOI: https://doi.org/10.1016/B978-0-323-91192-4.00012-2

grasp the intricate workings of biological systems (Li et al., 2018). Technology advancements in optics and imaging have made it possible for ML to analyze and track the trail of cells and even accomplish automated control of atomic force microscopes at the nanometer scale (Moen et al., 2019). ML may be employed as a support tool for human jobs to provide more dependable or efficient results. Additionally, ML may be proficient for tasks requiring the attribution of causation or establishing relationships between variables, which is beneficial for tackling problems that conventional modeling or experimentation cannot handle and generating new discoveries.

Notably, the area of biofuels has shown more interest in ML. To optimize a photobioreactor and microalgal production, Del Rio-Chanona et al. (2019) suggested a deep learning (DL) model based on a convolutional neural network (CNN), which significantly cut down on time required compared to conventional simulation calculations. To anticipate the best P50S50 biodiesel yield, Mujtaba et al. (2020) utilized an extreme learning machine (ELM) and a cuckoo search algorithm. The difference between the projected and experimental values was less than 0.26%. In order to forecast the cellulose enrichment factor and solid recovery of lignocellulosic biomass pretreated with ionic liquid solvents, Phromphithak et al. (2021) employed a random forest (RF), which comprised 23 factors. The regression coefficients (R^2) were, respectively, 0.94 and 0.84. More specifically, when it targets a microbial metabolic production process, it concentrates on how to maximize the yield of the desired products. ML models can offer guidance on various scales, whether gene-annotated strain planning to change metabolic pathways (Hannigan et al., 2019; Kumar et al., 2019) or fermentation operation parameter optimization to improve the production of metabolites (Zhang et al., 2020). As a result, the confluence of ML and bioenergy has grown significantly over the last 5 years, including microalgae monitoring and culture, energy material extraction and optimization, microbial community interaction analysis, and industrial production forecasting.

This chapter will present the status of the employment of AI in advancing biofuel production systems.

12.2 Machine learning in biofuel production systems

Many experiments are long-term studies that may need very complex measuring instruments. When theory-driven models run into significant difficulties due to the intricate nature of bioenergy conversion processes, data-driven models have been explored to avoid costly or time-consuming experiments and recapitulate the inner system to accomplish the goal of forecasts or optimization. Additionally, operations like image-activated cell sorting that theory-driven models are unable to execute may be completed by ML models (Isozaki et al., 2019; Nitta et al., 2018). Furthermore, most studies employed models, as indicated in Table 12.1. These models present a starting point for the development of AI-based systems.

Table 12.1 Studies employing model development.

Substrate	Study objective	Model employed	Reference
Sugarcane leaves	Optimize pretreatment	Box-Behnken design	Moodley and Kana (2017)
Corn cobs	Optimize pretreatment	Box Behnken design	Sewsynker-Sukai and Kana (2018)
Wheat straw	Optimize pretreatment	Central composite design	Vollmer et al. (2022)
Arachis hypogea shell	Enhance methane yields	ANFIS	Olatunji et al. (2022)
Elephant grass	Optimize pretreatment	Fuzzy logic	Godinho and Caneppele (2022)

12.2.1 Biological processes

In a moderate climate, biofuels may be created via biological conversion techniques. Although the fundamental assumptions supporting the AD projections based on operational parameters appear to be sound, other ingrained aspects will have a significant impact on the model's performance. The development of the microbial population, which brings about considerable alterations in microbial metabolism, is elusive in the long-term AD process. The final findings include some variety even with the same procedure, tools, and environment, including the microbial structure, which may alter periodically (Gruszka Vendruscolo et al., 2020).

When data were derived from biological processes, models' capacity to generalize was weak (Sydney et al., 2020; Wang et al., 2020), which may have been due to the intricacies of biological systems as well as the constraints and variance in the data inputs. The data set used by Wang et al. (2020) to forecast methane yield values from operational factors was compiled from peer-reviewed literature. These results were produced using the same AD reactor design. The small quantity of data points, however, could make these ML models less accurate. In an up-flow anaerobic sludge blanket reactor, methane yields were calculated by Antwi et al. (2017). An artificial neural network (ANN) model performed better ($R^2 = 0.9793$) than a nonlinear regression model ($R^2 = 0.9262$) in terms of performance. Since the experiment was ongoing, the results remained stable at various points. The results demonstrated the viability of modeling wastewater treatment facilities using ML. Hosseinzadeh et al. (2020) integrated ANN and fuzzy logic (FL) to estimate electrochemical activity in a microbial electrolysis cell. The results showed that the adaptive network built on the fuzzy inference system outperformed the ANN, which was in line with Rego et al. (2018) findings.

The possible applications of neural fuzzy models for breakthrough curve prediction were also mentioned by Calero et al. (2018). The functional resistance and durability of microbial fuel cells were assessed by Lesnik et al. (2020). The data limited the model's effectiveness, yet it gave insight into the genomic-based biological conversion process. On the other hand, Gruszka Vendruscolo et al. (2020) used principal component analysis (PCA) and RF algorithms to examine the significance of every functional group in a microbial community to the electrochemical activity of a microbial electrolysis cell. In addition to the issue of inadequate data brought on by experimental problems, there are issues in gathering data in a legitimate area. The performance of ML models might be improved by focusing more on flexibility, complementarity, and limitations between them. Not to mention, a more exacting experimental design would promote the production of high-level-quality data and coverage of the appropriate area.

12.2.2 Thermochemical processes

Despite requiring a large amount of energy, thermochemical conversion proceeds significantly more quickly than biological conversion. Furthermore, the numerical output is comparatively consistent and only dependent on the reaction parameters in the event of identical configuration settings and no intervention in the thermochemical reaction process. The result is that the trained ML models may make predictions with high accuracy and help create new plans.

When Mujtaba et al. (2020) integrated an ELM model with a cuckoo search method, they created a P50S50 with the increased cold flow and decreased friction (reduced by 12.37% compared to commercial diesel). In another study, Li et al. (2018) paired an ANN model with a genetic algorithm (GA) and were successful in catalyzing the conversion of discarded peanut shells into methyl levulinate. Both studies discovered that such hybrid models performed better than a response surface method (RSM).

Similar findings were concluded by Patil et al. (2017), who observed that an ANN outperformed RSM and support vector machine (SVM) in the yield of arginine deiminase by ultrasonic disruption. Based on published data, Zhu et al. (2019) estimated the yield and carbon levels of biochar and found that structural information resulted in a better R^2 (by 5%) than elemental information. However, Xing et al. (2019) found that elemental analysis and chemical contents exhibit a good association ($R^2 > 0.93$). According to these studies, pyrolysis conditions and biomass properties were crucial elements. Under various operating circumstances, engines adapt to biofuels to varying degrees, and the percentage of biofuel blends may be adjusted to optimize engine performance. The drawbacks of conventional trial-and-error-based calibration may be solved using ML, which can routinely determine the best solution. The optimized engine achieves lower exhaust emission characteristics and a greater biofuel consumption efficiency (Silitonga et al., 2018). In another study, an air-fuel ratio controller developed on an ELM model surpassed the engine's built-in controller (Wong and Wong, 2018). Furthermore, a kernel-based ELM model could enhance forecast precision even further (Wong and Wong, 2017; Zhao et al., 2021). The efficacy of ML in biological conversion is thus inferior to thermochemical conversion since biological conversion procedures take a long time and microorganisms' or enzymes' actions are very nonlinear and complicated. While ML overshadowed the forecasts for biological conversion, it produced perfect results in thermochemical optimization procedures.

12.3 Artificial intelligence employment in lignocellulosic biomass pretreatment

There are many pretreatment techniques that attempt to weaken the lignocellulosic structure's resistance to enzymatic saccharification and microbial fermentation. These techniques include using acid, alkali, inorganic salt, organosolv, ionic liquids, and microwave irradiation (Zabed et al., 2017). Due to its high catalytic activity, lower cost, and minimal toxicity, inorganic salt pretreatment is now proving to be a successful pretreatment (Moodley and Kana, 2017). Additionally, several studies have looked at the catalytic action of salt when combined with an acid, an alkali, or a surfactant to improve enzymatic hydrolysis (Raghavi et al., 2016; Kim et al., 2014; Moodley and Kana, 2017). Significant interest is also growing in microwave irradiation as a substitute heating source for lignocellulosic pretreatment (Aguilar-Reynosa et al., 2017). The lignocellulosic material is molecularly disrupted by microwave processing, resulting in fiber fragmentation (Diaz et al., 2015). Furthermore, microwave heating provides many benefits over traditional steam heating, including quicker heating, quicker response times, and energy economy (Aguilar-Reynosa et al., 2017).

Most of the time, modeling and optimization procedures are used to identify the ideal pretreatment conditions that will release the most sugar (Sindhu et al., 2016). However, since numerous experimental runs exist, modeling pretreatment is expensive and time-consuming. Prior to the optimization phase, thorough preliminary screenings are often conducted to identify an adequate navigation space. By eliminating many time-consuming activities required before the bioconversion of lignocellulose to biofuels, creating an intelligent predictive tool might considerably improve the workflow. ANN can recognize data patterns. In order to create highly effective models of complicated nonlinear processes, ANNs may be trained using experimental data (Desai et al., 2008). Many industries have used ANN models to precisely characterize connections between the input and output variables (da Silva Bispo et al., 2017; Rorke et al., 2017). This makes it possible to utilize ANN as a tool for virtual experimentation to estimate process parameters in real time that are difficult to monitor (Gonzaga et al., 2009).

Moodley et al. (2019) developed an ANN prediction model by pooling data from several pretreatment studies. The model was trained using the backpropagation approach with the goal of obtaining the lowest net error on the validation data set while avoiding overtraining. Based on validation data, coefficients of determination (R^2) were derived for each model and used to assess the model's correctness. Using the remaining 25% of the data sets that were not included in the training process, the constructed ANN models were evaluated on their ability to predict the creation of reducing sugar. Models based on steam and microwave technology produced coefficients of determination (R^2) of 0.97. The prediction trend line was followed by the vast majority of the data points, demonstrating improved accuracy in estimating the reduced sugar output under novel process conditions. By acting as an early preliminary screening of process parameters, as indicated in Fig. 12.1, the relatively high prediction accuracy of the generated models may improve pretreatment

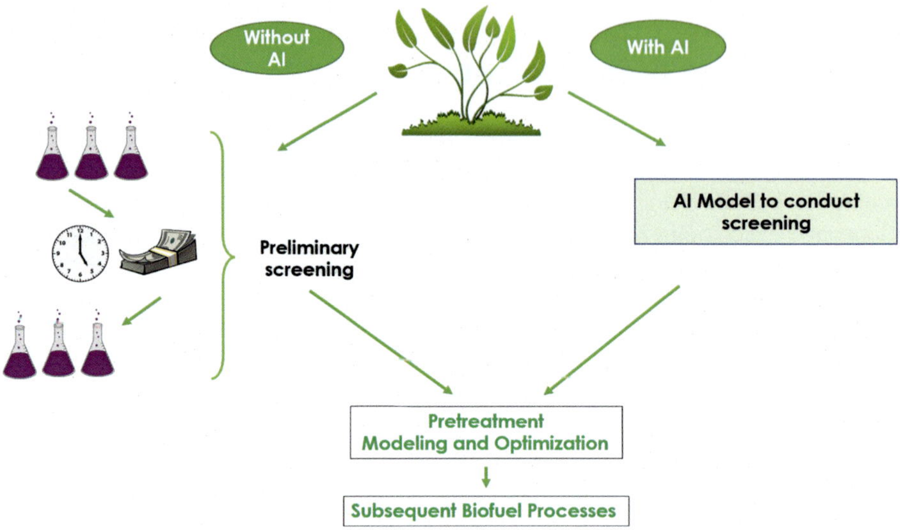

Figure 12.1 Schematic diagram illustrating the workflow of lignocellulosic pretreatment with and without artificial intelligence.

design studies by lowering the number of tests, saving reagents, time, and process development expenses.

12.4 Artificial intelligence employment in pretreatment inhibitor profile analysis

As discussed earlier, dilute, or concentrated acids and alkali solutions combined with heat energy are common pretreatment techniques. Unwanted by-products are often produced due to the extreme pretreatment conditions needed to partly disrupt the lignocellulosic structure (Cavka and Jönsson, 2013). The released by-products may adversely affect the enzymatic hydrolysis and fermentation processes (Zha et al., 2012). Depending on the makeup of the lignocellulosic material utilized and the severity of the pretreatment process conditions, different amounts of aromatic compounds, furan-derivatives, and aliphatic acids—fermentation inhibitors—can be present (Cavka and Jönsson, 2013). A more extended lag phase, slower microbial development, lower cell density, and decreased productivity are some of the inhibitory effects seen during fermentation (Zha et al., 2012). To improve microbial fermentation, several research studies have focused on the elimination or decrease of common inhibitors (Jönsson et al., 2013). Pretreatment difficulties often surface during the fermentation or postprocessing phases of bioprocesses. However, the detection and quantification of lignocellulosic pretreatment products is usually restricted to sugars and generally reported fermentation inhibitors owing to the diversity and number of by-products created. It can cause the adverse effects of unknown substances to go unnoticed.

Another significant barrier to the comprehensive evaluation of different intermediates has been the significant expense of the equipment necessary for these studies (Zhang et al., 2005), as well as the time-consuming sample planning procedures and lengthy run durations associated with some of these approaches (Humpula et al., 2011). Due to a lack of equipment, the expenses involved with the direct quantification of vast volumes of data may become overwhelming at both the research and development and production phases (Zhang et al., 2005). To estimate process parameters that are difficult and expensive to monitor, ANN may be employed as a virtual sensor (soft-sensor) (Gonzaga et al., 2009).

Soft sensors have made it possible to monitor real-time bioprocesses in an indirect manner. The fast working times of soft sensors also make it possible to get real-time readings. For example, Herrera and Filho (2013) reported a prediction accuracy of 0.998 when predicting the rate of product creation in the manufacture of bioethanol utilizing secondary data of pH, turbidity, CO_2 flow rate, and temperature. They used a hybrid model made up of ANN and mass balance.

Rorke et al. (2017) was a landmark study that investigated the development of an ANN soft sensor to analyze the inhibitor profile from the microwave-assisted alkaline and acid-pretreated hydrolysate. A four-factor Box—Behnken design was employed for the pretreatment, generating 58 experimental runs per pretreatment type. These were conducted in replicate. The pretreated hydrolysate was then analyzed via gas chromatography-mass spectrometry (GS-MS) for a profile of inhibitors. Using an ensemble neural network, an intelligent model was created to estimate the kind and proportion of volatile chemicals after microwave-assisted sorghum leaves (SL) pretreatment. Three multilayer perceptron ANNs made up the committee. Each committee member's topology had a single input layer with five neurons, two hidden layers with thirteen neurons each, and a single output layer with 21 neurons (5—13—13—21). The inputs included acid concentration, alkali concentration, microwave duration, microwave intensity, and solid-to-liquid (S:L) ratio. The outputs included reducing sugar and glucose (in g/L) and furfural, 1-hydroxy-2-propanone, 5-methyl furfural, acetic acid, formic acid, 2-phenyl acetaldehyde, citrannic anhydride, guaiacol, phenol, 4-ethyl guaiacol and 4-hydroxy benzaldehyde. The experimental data set was split into two parts: 25% for validation and 75% for training. For the hidden layer, a logistic transfer function was used. The network was trained using a backpropagation approach to minimize net error on the validation data set while avoiding overtraining or memorization. After 2500 training iterations, a net error value of 0.026 was obtained on the validation data set. Regression analysis on predicted and observed process outputs were used to evaluate the accuracy of the intelligent model, and coefficients of determination (R^2) were determined for each model output to show the model's accuracy in predicting the proportions of volatile chemicals created.

A committee of ANN models was created to forecast the relative percentages of 21 volatile chemicals from microwave-assisted acid and alkali-pretreated SLs.

Instead of employing a single multilayer perceptron, the technique used an ensemble of neural networks to improve network prediction performance via ensemble averaging (EA). Piotrowski et al. (2016) describe the beneficial effect of EA on ANN model prediction accuracy. The developed committee model's precision was tested by using it to forecast the relative fraction of volatile compounds from 15 experimental runs of microwave-assisted acid and alkali pretreatments that had not previously been exposed to the committee. It was done after training with 75% of the experimental data. The 21 compounds produced a range of output values with coefficients as high as 0.93. In addition, the profile of volatile chemicals and the pretreatment conditions were found to be significantly nonlinear. The conclusions from Rorke et al. (2017) indicate that the design of lignocellulosic pretreatment regimens for the production of biomaterials and biofuels will significantly benefit from ANN-based soft sensors to minimize the production of fermentation inhibitors, thus enhancing biofuel processes.

12.5 Impact of artificial intelligence on lignocellulosic biofuel production systems

The two case studies describe above are indications that research is underway to integrate AI into biofuel processes. Since pretreatment is an integral part of lignocellulosic biofuel production, implementing an AI system into the process can significantly enhance the workflow in the initial upstream stages. Additionally, the AI system capable of acting as a soft sensor in detecting fermentation inhibition byproducts is extremely useful in biofuel. This system could further streamline the key decisions in the upstream stage. AI systems such as these have the potential to enhance productivity and process yields, thus enabling smoother downstream processing. In light of the many bottlenecks and limitations in biofuel production processes, AI systems present the potential to ease these hindrances, thereby enhancing the outlook of these processes.

12.6 Conclusions and future perspectives

Although traditional ML techniques have been used to address issues with bioenergy conversion, there is still space to broaden the models' applicability and boost their effectiveness. Many studies have demonstrated the ability of DL to address challenging scientific issues. DL has shown outstanding accomplishments while dealing with challenging issues in different disciplines. The foundation of DL is the expansion of the size of conventional three-layer neural networks by adding additional layers and more intricate interactions between layer nodes. Convolutional and recurrent layers and extra pooling processes may be included as additional kinds of layers (beyond fully connected layers) to reduce the network size. Although these layers may be understood in terms of the conventional node-activation framework, they lead to more flexibility in the operations that can be carried out (regression, detection, and classification), as well as the kinds of data that can be handled.

For instance, a convolutional layer (specified by kernels or windows) may extract features to identify a more significant or condensed interpretation of inputs. Recurrent layer networks are effective for sequential data applications, such as time series analysis or natural language processing. Finding a methodical approach for their design is one of the biggest obstacles since DL networks have so many different architectural options. Currently, most of it is accomplished via trial and error and previous knowledge. Transfer learning may significantly improve the performance of DL networks by allowing networks to be pretrained on comparable data before being applied to a particular job. DL networks, as opposed to traditional ML techniques, are more appropriate for various complicated biofuel conversion processes because of their diversity. Several DL techniques have been used to describe morphological changes or object traces at the cellular level (Isozaki et al., 2019; Moen et al., 2019). The method feeds imaging data to a DL model when employing biotechnology to accurately identify microalgae cells or create biofuel. Examples include image segmentation for dead or living microalgae cell analysis and the use of augmented microscopy for more detailed image data. Another crucial subject is modeling time series in the bioenergy conversion process. While DL can complete increasingly challenging tasks at different geographical and temporal dimensions, it generally needs a large quantity of data that is either unavailable or costly and time-consuming to acquire. Researchers in the community need to work together more to collect and complete massive datasets so that they may build a database for evaluating new algorithms.

Additionally, consideration should be given to adding additional variables to datasets to perform more jobs. Alternative variables that are readily accessible and implicitly relevant to the outputs may be explored when the variables directly connected to the situation at hand are difficult to assess. There are several methods available right now because of the rapid growth of ML and the boom in interest and applications it has led to. It is possible to use ensemble learning to take advantage of learner diversity. The diversity of viewpoints causes the formation of multiview learning or multimodal learning. There are still some significant obstacles at the ML and biodiesel junction. The development of biofuel technology may be accelerated by overcoming these obstacles.

Abbreviations

AD anaerobic digestion
ADM1 Anaerobic Digestion Model No. 1
AI artificial intelligence
ANN artificial neural network
CNN convolutional neural network
DL deep learning

EA	ensemble averaging
ELM	extreme machine learning
FL	fuzzy logic
GA	genetic algorithm
GS-MS	gas chromatography-mass spectrometry
ML	machine learning
PCA	principal component analysis
RF	random forest
SL	sorghum leaves
S:L	solid-to-liquid ratio
SVM	support vector machine

References

Aguilar-Reynosa, A., Romani, A., Rodriguez-Jasso, R.M., Aguilar, C.N., Garrote, G., Ruiz, H.A., 2017. Microwave heating processing as an alternative of pretreatment in second generation biorefinery: an overview. Energy Convers. Manage. 136, 50–65.

Antwi, P., Li, J., Boadi, P.O., Meng, J., Shi, E., et al., 2017. Estimation of biogas and methane yields in an UASB treating potato starch processing wastewater with backpropagation artificial neural network. Bioresour. Technol. 228, 106–115.

Batstone, D.J., Keller, J., Angelidaki, I., Kalyuzhnyi, S.V., Pavlostathis, S.G., et al., 2002. The IWA Anaerobic Digestion Model No 1 (ADM1). Water Sci. Technol. 45 (10), 65–73.

Calero, M., Ĭáñez-Rodríguez, I., P̕erez, A., Martín-Lara, M.A., Bĭazquez, G., 2018. Neural fuzzy modelization of copper removal from water by biosorption in fixed-bed columns using olive stone and pinion shell. Bioresour. Technol. 252, 100–109.

Cavka, A., Jönsson, L.J., 2013. Detoxification of lignocellulosic hydrolysates using sodium borohydride. Bioresour. Technol. 136, 368–376.

da Silva Bispo, V.D., Scheid, C.M., Calçada, L.A., da Cruz Meleiro, L.A., 2017. Development of an ANN-based soft-sensor to estimate the apparent viscosity of water-based drilling fluids. J. Petrol. Sci. Eng. 150, 69–73.

Del Rio-Chanona, E.A., Wagner, J.L., Ali, H., Fiorelli, F., Zhang, D., et al., 2019. Deep learning-based surrogate modeling and optimization for microalgal biofuel production and photobioreactor design. AICHE J. 65 (3), 915–923.

Derbal, K., Bencheikh-lehocine, M., Cecchi, F., Meniai, A.-H., Pavan, P., 2009. Application of the IWA ADM1 model to simulate anaerobic co-digestion of organic waste with waste activated sludge in mesophilic condition. Bioresour. Technol. 100 (4), 1539–1543.

Desai, K.M., Survase, S.A., Saudagar, P.S., Lele, S.S., Singhal, R.S., 2008. Comparison of artificial neural network (ANN) and response surface methodology (RSM) in fermentation media optimization: case study of fermentative production of scleroglucan. Biochem. Eng. J. 41 (3), 266–273.

Diaz, A.B., de Souza Moretti, M.M., Bezerra-Bussoli, C., Nunes, C.C.C., Blandino, A., da Silva, R., et al., 2015. Evaluation of microwave-assisted pretreatment of lignocellulosic biomass immersed in alkaline glycerol for fermentable sugars production. Bioresour. Technol. 185, 315–323.

Diez-Olivan, A., Del Ser, J., Galar, D., Sierra, B., 2019. Data fusion and machine learning for industrial prognosis: trends and perspectives towards Industry 4.0. Inf. Fusion. 50, 92–111.

Dral, P.O., 2020. Quantum chemistry in the age of machine learning. J. Phys. Chem. Lett. 11 (6), 2336–2347.

Feng, D., Guo, X., Lin, R., Xia, A., Huang, Y., et al., 2021. How can ethanol enhance direct interspecies electron transfer in anaerobic digestion? Biotechnol. Adv. 52. Available from: https://doi.org/10.1016/j.biotechadv.2021.107812.

Godinho, E.Z., Caneppele, F.D.L., 2022. Fuzzy system in a ferrous sulfate pretreatment of elephant grass. Pesqui. Agropecuária Trop. 52.

Gonzaga, J.C.B., Meleiro, L.A.C., Kiang, C., Filho, M., 2009. ANN-based soft-sensor for real-time process monitoring and control of an industrial polymerization process. Comput. Chem. Eng. 33 (1), 43–49.

Gruszka Vendruscolo, E.C., Mesa, D., Rissi, D.V., Meyer, B.H., Pedrosa, F.d.O., et al., 2020. Microbial communities network analysis of anaerobic reactors fed with bovine and swine slurry. Sci. Total Environ. 742, 140314.

Hannigan, G.D., Prihoda, D., Palicka, A., Soukup, J., Klempir, O., et al., 2019. A deep learning genome-mining strategy for biosynthetic gene cluster prediction. Nucleic Acids Res. 47 (18), e110.

Herrera, W.E., Filho, R.M., 2013. Development of a monitoring hybrid system for bioethanol production. Chem. Eng. Trans. 32, 943–948.

Hosseinzadeh, A., Zhou, J.L., Altaee, A., Baziar, M., Li, D., 2020. Effective modelling of hydrogen and energy recovery in microbial electrolysis cell by artificial neural network and adaptive network-based fuzzy inference system. Bioresour. Technol. 316, 123967.

Humpula, J.F., Chundawat, S.P.S., Vismeh, R., Jones, A.D., Balan, V., Dale, B.E., 2011. Rapid quantification of major reaction products formed during thermochemical pretreatment of lignocellulosic biomass using GC–MS. J. Chromatogr. B 879, 1018–1022.

Isozaki, A., Mikami, H., Hiramatsu, K., Sakuma, S., Kasai, Y., et al., 2019. A practical guide to intelligent image-activated cell sorting. Nat. Protoc. 14 (8), 2370–2415.

Jönsson, L.J., Alriksson, B., Nilvebrant, N., 2013. Bioconversion of lignocellulose: inhibitors and detoxification. Biotechnol. Biofuels 6, 16.

Kim, T.H., Oh, K.K., Ryu, H.J., Lee, K.-H., Kim, T.H., 2014. Hydrolysis of hemicellulose from barley straw and enhanced enzymatic saccharification of cellulose using acidified zinc chloride. Renew. Energy 65, 56–63.

Kim, H.K., Min, S., Song, M., Jung, S., Choi, J.W., et al., 2018. Deep learning improves prediction of CRISPR-Cpf1 guide RNA activity. Nat. Biotechnol. 36 (3), 239–241.

Kumar, S., Dangi, A.K., Shukla, P., Baishya, D., Khare, S.K., 2019. Thermozymes: adaptive strategies and tools for their biotechnological applications. Bioresour. Technol. 278, 372–382.

Lesnik, K.L., Cai, W., Liu, H., 2020. Microbial community predicts functional stability of microbial fuel cells. Environ. Sci. Technol. 54 (1), 427–436.

Li, Y., Wu, F.-X., Ngom, A., 2018. A review on machine learning principles for multi-view biological data integration. Brief. Bioinform. 19 (2), 325–340.

Mehta, P., Bukov, M., Wang, C.-H., Day, A.G.R., Richardson, C., et al., 2019. A high-bias, low-variance introduction to machine learning for physicists. Phys. Rep.-Rev. Sec. Phys. Lett. 810, 1–124.

Moen, E., Bannon, D., Kudo, T., Graf, W., Covert, M., et al., 2019. Deep learning for cellular image analysis. Nat. Meth. 16 (12), 1233–1246.

Moodley, P., Kana, E.G., 2017. Microwave-assisted inorganic salt pretreatment of sugarcane leaf waste: effect on physiochemical structure and enzymatic saccharification. Bioresour. Technol. 235, 35–42.

Moodley, P., Rorke, D.C., Kana, E.B.G., 2019. Development of artificial neural network tools for predicting sugar yields from inorganic salt-based pretreatment of lignocellulosic biomass. Bioresour. Technol. 273, 682–686.

Mujtaba, M.A., Masjuki, H.H., Kalam, M.A., Ong, H.C., Gul, M., Farooq, M., et al., 2020. Ultrasound-assisted process optimization and tribological characteristics of biodiesel from palm-sesame oil via response surface methodology and extreme learning machine - cuckoo search. Renew. Energy 158, 202–214.

Nitta, N., Sugimura, T., Isozaki, A., Mikami, H., Hiraki, K., et al., 2018. Intelligent image activated cell sorting. Cell 175 (1), 266–276.

Olatunji, K.O., Ahmed, N.A., Madyira, D.M., Adebayo, A.O., Ogunkunle, O., Adeleke, O., 2022. Performance evaluation of ANFIS and RSM modeling in predicting biogas and methane yields from *Arachis hypogea* shells pretreated with size reduction. Renew. Energy 189, 288–303.

Patil, M.D., Dev, M.J., Tangadpalliwar, S., Patel, G., Garg, P., et al., 2017. Ultrasonic disruption of *Pseudomonas putida* for the release of arginine deiminase: kinetics and predictive models. Bioresour. Technol. 233, 74–83.

Phromphithak, S., Onsree, T., Tippayawong, N., 2021. Machine learning prediction of cellulose-rich materials from biomass pretreatment with ionic liquid solvents. Bioresour. Technol. 323, 124642.

Piotrowski, A.P., Napiorkowski, J.J., Osuch, M., Napiorkowski, M.J., 2016. On the importance of training methods and ensemble aggregation for runoff prediction by means of artificial neural networks. Hydrol. Sci. J. 61 (10), 1903–1925.

Raghavi, S., Sindhu, R., Binod, P., Gnansounou, E., Pandey, A., 2016. Development of a novel sequential pretreatment strategy for the production of bioethanol from sugarcane trash. Bioresour. Technol. 199, 202–210.

Rego, A.S.C., Valim, I.C., Vieira, A.A.S., Vilani, C., Santos, B.F., 2018. Optimization of sugarcane bagasse pretreatment using alkaline hydrogen peroxide through ANN and ANFIS modelling. Bioresour. Technol. 267, 634–641.

Reichstein, M., Camps-Valls, G., Stevens, B., Jung, M., Denzler, J., et al., 2019. Deep learning and process understanding for data-driven Earth system science. Nature 566 (7743), 195–204.

Rorke, D.C.S., Suinyuy, T.N., Gueguim Kana, E.B., 2017. Microwave-assisted chemical pretreatment of waste sorghum leaves: process optimisation and development of an intelligent model for determination of volatile compound fractions. Bioresour. Technol. 224, 590–600.

Sewsynker-Sukai, Y., Kana, E.G., 2018. Microwave-assisted alkalic salt pretreatment of corn cob wastes: process optimization for improved sugar recovery. Ind. Crop. Products 125, 284–292.

Silitonga, A.S., Masjuki, H.H., Ong, H.C., Sebayang, A.H., Dharma, S., et al., 2018. Evaluation of the engine performance and exhaust emissions of biodiesel-bioethanol diesel blends using kernel-based extreme learning machine. Energy 159, 1075–1087.

Sindhu, R., Binod, P., Pandey, A., 2016. A novel sono-assisted acid pretreatment of chilli post harvest residue for bioethanol production. Bioresour. Technol. 213, 58–63.

Stokes, J.M., Yang, K., Swanson, K., Jin, W., Cubillos-Ruiz, A., et al., 2020. A deep learning approach to antibiotic discovery. Cell 181 (2), 475–483.

Sydney, E.B., Duarte, E.R., Martinez Burgos, W.J., de Carvalho, J.C., Larroche, C., et al., 2020. Development of short chain fatty acid-based artificial neuron network tools applied to biohydrogen production. Int. J. Hydrog. Energy 45 (8), 5175–5181.

Vollmer, N.I., Driessen, J.L., Yamakawa, C.K., Gernaey, K.V., Mussatto, S.I., Sin, G., 2022. Model development for the optimization of operational conditions of the pretreatment of wheat straw. Chem. Eng. J. 430, 133106.

Wang, L., Long, F., Liao, W., Liu, H., 2020. Prediction of anaerobic digestion performance and identification of critical operational parameters using machine learning algorithms. Bioresour. Technol. 298, 122495.

Weinrich, S., Mauky, E., Schmidt, T., Krebs, C., Liebetrau, J., et al., 2021. Systematic simplification of the Anaerobic Digestion Model No. 1 (ADM1) – laboratory experiments and model application. Bioresour. Technol. 333, 125104.

Weinrich, S., Nelles, M., 2021. Systematic simplification of the Anaerobic Digestion Model No. 1 (ADM1) - model development and stoichiometric analysis. Bioresour. Technol. 333, 125124.

Wong, K.I., Wong, P.K., 2017. Optimal calibration of variable biofuel blend dual injection engines using sparse Bayesian extreme learning machine and metaheuristic optimization. Energy Conv. Manag. 148, 1170–1178.

Wong, K.I., Wong, P.K., 2018. Adaptive air-fuel ratio control of dual-injection engines under biofuel blends using extreme learning machine. Energy Conv. Manag. 165, 66–75.

Xing, J., Luo, K., Wang, H., Fan, J., 2019. Estimating biomass major chemical constituents from ultimate analysis using a random forest model. Bioresour. Technol. 288, 121541.

Zabed, H., Sahu, J.N., Suelu, A., Boyce, A.N., Faruq, G., 2017. Bioethanol production from renewable sources: currently perspectives and technological progress. Renew. Sustain. Energy Rev. Available from: https://doi.org/10.1016/j.rser.2016.12.076.

Zha, Y., Muilwijk, B., Coulier, L., Punt, P.J., 2012. Inhibitory compounds in lignocellulosic biomass hydrolysates during hydrolysate fermentation processes. Bioprocess. Biotech. 2 (1), 1–11.

Zhang, H., Zouaoui, Z., Lennox, B., 2005. A comparative study of soft-sensing methods for fed-batch fermentation processes. IFAC Proc. Volumes 38 (1), 55–60.

Zhang, D., Savage, T.R., Cho, B.A., 2020. Combining model structure identification and hybrid modelling for photo-production process predictive simulation and optimisation. Biotechnol. Bioeng. 117 (11), 3356–3367.

Zhao, Y., Li, Y., Fan, D., Song, J., Yang, F., 2021. Application of kernel extreme learning machine and Kriging model in prediction of heavy metals removal by biochar. Bioresour. Technol. 329, 124876.

Zhu, X., Li, Y., Wang, X., 2019. Machine learning prediction of biochar yield and carbon contents in biochar based on biomass characteristics and pyrolysis conditions. Bioresour. Technol. 288, 121527.

Integrated biorefineries: The path forward

13

Yeshona Sewsynker-Sukai[1], *Anthea N. David*[2], *Daneal C.S. Rorke*[2], *Isaac A. Sanusi*[1], *Gabriel S. Aruwajoye*[2], *Edson L. Meyer*[1] *and Evariste B. Gueguim Kana*[2]
[1]Fort Hare Institute of Technology, University of Fort Hare, Alice, South Africa, [2]School of Life Sciences, University of KwaZulu-Natal, Pietermaritzburg, South Africa

13.1 Introduction

The depletion of petroleum-based sources due to the increasing global population and the detrimental effect on the environment has intensified the search for suitable alternatives. Biofuels produced from microorganisms are being explored globally as the race towards alternative energy sources continues (David et al., 2020). While biofuels are renewable, sustainable, and reduce harmful effects on the environment, their production processes are currently expensive and generate low yields. Therefore integrated biorefineries have emerged as innovative multistaged systems for developing high-value bioproducts and biofuels (Pinales-Márquez et al., 2021; Patel and Shah, 2021). Remarkably, the biorefining of renewable lignocellulosic biomass (LCB) feedstocks has garnered significant attention towards achieving a circular bioeconomy (Banu et al., 2021). Unlike agricultural crops, LCB is a waste that is produced in large quantities and does not hinder food security, thus presenting an ideal feedstock for biorefinery systems (David et al., 2020; Ray, 2021). Wastes generated from corn, sugarcane, sorghum, rice, and wheat processing industries are examples of LCB that have previously been used for the microbial production of industrially relevant value-added compounds and biofuels (Moodley et al., 2020; David et al., 2020).

Three main steps, biomass pretreatment, enzymatic hydrolysis, and fermentation, are involved in lignocellulosic bioprocesses (such as bioethanol production). The pretreatment step is crucial since it unwinds the resistant lignocellulosic structures such as lignin and hemicellulose for cellulose release. The enzymatic hydrolysis step is essential for converting the complex polysaccharides to fermentable monosaccharides, which are metabolized into biofuels (David et al., 2020; Behera et al., 2022). Microbial biofuels and bioproducts can be produced using three process types: (1) separate hydrolysis and fermentation (SHF), (2) simultaneous saccharification and fermentation (SSF), and (3) simultaneous saccharification and fermentation with prehydrolysis (PSSF) (Sewsynker-Sukai and Gueguim Kana, 2018). Nevertheless, the high process cost and low product yields continue to plague the commercialization of lignocellulosic biofuel production.

Advances in Lignocellulosic Biofuel Production Systems. DOI: https://doi.org/10.1016/B978-0-323-91192-4.00019-5

Therefore integrated lignocellulosic biorefineries are being investigated globally to overcome these challenges and promote the commercialization of biofuels.

More recently, microbial-derived bioproducts such as lactic acid, itaconic acid, and succinic acid from lignocellulosic substrates have been assessed as high-value compounds of industrial significance (Chen et al., 2019; Krull et al., 2017; Anwar et al., 2021). These value-added compounds possess versatile applications and serve as precursors for producing biopolymers, chemicals, and fuels (Chen et al., 2019; Yang et al., 2020; Akhtar and Idris, 2017). Integrated biorefineries demonstrate the potential to alleviate the bottlenecks typically associated with biofuel production processes, with a different goal to achieve complete valorization of waste-based substrates and transform the landscape of lignocellulosic processes. Although biorefineries represent what could soon be ideal systems, several aspects must first be addressed to promote their feasibility at an industrial scale level.

This chapter provides an in-depth evaluation of integrated biorefineries by detailing the chemical and biological processes involved in forming these systems. First, feedstocks, pretreatment methods, and pretreatment selection criteria for biorefineries are outlined. Subsequently, the different fermentation process types are compared, followed by the microbial fermentative production of bioethanol, biohydrogen, and biogas. Then, microbial-derived high-value products (lactic acid, itaconic acid, and succinic acid) and value-added products (benzaldehyde, biochar, and bio-oil) are briefly outlined. Next, the innovations in lignocellulosic biorefineries are briefly highlighted. Additionally, the life cycle assessment (LCA) and techno-economic analysis of biorefineries are discussed. Lastly, the current challenges, trends, and future perspectives toward enhancing integrated biorefineries are presented.

13.2 Feedstocks for biorefineries

The evolution of integrated biorefineries has unraveled the conceptualization of biomass-based conversion processes analogous to petrochemical refineries. In this context, these substrates can be employed in a biorefinery system where several bioproducts are produced in a process. Interestingly, biomass contains an elementary composition of carbon, hydrogen, and oxygen similar to fossil resources, with the advantage of being renewable and abundant. Therefore several substrates have been extensively investigated to produce bioproducts ranging from biofuels, bioenergy, and commodity chemicals.

13.2.1 Lignocellulosic substrates

LCBs, such as postharvest forestry and agricultural residues, are an abundant resource available for the generation of biofuels and value-added products, with their global production reaching approximately 120 billion tons per annum (Abraham et al., 2020). Lignocellulose continues to attract global interest due to the broad areas of application and variety of conversion methods available (Chen, 2015). The primary composition of lignocellulose involves an intricate complex of the polysaccharides hemicellulose and cellulose and the phenolic polymer lignin. This polymer contributes significantly to the strength and rigidity of plant cell walls and protects hemicellulose and cellulose from enzymatic and microbial degradation (Yoo et al., 2020). Cellulose and hemicellulose contribute between 40%–60% and 20%–35%, respectively, to lignocellulose weight and are composed of 500–1400 D-glucose units (cellulose) and 100–200 glucose units (hemicellulose) (Zoghlami and Paës, 2019), while lignin makes up 5%–40% of lignocellulose and contributes significantly to the recalcitrance (Yoo et al., 2020). There is a wide variety of lignocellulose resources, so variation in lignocellulosic composition is high (Table 13.1). Efficient conversion remains a research hotspot as a "one size fits all" approach results in low hydrolysis rates and, subsequently, low product yields. Efforts to valorize different lignocellulosic substrates and their components (lignin, hemicellulose, and cellulose) to reduce this waste and increase its value may be achieved by adopting a biorefinery concept.

13.2.2 Lignocellulose-starch substrates

Presently, significant efforts are being driven toward biofuel generation from renewable sources. These renewable sources include sugar crops, starch crops, and lignocellulosic and lignocellulose-starch biomass (LCSB) (Mojović et al., 2009; Aruwajoye et al., 2017). For all the substrates previously mentioned, biofuel

Table 13.1 Lignocellulosic composition of commonly used substrates.

	Cellulose (% w/w)	Hemicellulose (% w/w)	Lignin (% w/w)	Reference
Pinus spp. sawdust	52.7 ± 2.8	12.9 ± 2.6	26.7 ± 0.9	Morales-Máximo et al. (2021)
Miscanthus sinensis	37.7 ± 3.8	22.9 ± 3.7	17.4 ± 1.3	Xu et al. (2020)
Miscanthus floridulus	36.3 ± 2.6	22.0 ± 3.6	16.9 ± 1.2	Xu et al. (2020)
Miscanthus sacchariflorus	39.3 ± 3.1	26.4 ± 3.7	18.1 ± 1.4	Xu et al. (2020)
Sugarcane bagasse	35.2 ± 0.9	24.5 ± 0.6	22.2 ± 0.1	Rezende et al. (2011)
Sweet sorghum bagasse	36.1 ± 1.1	31.2 ± 0.8	24.8 ± 2.0	Camargo et al. (2019)
Corn stover	36.5 ± 2.1	22.1 ± 0.9	18.8 ± 1.7	Yang et al. (2016)
Wheat straw	38.2 ± 0.1	29.0 ± 0.1	15.7 ± 0.0	Tufail et al. (2018)
Eucalyptus grandis sawdust	43.6 ± 2.1	11.1 ± 0.5	30.5 ± 0.5	Rochón et al. (2022)

production is only achievable through fermentative processes that utilize simple sugars released from the disintegration of polymeric carbohydrates embedded in the biomass. Currently, LCBs are generally preferred as substrates for biofuel production, despite the progress of this technology still being in the developing stages. Up to 70% of LCB are usually agricultural wastes generated as by-products of food preparation, farming activities, and the remains of crops postharvesting (Agamuthu, 2009; Balan, 2014; Nagendran, 2011). The preference for LCB over sugar and starch crops is founded on their lack of interference with food availability and the preservation of food security (Ray and Ramachandran, 2018). Although LCBs are considered wastes, they consist of highly organized structural carbohydrates—hemicellulose and cellulose often intertwined with lignin (Mosier et al., 2005).

On the other hand, LCSB have starch as an additional and major polysaccharide component in their composition, which confers the "starch-based" characteristic. Various LCSB substrates have been reported, including cassava peels, leaves, stems and bagasse, sweet potato peels and bagasse (Aruwajoye et al., 2017; Ray and Naskar, 2008; Chohan et al., 2020; Ojewumi et al., 2018; Pooja and Padmaja, 2015). The compositional analysis of LCSB reveals a starch composition in the range of 2%–66% depending on the type of source crop (Table 13.2), nature of the postharvesting activity, and other environmental factors involved during their generation (Aruwajoye et al., 2020). Starch, the reference component of LCSB, is a giant polymer of monomeric sugars, usually in the form of amylose and amylopectin chains. While α-1–4 glycosidic bonds link the amylose polymer, the amylopectin contains α-1–6 bonds in addition to the presence of α-1–4. Often, the extent of cleavage of the starch bonds during the pretreatment of LCSB significantly determines the level of the simple sugar yield for fermentative processes.

13.3 Overview of pretreatment

In lignocellulosic biorefineries, the general purpose of pretreatment is to unlock the lignin framework and expose the carbohydrate polymers to enzymatic hydrolysis. The highly ordered structure of lignin acts as a physical barrier that protects cellulose and hemicellulose from being biologically or chemically degraded into their metabolically favored form (Kumar et al., 2020a,b). To reduce these limitations, lignocellulosic pretreatment aims to modify the complex structures and improve the biodegradability of the biomass by removing lignin, reducing cellulose crystallinity, and increasing the reactive cellulose surface area and porosity (Kim, 2013). These key LCB pretreatment approaches are necessary prior to its exposure to downstream biological processes such as enzymatic saccharification and fermentation. Various pretreatment methods such as physical, biological, chemical, and a combination of these different strategies have been implemented.

Physical pretreatment usually involves mechanical, hot water, and radiation-based strategies that reduce the particle size of the biomass, expand the contact surface area, and decrease the degree of polymerization and cellulose crystallinity

Table 13.2 Different lignocellulos-starch biomass types with their composition.

LCSB substrate	Starch (%)	Cellulose (%)	Hemicellulose (%)	Lignin (%)	References
Cassava stem	15.0	22.8	28.8	22.1	Pooja and Padmaja (2015)
Cassava peels	2–45	9–18	7–28	1–10	Pooja and Padmaja (2015); Aruwajoye et al. (2017)
Potato peels	20–42	4–34	5–10	4–33	Ben Taher et al. (2017); Hijosa-Valsero et al. (2018); Chohan et al. (2020)
Sweet potato peels	32.05	13.31	13.32	8.15	Ray and Naskar (2008); Mithra and Padmaja (2016)
Pumpkin waste	65.3	NR	NR	NR	Chouaibi et al. (2020)
Pumpkin peels	24.61	21.05	17.74	10.66	Mithra and Padmaja (2017)
Elephant foot yam peels	28.96	15.63	14.00	7.01	Mithra and Padmaja (2016)
Tannia peels	30.46	17.32	14.48	8.26	Mithra and Padmaja (2016)
Greater yam peels	28.84	18.02	20.02	6.72	Mithra and Padmaja (2017)
Beetroot peels	27.13	18.94	19.17	3.87	Mithra and Padmaja (2017)
Banana peels	36.56	22.40	15.19	10.55	Mithra and Padmaja (2017)

Note: The starch, cellulose, hemicellulose and lignin are represented in wt.% and are determined from dry biomass. *NR*: not reported.

(Kumar et al., 2020a). Several approaches have been discovered, including grinding, extrusion, sonication, steam explosion, microwave, and gamma rays. Even though physical methods are environmentally friendly and produce little to no toxic by-products, these regimes are limited to low lignin-containing substrates since they cannot remove lignin (Kumar et al., 2020a). Therefore these methods are often applied as a preparation step prior to or in conjunction with other pretreatments (Kumari and Singh, 2018).

On the other hand, biological pretreatment involves the action of microbial metabolism (fungi and bacteria) or bio-based products (enzymes) for the degradation of cellulose, hemicellulose, and lignin fractions within lignocellulosic feedstocks (Kumar et al., 2020a). The microbial degradation process involves two extracellular systems: the hydrolytic and ligninolytic (Wagner et al., 2018). The hydrolytic system is required to break down cellulose and hemicellulose, while the ligninolytic system is involved in depolymerizing lignin using enzymes such as peroxidases and laccases (Baruah et al., 2018). Biological pretreatments are known for low energy requirements, cost efficiency, easy downstream processing, little to no waste generated, and reduced enzymatic and fermentation inhibitors (Abraham et al., 2020). However, despite these advantages, this type of pretreatment requires long process times and continuous records of microbial growth, making it a tedious and inefficient process (Kumari and Singh, 2018).

Chemical pretreatments are most often adopted and mediated by either acid, alkaline, inorganic salts, ionic liquids, or organosolvents. These strategies function by engaging in chemical reactions within the aqueous solution, resulting in delignification and hemicellulose solubilization. The mechanism of action and the intended outcome depend significantly on the type of chemical used and the operating conditions during the pretreatment process (Abraham et al., 2020). Singhvi et al. (2014) reported that chemical pretreatments tend to improve the enzyme digestibility of the biomass and enhance the overall yield of fermentable sugars by approximately 80% in comparison to pretreatments that are void of chemical supplementation. Additionally, these methods are energy efficient, stable, and flexible with the choice of substrate. Although chemical pretreatment regimes offer numerous advantages, these methods are usually costly, may cause corrosion of reaction vessels, and release inhibitory compounds that affect downstream processes. Consequently, the chemicals used for pretreatment must be carefully selected and evaluated for commercial production of the desired products.

Several pretreatment studies have focused on a combination of physical, biological, and chemical strategies, with alkaline-based methods shown to be the most promising (Woiciechowski et al., 2020). Nevertheless, alkaline methods are costly and energy-intensive and may produce undesirable inhibitor compounds due to their harsh operating conditions (Balat et al., 2008; Hendriks and Zeeman, 2009). Recently, waste-based pretreatment strategies such as paper wastewater and green liquor dregs have been proposed to alleviate limitations associated with high energy, chemical costs, and water consumption (David et al., 2020; Sewsynker-Sukai et al., 2020).

13.4 Pretreatment selection criteria for microbial-derived products in biorefineries

The selection of a specific lignocellulosic pretreatment method, although frequently determined based on costs, energy, and sugar yield, as stated previously, has also shown to be dependent on the target fermentation product to be produced. For instance, C5 sugars are derived from heteropolymer hemicellulose. Therefore a pretreatment that solubilizes this lignocellulosic structure must be applied. However, the application of C5 sugars in microbial bioprocesses remains limited since only a few microorganisms can metabolize sugars, such as xylose. Nevertheless, xylose has been widely used as an artificial sweetener in the food industry but may also be utilized for microbial biohydrogen production using the dark fermentation process (Mafuleka and Gueguim Kana, 2015).

On the other hand, the cellulose polymer gives rise to C6 sugars such as glucose, which is most often used in fermentation processes since microbial metabolic pathways typically follow glycolysis. Thus pretreatment methods that reduce cellulose crystallinity to enhance its enzymatic digestibility are selected. An array of microbially produced value-added compounds may be generated from glucose, such as biofuels (bioethanol, biohydrogen, biogas) and value-added products (lactic acid and itaconic acid), among several others.

Even with the selection of specific pretreatments that target cellulose and hemicellulose structures for the production of a particular compound, the aromatic lignin polymer causes a significant hindrance by acting as a resistant barrier that prevents and reduces hydrolysis of cellulose and hemicellulose structures. Because of this, the selected pretreatment must ideally be capable of degrading lignin and subsequently target cellulose, hemicellulose, or both. Apart from the desired value-added compound to be produced, the basis for selecting a pretreatment method should also consider the fermentation and/or enzymatic inhibitor compounds. These inhibitory compounds are formed as by-products of pretreatment and are primarily determined by the lignocellulosic feedstock composition and the severity and type of pretreatment method utilized (Rorke and Gueguim Kana, 2016). Some inhibitory compounds include aliphatic acids, phenolic compounds, and furan derivatives. More specifically, aliphatic acids (acetic acid) and furan aldehydes (furfural and 5-hydroxymethylfurfural) are inhibitors that are commonly found in pretreatment lignocellulosic hydrolysates and, to a lesser extent, certain phenolic compounds (2-methoxy-4-vinyl phenol and 4-hydroxy-benzaldehyde). The formation of acetic acid during chemical (acid, alkaline, oxidation, kraft pulping) and/or hydrothermal pretreatment occurs as a result of the degradation of acetyl and ester bonds in hemicellulose. Furan derivatives such as furfural and 5-hydroxymethylfurfural (HMF) form when carbohydrates (pentose and hexose) are degraded during acid- and/or hydrothermal-based pretreatments (Harmsen et al., 2010).

Moreover, under prolonged severe pretreatment conditions, including long reaction time and high temperature and/or acid concentration, HMF is further degraded to other inhibitory compounds such as levulinic and formic acids (Fengel and Wegener, 1989). On the other hand, phenolic compounds are formed when ether bonds in lignin are fragmented due to alkaline pretreatment. Inhibitor compounds negatively impact the cellulase enzyme and microbial fermentation process by interfering with enzyme activity and cell metabolism; therefore, their removal is imperative (Gupta et al., 2017). Generally, the application of harsh pretreatment conditions, such as high temperatures and high chemical concentration, in an attempt to improve fermentable sugar yields leads to inhibitor compound formation. Various strategies have previously been evaluated to alleviate the detrimental effects of lignocellulosic fermentation inhibitors, which include preventing their formation, converting them to a nontoxic form, or detoxifying the pretreatment hydrolysate (Taherzadeh et al., 2000). Furthermore, the selected pretreatment must be cost-effective from a chemical and energy perspective, eco-friendly, and lead to a high sugar and fermentation yield.

13.5 Microbial fermentation

Biowastes have become a well-recognized carbon source for their conversion into biofuels and platform chemicals with a wide range of applications in various industries. The monomeric sugars released following pretreatment and enzymatic hydrolysis of these biowastes become the intermediates in the overall process channeled towards microbial metabolism to produce value-added products. Microbial conversion processes depend on fermentation or anaerobic digestion (AD) by specific microorganisms, including bacteria, yeasts, fungi, and microalgae (Gomez et al., 2021). The preferred microorganisms required for microbial product formation must be able to metabolize broader substrate spectrums, possess high specific growth rates, increased substrate to substrate, higher substrate to biomass yield, higher overall tolerances, and improved metabolic fluxes (Höffner and Barton, 2014). Several microorganisms may utilize a single hydrolyzed biomass stream to produce various bioproducts, significantly boosting the economics of a biorefinery. However, the hydrolyzed biomass may contain a mixture of both hexose (glucose, fructose, mannose) and pentose sugars (xylose, arabinose) depending on the type and composition of the feedstock (Raud et al., 2019). Most microorganisms require hexoses, mainly glucose, to carry out their metabolic processes towards product formation. At the same time, pentose metabolism is restricted to a few native strains with relatively low yields and high sensitivity to process conditions (temperature, pressure, pH) (Mosier et al., 2005). Therefore choosing a suitable inoculum source is imperative to the productivity and functionality of a biorefinery process. For this reason, various microbial strains may be introduced into the system to produce different bioproducts, while in other cases, a single microorganism source with the dual capability of utilizing both pentose and hexose sugars is employed (Moodley, 2021).

13.6 Lignocellulosic fermentation process type

Cellulosic product formation occurs following three key steps:

1. LCB pretreatment,
2. enzymatic saccharification, and
3. microbial fermentation.

The enzymatic hydrolysis stage has been deemed a fundamental step in the bioproduct formation process because the fermentable sugars that will undergo fermentation are released during this step. Microbial bioproduct formation using lignocellulosic wastes can be produced using three different process types:

1. SHF,
2. SSF and,
3. simultaneous saccharification and fermentation with a prehydrolysis step (PSSF) (Carrillo-Nieves et al., 2017).

The above-mentioned process has its merits and bottlenecks that are discussed below.

13.6.1 Separate hydrolysis and fermentation

SHF is a method by which enzymatic hydrolysis and fermentation are performed sequentially. The process principle of SHF includes the separation of hydrolysis and fermentation of the reactions in separate units. In SHF, both hydrolysis and fermentation are carried out at their optimal temperatures for hydrolysis and fermentation. Within SHF systems, enzymatic saccharification of pretreated LCB is carried out first at the optimal temperature of the saccharifying enzyme (40°C−50°C); fermenting microbes (30°C−40°C) are then added to ferment the saccharified sugar at its optimal process conditions (David et al., 2020). Since enzymatic hydrolysis is performed at optimal temperature, SHF may require a lower quantity of saccharifying enzymes than the SSF process but achieves a higher saccharification efficiency (Binod et al., 2010). The SHF system, however, requires two independent reactors for the different reactions (enzymatic hydrolysis and fermentation), thus placing an additional cost on equipment. Moreover, the sugar accumulation during the hydrolysis stage may result in carbohydrate feedback inhibition effects on cellulolytic enzymes, reducing their hydrolyzing capability and efficiency (Koppram et al., 2013; Binod et al., 2010).

13.6.2 Simultaneous saccharification and fermentation

SSF is a process that combines enzymatic hydrolysis with instantaneous fermentation to obtain desired bioproducts (Ballesteros et al., 2004). This method is based on the use of the enzyme complex for hydrolyzing cellulose and sugar production (Guilherme et al., 2019). Microorganisms simultaneously ferment the released sugars to produce a particular product. The SSF process has several advantages over other process types. Compared to SHF, some advantages are using a single vessel for fermentation and saccharification, reduction in residence times, and the capital costs of the process (Pratto et al., 2020). Another prominent advantage is the reduction of inhibitory compounds from enzymatic hydrolysis, which improves the process's overall performance (Shi et al., 2019). Due to these advantages, SSF has been widely employed for the production of biofuels such as ethanol and butanol from lignocellulosic and starchy feedstocks (Izmirlioglu and Demirci, 2012;

Sewsynker-Sukai and Gueguim Kana, 2018; Moodley and Gueguim Kana, 2019; Shi et al., 2019; Sanusi et al., 2021).

Furthermore, the SSF process eliminates the inhibition of saccharifying enzymes that could occur from high initial fermentable sugar concentration because the resulting sugars are converted to products by fermenting microorganisms. However, the SSF process has a significant disadvantage compared to the SHF process since the optimum temperature and pH for microbial fermentation differ from that for the enzymatic hydrolysis step (Olofsson et al., 2008). It is, therefore, necessary to find an equilibrium point where the process works optimally for both the enzymatic and fermentation processes to avoid adverse effects on either reaction (Binod et al., 2010).

13.6.3 Simultaneous saccharification and fermentation with a prehydrolysis step

The PSSF processes have been introduced to enhance the efficiency of SSF processes by including a short prehydrolysis stage as a prior technique. Prehydrolysis before simultaneous saccharification and fermentation is usually performed by subjecting the saccharifying enzymes to their optimal temperatures for a short period. The optimal temperatures during prehydrolysis are often at elevated levels ($>40°C$), which are higher than the selected conditions for the SSF when the microbe is introduced. This elevated temperature reduces the initial viscosity of the medium and allows for optimal performance of the saccharifying enzyme for a short time, leading to a higher bioethanol yield (He et al., 2016; Zhu et al., 2015). However, despite the short time usually involved in prehydrolysis stages, higher process costs are generally experienced compared to direct SSF because of the demands of the additional stage, energy, and operational units needed (Wingren et al., 2003). Besides the disadvantage of additional process costs, comparative studies and kinetic evaluation of PSSF and SSF processes on some substrates have revealed a lack of significant difference in their bioethanol concentrations (Aruwajoye et al., 2020; Sewsynker-Sukai and Gueguim Kana, 2018). Consequently, to achieve process profitability from PSSF, it may be necessary to consider critical factors such as substrate type, yeast strain, and other fermentation conditions.

13.7 Lignocellulosic biofuel production

The microbial production of biofuels from LCB has become a global research hotspot in the search for sustainable processes. Bioethanol, biohydrogen, and biogas are being exhaustively studied for supplementing current energy sources, and these biofuels are discussed below.

13.7.1 Bioethanol

Bioethanol is one example of liquid biofuels that can be obtained from LCB. Currently, bioethanol is one of the preferable liquid biofuels for the replacement of fossil fuels. The significant advantages of bioethanol over fossil fuels include its renewability, high octane rating, eco-friendly nature, and lower emissions (Sutjahjo, 2018). Lignocellulosic bioethanol can be produced in four stages: pretreatment, enzymatic hydrolysis, fermentation, and distillation/product recovery. Bioethanol product formation during the fermentation of LCB usually occurs in the cell of the fermentative microbe. The most common fermentative microbes for lignocellulosic bioethanol fermentation are *Zymomonas mobilis* and *Saccharomyces cerevisiae*. *S. cerevisiae*, with the greater preference, has highly specialized systems for converting glucose to ethanol (Behera et al., 2022). The specialized pathway for converting glucose to ethanol in *S. cerevisiae* is glycolysis, which occurs in the microorganism's cytoplasm. During glycolysis, the glucose is catabolized by enzymes in multiple steps, resulting in the formation of pyruvate and bioethanol (Bai et al., 2008; Zabed et al., 2017). Various lignocellulosic substrates have previously been used for bioethanol production with different pretreatments and process types (Table 13.3).

13.7.2 Biohydrogen

The biological production of hydrogen from lignocellulose offers an environmentally friendly alternative to the current thermomechanical and chemical processes. It is a clean-burning fuel that exhibits a high energy density of between 120 and 142 MJ/kg and is converted to water when utilized for electricity as a fuel cell (McCay and Shafiee, 2020). Hydrogen production can be carried out using several approaches, including dark fermentation, microbial electrolysis, photosynthetically photofermentation, and biophotolysis (Fig. 13.1). The primary mechanism of microbial hydrogen formation is proton reduction catalyzed by hydrogenase or nitrogenase (Hallenbeck and Liu, 2016). Dark fermentation has been the most extensively studied technology for hydrogen production as it exhibits a higher treatment capacity for various waste streams than photo-fermentation, including LCB. Additionally, dark fermentation does not require light and has a low energy requirement, is a simpler process, and exhibits increased stability (Mohan et al., 2011). Hydrogen is often produced by a mixed culture of facultative and obligate anaerobic microorganisms by carbohydrate conversion to acetyl-CoA through the glycolytic pathway. It produces organic acids such as butyrate and acetate, CO_2 and H_2 (Ghimire et al., 2015). Theoretically, a maximum of 4 mol H_2/mol glucose can be produced when acetate is the sole organic acid produced from glucose. However, a much lower yield of 2 mol H_2/mol glucose is achievable when butyrate is the favored organic acid. In lignocellulosic systems where both organic acids are produced, a theoretical yield of 2.5 mol H_2/mol glucose can be obtained (Dalena et al., 2017).

To achieve maximal yields, factors that affect productivity, such as pH, operational temperature, substrate type, inoculum source, and retention time, should be

Table 13.3 Bioethanol produced from different lignocellulosic substrates.

Microorganism	Substrate	Pretreatment conditions	Process type	Process conditions	Ethanol concentration	Reference
Saccharomyces cerevisiae PTCC 5052	Wheat straw	1:5 ratio biomass to [TEA][HSO_4] solvent, 0.5 h, 130°C	SHF	3 g/L microorganism concentration, 5% SL, 28 FPU (filter paper units)/g Cellulase CelluMax GFL enzyme, 30°C, 96 h	52.84%	Ziaei-Rad et al. (2021)
S. cerevisiae PTCC 5052	Wheat straw	1:5 ratio biomass to [TEA][HSO_4] solvent, 3 h, 130°C	SHF	3 g/L microorganism concentration, 5% SL, 28 FPU/g Cellulase CelluMax GFL enzyme, 30°C, 96 h	84.34%	Ziaei-Rad et al. (2021)
S. cerevisiae CCUG 53310	Rice straw	5% SL, 0.5 M Na_2CO_3, 3 h, 93°C	SHF	5 g/L microorganism concentration, 20% SL, 10 FPU/g Cellic CTec2 and Cellic HTec2, 37°C, 120 h	55.1 g/L	Molaverdi et al. (2019)
S. cerevisiae CCUG 53310	Rice straw	5% SL, 0.5 M Na_2CO_3, 10 h, 93°C	SHF	5 g/L microorganism concentration, 20% SL, 10 FPU/g Cellic CTec2 and Cellic HTec2, 37°C, 120 h	67.7 g/L	Molaverdi et al. (2019)
S. cerevisiae	Corn cobs	5% SL, 0.95 g/L $FeSO_4$ and 29.8 g/L H_2O_2, 30 min, 35°C, 0.75% NaOH, 1 h, 90°C	PSSF	5 g/L,[a] 10%,[b] 15 FPU/g[c] Spezyme CP cellulase, 48 h	19.3 g/L	He et al. (2016)

S. cerevisiae	Bamboo	10% SL, 0.5% NaOH, 2 h, 170°C	SHF	0.027 g/mL,[a] 5%,[b] 15 FPU/g[c] cellulase and 30 IU xylanase, 35°C, 24 h	4.8 g/L	Yang et al. (2019)
Escherichia coli FBR5	Wheat straw	86 g/L SL, 0.5% H_2SO_4, 10 min, 160°C	SSF	5% (v/v),[a] 8.6%,[b] 150 μl/g[c] Celluclast, Novozym188 and ViscoStar cocktail, 35°C, 24 h	17.4 g/L	Saha et al. (2011)
Scheffersomyces stipitis NRRL-Y7124	Sugarcane bagasse	5% SL, 0.3 mol/L NaOH, 20 min, 60°C	SSF	5 g/L,[a] 11%,[b] 20 FPU/g[c] Cellic CTec2, 30°C, 96 h	17.26 g/L	Hilares et al. (2017)
Saccharomyces cerevisiae BY4743	Sugarcane bagasse	9.69% SL, 1.73 M $ZnCl_2$, 1.36 M NaOH, 30 min, 121°C	SSF	0.06 g/L,[a] 10%,[b] 100 U/g[c] Cellic CTec2, 39°C, 60 h	4.88 g/L	Jugwanth et al. (2020)

Note: a = biomass concentration, b = substrate solid loading, c = enzyme loading. *SHF*: Separate hydrolysis and fermentation; *SSF*: simultaneous saccharification and fermentation; *PSSF*: simultaneous saccharification and fermentation with prehydrolysis; *SL*: solid loading; *[TEA][HSO_4]*: triethylamine hydrogen sulfate.

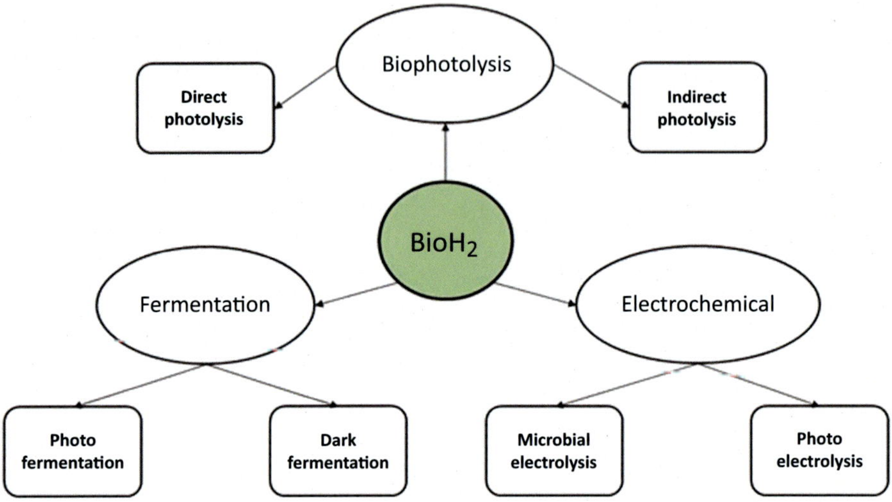

Figure 13.1 Biohydrogen production pathways from biomass.

considered and optimized (Ghimire et al., 2015). Some drawbacks of dark fermentation are low product yields, high by-product formation, and low substrate conversion efficiency (Osman et al., 2020).

LCB has been exhaustively used for biohydrogen production (Table 13.4). But the low yields remain a significant bottleneck. Recent studies have also focused on using industrial residues such as paper mill sludge to produce biohydrogen (Rorke et al., 2021; Moreno-Dávila et al., 2017).

13.7.3 Biogas

Biogas is an end product of the anaerobic microbiological conversion of organic material (Lin et al., 2016). It is a gaseous mixture of methane (55%–75%), carbon dioxide (CO_2) (25%–45%), and hydrogen (5%–10%) (de Mes et al., 2003; Kamali et al., 2016), which can be utilized for heating, upgraded to a higher quality (natural gas alternative) or used in the co-generation of electricity and heat (de Mes et al., 2003). AD is a process used for the generation of biogas from organic waste and wastewater and can be subdivided into four main characteristically unique phases (Angelidaki et al., 2011; Manyi-Loh et al., 2013; Zhang et al., 2016; Manchala et al., 2017) as seen in Fig. 13.2:

1. Hydrolysis converts complex biopolymers such as carbohydrates, lipids, and proteins to soluble organic compounds such as sugars, fatty acids, and amino acids. Hydrolysis is achieved by extracellular enzymes such as cellulases, proteases, lipases, and amylases, which are secreted by hydrolytic bacteria into the bulk liquid.
2. Acidogenesis—the anaerobic acid-producing process in which soluble organic compounds are converted to organic acids such as volatile fatty acids (VFAs) (butyrate, acetate, etc.), hydrogen, and carbon dioxide. The substrate serves as both electron donor and acceptor

Table 13.4 Biohydrogen produced from different lignocellulosic biomass and industrial waste sources.

Lignocellulosic substrate	Hydrogen yield (mol H_2/mol glucose)	Hydrogen yield (mL/g biomass)	Reference
Corn stover	1.5–3.0	–	Cheng et al. (2011)
Rice straw	0.8	–	Cheng et al. (2011)
Sugarcane bagasse	1.7	–	Cheng et al. (2011)
Potato steam peels	2.4–3.8	–	Cheng et al. (2011)
Sweet sorghum	–	15.1–127.3	Cheng et al. (2011)
Corn cobs	–	49.4	Zhang et al. (2021)
Sugarcane bagasse	2.3	–	Łukajtis et al. (2018)
Wheat straw	–	133.6	Zhu et al. (2022)
Corn stover	1.5–2.2	–	Yadav et al. (2020)
Sweet sorghum bagasse	1.3–2.6	–	Panagiotopoulos et al. (2010)

for obligate and facultative anaerobic bacteria within the species of *Clostridium*, *Propionibacterium*, *Lactobacillus*, etc.

3. Acetogenesis—organic acids [VFAs and long-chain fatty acids (LCFAs)] are converted to methane formation precursors such as acetate, hydrogen, and carbon dioxide. It can be achieved by hydrogen-utilizing acetogens, strictly anaerobic bacteria that use the acetyl-CoA pathway as the primary mechanism for reductive synthesis of acetyl-CoA from carbon dioxide. Organic acids are also oxidized to acetate by hydrogen-producing acetogens. These acetogens are obligate users of hydrogen ions and carbon dioxide as electron acceptors. Therefore the unfavorable energetics of the process limits them as energy derivation for growth is only possible when product concentration is low—creating an obligatory dependence on hydrogen scavengers such as sulfate reducers to remove the product. This conversion is, therefore, efficiently inhibited by high concentrations of products such as hydrogen ($>$ 10–4 atm) in the system.
4. Methanogenesis—acetate, carbon dioxide, and hydrogen are converted to methane by methanogenic bacteria such as acetoclastic/acetotrophic methanogens, which split acetate into methane and carbon dioxide, and hydrogenotrophic methanogens, which use hydrogen as an electron donor and carbon dioxide as an electron acceptor, thus producing methane.

In light of dwindling energy reserves, AD has gained significant interest, particularly in its potential for biogas production and solids reduction (Kamali et al.,

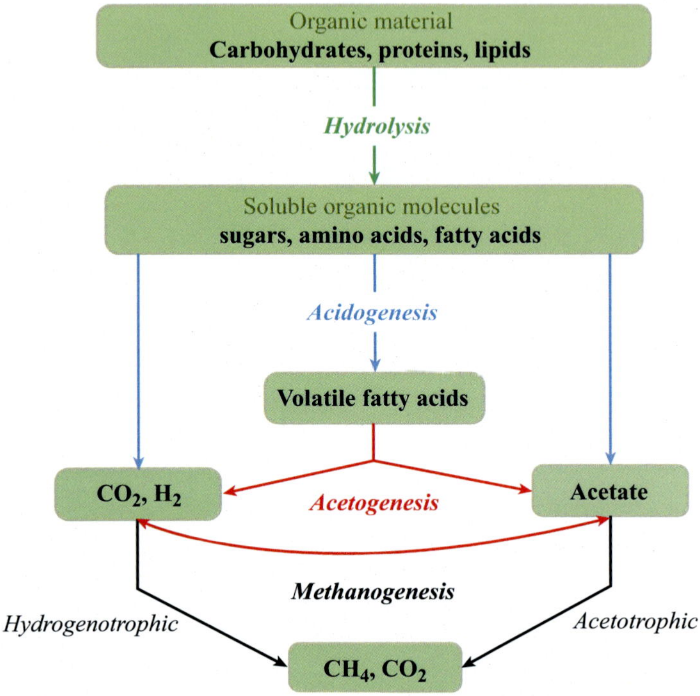

Figure 13.2 Schematic diagram of the anaerobic digestion process.

2016). Numerous advantages associated with the AD of organic wastes and LCB include (de Mes et al., 2003; Olatunji et al., 2021):

1. Low energy requirements of between 0.05 and 0.1 kWh/m^3 at mesophilic temperatures.
2. Significant retention of fertilizer nutrients (nitrogen, phosphate, and potassium).
3. Relatively low construction costs.
4. Stabilization of raw wastes.
5. Lower space requirements in comparison to conventional systems.

The efficiency of the AD process is dependent on start-up conditions and environmental factors such as a neutral pH, anaerobic conditions, and a suitable operating temperature, among others (Schirmer et al., 2014). Additionally, inhibitory compounds such as ammonia, hydrogen, sulfide, and VFAs can inhibit methanogens during AD (Zhang et al., 2016). Therefore, various methods have been applied to reduce inhibitory effects, such as gas stripping to reduce ammonia levels and ferric chloride to remove hydrogen sulfide (H_2S) and sulfate to reduce the inhibitory effect of propionate (Zhang et al., 2016).

Biogas is often produced to further upgrade the gas for use in applications requiring high methane concentrations. The content of the various gases present in biogas is determined by the nature of the substrate used as well as the pH of the process. The energy content of 50.4 MJ/kg—CH_4 (36 MJ/m^3—CH_4 at STP conditions) has been reported for methane by Angelidaki et al. (2018). Therefore any additional gases present are considered pollutants and will reduce the energy content of biogas. Additionally, the presence of compounds like siloxanes can result in the generation of sticky residues in combustion engines, which may cause machinery malfunctions (Angelidaki et al., 2018). Various methods have been implemented to remove impurities, including the application of bottom ash from municipal solid waste incineration to adsorb CO_2 and H_2S from the resultant biogas stream, the removal of H_2S by biological oxidation by aerobic sulfate oxidizers and the conversion of CO_2 to methane by reacting it with H_2 (Angelidaki et al., 2018). Numerous LCB sources have previously been used for biogas production (Table 13.5). The AD process for biogas production is highly versatile compared to bioethanol and biohydrogen since the microbes can break down complex substrates without pretreatment (Abraham et al., 2020).

13.8 Microbial high-value products from lignocellulosic biomass

LCB accounts for many nonfood-based portions, including leaves, stalks, and husks that are either landfilled or burnt in fields (Kee et al., 2020). It poses a significant hindrance to environmental and health considerations; therefore utilizing the residual waste components to produce biofuel and other bioproducts will add value and aid in an effective waste management solution (Manna et al., 2018). This "waste to wealth" concept enables the manipulation of microorganisms to utilize these lignocellulosic residues as a carbon source to produce end products that can be tailored to suit global demands. To date, various studies have been carried out on the conversion of hydrolyzed LCB into biofuels (bioethanol, biohydrogen, biogas, etc.) and

Table 13.5 Biogas produced from different lignocellulosic feedstocks.

Lignocellulosic biomass	CH_4 potential (L CH_4/kg VS)	Reference
Corn	291–338	Sawatdeenarunat et al. (2015)
Wheat	351–378	Sawatdeenarunat et al. (2015)
Grass	286–324	Sawatdeenarunat et al. (2015)
Fodder beet	398–424	Sawatdeenarunat et al. (2015)
Rice straw	152–263	Mustafa et al. (2016)
Wheat straw	251	Schroyen et al. (2015)
Miscanthus giganteus	285–333	Wahid et al. (2015)
Corn straw	325.7	Fu et al. (2015)
Pulp and paper sludge	225	Granström and Montelius (2014)
Reed biomass	188	Lizasoain et al. (2016)

VS: Volatile solids.

value-added chemicals (lactic acid, itaconic acid, succinic acid, etc.) (De Bhowmick et al., 2018). Nevertheless, the production of biofuels has shown to be expensive with low yields, thus rendering it unfeasible on a large scale. Therefore in recent times, the lignocellulosic bioprocess trajectory focuses on producing a high-value product of commercial relevance (Fig. 13.3), followed by biofuel production. Some examples of valuable bioproducts that are imperative for the commercial sector and can be produced from LCB in biorefinery systems include lactic acid, itaconic acid, and succinic acid.

13.8.1 Lactic acid

Lactic acid, a fermentation product of *Lactobacillus* species and other genetically engineered microbial strains, is an important product for the food, chemical, cosmetic and pharmaceutical industries. This bioproduct is naturally produced using *Levilactobacillus brevis*, *Lactiplantibacillus plantarum* and *Lacticaseibacillus rhamnosus* (Zhang and Vadlani, 2015; Cui et al., 2011). Lignocellulosic-derived sugar to lactic acid has been driven by the abundance and cost-effectiveness of feedstock and the tolerance and selectivity of the microorganism to enhance both the economic and environmental aspects of the bioprocess (Banu et al., 2021). Lactic acid production from LCB is increasingly gaining popularity. For example, Cizeikiene et al. (2018) reported a lactic acid concentration of 101.8 g/kg from pretreated wheat straw using *Fructilactobacillus sanfranciscensis* MW15. Similarly, previous studies have observed lactic acid concentrations of 62 and 92.5 g/L when using pretreated beechwood (Karnaouri et al., 2020) and rice straw (Chen et al., 2019), respectively, as a carbon source. In an earlier study, Ray et al. (2009)

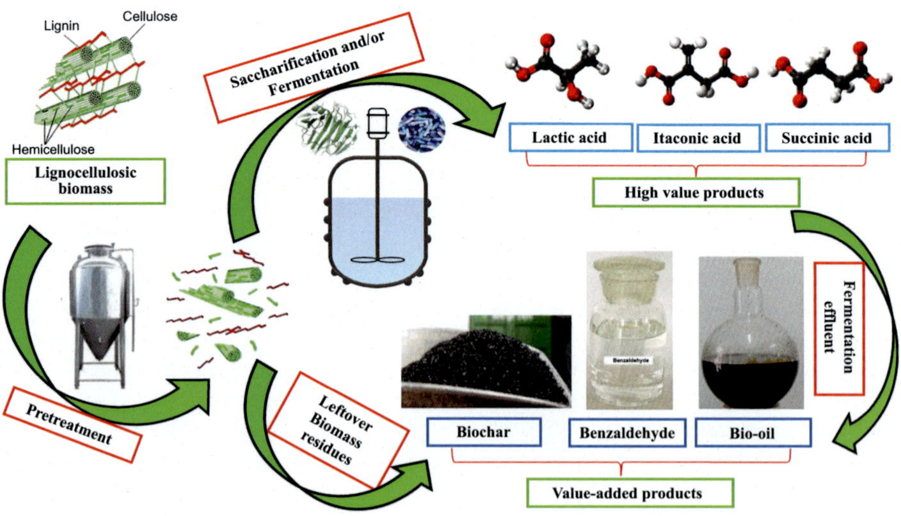

Figure 13.3 Integrated biorefinery process.

reported the production of 29.86 g of (L +) LA from 100 g of cassava bagasse (containing 60% starch) using *Lp. plantarum* MTCC 1407 as the fermenting organism.

13.8.2 Itaconic acid

Itaconic acid is an important precursor for polymers, chemicals, and fuels (Krull et al., 2017; Zhao et al., 2018). *Aspergillus terreus*, *A. itaconicus*, and *Pseudozyma antarctica* are microbes that naturally produce itaconic acid (Krull et al., 2017; Zhao et al., 2018; Kocabas et al., 2014). Other itaconic acid-producing microbes include *Ustilago maydis* and *Escherichia coli* (Maassen et al., 2014; Okamoto et al., 2014). Itaconic acid production using LCB as a substrate has gained prominence, with several aspects still to be investigated to maximize yields. Krull et al. (2017) and Yang et al. (2020) assessed itaconic acid production from alkaline pretreated wheat chaff and alkaline pretreated bamboo residues using *A. terreus* and observed maximum concentrations of 27.7 and 22.43 g/L, respectively.

13.8.3 Succinic acid

Succinic acid has versatile applications and can be used in the food, biodegradable polymer, pharmaceutical, and cosmetic industries (Yan et al., 2014). Major succinic acid-producing microbes include *Actinobacillus succinogenes*, *Mannheimia succiniciproducens*, and *Anaerobiospirillum succiniciproducens* (Guettler et al.,1999; Hong et al., 2004). Genetically engineered *E. coli* and *S. cerevisiae* have also been documented for succinic acid production (Yan et al., 2014; Thakker et al., 2013). Succinic acid from agricultural wastes such as LCB has been reported in the last few years (Zheng et al., 2010; Akhtar and Idris, 2017; Akhtar et al., 2020; Anwar et al., 2021). For example, Zheng et al. (2010) evaluated succinic acid production using corn stover and *A. succinogenes* as the substrate and inoculum source, respectively and obtained a yield of 47.4 g/L. In recent studies, Akhtar et al. (2020) and Anwar et al. (2021) reported succinic acid concentrations of 42.9 and 65.2 g/L from oil palm empty fruit bunch, respectively.

13.9 Value-added products from lignocelluloses: benzaldehyde, biochar, and bio-oil

Value-added products such as benzaldehyde, biochar, and bio-oil have gained popularity. They can be produced as a first, second, or third product for the complete valorization of the feedstock within integrated biorefineries. These bioproducts are briefly outlined below.

13.9.1 Benzaldehyde

Benzaldehyde is a bioproduct that can be obtained from the fermentation of different substrates by fermenting microbes (*Lp. plantarum*, *Pseudomonas putida*, and *Phanerochaete chrysosporium*) (Jensen et al., 1994; Nierop Groot and DeBont, 1998; Norliza and Ibrahim, 2005). For instance, benzaldehyde was obtained from the cultivation of *Rhizopus oligosporus* using mixed substrate (soya bean meal and rice husks). The maximum benzaldehyde concentration of 5.47 mg/g was obtained after 96 h (Norliza and Ibrahim, 2005). The biological-based benzaldehyde is a desirable flavoring agent next to vanillin (Welsh et al.,1989). Other flavoring compounds aside from benzaldehyde include vanillin, lactones, methyl ketones, alcohols, esters, terpenes, and cinnamaldehyde. These have been found to have industrial potential with commercial value. The biological-based benzaldehyde can also be obtained from plants as a glycoside (Norliza and Ibrahim, 2005). Chemical extraction of benzaldehyde results in the formation of toxic byproducts, limiting its usefulness. Therefore the production of biological-based benzaldehyde through fermentation processes is preferred. Furthermore, the low operational cost of fermentation, high productivity, low wastewater output, and improved product recovery has made fermentative benzaldehyde production attractive for commercialization (Bhattacharyya et al., 1998).

13.9.2 Biochar

Biochar is another bioproduct produced by heating choice biomass in the presence or partial absence of oxygen. Thermochemical conversion, such as pyrolysis, hydrothermal carbonization, gasification, and torrefaction, are standard techniques for biochar production (Lin et al., 2016; Pang, 2019; Yaashikaa et al., 2020). To obtain a high yield of biochar, the appropriate technique should be chosen as well as taking into consideration the biomass type (solid wastes, animal wastes, wood biomass, and agricultural residues), heating rate, carbonization temperature, biomass pretreatment, particle size and residence time (Yaashikaa et al., 2020). These conditions are vital because they affect the physical and chemical properties of the biochar produced. Biomasses are composed of cellulose, hemicellulose, and lignin. These components can be transformed into biochar under different reaction conditions and mechanisms (aromatization, condensation, repolymerization, crosslinking of intermediates, polymerization, and depolymerization) (Huang et al., 2012; Mu et al., 2013).

Biochar is usually characterized to determine its structural, elemental composition, surface area, porosity, and surface functional groups (Brewer et al., 2014). The techniques used to characterize biochar include scanning electron microscopy, Fourier transform infrared spectroscopy, X-ray diffraction, thermogravimetric analysis, nuclear magnetic resonance spectroscopy, Brunauer–Emmett–Teller and Raman spectroscopy (Yaashikaa et al., 2020). The stability of biochar during usage relates to its carbon structure since biochar's main constituent is the carbon structure (biochar's instability might be partly due to damage to the structural carbon)

(Huang et al., 2019). In addition, the stability of biochar should be considered due to environmental concerns; hence, it is important to determine in detail the toxicity effect(s) of produced biochar on the environment.

Biochar can play an important role in pollutant removal from the environment (soil and aqueous). It depends on porosity, functional group, surface area, pH, and hydrophobic nature. The eco-friendliness, inexpensive, and ease of biochar preparation from various biomass types make biochar an intensive area of interest. Biochar can also be used for other applications such as catalysts, wastewater treatment, composting, energy storage, carbon sequestration, and soil amendment (Yaashikaa et al., 2020).

13.9.3 Bio-oil

Bio-oil is a liquid fuel obtained through thermochemical processes (pyrolysis and hydrothermal liquefaction) from biomass materials, mainly from LCB. Biomass for bio-oil production includes crop residues, algal biomass, municipal wastes, and agricultural and forestry byproducts (Fu et al., 2009). Bio-oil is a promising feedstock to replace fossil fuel for engine generation. It is because it has a high energy density, is easy to store compared to gaseous products, is carbon neutral, biodegradable, and is environmentally friendly (Hu et al., 2008; Fu et al., 2009).

13.10 Innovations in lignocellulosic biorefineries

The industrial development of the global economy has presented unprecedented concerns regarding the over-utilization of nonrenewable energy sources and pressurized regulatory environmental guidelines (Usmani et al., 2021). Biomass-based energy can offset these challenges due to its sustainability, reduced greenhouse gas emissions, and abundance. LCB is a major renewable carbon source that may support carbon sequestration and the production of bioenergy and biomaterials (Banu et al., 2021). With significant interest attached to lignocellulosic bioproducts, the economic value of these waste residues may be maximized by potential conversion into multiple products through the valorization process of an integrated biorefinery approach (Thomsen, 2005). These links waste resources and the recovery of valuable products in closed-loop cycles, thus contributing to a sustainable environment (Nizami et al., 2017; Harrison et al., 2016). Feedstock contributes about 60% of the total costs of a typical biorefinery system (Botero et al., 2017). A constant and cost-effective supply of renewable and reliable feedstocks (such as lignocellulosic, microalgae, municipal solid waste, and wastewaters) is very important to implementing the biorefinery system (Cherubini et al., 2009). On the other hand, biorefinery platforms are intermediaries that link feedstocks and products (Cherubini et al., 2009). Well-known biorefinery platforms include sugar, biogas, carbon-rich chains, syngas, and plant product platforms (Harrison et al., 2016). The processing of these platforms involves a wide range of technological processes, including biochemical, thermochemical, mechanical, and chemical (Cherubini et al., 2009). Furthermore, biorefinery has been classified using various standpoints such as feedstock, platform, and processes (Harrison et al., 2016). The primary classifications are first, second, and third-generation biorefineries (Harrison et al., 2016).

1. First-generation biorefineries are based on "a single raw material, a single major product" systems (Cherubini et al., 2009; Harrison et al., 2016).
2. In the second-generation biorefinery, a single raw material or feedstock, often lignocellulosic materials, is converted to a wide range of products (Cherubini et al., 2009).
3. Third-generation biorefineries are a complex use of varieties of feedstocks and their transformation to generate multiple bioproducts (Harrison et al., 2016). A typical third-generation category is wastewater biorefinery (Harrison et al., 2016). Using wastewater as substrate minimizes waste emissions and promotes environmental sustainability (Harrison et al., 2016).

Particularly, the production of biofuels and value-added compounds from LCB faces significant economic challenges that need to be addressed by searching for low-cost feedstocks and pretreatments, suitable microbial inocula, robust enzyme cocktails, and bioprocess optimization. Recently, lignocellulosic biorefineries have emerged as potentially profitable systems due to the production of multiple products (low- and high-value) from a single waste stream. The biorefinery approach helps alleviate some economic barriers but still requires enhancement for its successful commercialization at an industrial scale. Biorefinery research is underway and has previously been documented (Banu et al., 2021). For example, in an earlier report, Kaparaju et al. (2009) investigated a biorefinery system in which biohydrogen was produced as the mainstream product, while lactic acid, acetic acid, and bioethanol were generated as by-products using hydrothermally pretreated coffee waste as a substrate.

Similarly, Kocabas et al. (2014) investigated the biorefinery concept by the production of xylanase enzymes and itaconic acid from corn cobs in a two-step fermentation and demonstrated yields of 70 IU/mL (xylanase) and 18 g/L (IA). Kuglarz et al. (2016) used hemp as a substrate for bioethanol and succinic acid production in a biorefinery system, while the latter authors also generated biogas. In a recent study, Moodley and Gueguim Kana (2019) assessed the kinetics of bioethanol production from sugarcane leaf wastes and proposed using effluent as animal feed. Other wastewater biorefinery applications consist of microalgal wastewater biorefinery. Microalgal biomass could be a source of multivaluable bioactive compounds in bioenergy and bioproducts recovery (Odjadjare et al., 2017). Microalgal wastewater biorefinery could transform microalgal biomass into a wide range of products such as biohydrogen bioethanol, biogas, biodiesel, polyhydroxyalkanoate (bioplastic), polyhydroxybutyrate, and animal feeds (Liu et al., 2020).

13.11 Life cycle assessment

LCA is essential in bioprocessing to increase economic sustainability and environmental friendliness. To achieve this, it is vital to determine the interconnectivity of every component of the process from the very beginning (Khang et al., 2017). The LCA has been used in different endeavors; these include energy, chemistry, food, and agriculture (Nemecek et al., 2011; Restianti and Gheewala, 2012; Garrett and Ronde, 2013). LCA as an assessment tool could be obtained by following four assessment steps:

1. scope and goal,
2. inventory analysis,
3. impact assessment, and
4. interpretation (ISO, 2006; Khang et al., 2017).

Equally, the LCA process can be considered as the environmental management accounting for all aspects of ecological releases, an array of environmental interactions, and resource use associated with the complete production process (Norris, 2006; Khang et al., 2017). These include various standpoints, from the extraction of feedstocks, the release of production energy, and recycling and disposing of waste. However, LCA is a vital tool to compare the different aspects, but not a complete exact, precise assessment. Highly reliable standard methodologies (International Organization for Standardization) are used in LCA (ISO, 2006). LCA as a tool can be implemented mainly in four analytical approaches: LCA scope and goal, LCA inventory analysis, Life cycle impact assessment, LCA interpretation. Consequently, final decisions and recommendations can be made.

13.12 Technoeconomic analyses

The process efficiency of many bioprocesses has shown that their industrial implementation is desirable and potentially feasible. Nevertheless, the economic implication of their industrial potential should be assessed through technoeconomic analyses (TEA) to determine the overall cost and identify the steps contributing to the highest expenses (Juneja et al., 2019). The initial costing assessment of a biological process usually contains an investment estimation detailing the capital, operating costs, and investment profitability. This economic evaluation is expected to include research and development (R and D) expenditures for both successful and unsuccessful products. The R and D spending is crucial to the overall cost assessment. Hence, engineers and scientists need to develop effective process designs and methodologies that can cost-effectively eliminate projects or ideas with the potential to be unsuccessful (Koutinas et al., 2016). From an economic perspective, achieving this at the early stages of product and process development is desirable. It is because the probability of new projects never reaching commercialization is over 90%. Such assessment would assist investors in decision-making as well as the implementation of the chosen project. Capital investment for an innovative bioprocess plant will include fixed capital, start-up capital, operational capital, and certification cost (Peters and Timmerhaus, 1991; Seider et al., 1999). The capital investment cost is best estimated by taking note of the wide range of multiplier values (such as equipment, maintenance, heating, ventilation, and air conditioning systems of processing space, laboratory, and office area). While the operational cost also covers a wide range of recurrent expenses such as raw materials, labor, energy cost, utility bills, waste disposal, wages, and salary. Process costing or TEA should result in a cost-competitive building block required for the project commercialization or industrial scale-up (Panda et al., 2018).

Furthermore, a TEA of bioprocesses is estimated using literature-cited (primary or secondary) data, conventional process design brochures, technical pamphlets, market prices, process flowsheets, and sensitivity analysis. Previous studies that are generally selected are based on data with high: process efficiencies, productivity, yield, and high substrate conversion rate (Koutinas et al., 2016). The reliable assessment might require the evaluation of different feedstocks or techniques to obtain the least expensive approach. Some of these feedstocks are available in substantial quantities and regarded as waste; hence, they are economically attractive. Moreover, there is a growing interest in their utilization for the production of various bioproducts such as bioethanol, biohydrogen, biogas, bioplastics, lactic acid, 2,3-butanediol, 1,3-butadiene, butanol, benzaldehyde, and bio-oil (Norliza and Ibrahim, 2005; Sriroth et al., 2000; Koutinas et al., 2016; Sivamani et al., 2018). The sensitivity assessment is equally crucial in technoeconomic assessment, which examines the impact of various key factors on the process. It reveals the variation in production cost associated with the influence of these fundamental factors and how each potentially alters the cost of production individually or interactively (Ong et al., 2012). Operational sensitivity analysis is performed on different process scenarios to show the influence of the diverse operation parameters (substrate type, temperature, pH, process time, enzyme, solvent concentrations, and product yields) to determine the most sensitive parameter in the process (Ong et al., 2012). Afterward, benefit/cost analysis would be considered to determine the most profitable approach for the best economic return.

13.13 Present challenges of integrated biorefineries

Despite the significant advances in developing lignocellulosic-based biorefineries, several aspects still need to be addressed related to the cost, resources, and energy.

1. For instance, the choice of feedstock plays a key role. It can prove detrimental to the entire biorefinery system when it is seasonal-dependent or affected by fluctuating weather conditions. This disrupts the steady year-round supply necessary for biorefinery operations (Oke et al., 2016).
2. In addition, other limitations of these feedstocks are also linked to the requirement for pretreatment processes, which is an integral step in degrading the tightly interwoven, recalcitrant matrice of LCB for enhanced accessibility to cellulose (Patel and Shah, 2021). The pretreatment process accounts for approximately 40% of the expenses within the biorefinery, thus posing a significant shortcoming (Kucharska et al., 2018). These costs include pretreatment chemicals, compatible reactors, energy resources, detoxification

technologies, and biomass separation processing, amongst others (Sewsynker-Sukai et al., 2020).

3. Another major concern is the high water consumption, a finite resource that remains a constraint in developing biorefineries. Lignocellulosic feedstocks are water-insoluble; therefore, substantial quantities of water are needed to enhance mass transfer during pretreatment, enzymatic hydrolysis, and fermentation processes (Sewsynker-Sukai et al., 2020).
4. Like the pretreatment process, enzymatic hydrolysis reactions play a key role in the biorefinery system by releasing sugars essential for microbes during fermentation but also face shortcomings. These may include: (1) costly enzymes that are negatively affected by various factors (inhibitor compounds, pH, temperature), (2) long process times, (3) low saccharification rates, and (4) reduced enzymatic activity as time progresses (Aguilar-Reynosa et al., 2017; Sewsynker-Sukai and Gueguim Kana, 2018).
5. Apart from the bottlenecks associated with pretreatment and enzymatic hydrolysis reactions, bioprocess development and optimization present their limitations. Of main concern is the lack of understanding of the network of metabolic fluxes for microbial cell growth and product formation within bioprocesses, which remains a limiting factor for optimization and scale-up (David et al., 2020; Sewsynker-Sukai and Gueguim Kana, 2018). The scarcity of information related to the microbial fermentation process has led to (a) poor sugar consumption by the fermenting microbes, (b) low product recovery, (c) high cost, and (d) energy-intensive bioprocesses (David et al., 2020).
6. Additionally, adequate research on LCA and TEA in lignocellulosic biorefinery processes has been scanty, which deters scale-up procedures and potential investors (Moodley, 2021).

13.14 Current status and future trends in integrated biorefineries

Integrated biorefineries aim to produce continuous streams of bioenergy and other valuable bioproducts while recycling all the waste outputs with little to no discharge, thus achieving a circular bioeconomy. The recent efforts to enhance integrated lignocellulosic biorefineries are leaning towards developing multistream processes that yield multiple products (Banu et al., 2021). For example, concerning substrate selection, either a single LCB substrate or a mixed feedstock approach can be adopted (Oke et al., 2016). While the single substrate method utilizes one feedstock type, the mixed substrate approach combines two selected wastes. The mixed feedstock approach has demonstrated the potential to reduce challenges associated with logistics issues (harvesting, collection, preprocessing, storage and transportation), intrinsic characteristics (low bulk density and nutrient-deficient biomasses), and extraneous factors (competing uses for the biomass and constant supply regardless of season and weather conditions) (Oke et al., 2016). Interestingly, the mixed biomass method has previously been used for the production of bioethanol (Ferreira et al., 2015), biohydrogen (Prakasham et al., 2009; Kim and Lee, 2010), biogas/methane (Lehtomäki et al., 2007; Kalra and Panwar. 1986; Li et al., 2014) polyhydroxyalkanoates (Sangkharak and Prasertsan, 2013; Shamala et al., 2012) and microbial enzymes (Shamala et al., 2012; Azin et al., 2007; Shamala and Sreekantiah, 1986). Further research on the mixed substrate approach in lignocellulosic biorefineries for complete valorization is recommended to negate feedstock limitations. Additionally, the pretreatment front has made significant strides with the appearance of green liquor dregs (David et al., 2020; Sewsynker-Sukai et al., 2020; Rorke et al., 2021) and paper wastewater (Sewsynker-Sukai et al., 2020) that are both generated from the paper industry as potential methods to replace chemicals and water, respectively. Furthermore, commercial biotechnology conglomerates have developed high-performing cellulase cocktail enzymes for LCB. They have demonstrated extraordinary saccharification efficiencies (Aguilar-Reynosa et al., 2017) and exhibit improved endo/exo activity ratio, resulting in lower end-product inhibition and higher tolerance to nonspecific binding (Ramos et al., 2015). Even with these breakthroughs that promote the implementation of these systems, effective isolation and procurement of the valuable aromatic lignin polymer remain an untapped goldmine in lignocellulosic biorefineries (Chio et al., 2019). The lignin macromolecule represents up to 30% of the LCB. It is usually an unexploited resource from a chemical perspective since it could potentially be used for the commercial production of plastics, pharmaceuticals, and paints. On the other hand, current trends in bioprocess development and optimization deal with evaluating fermentation process type (SHF, SSF, and PSSF) and kinetics of cell growth and product formation (David et al., 2020; Sewsynker-Sukai and Gueguim Kana, 2018; Moodley and Gueguim Kana, 2019), that may provide substantial information for translation to pilot scale. The bioprocess type has commonly been defined and associated with bioethanol production (David et al., 2020) but has also been applied to other biofuel processes (Rorke et al., 2021). The ideal biorefinery system should produce high-value commodities such as lactic acid, itaconic acid, and succinic acid, followed by biofuel production (bioethanol, biogas, and biohydrogen) using the effluent generated. Another innovative advancement in fermentation entails the use of industrial wastewaters (such as dairy, food, or pharmaceutical) that provide a dual purpose of (1) trace nutrients and (2) water sources to curb the use of this scarce resource in biorefinery systems. Other promising approaches include the supplementation of nanoparticles (Sanusi et al., 2021) and surfactants (Rorke et al., 2021) that improve both the enzymatic and fermentation processes. Utilization of industrial wastes from the food, agricultural and chemical sectors, in addition to the generation of multiple products, make provisions for: (1) the potential development of suitable LCA that provides significant insight into the environmental sustainability of the processes and (2) additional revenue streams that reduces the technoeconomic barriers of lignocellulosic biorefineries, thus promoting the circular bioeconomy approach. Innovations to speed up the circular bioeconomy encompass a range of different methods. More recently, virtual experimentation using machine learning models have increasingly been documented for several processes such as lignocellulosic pretreatment (Moodley and Gueguim Kana, 2019; Rorke and Gueguim Kana, 2016) and biofuel production (Sewsynker et al., 2015; Sewsynker and Gueguim Kana, 2016; Sewsynker-Sukai and Gueguim Kana, 2017). The convergence of machine learning models and biorefinery processes such as pretreatment, enzymatic hydrolysis, and microbial fermentation pave the way towards transitioning from resource-intensive, laborious, and time-consuming wet-lab experiments toward virtual environments. Artificial intelligence tools have gained popularity for application in biotechnological processes, which may be attributable to their effective encapsulation of biological systems known to be non-linear and dynamic. Artificial intelligence tools can extract functional relationships between the process inputs and corresponding outputs (product yields), thus allowing for the prediction of maximum productivity and reduced inhibitor yields. Experts can then assess the viability of the biorefinery process before scale-up procedures. However, this does not follow a straightforward approach but necessitates using robust artificial intelligence tools and large data sets to develop models that can equally predict defined chemicals and wastes.

References

Abraham, A., Mathew, A.K., Park, H., Choi, O., Sindhu, R., Parameswaran, B., et al., 2020. Pretreatment strategies for enhanced biogas production from lignocellulosic biomass. Bioresour. Technol. 301, 122725.

Agamuthu, P. 2009. Challenges and opportunities in agro-waste management: An Asian perspective. Inaugural Meeting of First Regional 3R forum in Asia 11(12): 4153–4158.

Aguilar-Reynosa, A., Romaní, A., Rodríguez-Jasso, R.M., Aguilar, C.N., Garrote, G., Ruiz, H.A., 2017. Comparison of microwave and conduction-convection heating autohydrolysis pretreatment for bioethanol production, Bioresour. Technol., 243. pp. 273–283.

Akhtar, J., Idris, A., 2017. Oil palm empty fruit bunches, a promising substrate for succinic acid production via simultaneous saccharification and fermentation. Renew. Energy 114, 917–923.

Akhtar, J., Hassan, N., Idris, A., Ngadiman, N.H.A., 2020. Optimization of simultaneous saccharification and fermentation process conditions for the production of succinic acid from oil palm empty fruit bunches. J. Wood Chem. Technol. 40 (2), 136–145.

Angelidaki, I., Karakashev, D., Batstone, D.J., Plugge, C.M., Stams, A.J.M., 2011. Chapter 16 – Biomethanation and its potential. Methods Enzymol. 494, 327–351.

Angelidaki, I., Treu, L., Tsapekos, P., Luo, G., Campanaro, S., Wenzel, H., et al., 2018. Biogas upgrading and utilization: current status and perspectives. Biotechnol. Adv. 36 (2), 452–466.

Anwar, N.A.K.K., Hassan, N., Yusof, N.M., Idris, A., 2021. High-titer bio-succinic acid production from sequential alkalic and metal salt pretreated empty fruit bunch via simultaneous saccharification and fermentation. Ind. Crop. Products 166, 113478.

Aruwajoye, G.S., Faloye, F.D., Gueguim Kana, E.B., 2017. Soaking assisted thermal pretreatment of cassava peels wastes for fermentable sugar production: process modelling and optimization. Energy Convers. Manag. 150, 558–566.

Aruwajoye, G.S., Sewsynker-Sukai, Y., Gueguim Kana, E.B., 2020. Valorisation of cassava peels through simultaneous saccharification and ethanol production: effect of prehydrolysis time, kinetic assessment and preliminary scale up. Fuel 278, 118351.

Azin, M., Moravej, R., Zareh, D., 2007. Production of xylanase by *Trichoderma longibrachiatum* on a mixture of wheat bran and wheat straw: optimization of culture condition by Taguchi method. Enzyme Microb. Technol. 40 (4), 801–805.

Bai, F., Anderson, W., Moo-Young, M., 2008. Ethanol fermentation technologies from sugar and starch feedstocks. Biotechnol. Adv. 26 (1), 89–105.

Balan, V., 2014. Current challenges in commercially producing biofuels from lignocellulosic biomass. Int. Sch. Res. Not. 2014.

Balat, M., Balat, H., Oz, C., 2008. Progress in bioethanol processing. Prog. Energy Combust. Sci. 34, 551–573.

Ballesteros, M., Oliva, J.M., Negro, M.J., Manzanares, P., Ballesteros, I., 2004. Ethanol from lignocellulosic materials by a simultaneous saccharification and fermentation process (SFS) with *Kluyveromyces marxianus* CECT 10875. Process. Biochem. 39 (12), 1843–1848.

Banu, J.R., Kavitha, S., Tyagi, V.K., Gunasekaran, M., Karthikeyan, O.P., Kumar, G., 2021. Lignocellulosic biomass based biorefinery: a successful platform towards circular bioeconomy. Fuel 302, 121086.

Baruah, J., Nath, B.K., Sharma, R., Kumar, S., Deka, R.C., Baruah, D.C., et al., 2018. Recent trends in the pretreatment of lignocellulosic biomass for value-added products. Front. Energy Res. 6, 141.

Behera, S.S., Saranraj, P., Ray, R.C., 2022. Microbial bioethanol fermentation technologies —recent trends and future prospects. In: Gómez Castro, F.I., Guti-Errez-Antonio, C. (Eds.), Biofuels and Biorefining. Elsevier, pp. 75–107.

Ben Taher, I., Fickers, P., Chniti, S., Hassouna, M., 2017. Optimization of enzymatic hydrolysis and fermentation conditions for improved bioethanol production from potato peel residues. Biotechnol. Prog. 33, 397–406.

Bhattacharyya, B.C., Tunga, R., Banerjee, R., 1998. Optimizing some factors affecting protease production under solid state fermentation. Bioprocess. Bioeng. 19, 187–190.

Binod, P., Sindhu, R., Singhania, R.R., Vikram, S., Devi, L., Nagalakshmi, S., 2010. Bioethanol production from rice straw: an overview. Bioresour. Technol. 101 (13), 4767–4774.

Botero, G., Restrepo, S., Cardona, A., 2017. A comprehensive review on the implementation of the biorefinery concept in biodiesel production plants. Biofuel Res. J. 4 (3), 691–703.

Brewer, C.E., Chuang, V.J., Masiello, C.A., Gonnermann, H., Gao, X., Dugan, B., et al., 2014. New approaches to measuring biochar density and porosity. Biomass Bioenergy 66, 176–185.

Camargo, D., Sydney, E.B., Leonel, L.V., Pintro, T.C., Sene, L., 2019. Dilute acid hydrolysis of sweet sorghum bagasse and fermentability of the hemicellulosic hydrolysate. Braz. J. Chem. Eng. 36 (1), 143–156. Available from: https://doi.org/10.1590/0104-6632.20190361s20170643.

Carrillo-Nieves, D., Ruiz, H.A., Aguilar, C.N., Illina, A., Parra-Saldivar, R., Torres, J.A., et al., 2017. Process alternatives for bioethanol production from mango stem bark residues. Bioresour. Technol. 239, 430–436.

Chen, H. (Ed.), 2015. Lignocellulose biorefinery engineering: An overview. Lignocellulose Biorefinery Engineering, pp. 1–17.

Chen, H., Huo, W., Wang, B., Wang, Y., Wen, H., Cai, D., et al., 2019. L-lactic acid production by simultaneous saccharification and fermentation of dilute ethylediamine pretreated rice straw. Ind. Crop Products 141, 111749.

Cheng, C., Lo, Y., Lee, K., Lee, D., Lin, C., Chang, J., 2011. Biohydrogen production from lignocellulosic feedstock. Bioresour. Technol. 102 (18), 8514–8523. Available from: https://doi.org/10.1016/j.biortech.2011.04.059.

Cherubini, F., Jungmeier, G., Wellisch, M., Willke, T., Skiadas, I., Van Ree, R., et al., 2009. Toward a common classification approach for biorefinery systems. Biofuels, Bioprod. Bioref. 3 (5), 534–546.

Chio, C., Sain, M., Qin, W., 2019. Lignin utilization: a review of lignin depolymerization from various aspects. Renew. Sustain. Energy Rev. 107, 232–249.

Chohan, N.A., Aruwajoye, G., Sewsynker-Sukai, Y., Gueguim Kana, E.B., 2020. Valorisation of potato peel wastes for bioethanol production using simultaneous saccharification and fermentation: process optimization and kinetic assessment. Renew. Energy 146, 1031–1040.

Chouaibi, M., Daoued, K.B., Riguane, K., Rouissi, T., Ferrari, G., 2020. Production of bioethanol from pumpkin peel wastes: comparison between response surface methodology (RSM) and artificial neural networks (ANN). Ind. Crop Products 155, 112822.

Cizeikiene, D., Juodeikiene, G., Damasius, J., 2018. Use of wheat straw biomass in production of L-lactic acid applying biocatalysis and combined lactic acid bacteria strains belonging to the genus Lactobacillus. Biocatal. Agric. Biotechnol. 15, 185–191.

Cui, F., Li, Y., Wan, C., 2011. Lactic acid production from corn stover using mixed cultures of *Lactobacillus rhamnosus* and *Lactobacillus brevis*. Bioresour. Technol. 102 (2), 1831–1836.

Dalena, F., Senatore, A., Tursi, A., Basile, A., 2017. Bioenergy production from second- and third-generation feedstocks. In: Dalena, F., Basile, A., Rossi, C. (Eds.), Bioenergy Systems for the Future. Woodhead publishing, Elsevier, United Kingdom, pp. 559–599.

David, A.N., Sewsynker-Sukai, Y., Sithole, B., Kana, E.B., 2020. Development of a green liquor dregs pretreatment for enhanced glucose recovery from corn cobs and kinetic assessment on various bioethanol fermentation types. Fuel 274, 117797.

De Bhowmick, G., Sarmah, A.K., Sen, R., 2018. Lignocellulosic biorefinery as a model for sustainable development of biofuels and value added products. Bioresour. Technol. 247, 1144–1154.

de Mes, T.Z.D., Stams, A.J.M., Reith, J.H., Zeeman, G., 2003. Methane production by anaerobic digestion of wastewater and solid wastes. In: Reith, J.H., Wijffels, R.H., Barten, H. (Eds.), Biomethane and Biohydrogen: Status and Perspectives of Biological Methane and Hydrogen Production. Dutch Biological Hydrogen Foundation, Petten, Netherlands, pp. 58–102.

Fengel, D., Wegener, G., 1989. Wood Chemistry, Ultrastructure, Reactions. Walter de Gruyter, Berlin.

Ferreira, P., da Silveira, F., dos Santos, R., Genier, H., Diniz, R., Ribeiro Jr., J., et al., 2015. Optimizing ethanol production by thermotolerant *Kluyveromyces marxianus* CCT 7735 in a mixture of sugarcane bagasse and ricotta whey. Food Sci. Biotechnol. 24 (4), 1421–1427.

Fu, Y., Hu, X., Xu, Y., Zhu, X., Jiang, S., 2009. Characterization of biodegradation of straw-based biomass-oil in aqueous culture conditions. Ind. Lubrication Tribol. 61, 0036–8792. 5, 277280.

Fu, S.F., Wang, F., Yuan, X.Z., Yang, Z.M., Luo, S.J., Wang, C.S., 2015. The thermophilic (55°C) microaerobic pretreatment of corn straw for anaerobic digestion. Bioresour. Technol. 175, 203–208.

Garrett, P., Ronde, K., 2013. Life cycle assessment of wind power: comprehensive results from a state-of-the-art approach. Int. J. Life Cycle Assess. 18, 37–48.

Ghimire, A., Frunzo, L., Pirozzi, F., Trably, E., Escudie, R., Lens, P.N.L., et al., 2015. A review on dark fermentative biohydrogen production from organic biomass: process parameters and use of by-products. Appl. Energy 144, 73–95.

Gomez, J.A., Höffner, K., Barton, P.I., 2021. Production of biofuels from sunlight and lignocellulosic sugars using microbial consortia. Chem. Eng. Sci. 239, 116615.

Granström, K., Montelius, J., 2014. Pre-treatment to enhance biogas yield from pulp and paper mill sludge. J. Chem. Chem. Eng. 8, 825–833.

Guettler, M.V., Rumler, D., Jain, M.K., 1999. *Actinobacillus succinogenes* sp. nov., a novel succinic-acid-producing strain from the bovine rumen. Int. J. Syst. Bacteriol. 49, 207–216.

Guilherme, A.D.A., Dantas, P.V.F., Padilha, C.E.D.A., Dos Santos, E.S., De Macedo, G.R., 2019. Ethanol production from sugarcane bagasse: use of different fermentation strategies to enhance an environmental-friendly process. J. Environ. Manag. 234, 44–51.

Gupta, R., Gautam, S., Shukla, R., Kuhad, R.C., 2017. Study of charcoal detoxification of acid hydrolysate from corncob and its fermentation to xylitol. J. Environ. Chem. Eng. 5, 4573–4582.

Hallenbeck, P.C., Liu, Y., 2016. Recent advances in hydrogen production by photosynthetic bacteria. Int. J. Hydrog. Energy 41 (7), 4446–4454.

Harmsen, P.F.H., Huijgen, W., Bermudez, L., Bakker, R., 2010. Literature review of physical and chemical pretreatment processes for lignocellulosic biomass. Wagening. UR-Food Biobased Research 1184.

Harrison, S., Johnstone-Robertson, M., Pott, R., Verster, B., Rumjeet, S., Nkadimeng, L., 2016. Towards wastewater biorefineries: integrated bioreactor and process design for combined water treatment and resource productivity. Water Res. Comm. Rep. 2580, 2016.

He, J., Zhang, W., Liu, X., Xu, N., Xiong, P., 2016. Optimization of prehydrolysis time and substrate feeding to improve ethanol production by simultaneous saccharification and fermentation of furfural process residue. J. Biosci. Bioeng. 122 (5), 563–569.

Hendriks, A.T.W.M., Zeeman, G., 2009. Pretreatments to enhance the digestibility of lignocellulosic biomass. Bioresour. Technol. 100 (1), 10–18.

Hijosa-Valsero, M., Paniagua-García, A.I., Díez-Antolínez, R., 2018. Industrial potato peel as a feedstock for biobutanol production. N. Biotechnol. 46, 54–60.

Hilares, R.T., Ienny, J.V., Marcelino, P.F., Ahmed, M.A., Antunes, F.A., da Silva, S.S., et al., 2017. Ethanol production in a simultaneous saccharification and fermentation process with interconnected reactors employing hydrodynamic cavitation-pretreated sugarcane bagasse as raw material. Bioresour. Technol. 243, 652–659.

Höffner, K., Barton, P.I., 2014. Design of microbial consortia for industrial biotechnology. In: Eden, M.R., Siirola, J.D., Towler, G.P. (Eds.), Computer Aided Chemical Engineering, 34. Elsevier, Cambridge, pp. 65–74.

Hong, S.H., Kim, J.S., Lee, S.Y., In, Y.H., Choi, S.S., Rih, J.K., et al., 2004. The genome sequence of the capnophilic rumen bacterium *Mannheimia succiniciproducens*. Nat. Biotechnol. 22, 1275–1281.

Hu, X., Ding, H., Xu, Y., Zhu, X., 2008. Isolation and identification of a strain for the degradation of biomass-oil and its degradation properties in soil. Res. Environ. Sci. 21, 1001–6929. 6 182186.

Huang, J., Liu, C., Tong, H., Li, W., Wu, D., 2012. Theoretical studies on pyrolysis mechanism of xylopyranose. Comput. Theor. Chem. 1001, 44–50.

Huang, M., Li, Z., Luo, N., Yang, R., Wen, J., Huang, B., et al., 2019. Application potential of biochar in environment: insight from degradation of biochar derived DOM and complexation of DOM with heavy metals. Sci. Total Environ. 646, 220–228.

ISO 14040:2006, 2006. Environmental Management—Life Cycle Assessment—Principles and Framework.

Izmirlioglu, G., Demirci, A., 2012. Ethanol production from waste potato mash by using *Saccharomyces cerevisiae*. Appl. Sci. 2, 738–753.

Jensen, K.A., Evans, K.M.C., Kirk, T.K., Hammel, K.E., 1994. Biosynthetic pathway for veratryl alcohol in the ligninolytic fungus *Phanerochaete chrysosporium*. Appl. Environ. Microbiol. 60, 709–714.

Jugwanth, Y., Sewsynker-Sukai, Y., Gueguim Kana, E.B., 2020. Valorization of sugarcane bagasse for bioethanol production through simultaneous saccharification and fermentation: optimization and kinetic studies. Fuel 262, 116552.

Juneja, A., Sharma, N., Cusick, R., Singh, V., 2019. Techno-economic feasibility of phosphorus recovery as a coproduct from corn wet milling plants. Cereal Chem. 96 (2), 380–390.

Kalra, M., Panwar, J., 1986. Anaerobic digestion of rice crop residues. Agric. Wastes 17 (4), 263–269.

Kamali, M., Gameiro, T., Costa, M.E.V., Capela, I., 2016. Anaerobic digestion of pulp and paper mill wastes – an overview of the developments and improvement opportunities. Chem. Eng. J. 298, 162–182.

Kaparaju, P., Serrano, M., Thomsen, A.B., Kongjan, P., Angelidaki, I., 2009. Bioethanol, biohydrogen and biogas production from wheat straw in a biorefinery concept. Bioresour. Technol. 100, 2562–2568.

Karnaouri, A., Asimakopoulou, G., Kalogiannis, K.G., Lappas, A., Topakas, E., 2020. Efficient D-lactic acid production by *Lactobacillus delbrueckii* subsp. bulgaricus through conversion of organosolv pretreated lignocellulosic biomass. Biomass Bioenergy 140, 105672.

Kee, S.H., Chiongson, J.B.V., Saludes, J.P., Vigneswari, S., Ramakrishna, S., Bhubalan, K., 2020. Bioconversion of agro-industry sourced biowaste into biomaterials *via* microbial factories–a viable domain of circular economy. Environ. Pollut. 116311.

Khang, D.S., Tan, R.R., Uy, O.M., Promentilla, M.A.B., Tuan, P.D., Abe, N., et al., 2017. Design of experiments for global sensitivity analysis in life cycle assessment: the case of biodiesel in Vietnam. Resour. Conserv. Recycl. 119, 12–23.

Kim, M.S., Lee, D.Y., 2010. Fermentative hydrogen production from tofu-processing waste and anaerobic digester sludge using microbial consortium. Bioresour. Technol. 101 (1, Supplement), S48–S52.

Kim, T.H., 2013. Pretreatment of lignocellulosic biomass. In: Yang, S.T., El-Enshasy, H.A., Thongchul, N. (Eds.), Bioprocessing Technologies in Biorefinery for Sustainable Production of Fuels, Chemicals, and Polymers. Wiley, New York, pp. 91–109.

Kocabas, A., Ogel, Z.B., Bakir, U., 2014. Xylanase and itaconic acid production by *Aspergillus terreus* NRRL 1960 within a biorefinery concept. Ann. Microbiol. 64 (1), 75–84.

Koppram, R., Nielsen, F., Albers, E., Lambert, A., Wännström, S., Welin, L., et al., 2013. Simultaneous saccharification and co-fermentation for bioethanol production using corncobs at lab, PDU and demo scales. Biotechnol. Biofuels 6, 2.

Koutinas, A.A., Yepez, B., Kopsahelis, N., Freire, D.M.G., de Castro, A.M., Papanikolaou, S., et al., 2016. Techno-economic evaluation of a complete bioprocess for 2,3-butanediolproduction from renewable resources. Bioresour. Technol. 204, 55–64.

Krull, S., Eidt, L., Hevekerl, A., Kuenz, A., Prüße, U., 2017. Itaconic acid production from wheat chaff by *Aspergillus terreus*. Process. Biochem. 63, 169–176.

Kucharska, K., Rybarczyk, P., Hołowacz, I., Łukajtis, R., Glinka, M., Kamiński, M., 2018. Pretreatment of lignocellulosic materials as substrates for fermentation processes. Molecules 23, 2937.

Kuglarz, M., Alvarado-Morales, M., Karakashev, D., Angelidaki, I., 2016. Integrated production of cellulosic bioethanol and succinic acid from industrial hemp in a biorefinery concept. Bioresour. Technol. 200, 639–647.

Kumar, A.A., Jain, K.K., Singh, B., 2020a. Process optimization for chemical pretreatment of rice straw for bioethanol production. Renew. Energy 156, 1233–1243.

Kumar, B., Bhardwaj, N., Agrawal, K., Chaturvedi, V., Verma, P., 2020b. Current perspective on pretreatment technologies using lignocellulosic biomass: an emerging biorefinery concept. Fuel Process. Technol. 199, 106244.

Kumari, D., Singh, R., 2018. Pretreatment of lignocellulosic wastes for biofuel production: a critical review. Renew. Sustain. Energy Rev. 90, 877–891.

Lehtomäki, A., Huttunen, S., Rintala, J.A., 2007. Laboratory investigations on co-digestion of energy crops and crop residues with cow manure for methane production: effect of crop to manure ratio. Resour. Conserv. Recycl. J. 51 (3), 591–609.

Li, J., Wei, L., Duan, Q., Hu, G., Zhang, G., 2014. Semi-continuous anaerobic co-digestion of dairy manure with three crop residues for biogas production. Bioresour. Technol. 156, 307–313.

Lin, Y., Ma, X., Peng, X., Yu, Z., Fang, S., Lin, Y., et al., 2016. Combustion, pyrolysis and char CO_2-gasification characteristics of hydrothermal carbonization solid fuel from municipal solid wastes. Fuel 181, 905–915.

Liu, Z., Wang, K., Chen, Y., Tan, T., Nielsen, J., 2020. Third-generation biorefineries as the means to produce fuels and chemicals from CO_2. Nat. Catal. 3, 274–288.

Lizasoain, J., Rinc_on, M., Theuretzbacher, F., Enguídanos, R., Nielsen, P.J., Potthast, A., et al., 2016. Biogas production from reed biomass: effect of pretreatment using different steam explosion conditions. Biomass Bioenergy 95, 84e91. Available from: https://doi.org/10.1016/j.biombioe.2016.09.021.

Łukajtis, R., Hołowacz, I., Kucharska, K., Glinka, M., Rybarczyk, P., Przyjazny, A., et al., 2018. Hydrogen production from biomass using dark fermentation. Renew. Sustain. Energy Rev. 91, 665–694. Available from: https://doi.org/10.1016/j.rser.2018.04.043.

Maassen, N., Panakova, M., Wierckx, N., Geiser, E., Zimmermann, M., Bölker, M., et al., 2014. Influence of carbon and nitrogen concentration on itaconic acid production by the smut fungus *Ustilago maydis*. Eng. Life Sci. 14, 129–134.

Mafuleka, S., Gueguim Kana, E.B., 2015. Modelling and optimization of xylose and glucose production from Napier grass using hybrid pre-treatment techniques. Biomass Bioenergy 77, 200–208.

Manchala, K.R., Sun, Y., Zhang, D., Wang, Z., 2017. Anaerobic digestion modelling. Adv. Bioenergy 2, 69–141.

Manna, M.C., Rahman, M.M., Naidu, R., Sahu, A., Bhattacharjya, S., Wanjari, R.H., et al., 2018. Bio-waste management in subtropical soils of India: future challenges and opportunities in agriculture. Adv. Agron. 152, 87–148.

Manyi-Loh, C.E., Mamphweli, S.N., Meyer, E.L., Okoh, A.I., Makaka, G., Simon, M., 2013. Microbial anaerobic digestion (bio-digesters) as an approach to the decontamination of animal wastes in pollution control and the generation of renewable energy. Int. J. Environ. Res. Public. Health 10, 4390–4417.

McCay, M.H. and Shafiee, S. 2020. Hydrogen. Future Energy, pp. 475–493.

Mithra, M.G., Padmaja, G., 2016. Compositional profile and ultrastructure of steam and dilute sulfuric acid pretreated root and vegetable processing residues. Curr. Biotechnol. 7, 288–301.

Mithra, M.G., Padmaja, G., 2017. Comparative alterations in the compositional profile of selected root and vegetable peels subjected to three pretreatments for enhanced saccharification. Int. J. Environ. Agricult. Biotechnol. 2 (4), 1732–1744.

Mohan, S.V., Mohanakrishna, G., Srikanth, S., 2011. Biohydrogen production from industrial effluents. Biofuels 499–524.

Mojović, L., Pejin, D., Grujić, O., Markov, S., Pejin, J., Rakin, M., et al., 2009. Progress in the production of bioethanol on starch-based feedstocks. Chem. Ind. Chem. Eng. Q. 15 (4), 211–226.

Molaverdi, M., Karimi, K., Mirmohamadsadeghi, S., 2019. Improvement of dry simultaneous saccharification and fermentation of rice straw to high concentration ethanol by sodium carbonate pretreatment. Energy 167, 654–660.

Moodley, P., 2021. Sustainable biofuels: opportunities and challenges. In: Ray, R.C. (Ed.), Sustainable Biofuels. Academic press, pp. 1–20.

Moodley, P., Gueguim Kana, E.B., 2019. Bioethanol production from sugarcane leaf waste: effect of various optimized pretreatments and fermentation conditions on process kinetics. Biotechnol. Rep. 22, e00329.

Moodley, P., Sewsynker-Sukai, Y., Gueguim Kana, E.B., 2020. Progress in the development of alkali and metal salt catalysed lignocellulosic pretreatment regimes: potential for bioethanol production. Bioresour. Technol. 310, 123372.

Morales-Máximo, M., García, C.A., Pintor-Ibarra, L.F., Alvarado-Flores, J.J., Velázquez-Martí, B., Rutiaga-Quiñones, J.G., 2021. Evaluation and characterisation of timber residues of *Pinus* spp. as an energy resource for the production of solid biofuels in an indigenous community in Mexico. Forests 12, 977. Available from: https://doi.org/10.3390/f12080977.

Moreno-Dávila, I.M.M., Herrera-Ramírez, E.B., Rodríguez-Garza, M.M., Garza-García, Y., Ríos-González, L.J., 2017. Bio-hydrogen production by SSF of paper industry wastes using anaerobic biofilms: a comparison of the use of wastes with/without pretreatment. Int. J. Hydrog. Energy 42 (51), 30267–30273.

Mosier, N., Wyman, C., Dale, B., Elander, R., Lee, Y.Y., Holtzapple, M., et al., 2005. Features of promising technologies for pretreatment of lignocellulosic biomass. Bioresour. Technol. 96 (6), 673–686.

Mu, W., Ben, H., Ragauskas, A., Deng, Y., 2013. Lignin pyrolysis components and upgrading—technology review. Bioenergy Res. 6, 1183–1204.

Mustafa, A.M., Poulsen, T.G., Sheng, K., 2016. Fungal pretreatment of rice straw with *Pleurotus ostreatus* and *Trichoderma reesei* to enhance methane production under solid-state anaerobic digestion. Appl. Energy 180, 661–671. Available from: https://doi.org/10.1016/j.apenergy.2016.07.135.

Nagendran, R. 2011. Agricultural Waste and Pollution. In: Waste: A handbook for management, pp. 341–355.

Nemecek, T., Dubois, D., Huguenin-Elie, O., Gaillard, G., 2011. Life cycle assessment of Swiss farming systems: I. Integrated and organic farming. Agric. Syst. 104, 217–232.

Nierop Groot, M.N., DeBont, J.A.M., 1998. Conversion of phenylalanine to benzadehyde initiated by anaminotransferase in *Lactobacillus plantarum*. Appl. Environ. Microbiol. 64 (8), 3009–3013.

Nizami, A.S., Rehan, M., Waqas, M., Naqvi, M., Ouda, O.K.M., Shahzad, K., et al., 2017. Waste biorefineries: enabling circular economies in developing countries. Bioresour. Technol. 241, 1101–1117.

Norliza, A.W., Ibrahim, C.O., 2005. The production of benzaldehyde by *Rhizopus oligosporus* USM R1 in a solid state fermentation (SSF) system of soy bean meal: rice husks. Malaysian J. Microbiol. 1 (2), 17–24.

Norris, G.A., 2006. Social impacts in product life cycles—towards life cycle attribute assessment. Int. J. Life Cycle Assess. 11, 97–104.

Odjadjare, E.C., Mutanda, T., Olaniran, A.O., 2017. Potential biotechnological application of microalgae: a critical review. Crit. Rev. Biotechnol. 37 (1), 37–52.

Ojewumi, M.E., Akwayo, I.J., Taiwo, O.S., Obanla, O.M., Ayoola, A.A., Ojewumi, E.O., et al., 2018. Bio-conversion of sweet potato peel waste to bioethanol using *Saccharomyces cerevisiae*. Int. J. Pharm. Phytopharmacol. Res. 8, 46–54.

Okamoto, S., Chin, T., Hiratsuka, K., Aso, Y., Tanaka, Y., Takahashi, T., et al., 2014. Production of itaconic acid using metabolically engineered *Escherichia coli*. J. Gen. Appl. Microbiol. 60 (5), 191–197.

Oke, M.A., Annuar, M.S.M., Simaran, K., 2016. Mixed feedstock approach to lignocellulosic ethanol production—prospects and limitations. Bioenergy Res. 9, 1189–1203.

Olatunji, K.O., Ahmed, N.A., Ogunkunle, O., 2021. Optimization of biogas yield from lignocellulosic materials with different pretreatment methods: a review. Biotechnol. Biofuels 14, 159.

Olofsson, K., Bertilsson, M., Lidén, G., 2008. A short review on SSF—an interesting process option for ethanol production from lignocellulosic feedstocks. Biotechnol. Biofuels 1 (7), 1–14.

Ong, H.C., Mahlia, T.M.I., Masjuki, H.H., Honnery, D., 2012. Life cycle cost and sensitivity analysis of palm biodiesel production. Fuel 98, 131–139.

Osman, A.I., Deka, T.J., Baruah, D.C., Rooney, D.W., 2020. Critical challenges in biohydrogen production processes from the organic feedstocks. Biomass Convers. Biorefinery. Available from: https://doi.org/10.1007/s13399-020-00965-x.

Panagiotopoulos, I.A., Bakker, R.R., de Vrije, T., Koukios, E.G., Claassen, P.A.M., 2010. Pretreatment of sweet sorghum bagasse for hydrogen production by *Caldicellulosiruptor saccharolyticus*. Int. J. Hydrog. Energy 35 (15), 7738–7747. Available from: https://doi.org/10.1016/j.ijhydene.2010.05.075.

Panda, S.K., Ray, R.C., Mshra, S.S., Kayitesi1, K., 2018. Microbial processing of fruit and vegetable wastes into potential biocommodities: a review. Crit. Rev. Biotechnol. 38, 1–16.

Pang, P., 2019. Advances in thermochemical conversion of woody biomass to energy, fuels and chemicals. Biotechnol. Adv. 37, 589–597.

Patel, A., Shah, A.R., 2021. Integrated lignocellulosic biorefinery: gateway for production of second generation ethanol and value added products. J. Bioresour. Bioprod. 6, 108–128.

Peters, M.S., Timmerhaus, K.D., 1991. Plant Design and Economics for Chemical Engineers. MacGraw-Hill, New York.

Pinales-Márquez, C.D., Rodríguez-Jasso, R.M., Araújo, R.G., Loredo-Treviño, A., Nabarlatz, D., Gullón, B., et al., 2021. Circular bioeconomy and integrated biorefinery in the production of xylooligosaccharides from lignocellulosic biomass: A review. Ind. Crop Products 162, 113274.

Pooja, N., Padmaja, G., 2015. Enhancing the enzymatic saccharification of agricultural and processing residues of cassava through pretreatment techniques. Waste Biomass Valoriz. 6 (3), 303–315.

Prakasham, R.S., Sathish, T., Brahmaiah, P., Rao, C.S., Rao, R.S., Hobbs, P.J., 2009. Biohydrogen production from renewable agri-waste blend: optimization using mixer design. Int. J. Hydrog. Energy 34, 6143–6148.

Pratto, B., Dos Santos-Rocha, M.S.R., Longati, A.A., Júnior, R.D.S., Cruz, A.J.G., 2020. Experimental optimization and techno-economic analysis of bioethanol production by simultaneous saccharification and fermentation process using sugarcane straw. Bioresour. Technol. 297, 122494.

Ramos, L.P., da Silva, L., Ballem, A.C., Pitarelo, A.P., Chiarello, L.M., Silveira, M.H.L., 2015. Enzymatic hydrolysis of steam-exploded sugarcane bagasse using high total solids and low enzyme loadings. Bioresour. Technol. 175, 195–202.

Raud, M., Kikas, T., Sippula, O., Shurpali, N.J., 2019. Potentials and challenges in lignocellulosic biofuel production technology. Renewable and Sustainable Energy Reviews. p. 111.

Ray, R.C., 2021. Sustainable Biofuels: Challenges and Opportunities. Academic Press, p. 502.

Ray, R.C., Naskar, S.K., 2008. Bio-ethanol production from sweet potato (*Ipomoea batatas* L.) by enzymatic liquefaction and simultaneous saccharification and fermentation. Dyn. Biotechnol. Process. Biochem. Mol. Biol. 2, 47–49.

Ray, R.C., Ramachandran, S., 2018. Bioethanol Production from Food Crops: Sustainable Sources, Interventions and Challenges. Academin Press, p. 460.

Ray, R.C., Sharma, P., Panda, S.H., 2009. Lactic acid production from cassava fibrous residue using *Lactobacillus plantarum* MTCC 1407. J. Environ. Biol. 30, 847–852.

Restianti, Y.Y., Gheewala, S.H., 2012. Life cycle assessment of gasoline in Indonesia. Int. J. Life Cycle Assess. 17, 402–408.

Rezende, C.A., de Lima, M.A., Maziero, P., deAzevedo, E.R., Garcia, W., Polikarpov, I., 2011. Chemical and morphological characterisation of sugarcane bagasse submitted to a delignification process for enhanced enzymatic digestibility. Biotechnol. Biofuels Bioprod. 4, 54. Available from: https://doi.org/10.1186/1754-6834-4-54.

Rochón, E., Cabrera, M.N., Scutari, V., Cagno, M., Guibaud, A., Martínez, S., et al., 2022. Co-production of bioethanol and xylosaccharides from steam-exploded eucalyptus sawdust using high solid loads in enzymatic hydrolysis: effect of alkaline impregnation. Ind. Crop. Products 175, 114253. Available from: https://doi.org/10.1016/j.indcrop.2021.114253.

Rorke, D.C.S., Gueguim Kana, E.B., 2016. Biohydrogen process development on waste sorghum (*Sorghum bicolor*) leaves: optimization of saccharification, hydrogen production and preliminary scale up. Int. J. Hydrog. Energy 41, 12941–12952.

Rorke, D.C.S., Lekha, P., Gueguim Kana, E.B. and Sithole, B.B. 2021. Surfactant-assisted green liquor dregs pretreatment to enhance the digestibility of paper mill sludge. Int. J. Hydrog. Energy.

Saha, B.C., Nichols, N.N., Qureshi, N., Cotta, M.A., 2011. Comparison of separate hydrolysis and fermentation and simultaneous saccharification and fermentation processes for ethanol production from wheat straw by recombinant *Escherichia coli* strain FBR5. Appl. Microbiol. Biotechnol. 92, 865.

Sangkharak, K., Prasertsan, P., 2013. Municipal wastes treatment and production of polyhydroxyalkanoate by modified two-stage batch reactor. J. Polym. Environ. 21 (4), 1009–1015.

Sanusi, A.I., Suinyuy, T.N., Gueguim Kana, E.B., 2021. Impact of nanoparticle inclusion on bioethanol production process kinetic and inhibitor profile. Biotechnol. Rep. 29, e00585.

Sawatdeenarunat, C., Surendra, K.C., Takara, D., Oechsner, H., Khanal, S.K., 2015. Anaerobic digestion of lignocellulosic biomass: challenges and opportunities. Bioresour. Technol. 178, 178–186. Available from: https://doi.org/10.1016/j.biortech.2014.09.103.

Schirmer, W.N., Jucá, J.F.T., Schuler, A.R.P., Holanda, S., Jesus, L.L., 2014. Methane Production in anaerobic digestion of organic waste from recife (Brazil) landfill: evaluation in refuse of different ages. Braz. J. Chem. Eng. 31 (2), 373–384.

Schroyen, M., Vervaeren, H., Vandepitte, H., Van Hulle, S.W., Raes, K., 2015. Effect of enzymatic pretreatment of various lignocellulosic substrates on production of phenolic compounds and biomethane potential. Bioresour. Technol. 192, 696–702. Available from: https://doi.org/10.1016/j.biortech.2015.06.051.

Seider, W.D., Seader, J.D., Lewin, D.R., 1999. Process Design Principles – Synthesis, Analysis, and Evaluation. John Wiley and Sons, New York.

Sewsynker, Y., Gueguim Kana, E.B., 2016. Intelligent models to predict hydrogen yield in dark microbial fermentations using existing knowledge. Int. J. Hydrog. Energy 41, 12929–12940.

Sewsynker, Y., Gueguim Kana, E.B.G., Lateef, A., 2015. Modelling of biohydrogen generation in microbial electrolysis cells (MECs) using a committee of artificial neural networks (ANNs). Biotechnol. Biotechnol. Equip. 29 (6), 1208–1215.

Sewsynker-Sukai, Y., Gueguim Kana, E.B., 2017. Does the volume matter in bioprocess model development? An insight into modelling and optimization of biohydrogen production. Int. J. Hydrog. Energy 42, 5780–5792.

Sewsynker-Sukai, Y., Gueguim Kana, E.B., 2018. Simultaneous saccharification and bioethanol production from corn cobs: process optimization and kinetic studies. Bioresour. Technol. 262, 32–41.

Sewsynker-Sukai, Y., David, A.N., Gueguim Kana, E.B., 2020. Recent developments in the application of kraft pulping alkaline chemicals for lignocellulosic pretreatment: potential beneficiation of green liquor dregs waste. Bioresour. Technol. 306, 123225.

Shamala, T.R., Sreekantiah, K.R., 1986. Production of cellulases and d-xylanase by some selected fungal isolates. Enzyme Microb. Technol. 8 (3), 178–182.

Shamala, T., Vijayendra, S., Joshi, G., 2012. Agro-industrial residues and starch for growth and co-production of polyhydroxyalkanoate copolymer and α-amylase by *Bacillus* sp. CFR-67. Braz. J. Microbiol. 43 (3), 1094–1102.

Shi, X., Liu, Y., Dai, J., Liu, X., Dou, S., Teng, L., et al., 2019. A novel integrated process of high cell-density culture combined with simultaneous saccharification and fermentation for ethanol production. Biomass Bioenergy 121, 115–121.

Singhvi, M.S., Chaudhari, S., Gokhale, D.V., 2014. Lignocellulose processing: a current challenge. RSC Adv. 4 (16), 8271–8277.

Sivamani, S., Chandrasekaran, A.P., Balajii, M., Shanmugaprakash, M., Hosseini-Bandegharaei, A., Baskar, R., 2018. Evaluation of the potential of cassava-based residues for biofuels production. Rev. Environ. Sci. Biotechnol. 17, 553–570.

Sriroth, K., Chollakup, R., Chotineeranat, S., Piyachomkwan, K., Oates, C.G., 2000. Processing of cassava waste for improved biomass utilization. Bioresour. Technol. 71, 63–69.

Sutjahjo, D.H., 2018. The characteristics of bioethanol fuel made of vegetable raw materials. IOP Conf. Ser.: Mater. Sci. Eng. 296 (1): 012019.

Taherzadeh, M.J., Gustafsson, L., Niklasson, C., Lidén, G., 2000. Physiological effects of 5-hydroxymethylfurfural on *Saccharomyces cerevisiae*. Appl. Microbiol. Biotechnol. 53 (6), 701–708.

Thakker, C., San, K.-Y., Bennett, G.N., 2013. Production of succinic acid by engineered *E. coli* strains using soybean carbohydrates as feedstock under aerobic fermentation conditions. Bioresour. Technol. 130, 398–405.

Thomsen, M.H., 2005. Complex media from processing of agricultural crops for microbial fermentation. Appl. Microbiol. Biotechnol. 68 (5), 598–606.

Tufail, T., Saeed, F., Imran, M., Arshad, M.U., Anjum, F.M., Afzaal, M., et al., 2018. Biochemical characterisation of wheat straw cell wall with special reference to bioactive profile. Int. J. Food Prop. 21 (1), 1303–1310. Available from: https://doi.org/10.1080/10942912.2018.1484759.

Usmani, Z., Sharma, M., Awasthi, A.K., Lukk, T., Tuohy, M.G., Gong, L., et al., 2021. Lignocellulosic biorefineries: the current state of challenges and strategies for efficient commercialization. Renew. Sustain. Energy Rev. 148, 111258.

Wagner, A.O., Lackner, N., Mutschlechner, M., Prem, E.M., Markt, R., Illmer, P., 2018. Biological pretreatment strategies for second-generation lignocellulosic resources to enhance biogas production. Energies 11 (7), 1797.

Wahid, R., Nielsen, S.F., Hernandez, V.M., Ward, A.J., Gislum, R., Jørgensen, U., et al., 2015. Methane production potential from *Miscanthus* sp.: effect of harvesting time, genotypes and plant fractions. Biosyst. Eng. 133, 71–80. Available from: https://doi.org/10.1016/j.biosystemseng.2015.03.005.

Welsh, F.W., Murray, W.D., Williams, R.E., 1989. Microbiological and enzymatic production off lavor and fragrance chemicals. Crit. Rev. Biotechnol. 9 (2), 105–169.

Wingren, A., Galbe, M., Zacchi, G., 2003. Techno-economic evaluation of producing ethanol from softwood: comparison of SSF and SHF and identification of bottlenecks. Biotechnol. Prog. 19 (4), 1109–1117.

Woiciechowski, A.L., Neto, C.J.D., Vandenberghe, L.P.D., Neto, D.P.D., Sydney, A.C.N., Letti, L.A.J., et al., 2020. Lignocellulosic biomass: acid and alkaline pretreatments and their effects on biomass recalcitrance –conventional processing and recent advances. Bioresour. Technol. . Available from: https://doi.org/10.1016/j.biortech.122848.

Xu, P., Cheng, S., Han, Y., Zhao, D., Li, H., Wang, Y., et al., 2020. Natural variation of lignocellulosic components in *Miscanthus* biomass in China. Front. Chem. 8, 595143. Available from: https://doi.org/10.3389/fchem.2020.595143.

Yaashikaa, P.R., Senthil Kumar, P., Varjanic, S., Saravanan, A., 2020. A critical review on the biochar production techniques, characterization, stability and applications for circular bioeconomy. Biotechnol. Rep. 28, e00570.

Yadav, M., Paritosh, K., Vivekanand, V., 2020. Lignocellulose to bio-hydrogen: an overview on recent developments. Int. J. Hydrog. Energy 45 (36), 18195–18210. Available from: https://doi.org/10.1016/j.ijhydene.2019.10.027.

Yan, D., Wang, C., Zhou, J., Liu, Y., Yang, M., Xing, J., 2014. Construction of reductive pathway in *Saccharomyces cerevisiae* for effective succinic acid fermentation at low pH value. Bioresour. Technol. 156, 232–239.

Yang, S., Zhang, Y., Yue, W., Wang, W., Wang, Y., Yuan, T., et al., 2016. Valorisation of lignin and cellulose in acid-steam-exploded corn stover by a moderate alkaline ethanol post-treatment based on an integrated biorefinery concept. Biotechnol. Biofuels 9, 238. Available from: https://doi.org/10.1186/s13068-016-0656-1.

Yang, H., Shi, Z., Xu, G., Qin, Y., Deng, J., Yang, J., 2019. Bioethanol production from bamboo with alkali-catalyzed liquid hot water pretreatment. Bioresour. Technol. 274, 261–266.

Yang, J., Xua, H., Jianga, J., Zhang, N., Xie, J., Zhao, J., et al., 2020. Itaconic acid production from undetoxified enzymatic hydrolysate of bamboo residues using *Aspergillus terreus*. Bioresour. Technol. 307, 123208.

Yoo, C.G., Meng, X., Pu, Y., Ragauskas, A.J., 2020. The critical role of lignin in lignocellulosic biomass conversion and recent pretreatment strategies: a comprehensive review. Bioresour. Technol. 301, 122784.

Zabed, H., Sahu, J., Suely, A., Boyce, A., Faruq, G., 2017. Bioethanol production from renewable sources: current perspectives and technological progress. Renew. Sustain. Energy Rev. 71, 475–501.

Zhang, Y., Vadlani, P.V., 2015. Lactic acid production from biomass-derived sugars via co-fermentation of *Lactobacillus brevis* and *Lactobacillus plantarum*. J. Biosci. Bioeng. 119 (6), 694–699.

Zhang, Q., Hu, J., Lee, D., 2016. Biogas from anaerobic digestion processes: research updates. Renew. Energy 98, 108–119.

Zhang, J., Xie, J., Zhang, H., 2021. Sodium hydroxide catalytic ethanol pretreatment and surfactant on the enzymatic saccharification of sugarcane bagasse. Bioresour. Technol. 319, 124171.

Zhao, M., Lu, X., Zong, H., Li, J., Zhuge, B., 2018. Itaconic acid production in microorganisms. Biotechnol. Lett. 40, 455–464.

Zheng, P., Fang, L., Xu, Y., Dong, J.J., Ni, Y., Sun, Z.H., 2010. Succinic acid production from corn stover by simultaneous saccharification and fermentation using *Actinobacillus succinogenes*. Bioresour. Technol. 101 (20), 7889–7894.

Zhu, J.Q., Qin, L., Li, W.C., Zhang, J., Bao, J., Huang, Y.D., et al., 2015. Simultaneous saccharification and co-fermentation of dry diluted acid pretreated corn stover at high dry matter loading: overcoming the inhibitors by non-tolerant yeast. Bioresour. Technol. 198, 39–46.

Zhu, J., Song, W., Chen, X., Sun, S., 2022. Integrated process to produce biohydrogen from wheat straw by enzymatic saccharification and dark fermentation. Int. J. Hydrog. Energy.

Ziaei-Rad, Z., Fooladi, J., Pazouki, M., Gummadi, S.N., 2021. Lignocellulosic biomass pretreatment using low-cost ionic liquid for bioethanol production: an economically viable method for wheat straw fractionation. Biomass Bioenergy 151, 106140.

Zoghlami, A., Paës, G., 2019. Lignocellulosic biomass: understanding recalcitrance and predicting hydrolysis. Front. Chem. 7. Available from: https://doi.org/10.3389/fchem.2019.00874.

Nanotechnology: Applications in biofuel production

14

Abhishek Joshi and Jaya Arora
Laboratory of Biomolecular Technology, Department of Botany, Mohanlal Sukhadia University, Udaipur, Rajasthan, India

14.1 Introduction

The cumulative global energy demand has increased the depletion of fossil fuels more than its creation in nature. It has been estimated that if the use of fossil fuel increase in such ways, all the known fossil fuel reservoirs will be entirely exhausted by the year 2050 (Shields-Menard et al., 2018). Another major concern associated with the extensive combustion of fossil fuels is the environmental damage caused by means of greenhouse gases (GHGs) emissions and global warming (Perera, 2018). In the last couple of decades, biofuels (i.e., biodiesel, bioethanol, biogas, biohydrogen, etc.) have been primarily considered a potential replacement option for fossil fuels because of their low GHGs emission, reduced carbon footprint, and sustainable nature (Joshi et al., 2019; Karaca and Dincer, 2020). Numerous attempts have been made to produce first-generation (1G), second-generation (2G), and third-generation (3G) biofuels using various raw materials like starch, sugar, vegetable oils, animal fats, lignocellulosic and algal biomass (Naik et al., 2010; Arora and Ramawat, 2013; Chowdhury and Loganathan, 2019; Joshi et al., 2020). In recent years, more emphasis has been given to the lignocellulosic biomasses (LCBs) as they are considered the cheapest and are likely the most abundant organic raw feedstock on the planet. It has been estimated that the global production of LCB is near 1.3×10^{10} metric tons per year. The United States of America, the leading biofuel-producing country in the world (Ray and Ramachandran, 2018), produces more than 400 metric tons of dry LCB per year (Valdivia et al., 2016). The high proportions of carbohydrate components (cellulose and hemicelluloses, around 70%) are also advantageous for their use in the production of biofuels, particularly bioethanol (Sharma et al., 2017; Joshi et al., 2018; Wang et al., 2020). Various biochemical, thermochemical, and microbiological methods (Joshi et al., 2022) have been implemented to convert LCBs into biofuels (Fig. 14.1). Most of these were reported to be promising ways, but at the same, they have certain limitations, most notably the cost factor associated with LCB conversion processes (Rodionova et al., 2017; Li et al., 2019).

Nanotechnology is a rapidly emerging field of research with numerous engineering and technological applications that can be used to improve the overall conversion

Advances in Lignocellulosic Biofuel Production Systems. DOI: https://doi.org/10.1016/B978-0-323-91192-4.00008-0

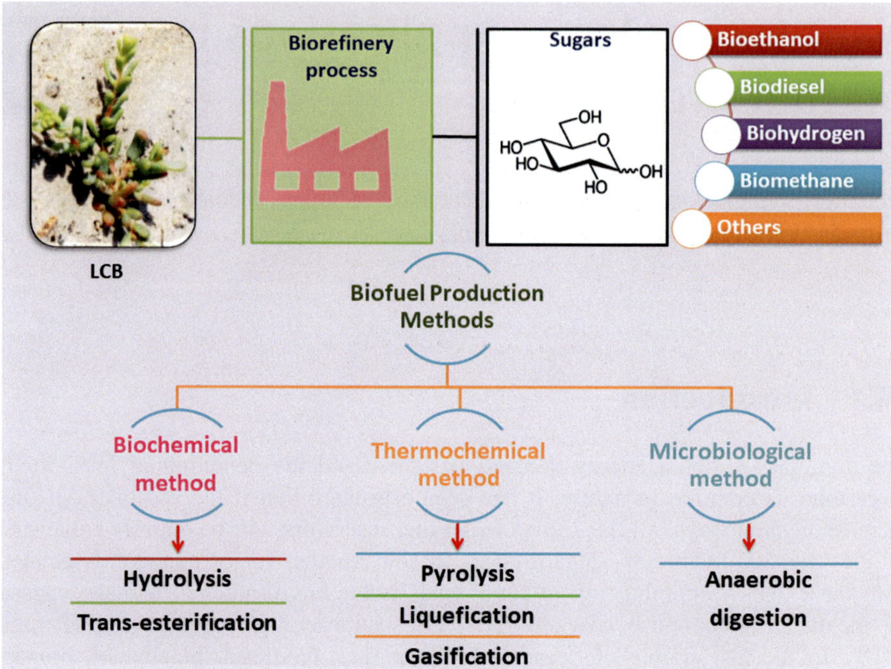

Figure 14.1 Schematic illustration of different methods of lignocellulose conversion to biofuels.

processes of LCB at a relatively low input cost (Singhvi and Kim, 2020). The available reports suggest that nanomaterials used for pretreatment and hydrolysis steps facilitate the molecular chemistry of LCB and release maximum monomeric and oligomeric sugars (Ingle et al., 2019). Immobilization of cellulases, hemicellulase and other catalytic enzymes on nanoscale material increased the rate of hydrolysis and sugar recovery. These immobilized nanocatalysts have been recovered and reused in different hydrolytic reactions, making the production process cost-effective (Salwan et al., 2020; Chamoli et al., 2020). Metal oxide nanocatalysts also play an important role in syngas production as they reduce tar formation in the gasification processes of LCB (Nanda et al., 2016). The increased liquid yield was also accomplished by applying a nanocatalyst of different acids and bases during biomass liquefaction (Khoo et al., 2020; Vasić et al., 2020). Both the nanoscale and subnanoscale instrumentation facilitates the understanding of cell wall ultrastructure and the enzymatic mechanism of LCB, which can be useful for the modification of lignocellulosic fiber at the molecular level (de Oliveira et al., 2017).

Therefore, the present chapter aims to review the application of nanotechnology in various biofuel production routes. The development of advanced technologies dealing with nanomaterial applications, particularly in preprocessing raw biomass and other fabrication stages of biofuel production like hydrolysis, transesterification, and gasification, has been critically reviewed.

14.2 Overview of analytical tools for nanotechnological consequence of biofuel production

Instrumentation or analytical tools are instrumental in understanding the various nanotechnological consequences of biofuel production, starting from transforming lignocellulose fiber to enzyme-nanomaterial interactions and product yield. Electron microscopy is extensively used to evaluate nanoscale materials. It covers techniques that can characterize various metals, polymers, and biological models' morphology, chemical constituents, and atomic structure (Khalil et al., 2006; Rajput et al., 2020). Several electron microscopy-based techniques, namely UV-vis spectroscopy, scanning electron microscopy, Transmission electron microscopy, atomic force microscopy, X-ray diffraction, Raman spectroscopy, time-of-flight secondary ion mass spectrometry, dynamic light scattering, X-ray photoelectron spectroscopy, selected area electron diffraction and Fourier transform infrared spectroscopy are now available to elucidate the microstructure of lignocellulosic fibers and enzymes adsorbed on/in the nanomaterials (Costa-Fernández et al., 2016). These sophisticated analytical tools explain in situ characterization of the structural and operational stability of immobilized enzymes on subsequent supports (Ganesan et al., 2009). Circular dichroism (CD) spectroscopy is also a promising technique for elucidating the impacts of carrier inclusion on the secondary structure of enzymes. Ji et al. (2010) employed CD spectrum analysis of lipase immobilized to carbon nanotubes (CNTs) and revealed that the CNTs immobilized lipase retained 62% of the α-helix structure of the native lipase.

14.3 Nanotechnological applications in bioethanol production

The first ever commercial concept of bioethanol production was brought after 1925 when Henry Ford stated that ethanol would be the "fuel of the future." Bioethanol is a well-recognized substitute for fossil fuels. Most developed and developing countries have enlarged their targets for bioethanol volumes and market shares daily. However, the existing bioethanol production process has certain limitations; nanotechnology could offer promising solutions to achieve cost-effective and process-efficient bioethanol yield. The following are only a very few examples (discussed in Sections 14.3.1–14.3.3) to illustrate the application of various nanotechnological approaches in the area ranging from preprocessing of raw LCB to different reactions used to produce bioethanol such as hydrolysis, fermentation, and detection (Fig. 14.2).

14.3.1 Nanotechnological applications in pretreatment of lignocellulosic biomass

Pretreatment is a crucial step for lignocellulosic biofuel production processes as the conversion of LCB is hindered by several structural and compositional complexities

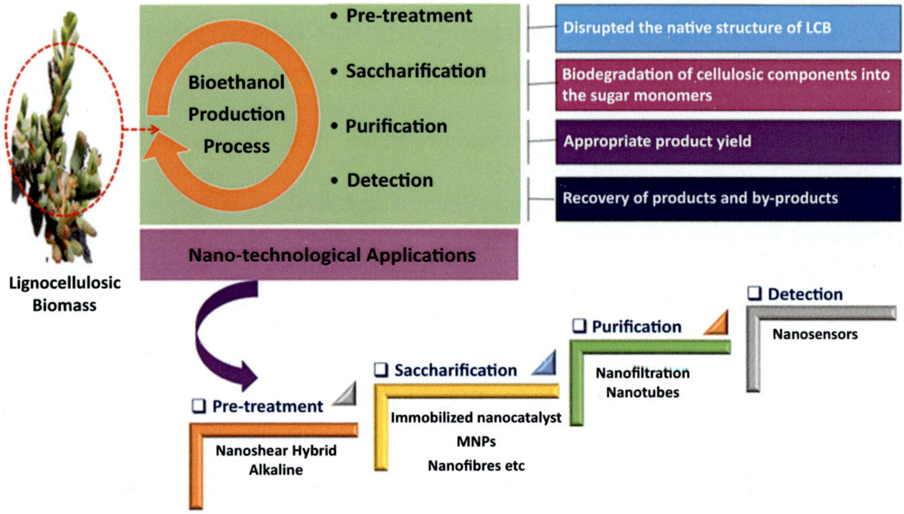

Figure 14.2 Schematic representation of the nanotechnology-based approaches used to produce bioethanol. *MNPs*: magnetic nanoparticles.

(i.e., cellulose crystallinity, degree of polymerization, hemicellulose, lignin sheathing, etc.). A pretreatment process disrupted the native structure of LCB to make it more accessible to hydrolytic enzymes that convert cellulosic components into sugar monomers (Joshi et al., 2021). In the past two decades, intensive research and development work has been carried out in this area. Several pretreatment methods expending diverse physical, chemical, and biological consortiums are available for degrading LCB (Mosier et al., 2005; Sindhu et al., 2016; Hassan et al., 2018). However, considering the concerns like high processing costs and the generation of inhibitor substances associated with these conventional methods, it is clear that there is a need to advance pretreatment procedures with novel, eco-friendly, and economically viable methods (Antunes et al., 2019). In this context, researchers have been involved the applications of diverse nanoparticles (NPs), nanofibers, and nanoscale tools like nanomixer to preprocess raw LCB (Ingle et al., 2019). NP_S are highly efficient in penetrating the cell wall of raw LCB because of their small structure and constructive interaction with biomolecules to release sugar monomers. It has been reported that the addition of NPs effectively changed the recalcitrant structural characteristics of the lignocellulosic matrix and reduced the inhibitory action of lignin. Several studies have been carried out using diverse NPs at the pretreatment stage to enhance the initial conversion of lignocellulosic components to sugar molecules (Rai et al., 2016; Leo and Singh, 2018; Singhvi and Kim, 2020). Earlier, the nanoshear hybrid alkaline (NSHA) technique that contains high-speed shear (within nanomixer) and chemical reagent synergy in the presence of milder thermal application provides broader prospects in the area of lignocellulose pretreatment (Pedersen and Iverseb, 2009; Lee

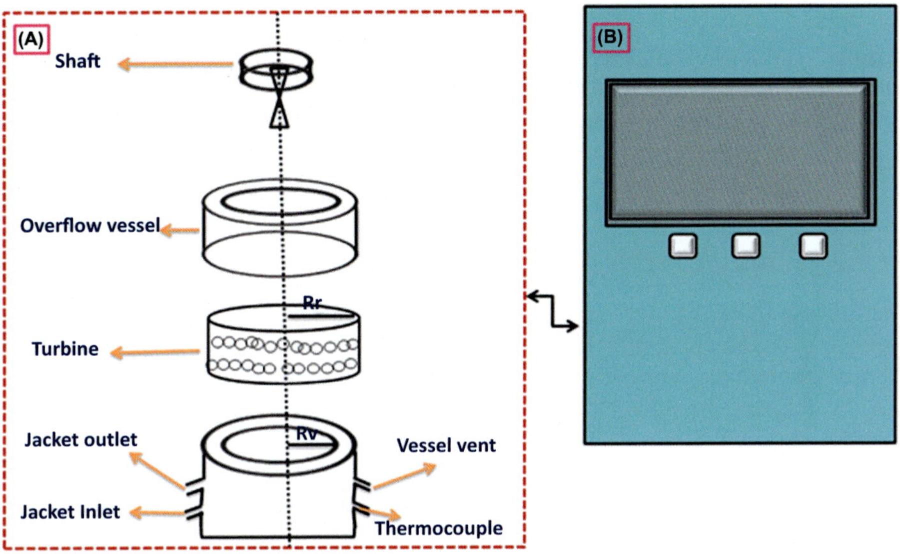

Figure 14.3 Schematic illustration of the Taylor–Couette modified reactor (nanomixer). (A) mixing unit and (B) control panel.

et al., 2012). This approach facilitates the exposition of cellulose and hemicellulose, even eliminating lignin within a very short period. Wang et al. (2013) proposed a novel NSHA pretreatment method for corn stover employing a modified Taylor–Couette reactor (12,500/s shear rate), (Fig. 14.3) with alkali (sodium hydroxide) at room temperature for 2 min. Pretreatment results showed fast and efficient removal of lignin and hemicellulose from raw biomass, leaving up to 82% of cellulose entities in the residual solid. In another study, Ji and Lee (2013) evaluated the impact of cationic polyelectrolyte (poly di allyl dimethyl ammonium chloride, PDAC) on NSHA pretreatment of corn stover employing fast processing conditions (~2 min, room temperature). Indeed, the addition of PDAC altered the cell wall morphology and stabilized lignin. It was also found that the PDAC applications enhance enzymatic hydrolysis yields.

14.3.2 Nanotechnological applications in lignocellulose hydrolysis

The cost of converting LCB to liquid biofuels is incurred in hydrolysis steps. One major reason is the high cost of lignocellulolytic enzymes, as they are produced by a microorganism that substantially adds capital and material expense. Furthermore, their operational properties like optimum temperature, pH, and cross-specificity also affect the process configuration and production cost. It is estimated that the cost contribution of the hydrolytic enzymes would be between $0.10 and $0.73/gal.

Nanotechnology could offer promising solutions to achieve cost-effective and process-efficient hydrolysis yields (sugar monomers). In the last decade, several potential and promising studies have been reported in this area. Many studies highlighted that the efficient hydrolysis of a typical LCB can be achieved mainly through (1) the addition of functionalized NPs, and (2) immobilization of various hydrolytic enzymes on nanomaterials. The applications of both these approaches are briefly discussed below (Sections 14.3.2.1 and 14.3.2.2).

14.3.2.1 Applications of functionalized nanoparticles

The utilization of different functionalized NPs in the hydrolysis step of LCB is a promising nanotechnological approach. In this approach, firstly, the NPs have been functionalized with varying groups like acids, silica, carbon, etc., then added into the reaction mixture along with the magnetic dynamism to obtain efficient hydrolysate yield (Pena et al., 2014a; Qi et al., 2018). Among the functionalization groups mentioned above, acid-functionalized magnetic NPs are the most preferably used to catalyze hydrolysis reactions (Su et al., 2015; Arora et al., 2020). As far as hydrolysis of LCB is concerned, Lai et al. (2011) were the pioneers who proposed the applications of sulfonated silica magnetic NPs as a catalyst in the hydrolysis of LCB. Further, Pena et al. (2012) demonstrated the applications of perfluoroalkyl sulfonic (PFS) and alkyl sulfonic acid-functionalized magnetic NPs for the hydrolysis of wheat straw. The results suggested that PFS and alkyl sulfonic (AS) acid-functionalized NPs showed 10-fold higher hydrolysis of hemicelluloses to oligosaccharides than the control. The applications of various functionalized NPs in the selective conversion of cellulose and hemicellulose into sugars were reviewed and summarized in Table 14.1. Apart from the reports mentioned above, some other studies elaborated on the efficacy of functionalized NPs in the hydrolysis of some commercially available compounds (i.e., cellobiose, starch, sucrose, etc.) which are directly or indirectly ensured in the LCB. Dhepe and Sahu (2010) studied the efficacy of sulfonic acid-modified mesoporous silica to hydrolyze sucrose and starch. The authors recorded promising efficacy of thus used acid-functionalized nanocomposites. Similarly, Peña et al. (2014b) demonstrated the catalytic efficacy of silica-coated NPs functionalized with propylsulfonic (PS) acid groups for cellobiose hydrolysis. The results obtained suggested promising efficacy of thus used functionalized NPs with a high degree of cellobiose conversion (up to 96%). Most recently, Ingle et al. (2020a,b) synthesized three different acid-functionalized magnetic NPs, which were further modified by applying silica coating, namely, alkyl sulfonic acid-functionalized magnetic NPs (Fe_3O_4−MNPs−Si−AS), butylcarboxylic acid-functionalized magnetic NPs (Fe_3O_4-MNPs-Si-BCOOH), and sulfonic acid-functionalized magnetic NPs (Fe_3O_4−MNPs−Si−SO_3H) and evaluate their catalytic efficiency for the cellobiose hydrolysis. The authors recorded promising efficacy of thus used functionalized NPs; moreover, they have also recovered and reused these nanocatalysts with the same efficacy up to the third cycle of hydrolysis.

Table 14.1 Applications of functionalized nanoparticles in the hydrolysis of lignocellulosic biomass.

Lignocellulosic substrate	Used nanomaterial	Summary	Reference
Corncobs	Carbon-based solid (C-SO_3H) acid catalyst using microcrystalline cellulose	Yielded 78.1% xylose with an enzymatic digestibility of up to 91.6% in 48 h	Qi et al. (2018)
Corn stover	Silica- propyl-sulfonic functionalized nanoparticles	90% glucose yield using 0.2 g of NPs	Pena et al. (2014a)
Jatropha hulls	Sulfonated magnetic carbonaceous acid NPs	47.2% and 35.6% yield of total reducing sugars and glucose, respectively	Su et al. (2015)
Plukenetia hulls		54.4% and 35.8% yield of total reducing sugars and glucose, respectively	
Rice husk and straw	Mesoporous silica NPs functionalized with xylanase	up to 35% reducing sugar production	Ariaeenejad et al. (2020)
Rice straw	Magnetic NPs functionalized with cellulase	Yielded high concentration of glucose (15 g/L)	Huang et al. (2015)
Sugarcane bagasse	Sulfonated magnetic carbonaceous acid NPs	79.8% and 58.3% yield of total reducing sugars and glucose, respectively	Su et al. (2015)
	Magnesium oxide NPs functionalized with xylanase (MgN-xyl)	1.82 and 1.91-fold increased reducing sugar and glucose, respectively, at 8°C	Dutta and Saha (2019)
Sugarcane straw	Alkyl sulfonic (AS) acid functionalized magnetic NPs (Fe_3O_4-MNPs-Si-AS), butyl carboxylic acid functionalized magnetic NPs (Fe_3O_4-MNPs-Si-BCOOH)	17.06 and 15.40 g/L sugar yield for Fe_3O_4-MNPs-Si-AS and Fe_3O_4- MNPs-Si-BCOOH, respectively	Ingle et al. (2020a)
Switchgrass	Activated carbon-supported sulfonic acid catalysts	Yielded up to 57.13% glucose	Ansanay et al. (2021)
Wheat straw	Perfluoroalkyl sulfonic (PFS) and AS acid-functionalized silica–cobalt ferrite NPs	Significantly higher amounts of oligosaccharides (46.3% and 45%, respectively) than the control (35.0%)	Pena et al. (2012)

14.3.2.2 Applications of nanomaterial-based enzyme immobilization

As earlier discussed, the major drawbacks associated with LCB hydrolysis are the cost, selectivity, and specificity of the enzymes involved in the process. Therefore researchers have used diverse nanomaterials [i.e., magnetic nanoparticles (MNPs), metal, metal oxide NPs, CNTs, etc.] as a carrier for immobilized hydrolytic enzymes, including cellulases and β-glycosidase (Cherian et al., 2015; Rai et al., 2016). These immobilized enzymes referred to as nanocatalysts, hold numerous advantages over the conventional biocatalyst or free enzymes because of higher surface area, higher stability, and lower toxicity (Rai et al., 2017). In addition, it has been evident that increased surface area and controlled pore size of nanomaterials permit higher enzyme loading and decrease the substrate mass transfer resistance. Also, the balanced chemistry of the surface of nanomaterials potentially helps to increase the efficacy and stability of the immobilized enzymes (Rai et al., 2019; Salwan et al., 2020). Recent developments focus on the use of MNPs for enzyme immobilization. They allow the repeated use of the same biocatalyst for several cycles because of their easy recovery by deploying a simple magnetic field (Huang et al., 2015; Gou et al., 2020; Chamoli et al., 2020). According to the reports summarized in Table 14.2, there is the possibility to use magnetic nanocatalyst for up

Table 14.2 Various nanomaterials used for the immobilization of different enzymes and their binding and catalytic efficacy.

Nanoparticles (NPs) or nanomaterials	Immobilized enzymes and immobilization method	Binding and catalytic efficiency after reuse	Reference
β-cyclodextrin-conjugated magnetic particles	Cellulase; salinization and reductive amidation	90%; 85% after 16 cycles	Huang et al. (2015)
Carbodiimide-activated magnetic NPs	Cellulase; covalent linkage	~90%	Jordan et al. (2011)
Carbodiimide-activated multi-walled carbon nanotubes (MWCNTs)	Cellulase; covalent linkage	85%, 75% after 6 cycles	Ahmad and Khare (2018)
Carboxylic acid functionalized MWCNTs	Cellulase; covalent linkage	98%	Abd and Nour (2012)
Chitosan-based magnetic microspheres	β-Glucosidase; covalent linkage	78.2%; 76.5% after 8 batches (total 256 h)	Zheng et al. (2013)

(*Continued*)

Table 14.2 (Continued)

Nanoparticles (NPs) or nanomaterials	Immobilized enzymes and immobilization method	Binding and catalytic efficiency after reuse	Reference
Chitosan-coated magnetic NPs	Cellulase; covalent linkage	100%; 80% after 15 cycles	Sanchez-Ramirez et al. (2017)
Functionalized MWCNTs	Cellulase; physical adsorption	97%; 52% after 6 cycles	Mubarak et al. (2014)
Gold (Au) NPs	Cellulase; physical adsorption	78% after 6 cycles	Mishra and Sardar (2015)
Magnetic iron NPs	β-Glucosidase; covalent linkage	89.78%; 85% after 10 cycles	Chamoli et al. (2020)
Magnetic iron oxide NPs	β-Glucosidase; covalent linkage	93%; more than 50% after 16 cycles	Verma et al. (2013)
Magnetite NPs	Laccase; covalent linkage	50% after 6 cycles	Gou et al. (2020)
Magneto-responsive grapheme nanosupports	Cellulase; covalent linkage	55%; 55% after 4 cycles	Gokhale et al. (2013)
Manganese dioxide (MnO_2) NPs	Cellulase; covalent linkage	75%	Cherian et al. (2015)
Mesoporous silica NPs	Xylanase; covalent linkage	90%; more than 80% after 9 cycles	Ariaeenejad et al. (2020)
	β-Glucosidase; covalent linkage	100%; 70% after 7 cycles	Sannino et al. 2020
Silver (Ag) NPs	Cellulase; physical adsorption	73% after 6 cycles	Mishra and Sardar (2015)
Titanium dioxide (TiO_2)−lignin hybrid support	Cellulase; physical interactions	80%; 90% after 10 cycles	Zdarta et al. (2017)
Vinyl and amino-functionalized mesoporous silica	Cellulase; physical adsorption	78%−100% and 72%−91% after first cycle	Hartono et al. (2010)
Zinc ferrite NPs	Cellulase; covalent linkage	53%;	Manasa et al. (2017)
	Cellulases; covalent linkage	88%; 50% after 5 cycles	

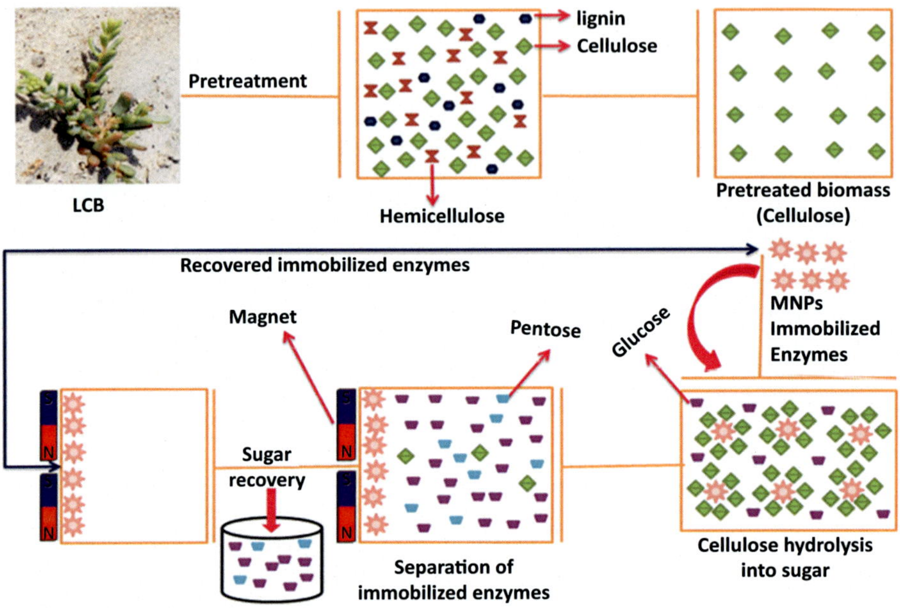

Figure 14.4 Schematic representation of enzyme immobilization on magnetic nanoparticles and their application in hydrolysis of lignocellulosic biomass. *MNPs*: magnetic nanoparticles; *LCB*: lignocellulosic biomass (a typical halophytes; *Suaeda fruticosa*).

to 16 cycles of LCB hydrolysis, and repeated use of highly expensive biocatalyst will reduce the cost and makes the process cost-effective and sustainable. Fig. 14.4 illustrates the schematic representation of enzyme immobilization on MNPs and their application in the hydrolysis of LCB.

14.3.3 Nanotechnological applications in bioethanol fermentation

Fermentation is crucial in the liquid biofuel production routes, especially bioethanol and biobutanol. In this process, the end product of LCB hydrolysis is commonly recognized as C6 (hexose; obtained from cellulose hydrolysis, i.e., glucose) and C5 (pentose; obtained from hemicellulose hydrolysis, i.e., xylose) sugar monomers can be converted into ethanol or butanol with the help of fermenting microorganism (Joshi et al., 2021). During the last couple of years, the application of nanomaterials for this purpose has attracted attention worldwide. Several reports demonstrated that applications of nanomaterials efficiently improve the fermentation yield by influencing either the conversion efficiency of microorganisms or gas–liquid mass transfer rate (Leo and Singh, 2018;

Prabhu et al., 2021; Singhvi et al., 2021). Earlier, Ivanova et al. (2011) reported nearly 91% of theoretical ethanol yield when used yeast cells were entrapped in a matrix of alginate/MNPs and immobilized on magnetite-containing chitosan cellulose-coated MNPs. The self-life of such MNPs was also examined and found to be more than 30 days at 4°C in saline.

Similarly, Kim et al. (2014) reported improvement in syngas fermentation in the presence of six different types of NPs, namely hydroxyl-functionalized single-walled carbon nanotubes (SWCNTs), iron oxide, silica (functionalized with methyl and isopropyl), alumina, and palladium on carbon and alumina. Among these, methyl-functionalized silica NPs displayed maximum bioethanol conversion efficacy (166.1%). Also, nearly 213.5% higher bioethanol production was obtained with the application of methyl-functionalized cobalt–ferrite–silica ($CoFe_2O_4@SiO_2-CH_3$) NPs on syngas fermentation (Kim and Lee, 2016). The reusability of these NPs was also examined and found to be more than five times with 90% fermentation efficiency. In addition, immobilization of metal NPs onto the nanosheet matrix can also help detect compounds during the bioethanol fermentation process (Pan et al., 2007). Lin et al. (2016) synthesized a heterojunction using the ultrathin two-dimensional (2D) polycrystalline ZnO nanosheets with uniform AgNPs coating by hydrothermal method to boost the surface reaction for ethanol detection. Similarly, Santos et al. (2016) developed a glassy carbon electrode using graphene oxide and copper NPs to determine total reducing sugars. As a result, the authors achieved better accuracy regarding the sugars and reusability of the equipped system.

14.4 Nanotechnological applications in biodiesel production

Biodiesel is known to be a promising renewal energy source, and it offers several advantages like less polluting combustion byproducts and a higher energy-to-mass ratio. Usually, it can be produced through transesterification reactions using different catalytic routes such as acid or base-catalyzed systems, enzyme-catalyzed systems, and deploying supercritical conditions (Aransiola et al., 2014; Patel et al., 2016). However, many limitations are associated with these conventional routes, like deactivation of the catalyst, saponification, and low reaction rate (Gebremariam and Marchetti, 2018; Mahlia et al., 2020). Therefore, over the last few years, implementing the nanobiotechnology approach is attracting consideration to increase biodiesel production and its fuel properties (Khoo et al., 2020). To date, several heterogeneous nano-based catalysts (either normal or functionalized), mainly associated with providing more significant surface area, high surface reactivity, reusability, etc., have been synthesized and introduced for the production of biodiesel (Wang et al., 2015;

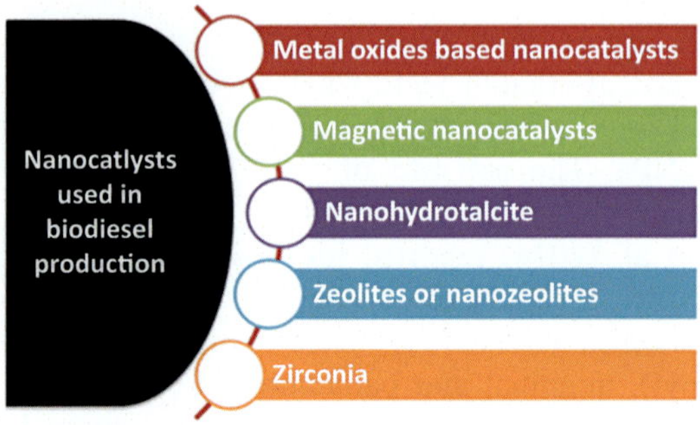

Figure 14.5 Overview of important nanocatalysts commonly used in biodiesel production.

Ingle et al., 2020c). The important heterogeneous nanocatalysts used in biodiesel production are illustrated in Fig. 14.5. Among these, metal-oxide-based nanocatalysts are considered the most promising and are widely studied in biodiesel production from diverse feedstock (Vasić et al., 2020; Ingle et al., 2020c). Based on the available reports employing heterogeneous nanocatalysts in biodiesel production processes, it was concluded that the number of required catalysts decreased up to 40% of the common catalyst. Also, the impacts of moisture and the amount of free fatty acids in feedstock on the bioconversion reactions were slighter than conventional catalytic reactions (Zhang et al., 2013; Ingle et al., 2020d). All the important studies related to the application of nanomaterials in biodiesel production, especially from lignocellulose-based feedstock, are summarized in Table 14.3.

In addition, several nanocatalysts have also been tested as nanoadditives to enhance the distinctive properties of biodiesel, such as viscosity, flash point density, cetane number, and some other combustions attributes. For example, Sajith et al. (2010) studied the influence of the inclusion of cerium oxide NPs on the major physicochemical properties and the performance of biodiesel. The authors reported a significant increase in viscosity and flash point and a decrease in hydrocarbon emission levels and NOx. A significant improvement in calorific value and cetane index has also been reported when diesel–promotion stearin wax biodiesel blends with zinc oxide (ZnO)-based NPs (Karthikeyan et al., 2014). Similarly, Aalam et al. (2015) reported enhanced cetane number and calorific value of iron oxide-based NPs blended diesel fuel. Nanofluids or nanoadditives are also promising for advancing the thermal brake efficiency of diesel engines. They are also helpful in resolving the emission (i.e., carbon monoxide emission, hydrocarbon emission, NO_x emission, etc.), combustion, and evaporation issues associated with biodiesel blends (Khond and Kriplani, 2016; Yaşar et al., 2019; Khoo et al., 2020).

Table 14.3 Applications of nanocatalysts in the production of biodiesel using lignocellulosic substrates.

Name of the substrate	Used nanocatalysts	Optimum operational conditions	Summary	Reference
Cotton seed oil	α-Fe_2O_3/AlOOH (γAl_2O_3)	T (60°C), methanol to oil molar ratio (6:1), catalyst (3 wt.%), the reaction rate (600 rpm), and reaction time (3 h)	Biodiesel yield from 94.1% to 100%, and catalyst shows high recyclability	Mohamed et al. (2020)
	Potassium fluoride (KF) impregnated CaO/NiO	T (60°C), methanol to oil molar ratio (1:15), catalyst (5 wt.%), and reaction time (4 h)	Biodiesel yield up to 99%, and catalytic activity was reduced after 4 cycles	Kaur and Ali (2014)
Jatropha Oil	Calcium oxide (CaO) nanocatalyst	Methanol to the oil molar ratio (5.15:1), catalyst ratio (0.02:1w/w), and reaction time (133.1 min)	98.54% biodiesel yield and catalyst reusability up to the 6 cycles	Anr et al. (2016)
	Nanocrystalline Li impregnated CaO	T (65°C) methanol to oil molar ratio (1:12), catalyst (1.75 wt.%)	Biodiesel yield up to 99%	Kaur and Ali (2011)
Musa balbisiana	Lithium-doped calcium oxide (Li−CaO)	T(65°C), methanol to the oil molar ratio (15:1), catalyst(4 wt.%), and reaction time (150 min)	Optimum biodiesel yield of nearly 97.8%, high catalytic efficiency up to 6 cycles of reuse	Kumar and Gupta (2020)
Palm oil	Titanium dioxide (TiO_2) and iron oxide (Fe_2O_3) NPs		Increased biodiesel yield (%) was found to be for TiO_2 (91.06%) and Fe_2O_3 (84.24%) in comparison to normal catalyst KOH (82.14%)	Prabhahar et al. (2021)
	TiO_2−ZnO	T(60°C), methanol to the oil molar ratio (1:6), catalyst(200 mg), reaction time (5 h)	Conversion efficiency (98%) and biodiesel yield (92%)	Madhuvilakkuand Piraman (2013)
Salicornia persica oil	CaO/Na-ZSM5	Deep eutectic solvents (12.9%), catalyst (14%) and reaction time (3 h)	Achieved 98.1% conversion into biodiesel	ShenavaeiZare et al. (2021)
Safflower oil			Achieved up to 97% conversion into biodiesel	

Apart from the nanocatalyst, a few studies have revealed the significance of CNTs in biodiesel production and their subsequent fuel characteristics. Singh and Bharj (2015) studied the effect of CNTs on fuel characteristics. The authors reported a significant improvement in engine performance, combustion attributes, and a reduction in emissions of toxic substances. Guana et al. (2017) used sulfonated multiwalled CNTs as a solid acid catalyst for biodiesel production. They revealed that the CNTs showed substantial catalytic activity for the transesterification of triglycerides due to apt interval porosity size, high acid, and dispersion sites.

14.5 Nanotechnological applications in biohydrogen production

Biohydrogen is considered the cleanest and most energy-efficient alternative among all the existing biofuels. It has the highest energy content of 142.0 kJ with nonpolluting combustion byproducts in the form of water vapors (Zilouei and Taherdanak, 2015; Taherdanak et al., 2016). Biohydrogen can be produced mainly through four different methods: (1) biophotolysis, (2) photofermentation, (3) dark-fermentation, and (4) use of microbial electrolysis. Moreover, the advantages of all these conventional methods have already been discussed in many research and review articles (Singh et al., 2015; Kumar et al., 2017; Bhatia et al., 2021). However, the major drawbacks associated with biohydrogen production are the slow production rate and low yields. It was observed that only 30%–35% of the substrate is converted into biohydrogen, and the residual part of the substrate (60%–65%) produces other bioproducts (Ren et al., 2016; Soares et al., 2020). Nowadays, the implementation of nanomaterials is attracting consideration in the area of biohydrogen production and bioproduct recovery. Hydrogenase is one of the crucial enzymes in biohydrogen production processes, especially dark fermentation. Fe and Ni-based NPs are mainly used for biohydrogen production as they have been proven to be the major cofactor for hydrogenase (Mishra et al., 2018; Yang and Wang, 2019; Sun et al., 2020). Besides these NPs, several other NPs such as Cu, Pd, Fe_2O_3, Ag, Co, Au, etc., have been used for improved biohydrogen production rate and yield under dark fermentation conditions (Beckers et al., 2013; Zhao et al., 2013; Mohanraj et al., 2016; Tawfik et al., 2019; Zhong et al., 2020). Table 14.4 summarizes the applications of various NPs/nanomaterials to produce biohydrogen from diverse lignocellulosic substrates.

Apart from these studies, lignocellulosic substrates based on nanocomposites have been reported to enhance biohydrogen generation and adsorption. For example, Schaefer et al. (2017) used rice straw-based activated carbons doped with SiC and noted increased hydrogen adsorption from 0.55 to 0.65 wt.% at 10 MPa and 298K. Similarly, Ding et al. (2018) developed a novel wheat straw cellulose (WSC)-based hydrogel and then applied it as a template for in situ preparation of Ni and Cu NPs for further utilization as a catalyst in biohydrogen generation. The authors reported an increased rate of biohydrogen generation when applying

Table 14.4 Applications of nanoparticles or nanomaterials in the production of biohydrogen using lignocellulosic substrates.

Name of the substrate	microorganism/ microbial enzymes/ inoculum	Used nanoparticles (NPs) or nanomaterials	Summary	Reference
Corn stover	Anaerobic sludge	Nickel ion and Ni^0 NPs	Additions of Ni^0 NPs at low levels (below 10 mg/L) were found to increase hydrogen production (40%) appreciably	Sun et al. (2020)
Cotton stalk	*Klebsiella* sp.	Zero-valent iron (Fe^0) and zero-valent nickel (Ni^0) NPs	Hydrogen yield enhanced higher than 20% upon optimum Fe^0 and Ni^0NPs (50 nm at a concentration of 30 mg/L) addition	Li et al. (2020)
Lignocellulosic sugars (glucose)	Anaerobic sludge	Nickel-doped magnetic carbon material	Hydrogen production (260 mL/g glucose) with an increase of up to 46.9%	Zhao et al. (2020)
	Enterobacter cloacae	Fe NPs (100 mg/L)	Hydrogen production (1.9 mol H_2/mol glucose) with an increase of up to 100%	Nath et al. (2015)
	Anaerobic sludge	Magnetite (Fe_3O_4) NPs (50 mg/L)	Maximum hydrogen yield of 12.97 mL H_2/g-volatile solids was obtained in batch experiments	Zhong et al. (2020)
Lignocellulosic grass	*Clostridium* sp.	Biochar and Fe^0 NPs	Hydrogen yield reached 50.6 mL/g-dry grass; it was 89.8% more than that for the control group	Yang and Wang (2019)
	Enterobacter and *Clostridium* sp.	Fe^0 NPs	The maximal hydrogen yield was 64.7 mL/g-dry grass and at Fe^0 NPs dosage of 400 mg/L, which was 73.1% higher compared with the control group	Yang and Wang (2018)
Palm oil mill effluent	*Bacillus anthraci*	Nickel oxide (NiO) and cobalt oxide (CoO) NPs	Hydrogen production yield is 151% (for NiO) and 167% (for CoO) higher than the control	Mishra et al. (2018)
Rice straw	*Bacillus* and *Clostridium*	Magnetic NPs	The addition of MNPs significantly improved the hydrogen production and yield, i.e., maximum H2 yield was 21.4 ± 3.2 and 60.6 ± 5.7 mL/g	Tawfik et al. (2019)

(*Continued*)

Table 14.4 (Continued)

Name of the substrate	microorganism/ microbial enzymes/ inoculum	Used nanoparticles (NPs) or nanomaterials	Summary	Reference
Sugarcane bagasse	Anaerobic sludge	Magnetite NPs	200 mg/L concentration of Magnetite NPs enhanced hydrogen yield up to 70%	Reddy et al. (2017)
		Nanotitanium dioxide (nanoTiO_2)	Maximum hydrogen production (101.5 mL/g volatile solids) was obtained at 1 g nanoTiO_2/L followed by UV irradiation (120 min) and acid hydrolysis (30 min)	Jafari and Zilouei (2016)
Sweet sorghum stove	*Trichoderma asperellum*	Chitosan coated Fe_3O_4-SiO_2-NPs.	Biohydrogen yield and production rates were 2.8 mol H_2/mol reducing sugar and 25 L H_2/L-day, respectively, in a continuous stirred tank reactor	Shanmugam et al. (2020)

WSC-based hydrogel-Ni (Cu) composites as catalysts. Additionally, these catalysts can be used up to 5 times with 100% conversion and 77.5% activity.

14.6 Conclusion and future perspective

The review discussed above demonstrated that the nanotechnological approach has a significant role in biofuel production routes starting from preprocessing raw LCB to other fabrication stages like hydrolysis, fermentation, transesterification, gasification, and detection. Nanomaterials or NPs can penetrate the cell wall of LCB and display virtuous constructive interaction with biomolecules that ensue in the LCB to release sugar monomers. Moreover, the implications of nanocatalysts on MNPs can largely satisfy the recovery and reusability of the immobilized enzymes and make the biofuel production process cost-effective and sustainable. Metal oxide nanocatalysts also play an important role in syngas production, while heterogeneous nano-based catalysts were successfully introduced to enhance biodiesel production and their subsequent fuel properties. However, significant progress has been made in this area still there is a prime necessity for further extensive studies on the following issues:

1. The thorough mechanism of nanomaterials or NPs degrading LCB structures.
2. Development of advanced nanosensors to detect the intermediate biomolecules and end products.
3. Synthesis of more versatile nanocatalysts.
4. Toxicity of NPs in fermentation microorganisms during the production of bio alcohols.
5. Effects of NPs on enzyme functionalities.

Abbreviations

AS alkylsulfonic
CNT carbon nanotubes
GHGs greenhouse gases
LCB lignocellulosic biomass
MNPs magnetic nanoparticles
NPs nanoparticles
PFS perfluoroalkylsulfonic

References

Aalam, S.C., Saravanan, C.G., PremAnand, B., 2015. Reduction of emissions from CRDI diesel engine using metal oxide nanoparticles blended diesel fuel. Int. J. Appl. Eng. Res. 10, 3865–3869.

Abd, R.M., Nour, A.H., 2012. Optimization of immobilized cellulase onto carbon nanotubes using response surface methodology. Int. J. Phys. Sci. 7, 841–849.

Ahmad, R., Khare, S.K., 2018. Immobilization of *Aspergillus niger* cellulase on multiwall carbon nanotubes for cellulose hydrolysis. Bioresour. Technol. 252, 72–75.

Anr, R., Saleh, A.A., Islam, M.S., Hamdan, S., Maleque, M.A., 2016. Biodiesel production from crude Jatropha oil using a highly active heterogeneous nanocatalyst by optimizing transesterification reaction parameters. Energy Fuels 30, 334–343.

Ansanay, Y., Kolar, P., Sharma-Shivappa, R., Cheng, J., Arellano, C., 2021. Pretreatment of switchgrass for production of glucose via sulfonic acid-impregnated activated carbon. Processes 9, 504.

Antunes, F.A.F., Chandel, A.K., Terán-Hilares, R., Ingle, A.P., Rai, M., Dos Santos Milessi, T.S., et al., 2019. Overcoming challenges in lignocellulosic biomass pretreatment for second-generation (2G) sugar production: emerging role of nano, biotechnological and promising approaches. 3 Biotech 9, 1–17.

Aransiola, E.F., Ojumu, T.V., Oyekola, O.O., Madzimbamuto, T.F., Ikhu-Omoregbe, D.I.O., 2014. A review of current technology for biodiesel production: state of the art. Biomass Bioenergy 61, 276–297.

Ariaeenejad, S., Jokar, F., Hadian, P., Ma'mani, L., Gharaghani, S., Fereidoonnezhad, M., et al., 2020. An efficient nano-biocatalyst for lignocellulosic biomass hydrolysis: xylanase immobilization on organically modified biogenic mesoporous silica nanoparticles. Int. J. Biol. Macromol. 164, 3462–3473.

Arora, A., Nandal, P., Singh, J., Verma, M.L., 2020. Nanobiotechnological advancements in lignocellulosic biomass pretreatment. Mater. Sci. Energy Technol. 3, 308–318.

Arora, J., Ramawat, K.G., 2013. Bioenergy resources of the Thar Desert. Biofuels 4, 617–633.

Beckers, L., Hiligsmann, S., Lambert, S.D., Heinrichs, B., Thonart, P., 2013. Improving effect of metal and oxide nanoparticles encapsulated in porous silica on fermentative biohydrogen production by *Clostridium butyricum*. Bioresour. Technol. 133, 109–117.

Bhatia, S.K., Jagtap, S.S., Bedekar, A.A., Bhatia, R.K., Rajendran, K., Pugazhendhi, A., et al., 2021. Renewable biohydrogen production from lignocellulosic biomass using fermentation and integration of systems with other energy generation technologies. Sci. Total Environ. 765, 144429.

Chamoli, S., Yadav, E., Saini, J.K., Verma, A.K., Navani, N.K., Kumar, P., 2020. Magnetically recyclable catalytic nanoparticles grafted with Bacillus subtilis β-glucosidase for efficient cellobiose hydrolysis. Int. J. Biol. Macromol. 164, 1729–1736.

Cherian, E., Dharmendirakumar, M., Baskar, G., 2015. Immobilization of cellulase onto MnO_2 nanoparticles for bioethanol production by enhanced hydrolysis of agricultural waste. Chin. J. Catal. 36, 1223–1229.

Chowdhury, H., Loganathan, B., 2019. Third-generation biofuels from microalgae: a review. Curr. Opin. Green. Sustain. Chem. 20, 39–44.

Costa-Fernández, J.M., Menéndez-Miranda, M., Bouzas-Ramos, D., Encinar, J.R., Sanz-Medel, A., 2016. Mass spectrometry for the characterization and quantification of engineered inorganic nanoparticles. Tr. Ana Chem. 84, 139–148.

de Oliveira, J.A.R., da Silva Martins, L.H., Komesu, A., Neto, J.M., 2017. Nanotechnology applications on lignocellulosic biomass pretreatment. In: Rai, M., da Silva, S. (Eds.), Nanotechnology for Bioenergy and Biofuel Production. Springer, Cham, pp. 19–37.

Dhepe, P., Sahu, R., 2010. A solid-acid-based process for the conversion of hemicellulose. Green. Chem. 12, 2153–2156.

Ding, J., Li, Q., Su, Y., Yue, Q., Gao, B., Zhou, W., 2018. Preparation and catalytic activity of wheat straw cellulose-based hydrogel-nanometal composites for hydrogen generation from $NaBH_4$ hydrolysis. Int. J. Hydrog. Energy 43, 9978–9987.

Dutta, N., Saha, M.K., 2019. Nanoparticle-induced enzyme pretreatment method for increased glucose production from lignocellulosic biomass under cold conditions. J. Sci. Food Agric. 99, 767–780.

Ganesan, A., Moore, B.D., Kelly, S.M., Price, N.C., Rolinski, O.J., Birch, D.J.S., et al., 2009. Optical spectroscopic methods for probing the conformational stability of immobilised enzymes. ChemPhysChem 10, 1492–1499.

Gebremariam, S.N., Marchetti, J.M., 2018. Economics of biodiesel production. Energy Convers. Manag. 168, 74–84.

Gokhale, A.A., Lu, J., Lee, I., 2013. Immobilization of cellulase on magnetoresponsive graphene nano-supports. J. Mol. Catal. B Enzym. 90, 76–86.

Gou, Z., Ma, N.L., Zhang, W., Lei, Z., Su, Y., Sun, C., et al., 2020. Innovative hydrolysis of corn stover biowaste by modified magnetite laccase immobilized nanoparticles. Environ. Res. 188, 109829.

Guana, Q., Lia, Y., Chenb, Y., Shia, Y., Gua, J., Lia, B., et al., 2017. Sulfonated multi-walled carbon nanotubes for biodiesel production through triglycerides transesterification. R.S.C. Adv. 7, 7250–7258.

Hartono, S.B., Qiao, S.Z., Liu, J., Jack, K., Ladewig, B.P., Hao, Z., et al., 2010. Functionalized mesoporous silica with very large pores for cellulase immobilization. J. Phys. Chem. C 114, 8353–8362.

Hassan, S.S., Williams, G.A., Jaiswal, A.K., 2018. Emerging technologies for the pretreatment of lignocellulosic biomass. Biores. Technol. 262, 310–318.

Huang, P.J., Chang, K.L., Hsieh, J.F., Chen, S.T., 2015. Catalysis of rice straw hydrolysis by the combination of immobilized cellulase from *Aspergillus niger* on β-cyclodextrin-Fe_3O_4 nanoparticles and ionic liquid. BioMed. Res. Int. 409103.

Ingle, A.P., Chandel, A.K., Antunes, F.A., Rai, M., da Silva, S.S., 2019. New trends in application of nanotechnology for the pretreatment of lignocellulosic biomass. Biofuels, Bioprod. Bioref. 13, 776–788.

Ingle, A.P., Chandel, A.K., Philippini, R., Martiniano, S.E., da Silva, S.S., 2020d. Advances in nanocatalysts mediated biodiesel production: a critical appraisal. Symmetry 12, 256.

Ingle, A.P., Philippini, R.R., de Souza Melo, Y.C., da Silva, S.S., 2020a. Acid-functionalized magnetic nanocatalysts mediated pretreatment of sugarcane straw: an eco-friendly and cost-effective approach. Cellulose 27, 7067–7078.

Ingle, A.P., Philippini, R.R., Rai, M., Silvério da Silva, S., 2020b. Catalytic hydrolysis of cellobiose using different acid-functionalised Fe_3O_4 magnetic nanoparticles. IET Nanobiotechnol. 14, 40–46.

Ingle, A.P., Philippini, R., Martiniano, S.E., da Silva, S.S., Chandel, A.K., 2020c. Application of metal oxide nanostructures as heterogeneous catalysts for biodiesel production. Advanced heterogeneous catalysts volume 1. Appl. Nano-Scale 261–289.

Ivanova, V., Petrova, P., Hristov, J., 2011. Application in the ethanol fermentation of immobilized yeast cells in matrix of alginate/magnetic nanoparticles, on chitosan-magnetite microparticles and cellulose-coated magnetic nanoparticles. Int. Rev. Chem. Eng. 3 (2), 289–299.

Jafari, O., Zilouei, H., 2016. Enhanced biohydrogen and subsequent biomethane production from sugarcane bagasse using nano-titanium dioxide pretreatment. Bioresour. Technol. 670–678.

Ji, P., Tan, H., Xu, X., Feng, W., 2010. Lipase covalently attached to multi-walled carbon nanotubes as an efficient catalyst in organic solvent. AIChE J. 56, 3005–3011.

Ji, S., Lee, I., 2013. Impact of cationic polyelectrolyte on the nanoshear hybrid alkaline pretreatment of corn stover: morphology and saccharification study. Bioresour. Technol. 133, 45–50.

Jordan, J., Kumar, C.S., Theegala, C., 2011. Preparation and characterization of cellulase-bound magnetite nanoparticles. J. Mol. Catal. B Enzym. 68 (2), 139–146.

Joshi, A., Kanthaliya, B., Arora, J., 2018. Halophytes of Thar desert: potential source of nutrition and feedstuff. Int. J. Bioassays 8, 5674–5683.

Joshi, A., Kanthaliya, B., Arora, J., 2019. Current scenario of potential renewable energy sources for sustainable development in India. J. Plant. Sci. Res. 35, 205–214.

Joshi, A., Kanthaliya, B., Arora, J., 2020. Halophytes: the nonconventional crops as source of biofuel production. In: Grigore, M.N. (Ed.), Handbook of Halophytes: From Molecules to Ecosystems Towards Biosaline Agriculture. Springer Nature, Switzerland AG, pp. 1–28. Available from: https://doi.org/10.1007/978-3-030-17854-3_126-1.

Joshi, A., Kanthaliya, B., Meena, S., Khan, F., Arora, J., 2021. Process consolidation approaches for cellulosic ethanol production. In: Ray, R.C. (Ed.), Sustainable Biofuels. Academic Press, pp. 43–72.

Joshi, A., Verma, K.K., D Rajput, V., Minkina, T., Arora, J., 2022. Recent advances in metabolic engineering of microorganisms for advancing lignocellulose-derived biofuels. Bioengineered 13 (4), 8135–8163. Taylor and Francis.

Karaca, A.E., Dincer, I., 2020. A new integrated solar energy based system for residential houses. Energy Convers. Manag. 221, 113112.

Karthikeyan, S., Elango, A., Prathima, A., 2014. Performance and emission study on zinc oxide nanoparticles addition with pomolion stearin wax biodiesel of CI engine. J. Sci. Ind. Res. 73, 187–190.

Kaur, M., Ali, A., 2011. Lithium ion impregnated calcium oxide as nano catalyst for the biodiesel production from Karanja and Jatropha oils. Renew. Energy 36, 2866–2871.

Kaur, M., Ali, A., 2014. Potassium fluoride impregnated CaO/NiO: an efficient heterogeneous catalyst for transesterification of waste cottonseed oil. Eur. J. Lipid Sci. Tech. 116 (1), 80–88.

Khalil, H.S.A., Alwani, M.S., Omar, A.K.M., 2006. Chemical composition, anatomy, lignin distribution, and cell wall structure of Malaysian plant waste fibers. BioResources 1, 220–232.

Khond, V.W., Kriplani, V.M., 2016. Effect of nanofluid additives on performances and emissions of emulsified diesel and biodiesel fueled stationary CI engine: a comprehensive review. Renew. Sustain. Energy Rev. 59, 1338–1348.

Khoo, K.S., Chia, W.Y., Tang, D.Y.Y., Show, P.L., Chew, K.W., Chen, W.H., 2020. Nanomaterial's utilization in biomass for biofuel and bioenergy production. Energies 13, 892.

Kim, Y.K., Lee, H., 2016. Use of magnetic nanoparticles to enhance bioethanol production in syngas fermentation. Bioresour. Technol. 204, 139–144.

Kim, Y.K., Park, S.E., Lee, H., Yun, J.Y., 2014. Enhancement of bioethanol production in syngas fermentation with *Clostridium ljungdahlii* using nanoparticles. Bioresour. Technol. 159, 446–450.

Kumar, G., Sivagurunathan, P., Sen, B., Mudhoo, A., Davila-Vazquez, G., Wang, G., et al., 2017. Research and development perspectives of lignocellulose-based biohydrogen production. Int. Biodeterior. Biodegradation 119, 225–238.

Kumar, U., Gupta, P., 2020. Modeling and optimization of novel biodiesel production from non-edible oil with *Musa balbisiana* root using hybrid response surface methodology along with African buffalo optimization. React. Kinet. Mech. Catal. 130, 875–901.

Lai, D.M., Deng, L., Guo, Q.X., Fu, Y., 2011. Hydrolysis of biomass by magnetic solid acid. Energy Environm Sci. 4, 3552–3557.

Lee, I., Wang, W., Ji, S., 2012. Device and Method for Pretreatment of Biomass. US Patent No. US20120036765 A1.

Leo, V.V., Singh, B.P., 2018. Prospectus of nanotechnology in bioethanol productions. In: Srivastava, N., Srivastava, M., Pandey, H., Mishra, P., Ramteke, P. (Eds.), Green Nanotechnology for Biofuel Production. Springer, Cham, pp. 129–139.

Li, P., Sakuragi, K., Makino, H., 2019. Extraction techniques in sustainable biofuel production: a concise review. Fuel Process. Technol. 193, 295–303.

Li, Y., Zhu, Q., Ding, P., You, S., Zhang, Q., Jiang, H., 2020. Effects of Fe0 and Ni0 nanoparticles on hydrogen production from cotton stalk hydrolysate using Klebsiella sp. WL1316: evaluation of size and concentration of the nanoparticles. Int. J. Hydrog. Energy 45, 6243–6253.

Lin, L., Liu, T., Zhang, Y., Liang, X., Sun, R., Zeng, W., et al., 2016. Enhancing ethanol detection by heterostructural silver nanoparticles decorated polycrystalline zinc oxide nanosheets. Ceram. Int. 42, 3138–3144.

Madhuvilakku, R., Piraman, S., 2013. Biodiesel synthesis by TiO_2–ZnO mixed oxide nanocatalyst catalyzed palm oil transesterification process. Bioresour. Technol. 150, 55–59.

Mahlia, T.M.I., Syazmi, Z.A.H.S., Mofijur, M., Abas, A.P., Bilad, M.R., Ong, H.C., et al., 2020. Patent landscape review on biodiesel production: technology updates. Renew. Sustain. Energy Rev. 118, 109526.

Manasa, P., Saroj, P., Korrapati, N., 2017. Immobilization of cellulase enzyme on zinc ferrite nanoparticles in increasing enzymatic hydrolysis on ultrasound-assisted alkaline pretreated *Crotalaria juncea* biomass. Indian J. Sci. Technol. 10, 1–7.

Mishra, A., Sardar, M., 2015. Cellulase assisted synthesis of nano-silver and gold: application as immobilization matrix for biocatalysis. Int. J. Biol. Macromol. 77, 105–113.

Mishra, P., Thakur, S., Mahapatra, D.M., Ab Wahid, Z., Liu, H., Singh, L., 2018. Impacts of nano-metal oxides on hydrogen production in anaerobic digestion of palm oil mill effluent—a novel approach. Int. J. Hydrog. Energy 43, 2666–2676.

Mohamed, M.M., Bayoumy, W.A., El-Faramawy, H., El-Dogdog, W., Mohamed, A.A., 2020. A novel α-Fe2O3/AlOOH (γ-Al2O3) nanocatalyst for efficient biodiesel production from waste oil: kinetic and thermal studies. Renew. Energy 160, 450–464.

Mohanraj, S., Anbalagan, K., Rajaguru, P., Pugalenthi, V., 2016. Effects of phytogenic copper nanoparticles on fermentative hydrogen production by *Enterobacter cloacae* and *Clostridium acetobutylicum*. Int. J. Hydrog. Energy. 41, 10639–10645.

Mosier, N., Wyman, C., Dale, B., Elander, R., Lee, Y.Y., Holtzapple, M., et al., 2005. Features of promising technologies for pretreatment of lignocellulosic biomass. Bioresour. Technol. 96, 673–686.

Mubarak, N.M., Wong, J.R., Tan, K.W., Sahu, J.N., Abdullah, E.C., Jayakumar, N.S., et al., 2014. Immobilization of cellulase enzyme on functionalized multiwall carbon nanotubes. J. Mol. Catal. B Enzym. 107, 124–131.

Naik, S.N., Goud, V.V., Rout, P.K., Dalai, A.K., 2010. Production of first and second generation biofuels: a comprehensive review. Renew. Sustain. Energy Rev. 14, 578–597.

Nanda, S., Reddy, S.N., Dalai, A.K., Kozinski, J.A., 2016. Subcritical and supercritical water gasification of lignocellulosic biomass impregnated with nickel nanocatalyst for hydrogen production. Int. J. Hydrog. Energy 41, 4907–4921.

Nath, D., Manhar, A.K., Gupta, K., Saikia, D., Das, S.K., Mandal, M., 2015. Phytosynthesized iron nanoparticles: effects on fermentative hydrogen production by *Enterobacter cloacae* DH-89. Bull. Mater. Sci. 38, 1533–1538.

Pan, X., Fan, Z., Chen, W., Ding, Y., Luo, H., Bao, X., 2007. Enhanced ethanol production inside carbon-nanotube reactor containing catalytic particles. Nat. Mater. 6, 507–511.

Patel, A., Arora, N., Sartaj, K., Pruthi, V., Pruthi, P.A., 2016. Sustainable biodiesel production from oleaginous yeasts utilizing hydrolysates of various non-edible lignocellulosic biomasses. Renew. Sustain. Energy Rev. 62, 836–855.

Pedersen, J.V., Iverseb, F.K., 2009. Devices and Methods for Discharging Pretreated Biomass From Higher to Lower Pressure Regions. Patent No. 2009147512 A2.

Peña, L., Hohn, K.L., Li, J., Sun, X.S., Wang, D., 2014b. Synthesis of propyl-sulfonic acid-functionalized nanoparticles as catalysts for cellobiose hydrolysis. J. Biomater. Nanobiotechnol. 5, 241.

Pena, L., Ikenberry, M., Hohn, K.L., Wang, D., 2012. Acid-functionalized nanoparticles for pretreatment of wheat straw. J. Biomater. Nanobiotechnol. 3, 342–352.

Pena, L., Xu, F., Hohn, K.L., Li, J., Wang, D., 2014a. Propyl-sulfonic acid functionalized nanoparticles as catalyst for pretreatment of corn stover. J. Biomater. Nanobiotechnol. 5, 8–16.

Perera, F., 2018. Pollution from fossil-fuel combustion is the leading environmental threat to global pediatric health and equity: solutions exist. Int. J. Environ. Res. Public Health 15, 16.

Prabhahar, R.S.S., Benitha, V.S., Nagarajan, J., 2021. Improved yield of palm oil biodiesel through nano catalytic transesterification. Mater. Today: Proc. . Available from: https://doi.org/10.1016/j.matpr.2021.03.473.

Prabhu, N., Karthikadevi, S., Gajendran, T., 2021. Comprehensive review of the prospectives and development for the production of bioalcohols using nanoparticles. Nanomaterials 379–391.

Qi, W., He, C., Wang, Q., Liu, S., Yu, Q., Wang, W., et al., 2018. Carbon-based solid acid pretreatment in corncob saccharification: specific xylose production and efficient enzymatic hydrolysis. ACS Sustain. Chem. Eng. 6, 3640–3648.

Rai, M., dos Santos, J.C., Soler, M.F., Marcelino, P.R.F., Brumano, L.P., Ingle, A.P., et al., 2016. Strategic role of nanotechnology for production of bioethanol and biodiesel. Nanotechnol. Rev. 5, 231–250.

Rai, M., Ingle, A.P., Gaikwad, S., Dussán, K.J., da Silva, S.S., 2017. Role of nanoparticles in enzymatic hydrolysis of lignocellulose in ethanol. In: Rai, M., da Silva, S. (Eds.), Nanotechnology for Bioenergy and Biofuel Production. Springer, Cham, pp. 153–171.

Rai, M., Ingle, A.P., Pandit, R., Paralikar, P., Biswas, J.K., da Silva, S.S., 2019. Emerging role of nanobiocatalysts in hydrolysis of lignocellulosic biomass leading to sustainable bioethanol production. Catal. Rev. 61, 1–26.

Rajput, V., Minkina, T., Sushkova, S., Behal, A., Maksimov, A., Blicharska, E., et al., 2020. ZnO and CuO nanoparticles: a threat to soil organisms, plants, and human health. Environ. Geochem. Health 42, 147–158.

Ray, R.C., Ramachandran, S. (Eds.), 2018. Bioethanol Production From Food Crops: Sustainable Sources, Interventions, and Challenges. Academic Press.

Reddy, K., Nasr, M., Kumari, S., Kumar, S., Gupta, S.K., Enitan, A.M., et al., 2017. Biohydrogen production from sugarcane bagasse hydrolysate: effects of pH, S/X, Fe2 + , and magnetite nanoparticles. Environ. Sci. Pollut. Control. Ser. 24, 8790–8804.

Ren, N.Q., Zhao, L., Chen, C., Guo, W.Q., Cao, G.L., 2016. A review on bioconversion of lignocellulosic biomass to H2: key challenges and new insights. Bioresour. Technol. 215, 92–99.

Rodionova, M.V., Poudyal, R.S., Tiwari, I., Voloshin, R.A., Zharmukhamedov, S.K., Nam, H.G., et al., 2017. Biofuel production: challenges and opportunities. Int. J. Hydrog. Energy. 42, 8450–8461.

Sajith, V., Sobhan, C.B., Peterson, G.P., 2010. Experimental investigations on the effects of cerium oxide nanoparticle fuel additives on biodiesel. Adv. Mech. Eng. 2, 581407.

Salwan, R., Sharma, A., Sharma, V., 2020. Nanomaterial-immobilized biocatalysts for biofuel production from lignocellulose biomass. In: Srivastava, M., Srivastava, N., Mishra, P.K., Gupta, V.K. (Eds.), Nanomaterials in Biofuels Research. Springer, Singapore, pp. 213–250.

Sanchez-Ramirez, J., Martinez-Hernandez, J.L., Segura-Ceniceros, P., Lopez, G., Saade, H., Medina-Morales, M.A., et al., 2017. Cellulases immobilization on chitosan-coated magnetic nanoparticles: application for *Agave Atrovirens* lignocellulosic biomass hydrolysis. Bioproc. Biosyst. Eng. 40, 9–22.

Sannino, F., Costantini, A., Ruffo, F., Aronne, A., Venezia, V., Califano, V., 2020. Covalent immobilization of β-glucosidase into mesoporous silica nanoparticles from anhydrous acetone enhances its catalytic performance. Nanomaterials 10, 108.

Santos, F.C.U., Paim, L.L., Silva, J.L.D., Stradiotto, N.R., 2016. Electrochemical determination of total reducing sugars from bioethanol production using glassy carbon electrode modified with graphene oxide containing copper nanoparticles. Fuel 163, 112–121.

Schaefer, S., Muñiz, G., Izquierdo, M.T., Mathieu, S., Ballinas-Casarrubias, M.L., González-Sánchez, G., et al., 2017. Rice straw-based activated carbons doped with SiC for enhanced hydrogen adsorption. Int. J. Hydrog. Energy. 42, 11534–11540.

Shanmugam, S., Krishnaswamy, S., Chandrababu, R., Veerabagu, U., Pugazhendhi, A., Mathimani, T., 2020. Optimal immobilization of *Trichoderma asperellum* laccase on polymer coated Fe_3O_4@ SiO_2 nanoparticles for enhanced biohydrogen production from delignified lignocellulosic biomass. Fuel 273, 117777.

Sharma, V., Joshi, A., Ramawat, K.G., Arora, J., 2017. Bioethanol production from halophytes of Thar desert: a "Green Gold. In: Basu, S.K., Zandi, P., Chalaras, S.K. (Eds.), Environment at Crossroads: Challenges. Dynamics and Solutions. Haghshenass Publishing, pp. 219–235.

ShenavaeiZare, T., Khoshsima, A., ZareNezhad, B., 2021. Production of biodiesel through nanocatalytic transesterification of extracted oils from halophytic safflower and salicornia plants in the presence of deep eutectic solvents. Fuel 302, 121171.

Shields-Menard, S.A., Amirsadeghi, M., French, W.T., Boopathy, R., 2018. A review on microbial lipids as a potential biofuel. Bioresour. Technol. 259, 451–460.

Sindhu, R., Binod, P., Pandey, A., 2016. Biological pretreatment of lignocellulosic biomass—an overview. Bioresour. Technol. 199, 76–82.

Singh, A., Sevda, S., Abu Reesh, I.M., Vanbroekhoven, K., Rathore, D., Pant, D., 2015. Biohydrogen production from lignocellulosic biomass: technology and sustainability. Energies 8, 13062–13080.

Singh, N., Bharj, R.S., 2015. Effect of CNT-emulsified fuel on performance emission and combustion characteristics of four stroke diesel engine. Int. J. Curr. Eng. Technol. 5, 477–485.

Singhvi, M.S., Deshmukh, A.R., Kim, B.S., 2021. Cellulase mimicking nanomaterial-assisted cellulose hydrolysis for enhanced bioethanol fermentation: an emerging sustainable approach. Green. Chem. . Available from: https://doi.org/10.1039/D1GC01239H.

Singhvi, M., Kim, B.S., 2020. Current developments in lignocellulosic biomass conversion into biofuels using nanobiotechology approach. Energies 13, 5300.

Soares, J.F., Confortin, T.C., Todero, I., Mayer, F.D., Mazutti, M.A., 2020. Dark fermentative biohydrogen production from lignocellulosic biomass: technological challenges and future prospects. Renew. Sustain. Energy Rev. 117, 109484.

Su, T.C., Fang, Z., Zhang, F., Luo, J., Li, X.K., 2015. Hydrolysis of selected tropical plant wastes catalyzed by a magnetic carbonaceous acid with microwave. Sci. Rep. 5, 1–14.

Sun, Y., Wang, Y., Yang, G., Sun, Z., 2020. Optimization of biohydrogen production using acid pretreated corn stover hydrolysate followed by nickel nanoparticle addition. Int. J. Energy Res. 44, 1843–1857.

Taherdanak, M., Zilouei, H., Karimi, K., 2016. The effects of Fe0 and Ni0 nanoparticles versus Fe2 + and Ni2 + ions on dark hydrogen fermentation. Int. J. Hydrog. Energy 1, 167–173.

Tawfik, A., El-Bery, H., Elsamadony, M., Kumari, S., Bux, F., 2019. Upgrading continuous H2 gas recovery from rice straw hydrolysate via fermentation process amended with magnetite nanoparticles. Int. J. Energy Res. 43, 3516–3527.

Valdivia, M., Galan, J.L., Laffarga, J., Ramos, J.L., 2016. Biofuels 2020: biorefineries based on lignocellulosic materials. Microb. Biotechnol. 9, 585–594.

Vasić, K., Hojnik Podrepšek, G., Knez, Ž., Leitgeb, M., 2020. Biodiesel production using solid acid catalysts based on metal oxides. Catalysts 10, 237.

Verma, M.L., Chaudhary, R., Tsuzuki, T., Barrow, C.J., Puri, M., 2013. Immobilization of β-glucosidase on a magnetic nanoparticle improves thermostability: application in cellobiose hydrolysis. Bioresour. Technol. 135, 2–6.

Wang, C., Zhang, X., Liu, Q., Zhang, Q., Chen, L., Ma, L., 2020. A review of conversion of lignocellulose biomass to liquid transport fuels by integrated refining strategies. Fuel Process. Technol. 208, 106485.

Wang, H., Covarrubias, J., Prock, H., Wu, X., Wang, D., Bossmann, S.H., 2015. Acid-functionalized magnetic nanoparticle as heterogeneous catalysts for biodiesel synthesis. J. Phys. Chem. C 119, 26020–26028.

Wang, W., Ji, S., Lee, I., 2013. Fast and efficient nanoshear hybrid alkaline pretreatment of corn stover for biofuel and materials production. Biomass Bioenergy 51, 35–42.

Yang, G., Wang, J., 2018. Improving mechanisms of biohydrogen production from grass using zero-valent iron nanoparticles. Bioresour. Technol. 266, 413–420.

Yang, G., Wang, J., 2019. Synergistic enhancement of biohydrogen production from grass fermentation using biochar combined with zero-valent iron nanoparticles. Fuel 251, 420–427.

Yaşar, A., Keskin, A., Yıldızhan, Ş., Uludamar, E., 2019. Emission and vibration analysis of diesel engine fuelled diesel fuel containing metallic based nanoparticles. Fuel 239, 1224–1230.

Zdarta, J., Jędrzak, A., Klapiszewski, Ł., Jesionowski, T., 2017. Immobilization of cellulase on a functional inorganic–organic hybrid support: stability and kinetic study. Catalysts 7, 374.

Zhang, X.L., Yan, S., Tyagi, R.D., Surampalli, R.Y., 2013. Biodiesel production from heterotrophic microalgae through transesterification and nanotechnology application in the production. Renew. Sustain. Energy Rev. 26, 216–223.

Zhao, W., Zhang, J., Zhang, H., Yang, M., Zang, L., 2020. Comparison of mesophilic and thermophilic biohydrogen production amended by nickel-doped magnetic carbon. J. Clean. Prod. 270, 122730.

Zhao, W., Zhang, Y., Du, B., Wei, D., Wei, Q., Zhao, Y., 2013. Enhancement effect of silver nanoparticles on fermentative biohydrogen production using mixed bacteria. Bioresour. Technol. 142, 240–245.

Zheng, P., Wang, J., Lu, C., Xu, Y., Sun, Z., 2013. Immobilized β-glucosidase on magnetic chitosan microspheres for hydrolysis of straw cellulose. Process. Biochem. 48, 683–687.

Zhong, D., Li, J., Ma, W., Xin, H., 2020. Magnetite nanoparticles enhanced glucose anaerobic fermentation for bio-hydrogen production using an expanded granular sludge bed (EGSB) reactor. Int. J. Hydrog. Energy. 45, 10664–10672.

Zilouei, H., Taherdanak, M., 2015. Biohydrogen from lignocellulosic wastes. In: Karimi, K. (Ed.), Lignocellulose-Based Bioproducts. Springer International Publishing, Switzerland, pp. 253–288.

Advances in combined biohydrogen and biomethane production processes from municipal-based lignocellulosic waste at small scale

15

Andrea Dell'Orto, Preshanthan Moodley and Cristina Trois
SARCHI Chair Waste and Climate Change, School of Engineering, University of KwaZulu-Natal, Durban, South Africa

15.1 Introduction

Lignocellulosic biomasses combine three different polymers forming a heterogeneous and recalcitrant matrix. While cellulose and hemicellulose are polysaccharides that can be decomposed into simple carbohydrates such as glucose and xylose, lignin is a methoxylated polyphenol that is severely recalcitrant and makes the walls of plant cells impermeable so that they can withstand the attacks of aggressive microorganisms and chemical compounds (Moodley et al., 2020). Lignocellulosic waste also has great potential for extensive, sustainable, and cost-effective energy production (Moodley and Trois, 2021).

Municipal solid wastes (MSWs) contain significant quantities of plant-derived carbohydrates, which have the potential to be exploited as a biomass source. Examples of municipal-based lignocellulosic waste (Fig. 15.1) are fruit and vegetable waste, garden refuse, and the organic fraction of MSW, which are not competing with the food or feed industry. Among energy crops, only the food plants produced through phytocapping can be considered a waste feedstock for a biorefinery because they grow on contaminated lands like landfill capping soil (Bolan et al., 2013). Therefore they are a suitable feedstock in countries like South Africa, where using maize for nonfeeding purposes is prohibited to preserve food security.

When landfilled, these materials emit significant amounts of greenhouse gases that become substantial contributors to climate change. Therefore the diversion from landfills and valorization in biorefinery processes has been investigated, concluding that municipal-based lignocellulosic waste is the ideal feedstock for the production of methane and hydrogen in second-generation biorefineries (Lin and Lu, 2021).

Advances in Lignocellulosic Biofuel Production Systems. DOI: https://doi.org/10.1016/B978-0-323-91192-4.00017-1

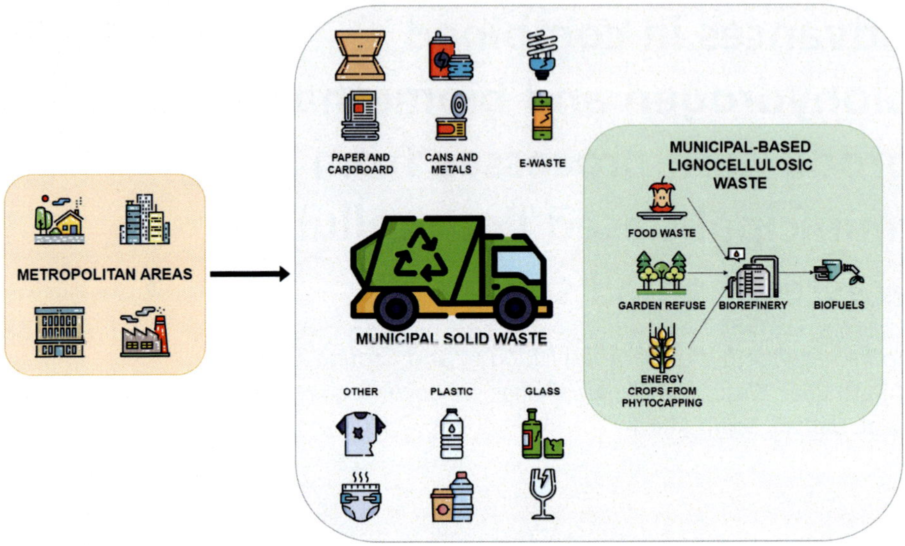

Figure 15.1 Municipal solid waste and municipal-based lignocellulosic waste.

Methane (CH_4) gas is commonly used in industrial chemical processes, as a fuel, or for electricity generation. Due to its hydrogen content, its combustion produces more energy per unit weight (55.7 kJ/g) than any other fossil fuel while also originating less carbon dioxide.

Hydrogen (H_2) is acknowledged as a secondary energy source or energy carrier which has more heat per mass (142.9 kJ/g) compared to any other fuel (Zheng et al., 2022). It may be used as a fuel in modified internal combustion engines generating fewer air pollutants than engines fueled by petroleum-based combustibles (Sharma et al., 2015). Moreover, water vapor is the only emission from vehicles powered by hydrogen fuel cells (Rosen and Koohi-Fayegh, 2016).

Conventional thermochemical hydrogen production systems rely upon fossil fuels, countering the advantages of clean combustion. On the other hand, biological methods depend on a widespread and readily available feedstock such as municipal-based lignocellulosic waste. These techniques can guarantee a clean, sustainable and continuous production of hydrogen independently from sun and wind (Nikolaidis and Poullikkas, 2017), while also addressing the critical issue of proper disposal of biodegradable waste.

This chapter provides an overview of biological processes such as anaerobic digestion (AD) and dark fermentation and explores the advances in the biological treatment of municipal-based lignocellulosic waste. Conventional single-stage AD is compared with a two-stage configuration, and innovative strategies for setting and controlling operational parameters are discussed. Moreover, the chapter presents the challenges and the techniques necessary to facilitate the biological degradation of complex lignocellulosic feedstocks.

15.2 Two-stage anaerobic digestion of municipal-based lignocellulosic waste

AD is a complex biochemical process that comprises four different phases: hydrolysis, acidogenesis, acetogenesis, and methanogenesis (Srisowmeya et al., 2020). Generally, the first two stages can be considered dark fermentation (production of hydrogen) whereas the last two stages are methanogenesis (production of methane). *Hydrolysis* consists of the degradation of particulate matter and complex organic polymers thanks to extracellular enzymes secreted by hydrolytic microorganisms. The resulting hydrolyzed substrate can cross bacterial cellular membranes and be broken down into volatile fatty acids (VFAs), low molecular weight alcohols such as ethanol, hydrogen and carbon dioxide during the second phase, called *acidogenesis* or *acid fermentation.* Thereafter, during *acetogenesis*, VFAs are converted into acetic acid and hydrogen, which in turn are used during *methanogenesis*, the last stage of AD. The final products are biogas containing methane and carbon dioxide and a nutrient-rich sludge called digestate which can have beneficial effects in crop cultivation.

15.2.1 Single- and two-stage anaerobic digestion

Conventional AD is operated in a single reactor, where all four stages are carried out simultaneously. Therefore such a configuration is set up to protect the weakest microorganisms (methanogens) and is not optimized for the diverse bacterial species that require different optimum conditions. For instance, microbes responsible for dark fermentation are usually spore formers and can thus survive harsh environmental conditions, particularly drastic changes in environment and pH. On the other hand, microbes involved in methanogenesis are extremely sensitive to slight changes in both temperature and pH. In order for these microbes to function at their optimum, their environmental conditions should be at their optimum values. Currently, the single-stage systems' advantages are related to simplified procedures and minimal equipment and operational costs. However, the reduced complexity of single-stage AD systems hampers their productivity. It exposes them to several drawbacks: incomplete degradation, reactor instability, inhibition of methanogenesis due to VFA accumulation, ammonia inhibition, inadequate buffering capacity, formation of toxic intermediates, and lack of tolerance for excessive organic loads (Srisowmeya et al., 2020). Moreover, when lignocellulosic biomass is used as a feedstock, the depolymerization of its complex matrix significantly hinders conventional AD, limiting the hydrolysis rate (Ferraro et al., 2020).

A two-stage AD system fulfills both needs of maximizing the efficiency of the limiting-rate step, hydrolysis and generating VFAs. Hydrolytic and acidogenic microorganisms are very diverse (obligate anaerobes, facultative anaerobes, or even aerobes) and more flexible than strict anaerobes such as acetogens and methanogens (Li and Fang, 2007). Hence, the physical separation between dark fermentation and methanogenesis gives both bacterial groups optimal conditions to thrive, resulting in increased energy recovery rates (De Gioannis et al., 2017). However, breaking up the stages also requires an additional buffer tank that will facilitate control of pH and organic loading rate while allowing for the removal of substrates recalcitrant to hydrolysis (Van et al., 2020).

15.3 Municipal-based lignocellulosic feedstocks for two-stage anaerobic digestion

Municipal-based lignocellulosic waste streams have been successfully exploited as feedstocks in two-stage AD (Table 15.1), with their potential for biohydrogen and biomethane production depending on biomass composition (volatile solids, VS, Table 15.2).

Substrates such as yard waste and food crops have significantly high carbon (C)-to-nitrogen (N) ratios and total solids, making them unsuitable for monodigestion. On the other hand, while FW fits the profile in terms of total solids and C/N content, the readily biodegradable compounds can lead to an excess of VFAs and the consequent acidification and failure of the process or make its management increasingly expensive and complex. The possibility of performing a codigestion of a mixture of these two waste categories has been recently explored to overcome these challenges.

15.3.1 Codigestion

Anaerobic codigestion is an efficient and cost-effective way to address the issues related to the monodigestion of suitable substrates and increase biohydrogen and biomethane yields. However, most studies focused on the codigestion of FW with waste-activated sludge, a system that has been extensively applied even at full-scale (Mata-Alvarez et al., 2011). Waste-activated sludge is the most common cosubstrate, also acting as inoculum, because of its modest concentration of organic waste, abundant trace elements, and significant alkalinity, which can help offset VFA accumulation in monodigestion of FW (Montecchio et al., 2019). In addition, the codigestion of different solid feedstocks guarantees the necessary balance in C/N ratio and nutrient content while protecting the fermentative process from acidification caused by an excess of readily available substrates.

Synergistic effects have been confirmed by recent research, resulting in methane production enhanced by eight times (280 mL CH_4 g/VS) compared to AD performed solely on FW (Patil et al., 2021) and 1.25−1.4 times to codigestion of fruits and vegetables wastes (FVWs) and waste-activated sludge (Ambrose et al., 2020; Montecchio et al., 2019). The codigestion of slaughterhouse waste with FVWs also yielded 74.2% more methane (776.3 mL CH_4 g/VS) than the monodigestion of FVWs (Miramontes-Martínez et al., 2021). Conversely, Gu et al. (2020) pointed out that codigestion of FW and sewage sludge led to addictive effects without boosting methane yields. The mesophilic codigestion produced 505.4−511.8 mL CH_4 g/VS, 1.02−1.10 times compared to the sum of monodigestion outputs, while the production from thermophilic codigestion was identical (0.99−1.01 times).

Few studies focused on the codigestion of two different municipal-based lignocellulosic feedstocks, which is necessary to balance the different C/N ratios of FW and

Table 15.1 Use of municipal-based lignocellulosic feedstocks in a two-stage anaerobic digestion.

Feedstock	Hydrogen yield	Methane yield	Reference
Food waste	8.6 ± 4.8 mL H_2/g VS d	428.3 ± 30.9 mL CH_4/g VS d	Baldi et al. (2019)
Food waste + garden waste	46.2 ± 0.9 mL H_2/g VS d	682 mL CH_4/g VS	Abreu et al. (2019)
Organic fraction municipal solid waste (OFMSW)	29.8 mL H_2/g VS	619 mL CH_4/g VS	Lavagnolo et al. (2018)
Garden waste (grass)	52 mL H_2/g VS	517 mL CH_4/g VS	Liczbiński and Borowski (2021)
Garden waste (leaves)	23 mL H_2/g VS	421 mL CH_4/g VS	Liczbiński and Borowski (2021)

VS: Volatile solids.

Table 15.2 Physicochemical characteristics of municipal-based lignocellulosic feedstocks suitable for stage anaerobic digestion.

Feedstock	TS (%)	VS (% TS)	pH	VFA (mg/L)	sCOD (mg/L)	C/N	Reference
Food waste	23.9 ± 0.1	91.3 ± 0.3	–	–	–	16.3 ± 0.2	Mu et al. (2020)
Food waste	22.6 ± 0.8	93.8 ± 0.10	3.7 ± 0.7	970 ± 127	2423 ± 164	21.6	Panigrahi et al. (2020)
Kitchen waste	19.99	90.2	–	–		17.3	Liu et al. (2012)
Vegetable/fruit residue	10.74	93.5	–	–	–	21.7	Liu et al. (2012)
Fruit and vegetable waste	3.7	96.78	5.0	–	29,100 ± 216	–	Ambrose et al. (2020)
Catering waste	14.30 ± 0.6	n.d.	3.91 ± 0.06	–	4980 ± 230	25 ± 1.05	Anjum et al. (2016)
Garden waste (grass)	12.8 ± 1.1	81.6				22.74	Liczbiński and Borowski (2021)
Garden waste (leaves)	92.9 ± 1.1	87.8				20.07	Liczbiński and Borowski (2021)
Yard waste	89.5 ± 0.6	87.3 ± 0.6	7.5 ± 0.1	364 ± 20	1281 ± 94	44.8	Panigrahi et al. (2020)
Yard waste	97.3 ± 0.1	91.9 ± 0.0	–	–	–	74.7 ± 4.0	Mu et al. (2020)
Wheat crop residues	99.2 ± 0.8	87.3	4.03 ± 0.02	–	2800 ± 240	76 ± 1.87	Anjum et al. (2016)
Sugar cane bagasse	95.3 ± 0.1	77.4	3.48 ± 0.05	–	3690 ± 250	85 ± 1.64	Anjum et al. (2016)

C/N: carbon/nitrogen; *COD*: chemical oxygen demand; *VFAs*: volatile fatty acids; *VS*: volatile solids; *TS*: total solids.

carbon-rich lignocellulosic biomasses and prevent acidification. An artificial neural network predicted that the codigestion of FW with garden waste would maximize methane yield at 270 mL CH_4 g/VS owing to optimizing the mixing ratio of substrates. In comparison, the monodigestion of FW would produce only 70 mL CH_4 g/VS (Casallas-Ojeda et al., 2021). Mu et al. (2020) evaluated the codigestion of FW, yard waste, and sewage sludge. According to their C/N ratios (74.7 ± 4.0, 16.3 ± 0.2, and 6.8 ± 0.1, for yard waste, FW, and sewage sludge, respectively); only FW would be an ideal substrate for AD, as confirmed by the biochemical methane potential tests performed on each feedstock. FW generated a considerably larger yield (526.8 ± 7.4 mL CH_4 g/VS) than yard waste (116.5 ± 1.9 mL CH_4 g/VS) and sewage sludge (254.6 ± 0.6 mL CH_4 g/VS). However, AD of FW experienced VFA accumulation that eventually stopped methanogenesis. Therefore the same biomasses were investigated as feedstocks for a semicontinuous codigestion. The more balanced composition in terms of C/N ratio, readily biodegradable substrates, trace elements, and buffering capacity led to a steady methane production while obtaining a maximum yield of 314.9 ± 1.7 mL CH_4 g/VS.

Song et al. (2021) focused on the codigestion of FW, garden waste, and tofu residue using wastewater treatment plant sludge as inoculum. This codigestion experienced a 9% positive synergistic effect compared to monodigestion of the single substrates, generating 370.54 mL CH_4 g/VS.

These results illustrated that, thanks to the synergistic effects of codigestion, AD of municipal-based lignocellulosic biomass shows promise as a feasible treatment method even in a semicontinuous mode (Table 15.3). Nevertheless, more studies need to be conducted to ascertain the ideal operating conditions and investigate the feasibility of a two-stage process for the combined production of hydrogen and methane, possibly at a larger scale. Only a single study (Abreu et al., 2019) focused on a mesophilic two-stage codigestion of FW and garden waste while using as inoculum a pure culture of *Caldicellulosiruptor saccharolyticus* in the dark fermentative reactor, resulting in a combined yield of 46.2 ± 0.9 mL H_2 g/VS and 682 mL CH_4 g/VS. Additionally, Zhu et al. (2022) performed a thermophilic two-stage codigestion of FW, corn straw, and chicken manure that doubled methane yield compared to monodigestion, leading to the production of 106 mL H_2 g/VS and 515.9 mL CH_4 g/VS.

15.3.2 Pretreatment

Despite the effectiveness of codigestion, pretreatment of lignocellulosic feedstocks is needed to ensure hydrolysis of the recalcitrant lignin matrix and release the carbohydrate content. There are three families of pretreatment, namely biological, physical and chemical. Biological pretreatment exploits the enzymes secreted by particular microorganisms or fungi to attack lignin or cellulose. Still, it requires very extended treatment times not acceptable in an industrial process like AD. Conversely, physical and chemical methods are extensively used to treat lignocellulosic feedstocks and increase their biodegradability. Widespread physical methods are thermal hydrolysis and microwave or ultrasound treatment, while biomasses can be chemically treated using acid or alkaline compounds, organic solvents, inorganic salts, and ionic liquids (Paudel et al., 2017) (Table 15.4).

Table 15.3 Codigestion of suitable feedstocks.

Feedstock 1	Feedstock 2	Inoculum	Reactor type	Reactor volume (L)	pH	T (°C)	HRT (d)	Yield	Reference
Catering waste	Wheat straw	Digestate from (AD) of organic waste	Batch	0.35	3.41–7.31	30°C ± 1°C	80	168 mL biogas/g VS	Anjum et al. (2016)
Catering waste	Sugar cane bagasse	Digestate from AD of organic waste	Batch	0.35	3.68–6.93	30°C ± 1°C	80	124 mL biogas/g VS	Anjum et al. (2016)
Food waste	Yard waste	Waste activated sludge WWTP	Batch	0.1	6.3–7.2	30°C	30	431 mL CH_4/g VS	Panigrahi et al. (2020)
Food waste	Yard waste	Anaerobic sludge WWTP	Semicontinuous	0.3	–	37°C	20	314.9 ± 1.7 mL CH_4/g VS	Mu et al. (2020)
Food waste	Garden waste + Tofu residue	Anaerobic sludge WWTP	Semicontinuous	0.25	–	38°C	40–50	370.54 mL CH_4/g VS	Song et al. (2021)
Food waste	Garden waste	*Caldicellulosiruptor saccharolyticus* + Anaerobic granular sludge	2-stage batch	0.16 + 0.27	n.d. + 7.0–7.2	70°C + 37°C	24 + 45	46.2 ± 0.9 mL H_2/g VS + 682 mL CH_4/g VS	Abreu et al. (2019)
Food waste	Corn straw, chicken manure	Digestate from AD of cow and chicken manure	2-stage batch	0.2	5.5 + 6.5	55°C ± 2°C	10 + 25	106 mL H_2/g VS + 515.9 mL CH_4/g VS	Zhu et al. (2022)

AD: anaerobic digestion; *HRT*: hydraulic retention time; *VS*: volatile solids; *WWTP*: waste water treatment plants.

Table 15.4 Physicochemical pretreatment methods for lignocellulosic biomass.

Pretreatment method	Feedstock(s)	Process conditions	Maximum yield	Effect of pretreatment	Reference
Hydrothermal	Food waste	140°C, 20 min	38.1 mL H_2/g VS−439.6 mL CH_4/g VS	+3.8% H_2 production; +168% CH_4 production	Ding et al. (2022)
Thermal + pressure hydrolysis	Fruit waste and garden waste	121°C, 0.1 MPa, 20 min	46.2 ± 0.9 mL H_2/g VS	+23%−46% COD solubilization	Abreu et al. (2019)
Thermal + pressure hydrolysis	Fruit and vegetable waste	120°C, 0.1 MPa, 20 min	23.53 mL H_2/g VS	+116% H_2 production	Abubackar et al. (2019)
Microwave	Yard waste + food waste	140°C, 2000W for 2 min	431 mL CH_4/g VS	+21%−28% COD solubilization; +29% in biogas production	Panigrahi et al. (2020)
Microwave + hydrogen peroxide (MW-H_2O_2)	Fruit and vegetable waste	80°C, 660 W for 2 min + 1% w/w H_2O_2/TS	276 mL CH_4/g VS (treated)	+33% COD solubilization; +25% biogas production	Ambrose et al. (2020)
Microwave + steam	Wheat straw + catering waste	800 J/g/min for 3 min +121°C for 40 min	168 mL CH_4/g VS	+28% COD solubilization, +85.2% COD removal	Anjum et al. (2016)
Microwave + steam	Sugar cane bagasse + catering waste	800 J/g/min for 3 min +121°C for 40 min	128 mL CH_4/g VS	+62% COD solubilization, +81.2% COD removal	Anjum et al. (2016)
Microwave	Lipid and food waste	1000 W for 50 s	738.63 mL CH_4/g VS	+14% CH_4 production	Yue et al. (2021)
Ultrasound	Lipid and food waste	500 W for 100 s	927.97 mL CH_4/g VS	+43% CH_4 production	Yue et al. (2021)
Dilute acid hydrolysis	Sugarcane leaf waste	4.90% HCl, 84.14 min, 99°C	18.6 mL H_2/g VS	93.15% hemicellulose removal	Moodley and Gueguim Kana (2018)

COD: chemical oxygen demand; *VS*: volatile solids.

15.3.2.1 Physical pretreatment

Conventional methods such as mechanical crushing destroy the lignocellulosic matrix without degrading its components or significantly reducing treatment costs. Therefore innovative techniques such as microwave/ultrasonic processing, gamma rays, or electron beam irradiation have been investigated to enhance the solubilization of recalcitrant substrates (Zheng et al., 2022).

Microwave-enhanced advanced oxidation is a promising technique for enhancing the solubilization of recalcitrant substrates such as lignin-rich lignocellulosic biomass. Panigrahi et al. (2020) report a 21%–28% increase in chemical oxygen demand (COD) solubilization of microwave-pretreated yard waste, resulting in a 29% higher biogas production yield (431 mL g/VS). When coupled with hydrogen peroxide chemical pretreatment, this technique improved hydrolysis (+33% COD solubilization) and biogas production (+25%, 276 mL CH_4 g/VS), increased the buffer ability of the system, and enhanced the quality of digestate (Ambrose et al., 2020).

Similar results originated from performing microwave and steam treatments to the wheat straw + catering waste and sugar cane bagasse + catering waste pairings, leading to a production of 168 mL CH_4 g/VS and 128 mL CH_4 g/VS, respectively (Anjum et al., 2016). Furthermore, a recent study on AD of lipid and FW proved that ultrasonic pretreatment was 25% more effective than microwave heating, leading to a methane yield of 927.97 mL CH_4 g/VS (Yue et al., 2021). The pulsed electric field is a technique commonly adopted in the medicine industry and food processing to break cell membranes and has the potential to be implemented on lignocellulosic biomass (Kovačić et al., 2021).

15.3.2.2 Thermal hydrolysis

Ding et al. (2022) performed a two-stage AD of hydrothermally pretreated FW, recording maximum yields of 38.1 mL H_2 g/VS and 439.6 mL CH_4 g/VS. While hydrogen production was similar when using untreated and treated FW, hydrothermally pretreated FW increased methane production by 168%, reducing HRT from 30 to 18 days.

Abreu et al. (2019) conducted thermal pretreatment while applying a pressure of 0.1 MPa, and reported a COD solubilization increased by 23%–46%, leading to a hydrogen yield of 46.2 ± 0.9 mL H_2 g/VS. Abubackar et al. (2019) confirmed the effectiveness of combined thermal and pressure hydrolysis, reporting a 116% increase in hydrogen production.

On the opposite temperature range, freezing and thawing pretreatment is another promising method that disrupts the cells enlarged by the temperature decrease (Chozhavendhan et al., 2020). When performed at $-80°C$ performed on peanut shells, the hydrogen yield increased by 2.4 times (Qi et al., 2020).

15.3.2.3 Chemical hydrolysis

Inorganic salt pretreatments using alkali or metal salts have been extensively investigated (Moodley et al., 2020). An enhancement to classic acid pretreatment is dilute acid hydrolysis, a method that can achieve greater hemicellulose solubilization. For example, when applied to sugarcane leaf waste before dark fermentation, it produced 18.6 mL H_2 g/VS thanks to the removal of over 93% of hemicellulose (Moodley and Gueguim Kana, 2018). Enzymatic pretreatment has also enhanced the hydrolysis rate and the hydrogen production from rice straw (Srivastava et al., 2017).

15.4 Current advances in two-stage anaerobic digestion systems

Two-stage AD is very sensitive to environmental conditions. In particular, pH is crucial to guarantee the stability of the digestion, with most studies focusing on the 5–6 range in the dark fermentative tank and 7–8 in the methanogenic reactor (Holl et al., 2022). VFAs are one of the by-products of dark fermentation, and their accumulation can lead to acidification and consequent process inhibition. Different strategies have been investigated to prevent VFA build-up and facilitate their extraction for bioplastic production (Asunis et al., 2022). In electrodialysis, the penetration of ionic compounds through selective ion exchange membranes is promoted by applying an electrical field to a solution (Chalmers Brown et al., 2020). Hassan et al. (2021) used this novel technique to remove excess VFA from the dark fermentative tank of a two-stage system, obtaining a 77.1% increase in biogas yields (33.68 mL H_2 g/VS and 43.94 mL CH_4 g/VS) and COD removal (86.7% with electrodialysis, compared to 73% of conventional two-stage AD). Microfiltration and ultrafiltration membranes have also been applied using immersed membrane bioreactors. However, the focus was on the valorization of extracted VFAs, not biogas production (Pervez et al., 2022).

Temperature controls the overall efficiency of the process. While mesophilic conditions are usually preferred to balance energy consumption and production rates, psychrophilic and thermophilic fermentation need further investigation, particularly in a double-stage configuration. Research on psychrophile conditions focused on exploiting acclimated biomass to perform conventional AD (Rajagopal et al., 2017; Xu et al., 2022).

On the other hand, Xiao et al. (2018) carried out a thermophilic (55°C) AD of FW in both single- and double-stage configurations while obtaining similar methane yields (0.459 ± 0.115 mL CH_4 g/VS and 0.477 ± 0.091 mL CH_4 g/VS, respectively). Moreover, thermophilic single-stage digestion of FW and sewage sludge did not lead to a significantly higher methane production than mesophilic conditions, increasing from 505.4 mL CH_4 g/VS to 511.8 mL CH_4 g/VS, even though it led to faster methane production (Gu et al., 2020). Consequently, a thermophilic second stage is not recommended, given the sensitivity of methanogenic bacteria to extreme conditions. Alternatively, the coupling of a hyperthermophilic (70°C) dark fermentation of FW and garden waste with a mesophilic (37°C) methanogenesis was successfully explored, obtaining 46.2 ± 0.9 mL H_2 g/VS and 682 mL CH_4 g/VS in the first and second stage, respectively (Abreu et al., 2019).

In a two-stage configuration, the microbial community involved in dark fermentation consists of wide-ranging bacteria responsible for different processes. However, undesired microorganisms compete with hydrogen-producing bacteria (HPB), acting like preprocessing inhibitors and reducing the H_2 yield. Among others, hydrogenotrophic methanogens, homoacetogens, propionate producers, sulfate-reducing bacteria, and nitrate-reducing bacteria are all included in a group called hydrogen-consuming bacteria (Bundhoo and Mohee, 2016). Additionally, lactic acid bacteria compete with hydrogen-consuming bacteria and are known inhibitors of dark fermentation (Elbeshbishy et al., 2017).

Therefore bacterial selection is paramount to controlling hydrogen-consuming bacteria in dark fermentation. A short hydraulic retention time (HRT) guarantees the washout of the growing methanogenic biomass while keeping HPB inside the dark fermentative reactor, thanks to the higher generation time of methanogens than HPB. When control of operational parameters is not enough, pretreating the inoculum ensures that nonsporulating methanogens will not endure the extreme conditions that HPB can withstand, thanks to their ability to produce endospores.

Heat shock treatment extensively removes lactic acid bacteria (Elbeshbishy et al., 2017). Inoculum is typically pretreated at 80°C−121°C for a duration between 10 min and 4 h (Baldi et al., 2019; Girotto et al., 2017; Lavagnolo et al., 2018; Moodley and Gueguim Kana, 2018). Other conventional pretreatment methods are chemical (with alkaline or acid compounds or inhibitors such as 2-bromoethanesulfonic acid and chloroform), microwave irradiation, or even aeration (Rafieenia et al., 2018).

Bioaugmentation is a novel approach to steer the fermentation towards specified pathways and avoid the proliferation of unwanted species. Both bacterial (Chin et al., 2020; Zhang et al., 2018) and fungal (Akyol et al., 2019; Shetty et al., 2020) augmentation have been successfully implemented to increase biogas yields in anaerobic digesters. Moreover, inoculation with lactate-producing bacteria successfully promoted methane production over hydrogen production and enhanced the overall energy recovery rate of a two-stage system fed with FW (García-Depraect et al., 2022).

Adding inorganic nanoparticles is another promising technique to improve biohydrogen efficiency (Bosu and Rajamohan, 2022). For example, nanocatalysts such as iron or aluminum oxides expedited and increased hydrogen production but need further investigation to understand the inhibition thresholds (Yörüklü et al., 2022).

15.5 Conclusion: two-stage anaerobic digestion is the future

AD of organic waste is a mature technology extensively used for biogas production to maximize the final methane yield obtained from the digestion of agricultural waste. However, the recent trend is to shift the focus from the final product to the valorization of intermediate by-products and to replace agricultural waste with municipal-based lignocellulosic waste as feedstock. In this novel approach, VFA and hydrogen can be recovered and valorized by splitting conventional single-stage AD into a two-stage process, in which a dark fermentative step precedes methanogenesis. This configuration ensures tailored conditions for HPB and methanogens while also improving hydrolysis, the rate-limiting step in AD, particularly when the feedstock utilized is lignocellulosic biomass, a complex of carbohydrate-rich polymers that need to be released from the recalcitrant lignin matrix that embraces them. For this purpose, the codigestion of different typologies of municipal-based lignocellulosic feedstocks, mainly food and garden waste, has been investigated at a laboratory scale. However, there is a clear need for further research on two-stage codigestion for combined hydrogen and methane production at the lab scale to fully understand the dynamics and synergistic aspects of the process and prepare for the scaling-up to a full-scale implementation of technology. Further emphasis must be placed on novel pretreatment methods for lignocellulosic biomasses.

Abbreviations

AD anaerobic digestion
C/N carbon-to-nitrogen
CH_4 methane
COD chemical oxygen demand
FVW fruit and vegetable waste
FW food waste
H_2 hydrogen
HPB hydrogen-producing bacteria
HRT hydraulic retention time
TS total solids
VFA volatile fatty acids
VS volatile solids
WWTP waste water treatment plants

References

Abreu, A.A., Tavares, F., Alves, M.M., Cavaleiro, A.J., Pereira, M.A., 2019. Garden and food waste co-fermentation for biohydrogen and biomethane production in a two-step hyperthermophilic-mesophilic process. Bioresour. Technol. 278, 180–186. Available from: https://doi.org/10.1016/j.biortech.2019.01.085.

Abubackar, H.N., Keskin, T., Arslan, K., Vural, C., Aksu, D., Yavuzyılmaz, D.K., et al., 2019. Effects of size and autoclavation of fruit and vegetable wastes on biohydrogen production by dark dry anaerobic fermentation under mesophilic condition. Int. J. Hydrog. Energy 44, 17767–17780. Available from: https://doi.org/10.1016/j.ijhydene.2019.05.106.

Akyol, Ç., Ince, O., Bozan, M., Ozbayram, E.G., Ince, B., 2019. Fungal bioaugmentation of anaerobic digesters fed with lignocellulosic biomass: what to expect from anaerobic fungus *Orpinomyces* sp. Bioresour. Technol. 277, 1–10. Available from: https://doi.org/10.1016/j.biortech.2019.01.024.

Ambrose, H.W., Philip, L., Suraishkumar, G.K., Karthikaichamy, A., Sen, T.K., 2020. Anaerobic co-digestion of activated sludge and fruit and vegetable waste: evaluation of mixing ratio and impact of hybrid (microwave and hydrogen peroxide) sludge pre- treatment on two-stage digester stability and biogas yield. J. Water Process. Eng. 37, 101498. Available from: https://doi.org/10.1016/j.jwpe.2020.101498.

Anjum, M., Khalid, A., Mahmood, T., Aziz, I., 2016. Anaerobic co-digestion of catering waste with partially pretreated lignocellulosic crop residues. J. Clean. Prod. 117, 56–63. Available from: https://doi.org/10.1016/j.jclepro.2015.11.061.

Asunis, F., Cappai, G., Carucci, A., De Gioannis, G., Dessì, P., Muntoni, A., et al., 2022. Dark fermentative volatile fatty acids production from food waste: a review of the potential central role in waste biorefineries. Waste Manag. Res. Available from: https://doi.org/10.1177/0734242X221103940.

Baldi, F., Pecorini, I., Iannelli, R., 2019. Comparison of single-stage and two-stage anaerobic co-digestion of food waste and activated sludge for hydrogen and methane production. Renew. Energy 143, 1755–1765. Available from: https://doi.org/10.1016/j.renene.2019.05.122.

Bolan, N.S., Thangarajan, R., Seshadri, B., Jena, U., Das, K.C., Wang, H., et al., 2013. Landfills as a biorefinery to produce biomass and capture biogas. Bioresour. Technol. 135, 578–587. Available from: https://doi.org/10.1016/j.biortech.2012.08.135.

Bosu, S., Rajamohan, N., 2022. Nanotechnology approach for enhancement in biohydrogen production—review on applications of nanocatalyst and life cycle assessment. Fuel 323, 124351. Available from: https://doi.org/10.1016/j.fuel.2022.124351.

Bundhoo, M.A.Z., Mohee, R., 2016. Inhibition of dark fermentative bio-hydrogen production: a review. Int. J. Hydrog. Energy 41, 6713–6733. Available from: https://doi.org/10.1016/j.ijhydene.2016.03.057.

Casallas-Ojeda, M., Soto-Paz, J., Alfonso-Morales, W., Oviedo-Ocaña, E.R., Komilis, D., 2021. Optimization of operational parameters during anaerobic co-digestion of food and garden waste. Env. Process. 8, 769–791. Available from: https://doi.org/10.1007/s40710-021-00506-2.

Chalmers Brown, R., Tuffou, R., Massanet Nicolau, J., Dinsdale, R., Guwy, A., 2020. Overcoming nutrient loss during volatile fatty acid recovery from fermentation media by addition of electrodialysis to a polytetrafluoroethylene membrane stack. Bioresour. Technol. 301, 122543. Available from: https://doi.org/10.1016/j.biortech.2019.122543.

Chin, K.L., Nurliyana, M.Y., H'ng, P.S., Lee, C.L., Go, W.Z., Khoo, P.S., et al., 2020. Effects of bacterial bio-augmentation on the methane potential from facultative digestion of palm oil mill effluent and empty fruit bunch. Waste Biomass Valor 11, 3407–3418. Available from: https://doi.org/10.1007/s12649-019-00680-3.

Chozhavendhan, S., Rajamehala, M., Karthigadevi, G., Praveenkumar, R., Bharathiraja, B., 2020. A review on feedstock, pretreatment methods, influencing factors, production and purification processes of bio-hydrogen production. Case Stud. Chem. Env. Eng. 2, 100038. Available from: https://doi.org/10.1016/j.cscee.2020.100038.

De Gioannis, G., Muntoni, A., Polettini, A., Pomi, R., Spiga, D., 2017. Energy recovery from one- and two-stage anaerobic digestion of food waste. Waste Manage 68, 595–602. Available from: https://doi.org/10.1016/j.wasman.2017.06.013.

Ding, L., Cheng, J., Qiao, D., Li, H., Zhang, Z., 2022. Continuous co-generation of biohydrogen and biomethane through two-stage anaerobic digestion of hydrothermally pretreated food waste. Energy Convers. Manage 268, 116000. Available from: https://doi.org/10.1016/j.enconman.2022.116000.

Elbeshbishy, E., Dhar, B.R., Nakhla, G., Lee, H.-S., 2017. A critical review on inhibition of dark biohydrogen fermentation. Renew. Sust. Energ. Rev. 79, 656–668. Available from: https://doi.org/10.1016/j.rser.2017.05.075.

Ferraro, A., Massini, G., Mazzurco Miritana, V., Rosa, S., Signorini, A., Fabbricino, M., 2020. A novel enrichment approach for anaerobic digestion of lignocellulosic biomass: process performance enhancement through an inoculum habitat selection. Bioresour. Technol. 313, 123703. Available from: https://doi.org/10.1016/j.biortech. 2020.123703.

García-Depraect, O., Martínez-Mendoza, L.J., Diaz, I., Muñoz, R., 2022. Two-stage anaerobic digestion of food waste: enhanced bioenergy production rate by steering lactate-type fermentation during hydrolysis-acidogenesis. Bioresour. Technol. 358, 127358. Available from: https://doi.org/10.1016/j.biortech.2022.127358.

Girotto, F., Lavagnolo, M.C., Pivato, A., Cossu, R., 2017. Acidogenic fermentation of the organic fraction of municipal solid waste and cheese whey for bio-plastic precursors recovery—effects of process conditions during batch tests. Waste Manage 70, 71–80. Available from: https://doi.org/10.1016/j.wasman.2017.09.015.

Gu, J., Liu, R., Cheng, Y., Stanisavljevic, N., Li, L., Djatkov, D., et al., 2020. Anaerobic co-digestion of food waste and sewage sludge under mesophilic and thermophilic conditions: focusing on synergistic effects on methane production. Bioresour. Technol. 301, 122765. Available from: https://doi.org/10.1016/j.biortech.2020.122765.

Hassan, G.K., Jones, R.J., Massanet-Nicolau, J., Dinsdale, R., Abo-Aly, M.M., El-Gohary, F. A., et al., 2021. Increasing 2-bio-(H_2 and CH_4) production from food waste by combining two-stage anaerobic digestion and electrodialysis for continuous volatile fatty acids removal. Waste Manage 129, 20–25. Available from: https://doi.org/10.1016/j.wasman. 2021.05.006.

Holl, E., Steinbrenner, J., Merkle, W., Krümpel, J., Lansing, S., Baier, U., et al., 2022. Two-stage anaerobic digestion: state of technology and perspective roles in future energy systems. Bioresour. Technol. 360, 127633. Available from: https://doi.org/10.1016/j.biortech.2022. 127633.

Kovačić, Đ., Rupčić, S., Kralik, D., Jovičić, D., Spajić, R., Tišma, M., 2021. Pulsed electric field: an emerging pretreatment technology in a biogas production. Waste Manage 120, 467–483. Available from: https://doi.org/10.1016/j.wasman.2020.10.009.

Lavagnolo, M.C., Girotto, F., Rafieenia, R., Danieli, L., Alibardi, L., 2018. Two-stage anaerobic digestion of the organic fraction of municipal solid waste – effects of process conditions during batch tests. Renew Energy 126, 14–20. Available from: https://doi.org/ 10.1016/j.renene.2018.03.039.

Li, C., Fang, H.H.P., 2007. Fermentative hydrogen production from wastewater and solid wastes by mixed cultures. Crit. Rev. Env. Sci. Technol. 37, 1–39. Available from: https://doi.org/10.1080/10643380600729071.

Liczbiński, P., Borowski, S., 2021. Effect of hyperthermophilic pretreatment on methane and hydrogen production from garden waste under mesophilic and thermophilic conditions. Bioresour. Technol. 335, 125264. Available from: https://doi.org/10.1016/j.biortech. 2021.125264.

Lin, C.-Y., Lu, C., 2021. Development perspectives of promising lignocellulose feedstocks for production of advanced generation biofuels: a review. Renew. Sust. Energ. Rev. 136, 110445. Available from: https://doi.org/10.1016/j.rser.2020.110445.

Liu, X., Wang, W., Gao, X., Zhou, Y., Shen, R., 2012. Effect of thermal pretreatment on the physical and chemical properties of municipal biomass waste. Waste Manag. 32 (2), 249–255. Available from: https://doi.org/10.1016/j.wasman.2011.09.027.

Mata-Alvarez, J., Dosta, J., Macé, S., Astals, S., 2011. Codigestion of solid wastes: a review of its uses and perspectives including modeling. Crit. Rev. Biotechnol. 31, 99–111. Available from: https://doi.org/10.3109/07388551.2010.525496.

Miramontes-Martínez, L.R., Rivas-García, P., Albalate-Ramírez, A., Botello-Álvarez, J.E., Escamilla-Alvarado, C., Gomez-Gonzalez, R., et al., 2021. Anaerobic co-digestion of fruit and vegetable waste: synergy and process stability analysis. J. Air Waste Manag. Assoc. 71, 620–632. Available from: https://doi.org/10.1080/10962247.2021.1873206.

Montecchio, D., Astals, S., Di Castro, V., Gallipoli, A., Gianico, A., Pagliaccia, P., et al., 2019. Anaerobic co-digestion of food waste and waste activated sludge: ADM1 modelling and microbial analysis to gain insights into the two substrates' synergistic effects. Waste Manage 97, 27–37. Available from: https://doi.org/10.1016/j.wasman.2019.07.036.

Moodley, P., Gueguim Kana, E.B., 2018. Comparative study of three optimized acid-based pretreatments for sugar recovery from sugarcane leaf waste: a sustainable feedstock for biohydrogen production. Eng Sci Technol Int. J. 21, 107–116. Available from: https://doi.org/10.1016/j.jestch.2017.11.010.

Moodley, P., Sewsynker-Sukai, Y., Gueguim Kana, E.B., 2020. Progress in the development of alkali and metal salt catalysed lignocellulosic pretreatment regimes: potential for bioethanol production. Bioresour. Technol. 310, 123372. Available from: https://doi.org/10.1016/j.biortech.2020.123372.

Moodley, P., Trois, C., 2021. 2 - Lignocellulosic biorefineries: the path forward. In: Ray, R.C. (Ed.), Sustainable Biofuels, Applied Biotechnology Reviews. Academic Press, pp. 21–42. Available from: https://doi.org/10.1016/B978-0-12-820297-5.00010-4.

Mu, L., Zhang, L., Zhu, K., Ma, J., Ifran, M., Li, A., 2020. Anaerobic co-digestion of sewage sludge, food waste and yard waste: synergistic enhancement on process stability and biogas production. Sci. Total Env. 704, 135429. Available from: https://doi.org/10.1016/j.scitotenv.2019.135429.

Nikolaidis, P., Poullikkas, A., 2017. A comparative overview of hydrogen production processes. Renew. Sust. Energ. Rev. 67, 597–611. Available from: https://doi.org/10.1016/j.rser.2016.09.044.

Panigrahi, S., Sharma, H.B., Dubey, B.K., 2020. Anaerobic co-digestion of food waste with pretreated yard waste: a comparative study of methane production, kinetic modeling and energy balance. J. Clean. Prod. 243, 118480. Available from: https://doi.org/10.1016/j.jclepro.2019.118480.

Patil, S.M., Kurade, M.B., Basak, B., Saha, S., Jang, M., Kim, S.-H., et al., 2021. Anaerobic co-digester microbiome during food waste valorization reveals Methanosaeta mediated methanogenesis with improved carbohydrate and lipid metabolism. Bioresour. Technol. 332, 125123. Available from: https://doi.org/10.1016/j.biortech.2021.125123.

Paudel, S.R., Banjara, S.P., Choi, O.K., Park, K.Y., Kim, Y.M., Lee, J.W., 2017. Pretreatment of agricultural biomass for anaerobic digestion: current state and challenges. Bioresour. Technol. 245, 1194–1205. Available from: https://doi.org/10.1016/j.biortech.2017.08.182.

Pervez, Md.N., Bilgiç, B., Mahboubi, A., Uwineza, C., Zarra, T., Belgiorno, V., et al., 2022. Double-stage membrane-assisted anaerobic digestion process intensification for production and recovery of volatile fatty acids from food waste. Sci. Total Env. 825, 154084. Available from: https://doi.org/10.1016/j.scitotenv.2022.154084.

Qi, N., Hu, X., Xin, X., Ye, S., Fu, Z., Zhao, X., 2020. Mechanisms of biohydrogen recovery enhancement from peanut shell by *C. guangxiense*: temperature pretreatment ranges from −80°C to 100°C. Bioresour. Technol. 304, 123026. Available from: https://doi.org/10.1016/j.biortech.2020.123026.

Rafieenia, R., Lavagnolo, M.C., Pivato, A., 2018. Pre-treatment technologies for dark fermentative hydrogen production: current advances and future directions. Waste Manage 71, 734–748. Available from: https://doi.org/10.1016/j.wasman.2017.05.024.

Rajagopal, R., Bellavance, D., Rahaman, Md.S., 2017. Psychrophilic anaerobic digestion of semi-dry mixed municipal food waste: for North American context. Process. Saf. Env. Prot. 105, 101–108. Available from: https://doi.org/10.1016/j.psep.2016.10.014.

Rosen, M.A., Koohi-Fayegh, S., 2016. The prospects for hydrogen as an energy carrier: an overview of hydrogen energy and hydrogen energy systems. Energy Ecol. Env. 1, 10–29. Available from: https://doi.org/10.1007/s40974-016-0005-z.

Sharma, S.K., Goyal, P., Tyagi, R.K., 2015. Hydrogen-fueled internal combustion engine: a review of technical feasibility. Int. J. Perform. Eng. 11, 491. Available from: https://doi.org/10.23940/ijpe.15.5.p491.mag.

Shetty, D., Joshi, A., Dagar, S.S., Kshirsagar, P., Dhakephalkar, P.K., 2020. Bioaugmentation of anaerobic fungus *Orpinomyces joyonii* boosts sustainable biomethanation of rice straw without pretreatment. Biomass Bioenergy 138, 105546. Available from: https://doi.org/10.1016/j.biombioe.2020.105546.

Song, Y., Meng, S., Chen, G., Yan, B., Zhang, Y., Tao, J., et al., 2021. Co-digestion of garden waste, food waste, and tofu residue: effects of mixing ratio on methane production and microbial community structure. J. Env. Chem. Eng. 9, 105901. Available from: https://doi.org/10.1016/j.jece.2021.105901.

Srisowmeya, G., Chakravarthy, M., Nandhini Devi, G., 2020. Critical considerations in two-stage anaerobic digestion of food waste – a review. Renew. Sust. Energ. Rev. 119, 109587. Available from: https://doi.org/10.1016/j.rser.2019.109587.

Srivastava, N., Srivastava, M., Kushwaha, D., Gupta, V.K., Manikanta, A., Ramteke, P.W., et al., 2017. Efficient dark fermentative hydrogen production from enzyme hydrolyzed rice straw by *Clostridium pasteurianum* (MTCC116). Bioresour. Technol. 238, 552–558. Available from: https://doi.org/10.1016/j.biortech.2017.04.077.

Van, D.P., Fujiwara, T., Leu Tho, B., Song Toan, P.P., Hoang Minh, G., Van, D.P., et al., 2020. A review of anaerobic digestion systems for biodegradable waste: configurations, operating parameters, and current trends. Env. Eng. Res. 25, 1–17. Available from: https://doi.org/10.4491/eer.2018.334.

Xiao, B., Qin, Y., Wu, J., Chen, H., Yu, P., Liu, J., et al., 2018. Comparison of single-stage and two-stage thermophilic anaerobic digestion of food waste: performance, energy balance and reaction process. Energy Convers. Manage 156, 215–223. Available from: https://doi.org/10.1016/j.enconman.2017.10.092.

Xu, X., Sun, Yong, Sun, Yongming, Li, Y., 2022. Bioaugmentation improves batch psychrophilic anaerobic co-digestion of cattle manure and corn straw. Bioresour. Technol. 343, 126118. Available from: https://doi.org/10.1016/j.biortech.2021.126118.

Yörüklü, H.C., Filiz, B.C., Figen, A.K., Özkaya, B., 2022. Screening of biohydrogen production based on dark fermentation in the presence of nano-sized Fe_2O_3 doped metal oxide additives. Int. J. Hydrog. Energy, Green. Hydrog. For. Glob. Decarbonization 47, 15383–15396. Available from: https://doi.org/10.1016/j.ijhydene.2022.03.148.

Yue, L., Cheng, J., Tang, S., An, X., Hua, J., Dong, H., et al., 2021. Ultrasound and microwave pretreatments promote methane production potential and energy conversion during anaerobic digestion of lipid and food wastes. Energy 228, 120525. Available from: https://doi.org/10.1016/j.energy.2021.120525.

Zhang, S., Chang, J., Liu, W., Pan, Y., Cui, K., Chen, X., et al., 2018. A novel bioaugmentation strategy to accelerate methanogenesis via adding *Geobacter sulfurreducens* PCA in anaerobic digestion system. Sci. Total Env. 642, 322–326. Available from: https://doi.org/10.1016/j.scitotenv.2018.06.043.

Zheng, Y., Zhang, Q., Zhang, Z., Jing, Y., Hu, J., He, C., et al., 2022. A review on biological recycling in agricultural waste-based biohydrogen production: recent developments. Bioresour. Technol. 347, 126595. Available from: https://doi.org/10.1016/j.biortech.2021.126595.

Zhu, X., Yellezuome, D., Liu, R., Wang, Z., Liu, X., 2022. Effects of co-digestion of food waste, corn straw and chicken manure in two-stage anaerobic digestion on trace element bioavailability and microbial community composition. Bioresour. Technol. 346, 126625. Available from: https://doi.org/10.1016/j.biortech.2021.126625.

Suitability and sustainability of lignocellulosic biomass from farm to biorefinery

16

Prasun Kumar[1], *Vinay Kumar*[2] *and Manas R. Swain*[3]
[1]Department of Chemical Engineering, Yeungnam University, Gyeongsan, Republic of Korea, [2]Department of Physiology and Cell Biology, Ohio State University Wexner Medical Centre, Columbus, OH, United States, [3]DBT-IOC Centre for Advance Bio-energy Research, Indian Oil Corporation, Faridabad, Haryana, India

16.1 Introduction

Plant biomass is considered a renewable energy source. It is an important fuel source in many countries, especially for cooking and heating (Carter et al., 2018). Biomasses are classified into several categories based on their origin and nature of subdivisions. Some of them are biomass feedstocks, energy crops, crop residues, forest deposits, wood processing residues, municipal wastes, etc. (Kumar et al., 2021). Recently, biomass feedstock has been used for second-generation fuel from nonfood feedstocks such as woody crops, lignocellulosic biomass (LCB), and agricultural wastes (Devi et al., 2021). Bioenergy production at the industrial level is evaluated based on the system, considering biomass resources, conversion technology, supply systems, and services found in bioenergy. The above-described parameters must be optimized to make the biorefinery operation socially sustainable. In terms of operations, harvesting, refining, and transporting biomass are the major challenges that must be addressed by managing the supply-chain operation. Producing biofuels requires the flow of biomass raw materials from supply sites to demand centers (Malladi and Sowlati, 2018). Along this route, biomass crosses certain facilities and experiences a variety of processing methods known as the biomass supply chain (BSC) (Razik et al., 2019). The different biomass resources, geographic distribution, and conversion technologies make the BSC more economical.

In most cases, the BSC is optimized by considering the locations of biorefineries and sizes and allocating biodegradable feedstock to those refineries by making the most of the profit over the entire supply chain. It includes feed stick supplies and fuel production. Furthermore, based on the biomass type, targeted end product, and the adopted bioconversion method, it may become necessary to add pretreatment, intermediate treatment, and blending facilities within the supply chain. The economic, environmental, and social performance and efficacy of BSC can be enhanced through optimal design (Zahraee et al., 2020). The BSC focuses on the economic aspect, which can be achieved by mixed—integrated linear programming

Advances in Lignocellulosic Biofuel Production Systems. DOI: https://doi.org/10.1016/B978-0-323-91192-4.00011-0

models for optimum design of the biorefinery supply chain. The decisions made about the design of BSC have increased due to the growing demand for utilizing renewable energy sources. In this chapter, an effort has been made to show the gap in biomass availability, storage, and supply chain.

16.2 Types of biomass, availability, and its biorefinery potential

Dedicated energy crops, agricultural residues, wood processing residues, municipal garbage, and wet waste are all examples of biomass feedstocks. Utilizing biomass in biorefineries generates more value over time and is more environmentally friendly than producing bioenergy and/or biofuels.

16.2.1 Biomass types

Plant-generated biomass can be classified into several groups depending on its origin, properties, nature, and usage. The prominent biomass types are elaborated on below.

16.2.1.1 Wood and wood processing waste

Wood biomass mainly comprises biomass harvested from forests, such as firewood, tree branches, bark, and sawdust. Wood waste is generated in wood processing facilities in sawmilling, veneer, trimming, and shaving of log woods. Sawdust, commonly generated from sizing, cutting, resawing, and edging, is the consequence of trimming and smoothing wood. Thus, sawdust, off-cuts, and barks are the most common biomass waste products produced by the wood processing industries.

16.2.1.2 Energy crops

Crops or plantations explicitly grown for fuel sources are considered energy crops. Mostly these crops are harvested for heat and power generation. Energy crops can be grouped based on their production target and on the type of feedstock that they provide for conversion. In addition, energy crops can be categorized into agricultural energy (e.g., giant reed, switchgrass, Miscanthus, rye, reed canary grass, etc.), short-rotation energy crops (e.g., reduced rotation forestry and coppice), and aquatic biomass including algal biomass.

16.2.1.3 Agricultural residues

The processing of crops produces residues having the potential to be used for energy generation. Residues from arable crops (such as stems, straw, or leaves) and organic materials from overproduction are two sources of agricultural residues for energy applications. Agricultural residues are abundant after the harvesting of crops and are widely available worldwide. Furthermore, agricultural biomass is herbaceous in nature, easier for processing, and relatively sustainable for biofuel production. Therefore biofuel production from agricultural residues directly or indirectly impacts the microegomaniacs of society.

16.2.1.4 Municipal solid wastes

Municipal waste is waste generated or produced by any individuals and households. Municipal wastes comprise both liquid and solid waste. Solid waste can be recycled after collection, but organic residues can be effectively used to enrich soil nutrients. However, the biogenic fraction of municipal waste can be considered renewable, mainly referring to cardboard, textiles, food, paper, wood, and yard trimming.

16.2.1.5 Industrial wastes

Waste produced during manufacturing at industries, factories, and mining units is considered industrial waste. Industrial wastes are major causes of environmental pollution and health hazard since they are dumped and discarded into the environment. Most food processing industries produce a large quantity of waste, such as starch/sugars, juice residues, etc., which can be utilized as a source of biomass for biogas production through anaerobic digestion. Similarly, paper and pulp industries generate many waste products that can be used for bioenergy production.

16.2.2 Biomass availability

The availability of biomass for energy is a contentious issue, especially illustrated by the debate between fuels versus food (Ray and Ramachandran, 2018). All agricultural biomass, energy crops, and wood biomass have seasonal availability, and narrowed biomass availability window may make the process more complicated for the BSC. Furthermore, biomass availability is difficult to quantify, especially in light of potential competition of biomass for food, fodder, materials, and energy.

16.2.3 Biorefinery potential of various biomass

Compared to its biobased competitors, fossil-based fuels and materials are more dependent on the price of oil. Therefore it is extremely difficult to predict the point at which bio-based products become cost-competitive with their fossil counterparts. However, using biomass in biorefineries will increase value creation over time and be more sustainable than producing only bioenergy and/or biofuels. Furthermore, contrary to the biofuels business, which is seeing rapid expansion in recent years due to numerous national regulations, the use of bio-based chemicals is supported by various methods but lacks clear policy drivers. But significant expansion in the use of lignocellulosic feedstocks will necessitate modifications to the architecture of the supply chain and perhaps even more effective socioeconomic and regulatory frameworks.

16.3 Harvesting and processing strategies

Woody plant biomass requires a unique alternative for harvesting and transportation. But other agricultural biomasses are harvested during the crop harvest and do not require special attention. In general, biomass harvest can be done with conventional or suppurate harvest operations. Although traditional harvesting systems are physically capable of harvesting and recovering wood biomass, they must be cost-effective. Large and highly mechanical harvesting equipment are generally more cost-effective than small-scale systems and give high productivity.

16.4 Biomass supply chain networks

BSC management planning design can be categorized into three major segments such as (1) strategic, (2) tactical, and (3) operational. Strategic decisions are classified into three general classes (a) facility-related, (b) biomass related, and (3) final product related. The facility-related strategic decision includes decisions regarding the technology, capacity, and location of different facilities, including supply collection, preprocessing, processing, and blending and distribution sites. The above points can specifically and separately determine each facility's location and capability (Gold and Seuring, 2011; Sharma et al., 2013). A pictorial presentation and brief descriptions regarding BSC are given in Fig. 16.1 and Table 16.1, respectively.

Notably, the location of the processing site directly affects the biomass feedstock allocation– the tactical and operational aspects of planning design. Competition in the region of BSC is influenced by the proximity of the sites for supply and processing. Since low bulk-density biomass feedstock takes up a lot of space, it raises the cost of transportation, especially over long distances. Similarly, the ideal location of facilities and capacity can lower the costs of the BSC (Awudu and Zhang, 2012). Even though there are numerous biomass sources, the lateral supply chain becomes less efficient as

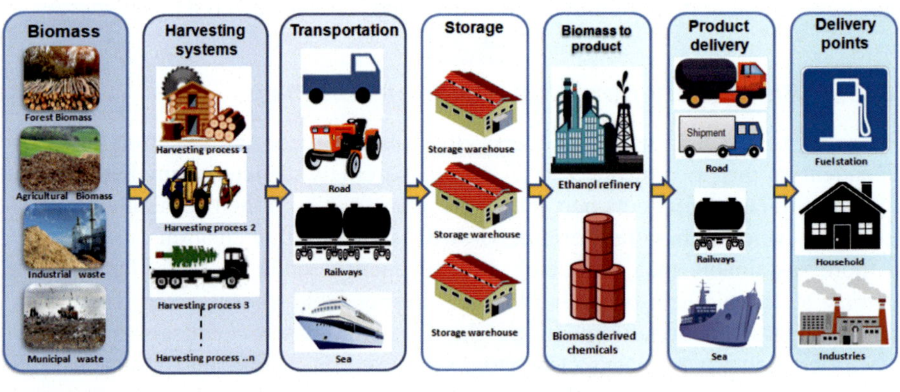

Figure 16.1 Framework of the supply chain of different biomass sources to second generation ethanol and derived chemicals from biomass and their destinations.

Table 16.1 Issues and challenges associated with biomass supply chain.

Harvesting and collection	• Postharvest biomass processing such as drying, baling, chipping • Discriminatory harvesting • Negative environmental impacts of harvesting machinery • Scattered biomass distribution and topographical constraints • Harvesting frequency and time
Storage	• Matching biomass supply and demand (agricultural crop growing cycles, weather fluctuations, etc., and the continuous bioenergy production required at the plant is a massive challenge for any storage systems) • Risky storage of biodegradable materials that are also prone to dry matter losses and quality deterioration • Sites, distance, and nature of facilities
Transport	• Infrastructure for transportation and governing laws • Higher density of transported biomass with optimized utilization of carrier vehicle • Social and environmental effects due to biomass transport
Pretreatment	• Drying and pelletization are required for the enhanced biomass energy value • Challenging torrefaction and pyrolysis
System design	• Individualization and system design optimization • High levels of interdependence and complexity in bioenergy production systems • Biomass cascade usage strategy • Look for managerial and technological improvements that can improve economic, environmental, and social effectiveness while reducing adverse effects
Technical and technological	• Biomass spatial and temporal variability • Variability in biomass composition • Conversion yields are impacted by biomass quality • Rather than maximizing residue extraction, current supply chains are constructed around high-value agricultural and forestry goods • High capital costs (CAPEX)
Financial	• The push for lower CAPEX (smaller plants) versus economies of scale (larger plants) • Technical difficulty restricting financial access • Threats posed by emerging technologies (insurability, performance, rate of return) • Increasing market volatility (energy and food markets)
Social	• Potential risks to health and safety

transportation, communication, and processing costs escalate (as a result of supplier competition). Depending on the distance between supply sites and production facilities, the kind of biomass, the production method, and the capacity of each facility, different production facilities will receive different amounts of biomass (Sharma et al., 2013).

16.5 Transport and storage

The transportation, storage, and logistics operations are primary unit operations in converting biomass feedstocks to fuels. These operations impacted major inputs on the final product delivery. So, these operations must be critically optimized for a sustainable biofuel refinery.

16.5.1 Storage in biomass supply chain

The biomass supply and demand must be well-matched for biomass storage to be effective. Because different biomass types have short harvesting seasons and are geographically dispersed over the area, storage is inherently necessary to maintain a steady supply of feedstock for bioenergy plants or biorefineries.

Storage challenges for LCB can be categorized into three major groups:

1. Need for storage of biomass throughout the BSC.
2. Cost of risks of storage.
3. Storage option.

Biomass storage is reasonably necessary to fulfill the biorefinery plant demand and manage a sustainable BSC. The short harvesting period and biomass distribution in scattered locations induce storage to ensure continued feedstock supply for bioenergy plants or biorefineries. The quicker the harvesting period emphasizes, the higher number of storage units needed as buffer capacity. Storage terminals can be the supply unit source for several other plants with insufficient storage capacity. Storage cost mainly depends on the location and type of storage.

Moreover, the volume of biomass to be stored and the duration of storage is the pivotal cost factors. During storage practice, biomass may lose its quality because of the loss of moisture and nutrients. Further, it is recommended to use close-type warehouses. If the storage facility is adjutant to the biorefinery, it may be used to dry the stored biomass using exhaust heat from the facility, which may improve the biomass quality during storage.

The location and kind of storage facility have the most significant impact on the storage choice (possibly close to the harvesting site, near the field or forest). The storage facility needs to be connected via appropriate road connectivity for proper automobile transportation. Open, roof-covered, and air-fan storage are just a few of the storage options. The environment and storage of biomass processing influence the kind of storage. Open or covered storage may be beneficial when biomass is compressed into bales since round bales can withstand exposure to specific degrees of rain and weather. However, while they can be stacked well, square bales must be kept in covered storage because they are somewhat vulnerable to weather damage.

16.5.2 Transport in biomass supply chain

The overall infrastructural framework where a bioenergy supply system operates constitutes its essential preconditions of functioning. However, developing countries may not be able to provide the basic transportation infrastructure and face a severe hurdle for affecting foreign investment in bioenergy systems.

Biomasses are generally available in low energy density. Therefore, biomass must be collected from broad areas depending on availability. The most popular and cost-effective method of transportation is by rail. However, there is always no rail connectivity between biomass collection centers and rail hubs. In this regard, intermediate road transportation is required with rail transportation, which collectively increases transportation costs. In some cases, biomass is received at ports with supporting rail infrastructure (Eksioglu et al., 2009).

Compared to other ways, road transportation is the one that is most frequently used to move biomass since it is readily available. Road transportation produces significantly high levels of greenhouse gas (GHG) emissions. Speed and distance significantly affect cost analysis for transporting a tonne of biomass via road. According to a survey, a maximum of 18 tonnes of biomass can be transported per load of 21 tonnes in most large articulated trucks. Varying feedstocks will have different road transport GHG emission factors because of changes in bulk density (Ghaderi et al., 2016).

Bulk ships with a capacity of 80,000 tonnes or more are used for the marine transportation of biomass. The GHG emissions from shipping are typically 40 times lower than trucks fueled with fossil fuels (Ghaderi et al., 2016).

16.6 Modeling approaches

The three primary classifications of modeling methods are– (1) mathematical programs, (2) multicriteria decision-making, and (3) heuristic and metaheuristic programs. In this respect, mathematical programming with more than 90% uses is the most practical approach in the BSC network design (BSCND) model. However, real-world problems are complicated and cannot be modeled by linear methods. As a result, each modeling strategy will be assessed and examined using features like objective type, number of objective functions, time periods, sustainability, uncertainty factors, and solution methods.

All requirements are met by mathematical programming models based on objective function restrictions and decision variables. The mathematical programming models employed in the chosen studies include linear, nonlinear, mixed integer linear, and mixed integer nonlinear programming models, considering the goal functions, decision variables, and constraints. A particular kind of mathematical programming called "linear programming" has a linear goal function and linear constraints. Every linear BSC network design model has a single aim and disregards sustainability. The cost of transpiration in the BSC is significantly reduced by site optimization, which uses mostly linear programming. Additionally, optimizing the facility's location reduces transportation expenses for the BSC. By combining the data from many aspects, Geographical Information Systems (GIS) can assist in determining the ideal location of the facilities in a particular area.

When there is a substantial price variability of biomass, a GIS-based method can find the best location of the processing sites and the best allocation of biomass based on marginal delivery costs and resource competition across facilities (Panichelli and Gnansounou, 2008). A GIS-based Environmental Decision Support System for the best logistical and strategic planning to create energy from woody biomass was presented by Frombo et al. (2009). Based on reducing overall costs, this system may optimize the decision factors for the annual biomass collection and the capacity of the biorefinery.

Biomass is sometimes expensive to procure since the locations for harvesting and processing are spread out over a large area (Möller and Nielsen, 2007). The shortest possible travel routes between the biomass harvesting sites and processing facility are determined by various algorithms (like Dijkstra) proposed in the literature. Ren et al. (2013) put forth a model for creating a sustainable bioethanol supply chain that considers various production technologies: biomass feedstocks, transportation methods, patterns, and disposal options. A few limitations impact the goal function of this model. The target function seeks explicitly to reduce the overall ecological impact.

Cundiff et al. (1997) devised a scenario-based multiperiod linear programming model to estimate the capacity of the collecting locations in a herbaceous biomass delivery system. This model accounts for the variability in biomass production levels brought on by changing environmental circumstances and irregular growth and harvesting times.

All or some of the decision variables in mixed integer linear programming (MILP) models, which feature linear objective functions and constraints, are integers. For cases where integer variables are required due to discrete phenomena, MILP models are helpful tools. Since the relevant binary choice variables may accurately represent the development of facilities in various capacities and production methods, the facility placement decision is an important consideration when utilizing MILP models. In these models, optimization is carried out either following economic, environmental, or social objectives or, in some circumstances, simultaneously with the economic, environmental, and social components (You et al., 2012; Yue et al., 2014). Nagel (2000) suggested a MILP model that, through the dynamic evaluation of economic performance and efficiency, may identify the most economical and ecological structure for supplying biomass. The assessment of producing heat and electricity from biomass at the regional level (Freppaz et al., 2004), the design and scheduling of supply chain processes to produce heat from biomass (Dunnett et al., 2007), the design and evaluation of waste biomass sustainable supply chains (Toka et al., 2008), and the strategic planning of bioethanol supply chain systems (Huang et al., 2010) are emphasized. According to Akgul et al. (2011), optimal bioethanol supply chain architecture includes figuring out the best harvesting and production rates, strategic planning for BSCs, and the best feedstock allocation under unpredictable site selections (Chen and Fan, 2012). To determine the estimated value of the biomass supply and the transportation distance, Geijzendorffer et al. (2008) coupled GIS and MILP. In addition to evaluating facility location, GIS can be utilized to choose the best path for the feedstock flow between facilities.

Most MILP models have solely taken the economic factor into account. Some authors have since concentrated on environmental optimization. For instance, Foo et al. (2013)

suggested some robust models for the synthesis of empty fruit bunch allocation networks in the regional BSC to generate energy from palm oil. Lam et al. (2010) presented a new method for regional energy clustering. The reduction of GHG emissions is the sole goal of both approaches. In most MILP models, economic and environmental objectives are typically represented by minimizing the overall supply chain costs and GHG emissions (You and Wang, 2011, 2012; Zamboni et al., 2009a,b). However, in real-world issues, it occasionally happens that all or some parameters change over time in accordance with a predictable tendency. Because it has the potential to influence planning decisions greatly, it is important to take this dynamic into account when defining problem parameters. In these circumstances, a multiperiod model is utilized to conform to the dynamic changes, with the time horizon being separated into numerous periods (Melo et al., 2009). This means that choices must be taken for each period regarding the placement and capacity of the facilities and the flows of biomass and biofuel. A multiobjective multiperiod model for developing and planning a bioethanol supply chain taking into account first- and second-generation bio-refineries has been presented by Giarola et al. (2011). Mele et al. (2011) proposed a multiperiod MILP model to create a supply chain for synthetic sugar-bioethanol that considers both economic and environmental goals. The key choices that need to be made for each time period in this model are the number and capacity of collection sites and biorefineries.

The strategic planning and feedstock allocation decisions of the bioenergy supply chain systems are linked to uncertainties in demand for biomass and bioethanol according to the mixed-integer stochastic programming model proposed by Chen and Fan (2012). For the biofuel supply chain design problem under varying demand conditions, Kim et al. (2011) proposed a scenario-based robust MILP model that could specify the ideal value of the feedstock flow between various supply chain facilities and the excellent values of the biorefinery-related strategic variables. To deal with the ambiguity and dynamism of biomass supply, biofuel pricing, crude oil demand, and production technology, Tong et al. (2014) suggested a multiperiod, stochastic MILP model.

The actual BSC operation has an impact on the environment in addition to the economic component (De Meyer et al., 2014). Environmental factors (caused by land, water, forest, and GHGs emissions) and social factors (caused by the development or deletion of job possibilities) are the most significant impact on BSC operation (Sabri and Beamon, 2000; Amigun et al., 2011; Duku et al., 2011).

Several entities are likely to change from one BSC site to another one. These entities consist of the following:

1. Supply facilities where the biomass feedstock is grown, harvested, or produced.
2. Locations where the biomass or the biofuel produced is collected and stored.
3. Processing plants where preprocessing operations such as size reduction, pelleting, and drying are carried out to make the necessary changes to assist storage and transportation.
4. Locations where intermediate products such as bio-oil are produced.
5. Sites where bioethanol is blended and mixed with other additives or used to synthesize gasoline.
6. Distribution sites from where the stored biofuel or bioenergy is transferred.
7. Demand centers where final products such as biodiesel, bioethanol, gasoline, and the energy produced from biomass are used.

16.7 Conclusions and future perspectives

Strategic choices are necessary for network design modeling. This chapter strongly emphasizes the strategic choices made in a BSC, which include facility-related, biomass-related, and ultimate product-related strategic choices. This study also analyses and groups the selected publications according to modeling approaches, uncertainties, solution methodology, sustainability, model features, entities, and data to highlight the gaps in the literature and provide research possibilities and future directions. Without considering all aspects of sustainability, researchers have mostly focused on single-feedstock, single-product, single-period, single-objective, and deterministic models. The majority of BSCND articles use approaches from mathematical programming.

Real-world issues are invariably complex, making it impossible to represent them using straightforward MILP techniques. Researchers are motivated to create MILP models for BSCND situations because of the simplicity of the linear model compared to the complexity of nonlinear models. Researchers may find it easier to tackle real-world issues by using nonlinear approaches that are appropriate for real-world problems rather than straightforward linear methods and more adaptable convex optimization methods as well as solution methodology. The small percentage of nondeterministic models suggest reluctance because of the computing challenges in their implementations. Stochastic (chance-constrained, probabilistic, two-stage, and scenario-based) and probabilistic programming are two methods for addressing uncertainty in BSCND models. This study shows that researchers closely examined stochastic approaches to handling uncertainties. However, fuzzy logic, robust optimization, interval approaches, and chaos theory are useful approaches that can be utilized to cope with data uncertainty.

Analyses suggest that future research in a stochastic approach to handling uncertainty should focus on multistage stochastic approaches and robust optimization techniques instead of typical stochastic programming models. The availability of biomass and the demand for biofuel are the two most important parameters discussed as unknown parameters in the most relevant articles. To be sustainable, a supply chain must consider economic, environmental, and social goals (Amigun et al., 2011; Duku et al., 2011).

References

Akgul, O., Zamboni, A., Bezzo, F., Shah, N., Papageorgiou, L.G., 2011. Optimization-based approaches for bio-ethanol supply chains. Ind. Eng. Chem. Res. 50 (9), 4927–4938.

Amigun, B., Musango, J.K., Stafford, W., 2011. Biofuels and sustainability in Africa. Renew. Sustain. Energy Rev. 15 (2), 1360–1372.

Awudu, I., Zhang, J., 2012. Uncertainties and sustainability concepts in biofuel supply chain management: a review. Renew. Sustain. Energy Rev. 16 (2), 1359–1368.

Carter, E., Shan, M., Zhong, Y., Ding, W., Zhang, Y., Baumgartner, J., et al., 2018. Development of renewable, densified biomass for household energy in China. Energy Sustain. Dev. 46, 42–52.

Chen, C.W., Fan, Y., 2012. Bio-ethanol supply chain system planning under supply and demand uncertainties. Transport. Res. Part. E: Logist. Transport. Rev. 48 (1), 150–164.

Cundiff, J.S., Dias, N., Sherali, H.D., 1997. A linear programming approach for designing a herbaceous biomass delivery system. Bioresour. Technol. 59 (1), 47–55.

De Meyer, A., Cattrysse, D., Rasinmäki, J., Van Orshoven, J., 2014. Methods to optimize the design and management of biomass-for-bio-energy supply chains: a review. Renew. Sustain. Energy Rev. 31, 657–670.

Devi, A., Niazi, A., Ramteke, M., Upadhyayula, S., 2021. Techno-economic analysis of ethanol production from lignocellulosic biomass—a comparison of fermentation, thermo catalytic, and chemocatalytic technologies. Bioprocess. Biosyst. Eng. 44 (6), 1093–1107.

Duku, M.H., Gu, S., Hagan, E.B., 2011. A comprehensive review of biomass resources and biofuels potential in Ghana. Renew. Sustain. Energy Rev. 15 (1), 404–415.

Dunnett, A., Adjiman, C., Shah, N., 2007. Biomass to heat supply chains: applications of process optimization. Process. Saf. Environ. Prot. 85 (5), 419–429.

Eksioglu, S.D., Acharya, A., Leightley, L.E., Arora, S., 2009. Analyzing the design and management of biomass-to-biorefinery supply chain. Comput. Ind. Eng. 57, 1342–1352.

Foo, D.C., Tan, R.R., Lam, H.L., Aziz, M.K.A., Klemeš, J.J., 2013. Robust models for the synthesis of flexible palm oil-based regional bio-energy supply chain. Energy 55, 68–73.

Freppaz, D., Minciardi, R., Robba, M., Rovatti, M., Sacile, R., Taramasso, A., 2004. Optimizing forest biomass exploitation for energy supply at a regional level. Biomass Bioenergy 26 (1), 15–25.

Frombo, F., Minciardi, R., Robba, M., Sacile, R., 2009. A decision support system for planning biomass-based energy production. Energy 34 (3), 362–369.

Geijzendorffer R., Annevelink E., Elbersen B., Smidt R., de Mol R.M., 2008. Application of a GIS-Bioloco tool for the design and assessment of biomass delivery chains. In Proceedings of the Sixteenth European Biomass Conference and Exhibition, 2–6 June, Valencia, Spain.

Ghaderi, H., Pishvaee, M.S., MoiniA, 2016. Biomass supply chain network design: an optimization-oriented review and analysis. Ind. Crop. Products 94, 972–1000.

Giarola, S., Zamboni, A., Bezzo, F., 2011. Spatially explicit multiobjective optimization for design and planning of hybrid first and second generation bio-refineries. Comput. Chem. Eng. 35 (9), 1782–1797.

Gold, S., Seuring, S., 2011. Supply chain and logistics issues of bio-energy production. J. Clean. Prod. 19 (1), 32–42.

Huang, Y., Chen, C.W., Fan, Y., 2010. Multistage optimization of the supply chains of biofuels. Transport. Res. Part. E: Logist. Transport. Rev. 46 (6), 820–830.

Kim, J., Realff, M.J., Lee, J.H., Whittaker, C., Furtner, L., 2011. Design of biomass processing network for biofuel production using an MILP model. Biomass Bioenergy 35 (2), 853–871.

Kumar, A., Adamopoulos, S., Jones, D., Amiandamhen, S.O., 2021. Forest biomass availability and utilization potential in Sweden: a review. Waste Biomass Valoriz. 12 (1), 65–80.

Lam, H.L., Varbanov, P.S., Klemeš, J.J., 2010. Optimization of regional energy supply chains utilizing renewable: P-graph approach. Comput. Chem. Eng. 34 (5), 782–792.

Malladi, K.T., Sowlati, T., 2018. Biomass logistics: a review of important features, optimization modeling and the new trends. Renew. Sustain. Energy Rev. 94, 587–599.

Mele, F.D., Kostin, A.M., Guillén-Gosálbez, G., Jiménez, L., 2011. Multiobjective model for more sustainable fuel supply chains: a case study of the sugar cane industry in Argentina. Ind. Eng. Chem. Res. 50 (9), 4939–4958.

Melo, M.T., Nickel, S., Saldanha-Da-Gama, F., 2009. Facility location and supply chain management—a review. Eur. J. Oper. Res. 196, 401–412.

Möller, B., Nielsen, P.S., 2007. Analyzing transport costs of Danish forest wood chip resources by means of continuous cost surfaces. Biomass Bioenergy 31 (5), 291–298.

Nagel, J., 2000. Determination of an economic energy supply structure based on biomass using a mixed-integer linear optimization model. Ecol. Eng. 16, 91–102.

Panichelli, L., Gnansounou, E., 2008. GIS-based approach for defining bioenergy facilities location: a case study in Northern Spain based on marginal delivery costs and resources competition between facilities. Biomass Bioenergy 32 (4), 289–300.

Ray, R.C., Ramachandran, S. (Eds.), 2018. Bioethanol Production from Food Crops: Sustainable Sources, Interventions, and Challenges. Academic Press.

Razik, A.H.A., Khor, C.S., Elkamel, A., 2019. A model-based approach for biomass-to-bioproducts supply chain network planning optimization. Food Bioprod. Process. 118, 293–305.

Ren, J., Manzardo, A., Toniolo, S., Scipioni, A., Tan, S., Dong, L., et al., 2013. Design and modeling of the sustainable bio-ethanol supply chain by minimizing the total ecological footprint in life cycle perspective. Bioresour. Technol. 146, 771–774.

Sabri, E.H., Beamon, B.M., 2000. A multiobjective approach to simultaneous strategic and operational planning in supply chain design. Omega 28 (5), 581–598.

Sharma, B., Ingalls, R.G., Jones, C.L., Khanchi, A., 2013. Biomass supply chain design and analysis: Basis, overview, modeling, challenges, and future. Renew. Sustain. Energy Rev. 24, 608–627.

Toka, A., Iakovou, E., Vlachos, D., 2008. Design of waste biomass supply chains for energy production. In Proceedings of the East European Doctoral Student Conference, p. 68.

Tong, K., Gong, J., Yue, D., You, F., 2014. Stochastic programming approach to optimal design and operations of integrated hydrocarbon biofuel and petroleum supply chains. ACS Sustain. Chem. Eng. 2 (1), 49–61.

You, F., Wang, B., 2011. Life cycle optimization of biomass-to-liquid supply chains with distributed–centralized processing networks. Ind. Eng. Chem. Res. 50 (17), 10102–10127.

You, F., Wang, B., 2012. Multiobjective optimization of biomass-to-liquids processing networks. Found. Comput. Process. Oper. 1 (7).

You, F., Tao, L., Graziano, D.J., Snyder, S.W., 2012. Optimal design of sustainable cellulosic bio-fuel supply chains: multiobjective optimization coupled with life cycle assessment and input–output analysis. AIChE J. 58 (4), 1157–1180.

Yue, D., You, F., Snyder, S.W., 2014. Biomass-to-bio-energy and biofuel supply chain optimization: overview, key issues and challenges. Comput. Chem. Eng. 66, 36–56.

Zahraee, S.M., Shiwakoti, N., Stasinopoulos, P., 2020. Biomass supply chain environmental and socio-economic analysis: 40-Years comprehensive review of methods, decision issues, sustainability challenges, and the way forward. Biomass Bioenergy 142, 105777.

Zamboni, A., Bezzo, F., Shah, N., 2009b. Spatially explicit static model for the strategic design of future bioethanol production systems. 2. Multiobjective environmental optimization. Energy Fuels 23 (10), 5134–5143.

Zamboni, A., Shah, N., Bezzo, F., 2009a. Spatially explicit static model for the strategic design of future bio-ethanol production systems. 1. Cost minimization. Energy Fuels 23 (10), 5121–5133.

Life-cycle analysis for sustainable biofuel production

17

Noé Aguilar-Rivera[1], Christian Michel-Cuello[2] and María Mercedes González Bernal[3]

[1]Faculty of Biological and Agricultural Sciences, Universidad Veracruzana, Córdoba, Veracruz, Mexico, [2]Multidisciplinary Academic Unit Central Zone, Autonomous University of San Luis Potosí, Rioverde, San Luis Potosí, México, [3]Universidad Industrial de Santander, Bucaramanga, Colombia

17.1 Introduction

Various factors such as the increase in population and the demand for fuels to satisfy energy requirements, especially in transport, and the decrease in both reserves and volumes of oil extraction have generated an energy crisis worldwide. The growing concern about the adverse effects and current and future consequences of global warming has driven the development of more efficient, economical, and less environmental impact technologies for power generation (Morone et al., 2020; Demırbas, 2017).

A viable alternative that takes several decades of development and research is the generation of fuels in liquid or gaseous forms from plants, animals, and waste biomass or generated from the application of microorganisms on sources such as CO_2. However, a rigorous analysis of the life cycle of such fuels is required to position them with an advantage over fossil fuels (Aro, 2016). According to their origin, biofuels are classified as first (1G), second (2G), and third generation (3G) fuels. However, its large-scale production is still challenging. It is essential to convert biomass into biofuels by developing novel technology to increase biofuel production to fulfill the current and future energy demand (Ambaye et al., 2021). The technological utilization of lignocellulosic biomass (LCB: cellulose, hemicellulose, and lignin) would play a significant role in the economic viability of 2G ethanol (Patel and Shah, 2021). Despite the advancements in 2G ethanol, they still represent less than 3% of total ethanol production. The environmental impacts of 2G ethanol production depend on feedstock availability, pretreatment methods, and conversion technology [fermentation, hydrolysis process, and separation (Sharma et al., 2020)].

Some authors differ in the classification of biofuels by generation and prefer to characterize them according to their biomass source and classify them as conventional biofuels (related to the 1G) and advanced biofuels (related to the 2G). The more detailed classification and examples can be seen in Table 17.1. The fuels classified as 1G applied conventional technologies such as fermentation and

Advances in Lignocellulosic Biofuel Production Systems. DOI: https://doi.org/10.1016/B978-0-323-91192-4.00003-1

Table 17.1 Examples of biofuels and their different classifications.

Classification by generation	Alternative classification	Biofuel	Characteristics	Raw material	Applications
1ª gen	Conventional	Bioethanol	Fermentation of sugars by yeasts, previous hydrolysis of starch by enzymes, and the CO_2 generated in the process can have different applications	Sugar cane, sugar beet, sweet sorghum, cassava, corn, sorghum, wheat and millet.	Commercial vehicles
		Biodiesel/pure vegetable oil	Obtained by esterification	Vegetable oils are extracted from seeds or fruit kernels of plants such as canola (rapeseed), soybean, oil palm, sunflower, and jatropha. Oil-rich and fatty wastes from the kitchen	Commercial vehicles based on conventional/ modified diesel engines
		Biogas	Decomposition of organic matter by microorganisms under anaerobic conditions by refining biomethane is obtained	Sewage sludge, animal manure, organic waste, municipal putrescible waste, and green or ensilaged biomass from energy crops such as corn or sorghum.	Production of electricity and heat, industrial processes. Compressed biomethane is used for transport
2ª gen	Advanced	Cellulosic ethanol	Conversion of cellulose and hemicellulose into fermentable sugars and subsequent alcoholic fermentation	Agricultural and forestry residues such as corn straw and stubble, forest crop residues, and paper processing	Commercial vehicles
		Hydrotreated vegetable oil (HVO)	Alternative esterification process	Residues of vegetable oils or animal fats	Substitute aviation fuels or diesel
		Biomass to liquids (BtL) diesel	First, gasified biomass to produce a synthesis gas (Syngas), impurities such as tar, suspended particles, and other contaminants are removed. By methods like Fischer-Tropsch (FT)or Mobil process, the syngas is catalytically converted into various hydrocarbon liquids	Biomass	Synthetic diesel for vehicles

		Synthetic natural gas (BioSNG)	Cellulose gasification processes	Forest residues	Used in a similar way to biomethane upgraded from biogas.
		Bio-based dimethyl ether (BioDME)	Methanol is produced from gasified biomass feedstock and then is converted to Bio-DME.	Gasified biomass	It is a gas at environmental temperature and pressure and burned as natural gas
		Biobutanol	Fermentation from biomass using specific microorganisms	Biomass	Due to its higher energy density, it is used as gasoline without modifications, with less corrosive power
		Algal biofuel	CO_2 sources, especially polluting substances or elements as raw materials	Certain varieties of single- or multicelled algae	Substitute potential for petroleum-based fuel
3a̲ gen		Biodiesel and ethanol at the level pilot plant	The production of its raw material fixes CO_2 in its photosynthetic process and reduces the emission of greenhouse gases	Evergreen grasses, fast-growing trees and plants, green algae, and blue-green algae	Emerging production technologies
4a̲ gen		Bioethanol	Takes raw material carbonic anhydride (CO_2) as raw material	Genetically modified bacteria	Technologies in theoretical phase cellular biology, photobiological solar fuels, and electro fuels

Sources: From Álvarez-Maciel (2009), World Bioenergy Association (2013), Siliceo (2014), Aro (2016), and Robak and Balcerek (2020).

esterification to obtain fuels from carbohydrates and oils from various plant varieties. Its primary disadvantage lies in the use of plants for food use (Guo and Tanaka, 2022). This dilemma over its use towards its consumption as food or its use as a source of fuel has, therefore, increased the price of food or even affected food security.

2G biofuels have emerged as an alternative to using food crops as raw material. Their technology is oriented towards using process residues and inedible plant parts with high cellulose contents, which are used as a highly fermentable substrate and prone to saccharification. This technology is economically viable but does not contribute to reducing greenhouse gas (GHG) emissions since it shares the technology of its production process with 1G biofuels.

The 3G biofuels adapt their transformation processes for use as the raw material of fast-growing plant species with high photosynthetic efficiency and high biomass production, such as Bermuda grass (*Cynodon dactylon*), sugarcane (*Saccharum officinarum* L.), corn (*Zea mays*), sorghum (*Sorghum* spp.) and amaranth (*Amaranthus* spp.). The diversification of these crops has resulted in an increase in the uptake of CO_2 by these plants and, consequently, a decrease in GHG. However, its negative effect is the displacement of food crops. The change in land use to open new crops promotes deforestation and removes their native or endemic species with ecological functionality, causing a significant environmental impact (Álvarez-Maciel, 2009). Therefore 3G biofuels have been envisioned, where microorganisms are deployed. For instance, since algae can form up to 200 times more biomass per area than terrestrial biomass, they hold great promise for future biofuel production on marginal land or in the ocean (Behera et al., 2014). The challenge for the 3G process is to facilitate the exposure of more cellulose and hemicelluloses for enzymatic hydrolysis named "green fractionation" of biomass by applying the principles of green chemistry for 4G ethanol production (Satari and Jaiswal, 2020; Robak and Balcerek, 2020).

Table 17.1 shows the productive characteristics of each generation and type of biofuels. These technologies have evolved to increase efficiency in the sustainable use of raw materials and improve available refining technologies. However, little research has been done about the environmental, social, and economic implications related to the contamination of each process.

17.2 Sustainability of biorefinery from biomass

The concepts "bioenergy," "biofuels," and "bioeconomy" are relatively recent. They have been the subject of controversy and discussion due to the constant instability in the geopolitics of oil. Biorefinery development is a viable, sustainable, and rational alternative to supply part of the current demand for energy(fuel), bioproducts, and commodities (Cortez et al., 2020). The sustainability of any biorefinery majorly depends on defining the feedstock (Du et al., 2019).

Agriculture based on the production of bioenergy crops has had a notable development in recent years. As mentioned above, it is related to negative effects such as its high consumption of fuels for agricultural work, higher requirements of agrochemicals than traditional crops, its impact on ecosystems, and competition for arable land with other agricultural and forest species. In addition, its transformation processes must present economic and environmental advantages over fossil fuel production processes. The generation number and alternative classification of fuels are based mainly on the characteristics and properties of their raw material. Other important aspects considered for the design and adaptation of production processes are their energy efficiency and the reduction of emissions compared to fossil fuels (Singh et al., 2022). For example, the process based on the use of sugar cane to produce ethanol presents the following values: the energy invested in the crop includes all that inherent to fertilizers, herbicides, pesticides, transport of materials and personnel, irrigation, and fuels, which it translates into that the energy required for its cultivation phase is 4,036,011 kcal/ha, taking a reference yield of 5700 L/ha, a cost of 708.07 kcal/L is obtained. Later the industrial process that integrates transport, water, distillation, concentration, electricity, and wastewater treatment; requires approximately 3898.68 kcal/L. Adding both production and processing values, an investment of 4606.75 kcal/L is obtained. The energy released by 1 L of ethanol is 5610 kcal, resulting in a yield of 22%. Furthermore, if the energy provided by co-products such as distillation and crop residues (cellulose, hemicellulose, and lignin) is integrated into the calculation as a positive value, the yield value reaches 88% (García, 2011). To produce the same biofuel, the analysis of CO_2 emissions generates about 0.253 kg/L in its cultivation and 1170 kg/L in its industrial process, totaling 1423 kg of CO_2/L. In contrast, gasoline production represents 2800 kg of CO_2/L through its exploration, drilling, extraction, and refining activities. The energy contribution ratio between gasoline and bioethanol is 1.0: 1.5; that is, it is necessary to apply 1.5 L of bioethanol to replace the energy provided by 1 L of gasoline. Therefore, the substitution of gasoline for bioethanol results in a 23% reduction in emissions, equivalent to 665 g of CO_2 for each liter of gasoline substituted by bioethanol obtained from sugar cane (García, 2011).

17.3 Life cycle analysis

It is important to analyze the impacts of environmental, social, and economic factors during the entire biofuel production process. For this, tools such as the life cycle analysis (LCA) are used, which analyze the potential environmental aspects and impacts during the whole life cycle of a product from the generation of its raw material, through its use, to its final disposal as waste/byproduct. It is also called a study from the cradle to the grave. This tool can be applied to determine the environmental cost of biofuel production materials, elements, and processes, and the different regulatory contexts of each country determine the methodologies and their reference values. There are established models that consider different factors to create analysis (Rathore et al., 2013; De Jong et al., 2017; Bonomi et al., 2018):

1. The greenhouse gases.
2. Regulated Emissions and Energy use in Transportation (GREET) (United States of America).
3. BioGrace (European Community).
4. GHGenius (Canada).
5. New EC (European Community).
6. Virtual Sugarcane Biorefinery (VSB, Brazil).

It is important to highlight that each model presents differences in data integration and methodological procedures. Therefore, they offer different results depending on the value assigned to the analyzed variables.

McManus and Taylor (2015) concluded the LCA has evolved from its origins in energy analysis in the 1960s and 1970s into a wide-ranging tool used to determine the impacts of products or systems on several environmental and resource issues. The approach has become more prevalent in research, industry, and policy.

For a few decades, several LCA studies have been conducted to measure the sustainability of ethanol production. However, research is contradictory when compared to each other in terms of methodological assumptions, technical and agricultural issues applied, bottlenecks and regions because it is not possible to compare the influence of each component of the methodology on the results with high certainty because various technical and agricultural-related parameters are also involved in the results of studies (Soleymani Angili et al., 2021).

Because LCAs are a transdisciplinary tool that can and are used by all professions but works best when conducted by stakeholders with diverse skills and backgrounds, and the life cycle interpretation is the final phase, in which the results of a Life Cycle Inventory or a Life Cycle Impact Assessment tools or both, are summarized and discussed as conclusions, recommendations, and a decision-making process following the goal and scope definition (Pinto et al., 2022; Brandao et al., 2022; Sieverding et al., 2020)

Biofuel generation from different types of feedstocks presents various characteristics and requirements. These requirements are the chemical composition of feedstocks, type of soil in which the biomass is grown, water supply, nutrient demand, agrochemicals, cultural management, environmental conditions (i.e., radiation, temperature, relative humidity), transport systems, and specialized transformation processes with specific inputs and outputs of matter and energy; finally, its transformation into energy using motors may require unique characteristics. Evidence from studies on the development of processes to obtain biofuels indicates some advantages but leaves aside negative consequences. Some examples point out the decrease in GHGs but do not consider that the uses of nitrogen fertilizers contribute to considerable emissions of toxic N_2O, which is why the importance of integrating environmental, political, social, and technological components is essential (Achinas et al., 2019)

Gupta and Verma (2015) discussed current technologies available for sustainable ethanol production from agroresidues such as corn, sugarcane, wheat, sugar beet, cassava, barley, and rye. They also mentioned other lignocellulosic materials for ethanol production such as corn stover, wheat, and cereal straw, poplar aspen, rice straw, bagasse, seaweed (*Sargassum spp.*), paper, cotton seed hairs, oil palm frond, coconut fiber, pineapple leaf, banana stem, softwood and hardwood sawdust, switchgrass, and jatropha waste. Maga et al. (2019) mentioned that to valorize several wastes, byproducts, and biomass, the researchers investigated new conversion technologies that could process the raw material and produce ethanol biofuel competitively and sustainably. However, several constraints, such as limited land availability and substantial restrictions on extending areas for energy plantations or collecting biomass due to competition with food crops, logistical requirements, and climate conditions, hinder large-scale biofuel production. Moreover, 2G ethanol promises lower environmental impacts than 1G ethanol. In this context, sugarcane is the most widely used source of biomass to produce 1G and 2G ethanol. It is carried out through two technological options:

1. *1G ethanol:* Sugarcane production system and trash harvesting from the crop field, cleaning, preparation, and milling, sugarcane juice treatment, sugar production, yeast fermentation of molasses, sugarcane juice, or both, yeast acid wash for recycling, distillation, and dehydration, and steam and power generation where bagasse and trash are used as fuel in steam and power generation.
2. *2G ethanol:* Sugarcane bagasse transportation from 1G to 2G plant from sugar mills, trash harvesting from the field (recovery of 50% of trash), loading and transportation, trash cleaning and milling, biomass pretreatment, on-site enzyme production, enzymatic hydrolysis, yeast propagation; C5/C6 sugar fermentation, distillation and dehydration, wastewater treatment, and steam and power generation from bagasse and trash.

To be economically feasible, a stand-alone 2G plant should be located nearby a 1G ethanol plant and consider the simultaneous production of several bioproducts derived from sucrochemistry, ethanolchemistry, and lignochemistry. LCA determines the advantages and disadvantages of each of them. Their analysis also guides permanent research for developing and generating new biofuels, bioproducts from collateral production, and public policies with stakeholders that guide technological development (Rathore et al., 2013). The integration of 2G ethanol production processes from sugarcane bagasse and trash in 1G biorefineries and artisanal methods (non centrifuged sugar) may improve their feasibility, mainly because integrated processes will require a lower investment because some operations like concentration, fermentation, distillation, storage, and cogeneration may be shared between 1G and 2G plants and byproducts utilization for bioproducts (Fig. 17.1).

According to Karp et al. (2021), three main stages can be recognized during the determination of boundaries in an LCA: the "cradle" stage is usually composed of cultivation, collection, and harvesting, passing through the "gate" stage (preprocessing, transport, hydrolysis, fermentation, and distillation), and the "grave" stage (transport and use). The choice of environmental indicators, the definition of system boundaries, that is, "well to the gate," "gate to wheel," "well to wheel," energy

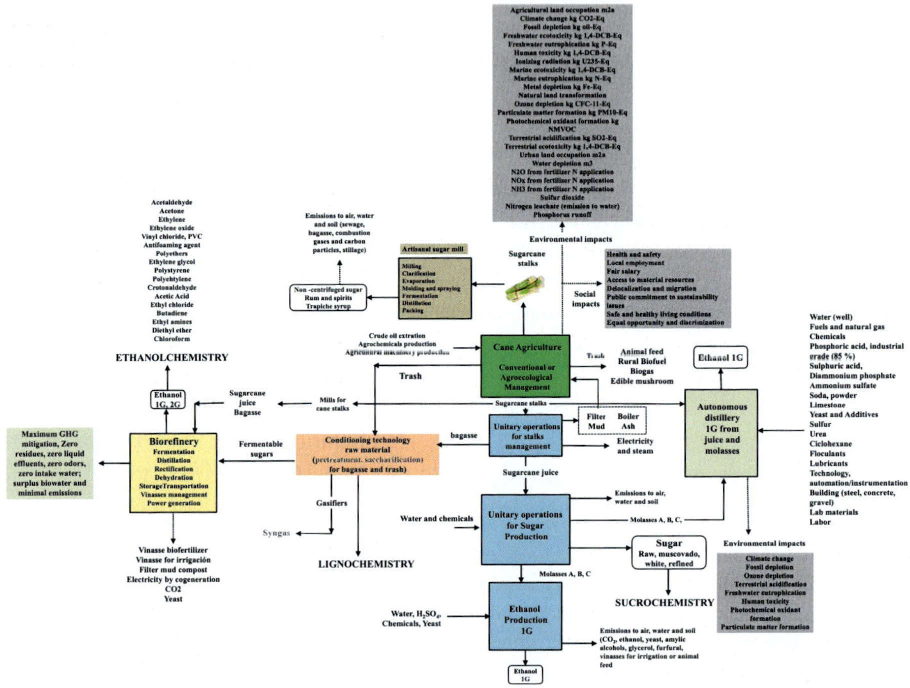

Figure 17.1 Life cycle analysis of first- and second-generation ethanol and productive integration for sustainability for the conventional sugar industry (Amores et al., 2013; Hun et al., (2017).
Source: Adapted from Amores, M. J., Mele, F. D., Jiménez, L., and Castells, F. 2013. Life cycle assessment of fuel ethanol from sugarcane in Argentina. Int. J. Life Cycle Assess., 18 (7), 1344–1357. and Hun, A. L. N., Mele, F. D., and Pérez, G. A. 2017. A comparative life cycle assessment of the sugarcane value chain in the province of Tucumán (Argentina) considering different technology levels. Int. J. Life Cycle Assess., 22(4), 502–515.

recovery, carbon emissions, and storage, among other differences in the basic assumptions, depend on several considerations, data limitations, and research regions.

A basic LCA framework consists of 14 environmental indicators: acidification, climate change, ecotoxicity for fresh aquatic water (Fecotox), aquatic eutrophication, terrestrial eutrophication, human toxicity (cancer effects), human toxicity (noncancer effects), ionizing radiation (human effects), land transformation, ozone depletion, particulate matter/respiratory inorganics, photochemical ozone formation, mineral and fossil resource depletion, and water resource depletion (Sacramento-Rivero et al., 2022; Onat, 2022).

Social indicators must be incorporated to extrapolate sustainability by addressing technology analysis by expanding the concept of environmental services (provided by farmers and land users) to socioenvironmental services. Here, social constraints are represented by rural development, more specifically by new sources of demand for agricultural products and access to energy in rural areas (classified as opportunities), negative implications for poverty levels due to the competition for resources (classified as a risk).

17.4 Social barriers to the use of biofuels

Currently, research on biofuels is aimed at increasing the efficiency of the production processes and diversifying their use by a greater number of countries. However, programs related to the production of biofuels must consider the economic and social implications, and although it would seem ironic, there are also environmental impacts to consider. Biofuel generation systems, in general, can generate a global problem based on three phenomena that interact with each other and add complexity: the rise in food prices that have the same origins as said fuels, competition for land and resources existing water sources, and deforestation to open new farmland (Prasad and Ingle, 2019; Viccaro et al., 2022).

The development of biofuel production processes fosters greater economic development in rural sectors of countries dependent on oil imports. For this, developing policies that promote job creation, greater interaction, and efficiency of the business environment should be considered, and above all, updating environmental protection protocols. The development of management policies for the entire biofuel production process includes research, the establishment of crops, their management, harvest, markets, operations, logistics, transfers, etc. It has been successful based on the projections of benefits in areas of interest to governments, such as agricultural production, reduction of GHG emissions, energy security, trade balances, rural development, and economic development opportunities. Some nations have established programs, legislation, and mandates to promote the use of biofuels and highlight their benefits, such as reducing their dependence on fossil fuels and their environmental impact (Leibensperger et al., 2021). However, there are limitations regarding the study and analysis of social barriers when implementing these technologies and processes. If not addressed, they can significantly limit their integration and implementation.

The needs and paradigms of the population are critical elements in establishing new policies or development trends. It is necessary to interact with them and take them into account. In the context of the development of renewable energies, the participation of society and its approach to the development and advancement of production processes, as well as the dissemination of the benefits and problems inherent to it, is of utmost importance. Authors such as Elaheh (2011), Awasthi et al. (2022), and Pandey et al. (2021) studied the characteristics of the social acceptance of new technology, such as those related to biofuel generation processes. For this, they point out the importance of actions such as:

1. A close relationship between experts on the subject and those responsible for formulating and implementing new technologies.
2. Wide public dissemination of the inherent information towards all the sectors involved and a positive but real vision of the new technology.

3. Actions that enrich the will to apply new technologies to society.
4. Avoid obstacles by the interest groups involved such as NGOs, local political representatives, institutions, companies, etc.

Social barriers have been identified that discourage farmers or producers from cultivating or harvesting raw materials using bioenergy and consumers from acquiring new biofuels. Various investigations carried out by Uslu et al. (2017), NRC (2011), and Ambaye et al. (2021) point out some social barriers not yet fully resolved to the integration of these technologies. Examples of these barriers are as follows:

1. Farmers dedicated to the production of raw materials or who in their production processes generate waste rich in cellulose, starch, or sugars, as well as the owners of plots with forest use, are unaware of the management practices of their materials and how to integrate them into the production system. The need to establish systems for collecting, transferring, and adapting agricultural raw material residues is observed. There is a lack of information to enter emerging markets and the need for sufficient economic incentives, as well as their duration and supporting policies.
2. Lack of infrastructure for collecting, conserving, and using biomass by farmers and producers.
3. There are technological, economic, and logistical barriers that hold back the biofuel industry must overcome. Farmers still consider that its benefits are not tangible and its production is not economically viable.
4. Lack of dissemination of information to farms and agricultural organizations on new crops of interest with greater efficiency for biofuel production. Such is the case of new varieties, affinity to environmental conditions, and ability to interact with other plant species of interest other than the production of biofuels such as food.
5. Social factors based on personal characteristics such as income, age, academic level, etc., that limit the disposition towards new technologies or risk-taking. For example, Jensen and Babcock (2007) and Ebi et al. (2020) indicate that farmers with higher income from nonagricultural businesses, higher academic levels, and a young age range are more willing and open to applying new technologies and crops.
6. Cultural barriers to integrating new elements into the beauty of the landscape are dominated by species of previous interest or endemic.
7. Size and characteristics of the exploitation area, demographic factors such as the age of the producers, and the willingness to join new markets and crops.
8. Lack of awareness of the importance of biofuel generation, its impact on sustainable agriculture and the environment, and the potential socioeconomic benefits.
9. The volatility of biofuel prices and their costly reversibility of the investment transmits a very high risk that hinders the conversion of a traditional crop and integration into a solid, productive system through the cultivation of bioenergetic raw materials.
10. Social ignorance of the new processes for producing 2G and 3G biofuels provokes the false belief that the price of food will increase or, where appropriate, shortages will be generated.
11. Public policies that encourage the development and production of biofuels present a high level of uncertainty and discourage investment in infrastructure.
12. The use of biofuel in transport is limited to certain economic levels that can acquire a vehicle adapted to its use; its use in cheaper and more accessible vehicles should be widespread.

Finally, the most significant social barrier is the lack of information about technological advances, transformation processes, advantages, and environmental benefits related to biofuel generation. Generally, this information remains in the circles of researchers and decision-makers; however, they are not always adequately shared with the bulk of society. Issues related to biofuels, bioenergy, and the reduction of global warming are favorably accepted by society, but the lack of dissemination of technical information prevents it from being considered as renewable energy. One reason originates the breadth of related concepts or terms, which generates confusion and disinterest. The new policies should be oriented to transmit more information, considering social factors and associated paradigms, thus achieving greater public acceptance (Ren et al., 2019; Mizik, 2021).

17.5 Analysis of existing technologies in the period 2019–21 (pandemic COVID-19)

The technologies and infrastructure to produce biofuels have been designed according to the available raw materials of the crops. In some cases, they take advantage of harvest or process residues and surplus food crops with potential; others are integrated crops without food use or developed on marginal lands with adequate adaptation. The biofuels with the highest degree of industrial development are bioethanol and biodiesel. As seen in Fig. 17.2, they use crops or oils as the raw material used as food for human consumption. This has generated changes in the patterns in agricultural production systems to try to satisfy the food and biofuel markets. To avoid the dilemmas of crop orientation, new processes have been developed to generate energy from unknown sources or raw materials (Umakanth et al., 2022).

New technologies are currently being applied to obtain 2G and 3G biofuels based on digestion, fermentation, saccharification, esterification processes, etc. Each one presents variations in the requirements of raw material in the conditions of the processes. The 2G takes advantage of agricultural and forestry residues

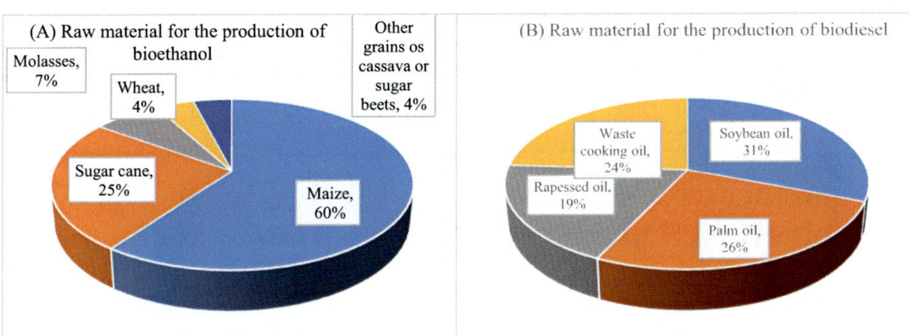

Figure 17.2 Raw material for bioethanol and biodiesel production (IRENA, 2019).

without risking food safety. The 3G can also take advantage of algae on a large scale with the benefit of its abundant availability and excellent absorption of CO_2 (Laurens and Nelson, 2020; Nguyen and Pham, 2020; Sadatshojaei et al., 2020)

The biomass used by 3G fuels has a considerable concentration of carbohydrates and lipids with different orientations, mainly for the production of bioethanol, biohydrogen, biodiesel, bio-oil, biomethane, bio-crude oil, etc. The different species of algae used can take advantage of wastewater gases and nutrients for the growth of their biomass. Its basic requirements are light, water, carbon dioxide, and nutrients such as phosphorus and nitrogen. It is possible to apply genetic modification techniques to increase the light absorption capacity, the concentration of lipids and carbohydrates, and maximize biomass yield (Malode et al., 2021).

Recent investigations of the last three years highlight the advances in the generation of biofuels from algae. Very optimistic forecasts indicate that significant production volumes will be reached from the year 2030. Even government agencies contemplate very optimistic scenarios, even considering that in that year, the processes that use algae as raw material will be up to 100 times more productive than the processes from traditional raw materials (DOE, 2018; Lackner, 2022). These expectations are based on the trends of advances in this type of bioenergy since 2008. Its large-scale production stands out since algae can be cultivated in large areas and very diverse environments. Numerous species and varieties of algae are suitable for its use with great potential for the production of lipids. Their tolerance to saline environments and apparent capacity to be cultivated in seawater eliminates the need to orient productive soils of other agricultural species for production. This characteristic also allows their cultivation on land without agricultural vocation or even wastewater. The subprocess of the massive cultivation of algae is vital for the process in general and represents almost half of the total production cost. Therefore the proper management of these organisms is critical. The stress-generating factors trigger the accumulation of lipids but reduce the photosynthetic processes, and thus their efficiency decreases. There is also concern about the environmental impacts caused by water and spaces and the release of elements such as nitrogen and phosphorus from applied fertilizers (IRENA, 2019).

It is necessary to deeply analyze the implications of the biofuel generation processes with a complex approach of all the factors involved, that is, characteristics of the raw material, current and future technologies, and their use either in engines, boilers, or turbines, etc. This forms a matrix where the environmental, socioeconomic, political, spatial, and technological implications that guide the continuity of development promoted by biofuels are integrated.

17.6 Impact of SARS CoV2 virus (COVID-19) on bioethanol production and progress to 2022

The agroindustrial wastes such as corn, wheat, cassava, sugar cane bagasse, etc., with high cellulose and starch concentration, are hydrolyzed and isomerized to obtain fermentable sugars and then fermented into bioethanol (2G) (Brosse et al.,

2011; Mohd-Azhar et al., 2017; Rezania et al., 2020). Some efforts have been aimed at developing technology that allows the use of marine microalgae as an alternative for obtaining bioethanol (3G) (Lum et al., 2013) due to its large amount of available biomass. However, the COVID-19 pandemic caused by the severe acute respiratory syndrome Coronavirus (SARS-CoV-2 virus) in 2020 caused the global demand for transportation fuels to decrease by approximately 8.5% compared to 2019. This was caused by the containment measures applied by governments, specifically in the restrictions on the movement of people and, therefore, the problems of logistics and commercial distribution worldwide. In this same period, the decrease in the demand for biofuels stood at 8.7% in 2020 compared to 2019. Fig. 17.3 shows the trend in the volume of bioethanol production worldwide in billion liters per year. Between 2019 and 2020, there was a notable decrease in bioethanol production, close to 8.4 billion liters per year. Although the declining trend is partially averted with the projected ethanol production for 2022, the production volumes recorded before the pandemic have not yet been reached. In addition, actions to counteract the advance of the pandemic, such as the use of ethanol as a hand sanitizer and disinfectants, somewhat compensate for the declining use as transport fuel (Hans et al., 2021; IEA, 2021).

The rapid worldwide spread of the COVID-19 pandemic from China at the end of 2019 and the beginning of 2020 caused numerous global phenomena. They significantly affected health, economic, environmental, social, and cultural contexts. One of the sectors that have been most affected by this pandemic is oil extraction. Due to the decrease in demand, a significant drop in the prices of this energy is observed. Other energy sources such as biofuels have also been affected by these

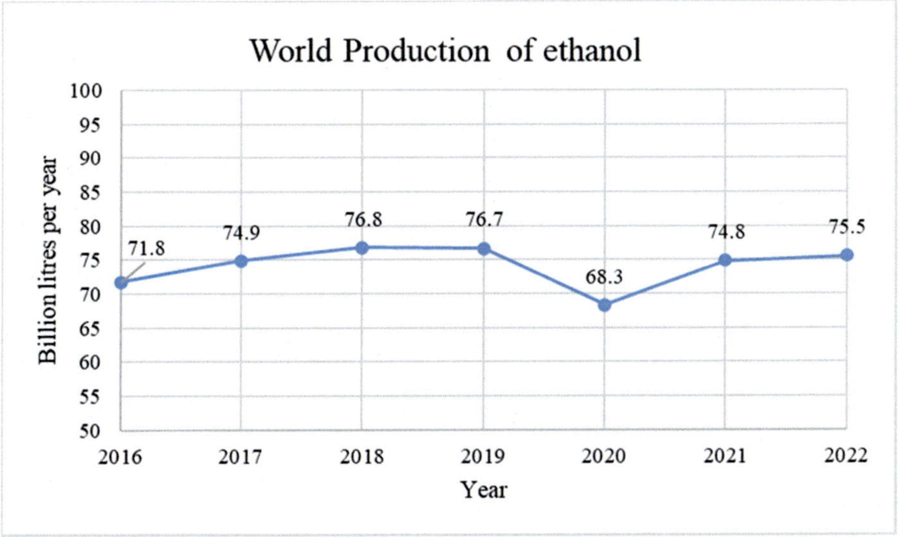

Figure 17.3 The volume of world ethanol production before and after the SARS-CoV-2 pandemic (IEA, 2021).

repercussions. In the case of bioethanol integrated with gasoline, its demand has decreased notably, causing the ethanol industry to reduce its production and, in extreme cases, to close its plants. Fig. 17.4 shows the behavior of bioethanol production in the most important regions and countries for the period from 2015 to 2020. There is a trend toward a considerable increase in production between 2015 and 2019; however, in 2020, with the spread of the COVID pandemic, the volume of production decreased on average to the production levels of 2016. It is important to note that only India increased (1.0%) bioethanol production from 2019 to 2020 despite restriction conditions.

Some countries like the United States of America allocate about 40% of their corn production to bioethanol. This decrease in 2020 gives us an idea of the magnitude of the effects on their production systems. In the case of biodiesel, a particular situation arises. The vehicles that require this fuel are used mainly as heavy trucks, agricultural vehicles, construction vehicles, or industrial applications. Although these activities have decreased, they do not significantly impact the reduction in biodiesel demand compared to gasoline. Some authors, such as Taheripour and Mintert (2020), reported that the decrease in the demand for diesel due to COVID-19 represents half the reduction rate of gasoline, and the results from it are more significant to indicate that the consumption of biodiesel only represents in 2019 about 3% of the volume of petroleum diesel produced. As a result of the drop in international oil prices, biofuels are now less used

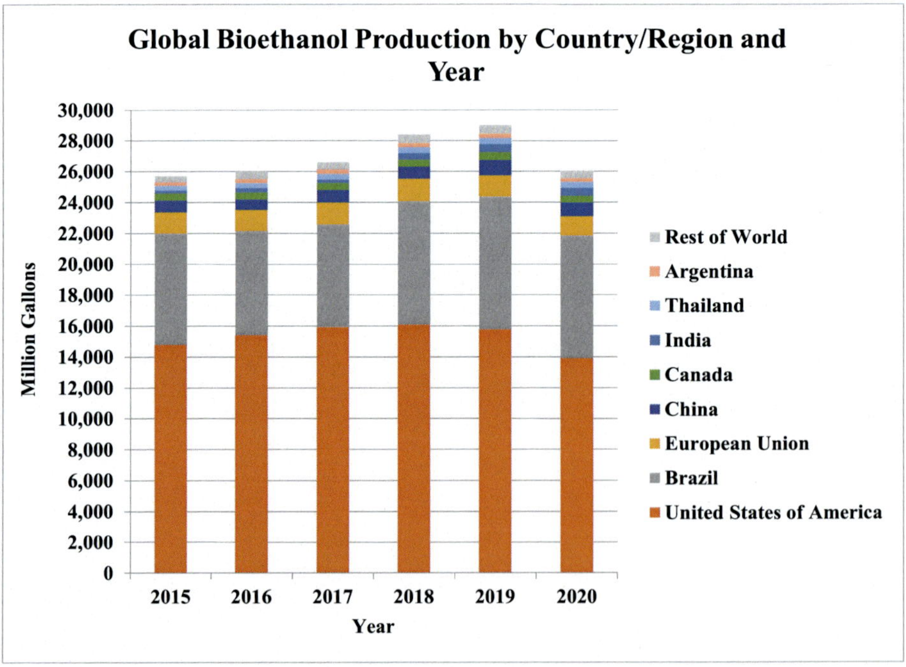

Figure 17.4 Decrease in bioethanol production by country in the period 2015–20 (RFA, 2020).

than fossil fuels (Elleby et al., 2020). As a collateral effect of the decrease in biofuel production, a decrease in the prices of its crops with bioenergy use, such as corn, sugar, soybeans, and those precursors of vegetable oils, is expected.

The decrease in energy cost for various agricultural processes such as irrigation, land preparation, planting, harvesting, and inputs such as fertilizers and fuels, acts as a stabilizer of the prices of basic foods and, in a certain way, mitigates the impacts on food and agriculture (Elleby et al., 2020; FAO, 2020).

Finally, it is necessary to highlight that the COVID-19 pandemic also indirectly affects the production of biofuels since the demand for this energy reacts to changes in the Gross Domestic Product. On average, this indicator fell between 4% and 8% during 2020 which caused a decrease in the consumption of gasoline and diesel, which in turn affects the demand for biofuels that are considered in its mix (Elleby et al., 2020).

17.7 Conclusions and future perspectives

Bioethanol obtained from various raw materials (crops, wastes, byproducts, etc.), biodiesel, and several biofuels technological processes and potential uses face challenges yet to be solved, such as environmental and socioeconomic impacts, user acceptance, economic competition with established fossil fuels, petroleum geopolitics, sustainability, etc. This last issue constitutes the main objective in which numerous researchers have worked for decades with various methodologies, scenarios, and multiple variables to determine bioethanol impacts in a precise and robust way. However, its measurement for decision-making by stakeholders depends on the region of evaluation, the technology used, available data, production and handling of raw materials from their origin, pretreatments, transformation options, management, etc. Therefore there are significant differences in sustainability between 1G, 2G, and 3G for ethanol, making a transdisciplinary approach highly needed and urgent to determine the most sustainable options. LCA, as a framework in contrast to conventional technologies, essentially follows an "end-of-pipe" system, which has been the methodological approach most used to assess sustainability and environmental impacts. However, there is still controversy in its use of looking for "green" or sustainable production due to the different evaluation scenarios and indicators that have restricted the development of universally applicable conclusions. Therefore any ethanol project must consider the evaluation of impacts with the integration of methodological frameworks such as ecological footprint (carbon, water, and energy), life cycle sustainability assessment, life cycle costing, cost–benefit analysis, full cost accounting, and multicriteria decision analysis.

References

Achinas, S., Horjus, J., Achinas, V., Euverink, G.J.W., 2019. A PESTLE analysis of biofuels energy industry in Europe. Sustainability 11 (21), 5981.

Álvarez-Maciel, C., 2009. Biofuels; historical-technological development, current markets and international trade. Econ. Inf. 359, 63–89.

Ambaye, T.G., Vaccari, M., Bonilla-Petriciolet, A., Prasad, S., van Hullebusch, E.D., Rtimi, S., 2021. Emerging technologies for biofuel production: a critical review on recent progress, challenges and perspectives. J. Environ. Manag. 290, 112627.

Amores, M.J., Mele, F.D., Jiménez, L., Castells, F., 2013. Life cycle assessment of fuel ethanol from sugarcane in Argentina. Int. J. Life Cycle Assess. 18 (7), 1344–1357.

Aro, E., M., 2016. From first generation biofuels to advanced solar biofuels. Ambio 45 ((Suppl.1), S24–S31.

Awasthi, M.K., Yan, B., Sar, T., Gómez-García, R., Ren, L., Sharma, P., et al., 2022. Organic waste recycling for carbon smart circular bioeconomy and sustainable development: a review. Bioresour. Technol. 360, 127620.

Behera, S., Singh, R., Arora, R., Sharma, N.K., Shukla, M., Kumar, S., 2014. Scope of algae as third generation biofuels. Front. Bioeng. Biotechnol. 2, 90. Available from: https://doi.org/10.3389/fbioe.2014.00090. Published online 2015 Feb 11.

Bonomi, A., Colling-Klein, B., Ferreira-Chagas, M., Dias-Souza, N., 2018. Technical report comparison of biofuel life cycle analysis tools. IEA Bioenergy. Braz. 94.

Brandao, M., Heijungs, R., Cowie, A.L., 2022. On quantifying sources of uncertainty in the carbon footprint of biofuels: crop/feedstock, LCA modelling approach, land-use change, and GHG metrics. Biofuel Res. J. 34, 1608–1616.

Brosse, N., Mohamad Ibrahim, M.N., Abdul Rahim, A., 2011. Biomass to bioethanol: initiatives of the future for lignin. ISRN Mater. Sci. 2011, 1–10. Available from: https://doi.org/10.5402/2011/461482.

Cortez, L.A.B., Baldassin, R., de Almeida, E., 2020. Energy from sugarcane. In: Santos, F., Rabelo, S., De Matos, M., Eichler, P. (Eds.), Sugarcane Biorefinery, Technology, and Perspectives. Academic Press, pp. 117–139.

De Jong, S., Antonissen, K., Hoefnagels, R., Lonza, L., Wang, M., Faiij, A., et al., 2017. Life-cycle analysis of greenhouse gas emissions from renewable jet fuel production. Biotechnol. Biofuels 10, 64. Available from: https://doi.org/10.1186/s13068-017-0739-7.

Demırbas, A., 2017. The social, economic, and environmental importance of biofuels in the future. Energy Sources B Econ. Plan. Pol. 12 (1), 47–55.

Du, H., Liu, W., Zhang, M., Si, C., Zhang, X., Li, B., 2019. Cellulose nanocrystals and cellulose nanofibrils based hydrogels for biomedical applications. Carbohydr. Polym. 209, 130–144. Available from: https://doi.org/10.1016/j.carbpol.2019.01.020.

Ebi, K.L., Harris, F., Sioen, G.B., Wannous, C., Anyamba, A., Bi, P., et al., 2020. Transdisciplinary research priorities for human and planetary health in the context of the 2030 agenda for sustainable development. Int. J. Environ. Res. Public Health 17 (23), 8890.

Elaheh, A. 2011. Social Acceptance of Bioenergy in Europe. Master's thesis. Environmental Management and Policy—Lund University, Lund, Sweden. 73 p.

Elleby, C., Pérez-Domínguez, I., Adenauer, M., Genovese, G., 2020. Impacts of the COVID-19 pandemic on the global agricultural markets. Environ. Resour. Econ. 76, 1067–1079. Available from: https://doi.org/10.1007/s10640-020-00473-6.

Food and Agriculture Organization (FAO). 2020. Food Outlook—Biannual Report on Global Food Markets. Rome. Available from: https://doi.org/10.4060/ca9509en.

García, C., 2011. Bioethanol, Biofuels, Energy or Food? Omniscellula, 25. Universitat de Barcelona, pp. 21–27.

Guo, J., Tanaka, T., 2022. Energy security versus food security: an analysis of fuel ethanol-related markets using the spillover index and partial wavelet coherence approaches. Energy Econ. 112, 106142.

Gupta, A., Verma, J.P., 2015. Sustainable bio-ethanol production from agro-residues: a review. Renew. Sustain. Energy Rev. 41, 550−567.

Hans, M., Lugani, Y., Chandel, A.K., Rai, R., Kumar, S., 2021. Production of first- and second-generation ethanol for use in alcohol-based hand sanitizers and disinfectants in India. Biomass Conv. Bioref. Available from: https://doi.org/10.1007/s13399-021-01553-3.

Hun, A.L.N., Mele, F.D., Pérez, G.A., 2017. A comparative life cycle assessment of the sugarcane value chain in the province of Tucumán (Argentina) considering different technology levels. Int. J. Life Cycle Assess. 22 (4), 502−515.

IEA. 2021. Renewable energy market update 2021, outlook for 2021 and 2022. Int. Energy Agency. Paris Fr. 29 p Con Acceso: August 15, 2022. Available in: https://www.iea.org/reports/renewable-energy-market-update-2021

International Renewable Energy Agency (IRENA), 2019. Advanced biofuels, what holds them back? Abu Dhabi, 92p.

Jensen, H.H., Babcock, B.A., 2007. Do biofuels mean inexpensive food is a thing of the past? Iowa Ag. Rev. 13 (3), 4−5. 11.

Karp, S.G., Medina, J.D., Letti, L.A., Woiciechowski, A.L., de Carvalho, J.C., Schmitt, C.C., et al., 2021. Bioeconomy and biofuels: the case of sugarcane ethanol in Brazil. Biofuels Bioprod. Bioref. 15 (2). Available from: https://doi.org/10.1002/bbb.2195.

Lackner, M., 2022. Third-generation biofuels: bacteria and algae for better yield and sustainability. Handbook of Climate Change Mitigation and Adaptation. Springer International Publishing, Cham, pp. 1947−1986. Available from: https://link.springer.com/content/pdf/10.1007/978-3-030-72579-2_90.pdf.

Laurens, L.M., Nelson, R.S., 2020. Sustainable technologies for seaweed conversion to biofuels and bioproducts. In: Torres, M.D., Herminia Dominguez, S.K. (Eds.), Sustainable Seaweed Technologies: Cultivation, Biorefinery, and Applications. Elsevier, pp. 643−661.

Leibensperger, C., Yang, P., Zhao, Q., Wei, S., Cai, X., 2021. The synergy between stakeholders for cellulosic biofuel development: perspectives, opportunities, and barriers. Renew. Sustain. Energy Rev. 137, 110613.

Lum, K.K., Kim, J., Gen, Lei, X., 2013. Dual potential of microalgae as a sustainable biofuel feedstock and animal feed. J. Anim. Sci. Biotechnol. 4 (53), 7p.

Maga, D., Thonemann, N., Hiebel, M., Sebastião, D., Lopes, T.F., Fonseca, C., et al., 2019. Comparative life cycle assessment of first-and second-generation ethanol from sugarcane in Brazil. The Int. J. Life Cycle Assess. 24 (2), 266−280.

Malode, S., Keerthi, K., Mascarenhas, R., Shetti, N., Aminabhavi, T., 2021. Recent advances and viability in biofuel production. Energy Convers. Manage.: X. Artic. June, 100070.

McManus, M.C., Taylor, C.M., 2015. The changing nature of life cycle assessment. Biomass Bioenergy 82, 13−26.

Mizik, T., 2021. Economic aspects and sustainability of ethanol production—a systematic literature review. Energies 14 (19), 6137.

Mohd-Azhar, S.H., Abdulla, R., Azma, J.S., Marbawi, H., Azlan, G.J., Mohd, F.A.A., et al., 2017. Yeasts in sustainable bioethanol production: a review. Biochem. Biophys. Rep. 10, 52−61.

Morone, P., Strzałkowski, A., and Tani, A. 2020. Biofuel transitions: an overview of regulations and standards for a more sustainable framework. Biofuels a More Sustain. Future, Elsevier. pp. 21–46.

National Research Council (NRC), 2011. Renewable Fuel Standard: Potential Economic and Environmental Effects of U.S. Biofuel Policy. The National Academies Press, Washington, DC. Available from: https://doi.org/10.17226/13105.

Nguyen, D.C., Pham, V.V., 2020. Pathway of sustainable fuel development with novel generation biofuels. AIP Conferemce Proceedings. 2235 (1), 020012. AIP Publishing LLC.

Onat, N.C., 2022. How to compare sustainability impacts of alternative fuel Vehicles? Transport. Res. D Transp. Environ. 102, 103129.

Pandey, P., Valkenburg, G., Mamidipudi, A., Bijker, W.E., 2021. "All we want, is to get rid of the straw": how biofuel policies need to be multiple. East. Asian Sci. Technol. Soc. Int. J. 15 (1), 4–23.

Patel, A., Shah, A.R., 2021. Integrated lignocellulosic biorefinery: gateway for production of second generation ethanol and value added products. J. Bioresour. Bioprod. 6 (2), 108–128.

Pinto, A.S., Elias, A.M., Furlan, F.F., Ribeiro, M.P., Giordano, R.C., Farinas, C.S., 2022. Strategies to reduce the negative impact of inhibitors in biorefineries: a combined techno-economic and life cycle assessment. J. Clean Prod. 345, 131020.

Prasad, S., Ingle, A.P., 2019. Impacts of sustainable biofuels production from biomass. In: Rai, M., Ingle, A. (Eds.), Sustainable Bioenergy: Advances and Impacts. Elsevier, pp. 327–346.

Rathore, D., Pant, D., Singh, A., 2013. A comparison of life cycle assessment studies of different biofuels. In: Singh, A., Pant, D., Olsen, S. (Eds.), Life Cycle Assessment of Renewable Energy Sources. Springer-Verlag, London, pp. 270–289.

Biofuels for a more sustainable. In: Ren, J., Scipioni, A., Manzardo, A., Liang, H. (Eds.), Future: Life Cycle Sustainability Assessment and Multi-Criteria Decision Making. Elsevier.

Rezania, S., Oryani, B., Cho, J., Talaiekhozani, A., Sabbagh, F., Hashemi, B., et al., 2020. Different pretreatment technologies of lignocellulosic biomass for bioethanol production: an overview. Energy 199, 117457. Available from: https://doi.org/10.1016/j.energy.2020.117457.

Renewable Fuels Association (RFA). 2020. Annu. Fuel Ethanol Prod. U.S. World Ethanol Prod. Available from: https://ethanolrfa.org/statistics/annual-ethanol-production/.

Robak, K., Balcerek, M., 2020. Current state-of-the-art in ethanol production from lignocellulosic feedstocks. Microbiol. Res. 126534.

Sacramento-Rivero, J.C., Mwampamba, T.H., Navarro-Pineda, F.S., Musule, R., García, C. A., Martínez-Bravo, R.D., et al., 2022. A methodological framework for assessing the sustainability of solid biofuels systems. Bioenergy Res. 1–23.

Sadatshojaei, E., Wood, D.A., Mowla, D., 2020. Third generation of biofuels exploiting microalgae. In: Inamuddin, Asiri, A. (Eds.), Sustainable Green Chemical Processes and their Allied Applications. Springer, Cham, pp. 575–588.

Satari, B., Jaiswal, A.K., 2020. Green fractionation of 2G and 3G feedstocks for ethanol production: advances, incentives and barriers. Curr. Opin. Food Sci. 37, 1–9.

Sharma, B., Larroche, C., Dussap, C.G., 2020. Comprehensive assessment of 2G bioethanol production. Bioresour. Technol. 313, 123630. Available from: https://doi.org/10.1016/j.biortech.2020.123630.123630. 2020 Oct.

Sieverding, H., Kebreab, E., Johnson, J.M., Xu, H., Wang, M., Grosso, S.J.D., et al., 2020. A life cycle analysis (LCA) primer for the agricultural community. Agron. J. 112 (5), 3788–3807.

Singh, A., Prajapati, P., Vyas, S., Gaur, V.K., Sindhu, R., Binod, P., et al., 2022. A comprehensive review of feedstocks as sustainable substrates for next-generation biofuels. BioEnergy Res. 1–18.

Soleymani Angili, T., Grzesik, K., Rödl, A., Kaltschmitt, M., 2021. Life cycle assessment of bioethanol production: a review of feedstock, technology and methodology. Energies 14 (10), 2939.

Taheripour, F., Mintert, J., 2020. Impact of COVID-19 on the biofuels industry and implications for corn and soybean markets. Coll. Agricult. 10. Purdue University.

Umakanth, A.V., Datta, A., Reddy, B.S., Bardhan, S., 2022. Biomass feedstocks for advanced biofuels: sustainability and supply chain management. Adv. Biofuel Technol. 39–72. Available from: https://doi.org/10.1016/B978-0-323-88427-3.00023-4.

Uslu, A., Detz, R.J., Mozaffarian, H., 2017. Barriers to advanced liquid biofuels and renewable liquid fuels of non-biological origin. Eur. Union. 50.

Viccaro, M., Caniani, D., Masi, S., Romano, S., Cozzi, M., 2022. Biofuels or not biofuels? The "Nexus Thinking" in land suitability analysis for energy crops. Renew. Energy 187, 1050–1064.

World Bioenergy Association (WBA), 2013. Biofuels for Transport. Available from: http://www.worldbioenergy.org/uploads/Factsheet%20-%20Biofuels%20for%20transport.pdf.

Section V

Public policy and circular economy

18 Impacts of public policies and stakeholders in the transition from the biofuel value chain to a circular bioeconomy: México as a case study

Noé Aguilar-Rivera[1] *and Luis Alberto Olvera-Vargas*[2]
[1]Faculty of Biological and Agricultural Sciences, Universidad Veracruzana, Córdoba, Veracruz, México, [2]Center for Research and Assistance in Technology and Design of the State of Jalisco (CIATEJ) Guadalajara, Jalisco, Mexico

18.1 Introduction

The analysis of the nature–society relationship and anthropogenic impacts are essential in developing the necessary knowledge of the complexity of environmental problems in constructing policies and alternatives for managing agroecosystems by stakeholders. That is why understanding the relationship between environment, society, and culture will allow us to understand in parallel the reason for global problems such as food, energy, raw materials, rural poverty, global warming, greenhouse gas (GHG) emissions, deforestation, loss of biodiversity, migration, drug trafficking and lack of respect for human rights.

The extraction and handling of raw materials, conditioning, manufacturing, and waste processing lead to pollution, degradation of ecosystems, and the need for large quantities of water and mainly energy from fossil fuels. To fulfill the growing demand, less accessible and efficient sources by industry or population will need to be explored, requiring more energy, causing more waste, disturbance, and deterioration of ecosystems. The negative impact on the environment is, therefore exponentially more significant, so technological options and the participation of decision-makers are needed within the transdisciplinarity of the circular economy (CE) (Tan et al., 2022).

18.2 Energy and circular economy

Energy is the basis of industrial and socioeconomic development and well-being. Since the 20th century, the primary source has been those derived from fossil

Advances in Lignocellulosic Biofuel Production Systems. DOI: https://doi.org/10.1016/B978-0-323-91192-4.00010-9

sources. However, bioenergy as a source has limitations on cost due to the lack of awareness, interest, community participation, and/or responsiveness (Pattanaik et al., 2019). Esparza et al. (2020) concluded it could be envisaged that the optimal approach will consist of an adequate balance between conventional energy sources (traditional end-of-pipe technologies) and emerging valorization technologies with bioenergy and waste biomass. Both strategies must coordinate to allow integral management that minimizes the impact on food security, wastage of natural resources, and environmental damage.

The successful conversion of biomass, agroindustrial by-products, wastes, and residues into a wide range of valuable chemicals, materials, and energy with economically feasible production costs, is the overall goal of sustainable biorefineries. But socioeconomic and political context, governance, public health, social macroeconomic policies, and institutional capacity are to be considered (Chandel et al., 2021).

However, the transition to renewable energy and bioproducts production requires many innovations, stakeholders, and various technologies to compete with fossil fuel technologies. The circular economy (CE) approach involves many uncertainties, mainly if the raw material is seasonal and highly variable due to uncontrollable factors such as climate change and rising production costs mean that operations management aims to balance risk management, operations, and crop optimization but due to its economic and environmental importance, especially regarding the generation of renewable energy or biofuels.

Sugarcane ethanol is the most representative world biofuel, mainly produced in Brazil. This is a result of governmental policies, stakeholder participation, extensive research, and technological development (Fig. 18.1).

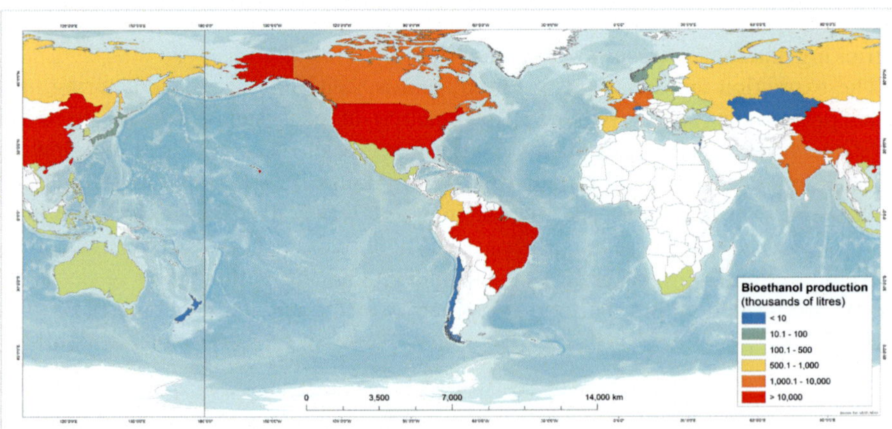

Figure 18.1 World ethanol production OECD (2019). (Agrienvironmental other indicators, ethanol production. Organization for Economic Co-operation and Development).
Source: From https://stats.oecd.org/viewhtml.aspx?datasetcode = AEI_OTHER.

Several researchers have been attracted by biomass or wastes, especially sugarcane, as a renewable energy source to produce bioethanol. Based on CE, the cornerstone of this approach is the biorefinery, in which renewable resources are intelligently converted to a large number of bioproducts, maximizing the valorization of the feedstocks. Innovation is a prerequisite to moving a fossil-based economy toward sustainable alternatives, and the viability of the bio-based economy depends on the integration between plant (green) and industrial biotechnology and other novel technologies (Gilani et al., 2020; Vanholme et al., 2013).

Ajanovic and Haas (2014) elaborated further on the use of biofuels:

1. To reveal the real future market value of biofuels, a CO_2-based tax system should be implemented for all types of fuels providing a neutral environmental incentive for competition between fossil and renewable fuels.
2. The research and development for all biofuels, especially 2G biofuels, must be intensified.

Fava et al. (2015) stated research and development areas for biowaste biorefinery could be focused on:

1. Collection strategy of biowaste streams from various sources.
2. Flexibility improvement of biowaste biorefinery to produce a wide range of biowaste streams.
3. Integration of the valorization processes and closing the loop to produce.
4. Products that are needed by the market.

Furthermore, sugarcane biomass can contribute to renewable energy by producing first generation(1G) and second generation (2G) ethanol. However, few studies have explored prioritizing barriers and stakeholders to the adoption of the CE even though it is estimated that the demand for ethanol in coming years will be supplied basically by two pathways: improvements in the use of existing feedstocks (corn and sugarcane) and an increase in feedstock production (biomass) in biorefineries. However, lignocellulose biomass (LCB) valorization through 2G bioethanol should account for its integration with territories and other productive chains along the raw materials. This can be accomplished by simulating the locations, production and demand of biorefineries to minimize the costs associated with these processes (Galanopoulos et al., 2020).

The production of 2G bioethanol from sugarcane bagasse has a complex technology, especially regarding the pretreatment step, which has the general objective of making carbohydrates, mainly cellulose and hemicelluloses, available for subsequent fermentation. This is primarily accomplished by separating lignin and disrupting the complex polysaccharide structure (Karp et al., 2021).

On the other hand, Hoque et al. (2020). Jesus et al. (2021), and Ali et al. (2021) concluded the next wave of vehicles hybrid and electric cars depends on an infrastructure that also requires a lot of investment. There are millions of gasoline vehicles; therefore, the use of electric energy in automobiles is likely to be slower and will not significantly impact the market in the next 10 years. An increase in ethanol demand is, therefore to be expected. Thus the energy, sugar, and alcohol sector can play a prominent role in the transition from the linear economy to the worldwide CE.

Therefore the outlook for biofuels remains mixed with uncertainty but intensifying the processes, the exploration of driving-change technologies, and the integration of wastes through biorefinery processes can lead to a more sustainable sugarcane processing industry (Sherpa et al., 2017).

18.3 Sugarcane as feedstock for ethanol biofuel

Worldwide, sugarcane is produced in 23,832,000 ha. America is the world's largest region in sugarcane production, with 12,618,746 ha. Brazil is far from all countries with 10,042,199 ha and México the third with 783,486 ha (FAOSTAT, 2022). However, sugar industries have some specific competitive challenges: (1) long cycle crop, (2) the high initial investment, (3) a resilient crop and less water demanding, (4) sugarcane growing is part of the history and culture in many countries, (5) small farmers living on sugarcane for generations, (6) sugarcane harvesting, logistics and sugar mills employ many people, (7) industry requires reliable and resilient infrastructure, (8) mills and factories require regular maintenance and external support from the government in many countries, (9) the worldwide sugar industry is regulated by various types of public policies and numerous stakeholders are immersed in its development, (10) Brazil the most important switch factor on the short/medium term, (11) the sugar industry have been cyclically impacted by low sugar prices, (12) production has been growing faster than consumption, (13) government intervention preventing low prices from balancing supply and demand, (14) Campaigns against sugar seriously hampering consumption, (15) labeling and advertisement regulations are increasing, (16) food and soft drink makers are reacting to public health policies, (17) reducing sugar content in their products, (18) launching new products with no sugar, (19) adapting to consumers' demand (20) advertising low or no sugar products, and (21) alternative sweeteners improve.

Sugar crop fields and sugar mills generate jobs for the rural population and valuable products for society. Besides, significant policy changes are needed to balance the market in the long run (diversification with ethanol 1G and 2G, cogeneration, alcohol for the pharmaceutical industry, bioplastics), reduce costs, and improve yields with sustainable management practices such as green harvest.

According to Bonsucro (2020) (https://www.bonsucro.com/the-impact-of-covid-19-on-sugarcane-sustainability/), the future of sugarcane research perspectives are:

1. Promote the use of sugar according to individual health requirements to increase global sugar demand.
2. The COVID-19 crisis represents an opportunity to raise awareness of the advantages of sugarcane as a biofactory to produce "clean" bioenergy, bioplastics, clothes, etc.
3. Certify traceability and sustainable practices with the use of blockchain.
4. Development in Industry 4.0 and Big Data algorithms.
5. Apply sales on digital platforms using social media and E-commerce.

6. Implement Best Management Practices in factory and field operations.
7. Review objectives in plant breeding, management systems, and biotechnology applications.
8. A closer look at soil biology, water and soil conservation, and buffer areas to preserve biodiversity, including surrounding communities and stakeholders.

Both industrial plants (1G and 2G) refer to modern plants with high-pressure boilers and low energy consumption, which, in addition to straw recovery, allow for high electricity production levels. The process development for 1G and 2G ethanol has, therefore, important impacts in multiple contexts, which must be considered intensely so that its application does not generate adverse consequences that counterpoise the advantages that the use of friendly technologies can develop with the environment, economic growth and the well-being of the population. Furthermore, the potential environmental benefits of alternatives for 1G and 2G ethanol production should be adequately quantified to support public policy formulation that will affect the future of climate and other environmental aspects of our planet. Therefore a lack of consensus on measurement methodologies and certification schemes, for example, BONSUCRO and agenda 2030, are some of the current key problems in correctly monitoring biofuel indicators, thus hindering their impact on sustainability policies (Table 18.1).

According to Neto et al. (2018) and Chandel et al. (2021), the 2G bioethanol has critical stages in the production process, which have been limiting for companies to achieve production on a commercial scale. Most sugarcane bagasse studies for ethanol production have been conducted in Brazil. Brazilian mills, such as Raízen and Granbio, have already used this raw material for commercial cellulosic ethanol production, mainly using sugar cane bagasse and straw. Besides, there are four major 2G commercial ethanol plants in the world Poet-DSM, BetaRenewables, Abengoa (plant bought by Synata Bio), and DuPont—two are Brazilian: GranBio and Raízen,

18.4 Mexican sugar industry boosting ethanol production by public policies

México had 783,486 ha under cane cultivation. Specifically, 47,265,093 tons of cane were harvested, and 5,278,320 tons of sugar and 12,111,980 L of ethanol were produced in fifteen sugarcane-producing states during harvest season 2019/2020. Although the cultivation and processing of sugarcane occur under adverse conditions with a cane yield of 62,891 t/ha and agroindustrial yield of 6737 t/ha, México has managed to position itself as a major producer of sugar in the world. The socioeconomic, political, environmental, and technical aspects create complex challenges that must be tackled to improve competitiveness. Sugarcane production in México has nearly 500 years of history and technological, socioeconomic, political, and productive agricultural and industrial activities in sugar plantations, artisanal Noncentrifuged facilities (trapiches), sugar mills, and distilleries. These have been studied by various disciplines such as engineering, anthropology, sociology, geography, history, law, and others by Mexican and foreign researchers, but the low level

Table 18.1 Research trending to consider in ethanol projects.

Issue	Conclusions	References
Integrated 1G and 2G ethanol	Synergic production reduces most environmental impacts except for global warming, photochemical ozone depletion, and resource depletion due to the mechanization of the planting and harvesting processes. The ethanol yield/ha increases since bagasse and trash are used to produce 2G ethanol. Consequently, 2G ethanol allows reduction of land use by approximately 30%.	Maga et al. (2019)
Combined use of lignocellulosic biomass	2G technology is not as mature as conventional 1G ethanol from sugarcane. However, 2G ethanol production could be significantly improved by the combined use of bagasse, straw, and tops compared to the use of bagasse alone.	Pereira et al. (2015)
Sustainability of 2G ethanol production	The sustainability of biofuel is severely affected by the technological route of its production. green solvents, heat and electricity generation through bioenergy, reutilization of water from evaporators, vinasse concentration, and the upgrading of lignin are strategic routes for developing sustainable 2G ethanol, minimization of the use of natural resources, effluent generation, and energy expenditure.	Vieira et al. (2020)
LCA of 2G ethanol	For the sustainable use of energy crops and LCB, the reduction of global warming potential is the main objective; however, it is necessary for processing to avoid increasing acidification, eutrophication, and photochemical oxidant formation impact categories because the feedstock provision is still complex.	Soleymani Angili et al. (2021)

(*Continued*)

Table 18.1 (Continued)

Issue	Conclusions	References
RenovaBio program and decarbonization credit (CBio)	This program aimed to reduce GHG emissions by 2030 and the development of the ethanol industry to increase profits and improve process efficiency in all of the production steps (from agriculture up to product distribution), reducing costs and improving environmental efficiency. To obtain their CBios, the ethanol producers will be certified through LCA using an appropriate tool called RenovaCalc.	Gonçalves et al. (2021) Chandel et al. (2021)
Sugarcane straw biorefinery	New applications and valorization of trash have emerged due to the substitution of the manual harvest of sugarcane, which led to a high generation of this substantially novel residue for applications in biorefinery.	Aguiar et al. (2021)
Valorization of wastes from sugarcane ethanol production	The sugarcane industry must embrace novel technologies for biomass fractionation and pyrolysis, microalgae cultivation for CO_2 recovery and vinasse treatment, CO_2 chemical fixation, energy generation through the anaerobic digestion of vinasse, and genetically improved fermenting yeast strains.	Sydney et al. (2020)
Process-based models for sustainability assessments of 2G sugarcane ethanol	2G ethanol provides an opportunity to intensify production, obtaining more energy per unit of the area cropped and potentially reducing the environmental footprint. However, integrating models and geographic information systems allows for regional assessments of the potential impacts of bioenergy production, contributing to the identification and promotion of sustainable pathways for cellulosic ethanol production.	Galdos et al. (2017)

(Continued)

Table 18.1 (Continued)

Issue	Conclusions	References
2G Ethanol technology	Technology learning curves will reduce environmental impacts and costs of 2G ethanol. Besides, increased sugarcane yields and the gradual introduction of more productive varieties, such as energy cane, are expected to affect 1G and 2G ethanol production processes.	Cavalett et al. (2017)

1G, First generation; *2G*, second generation; *GHGs*, greenhouse gases; *LCA*, life cycle analysis; *LCB*, lignocellulose biomass.

of coordination and holistic solutions among stakeholders, mainly farmers, biofuel producers, traders, sugar millers, technology and agrochemical companies, government, universities, researchers, shipping firms, logistics companies, PEMEX (Federal Mexican Oil Company), CFE (Federal Electric Commission), and users of fuel, industrial and beverage ethanol (Fig. 18.2).

Cane growing and milling in México focuses on producing raw, standard, white, and refined sugar and molasses for domestic consumption and raw sugar for export. The industry employs an estimated 440,000 people directly and through upstream and downstream multipliers, supporting a further 2.2 million indirect jobs. The sugarcane sector has been struggling for years to develop a national plan with stakeholders to improve the sustainability of the industry. However, it presents high production costs distributed in preparation at 8%, planting at 11%, crop management at 15%, inputs at 23%, harvest at 27%, social security at 4%, and other 12%. In addition, the structure of land ownership (<5 ha/capita), the lack of a grower unit to reduce harvest and transportation costs, and the lack of actual plans for diversifying rural and industrial production options around sugar cane and synergy with other agroindustrial crops as corn wastes, are some of the other concerns.

From the current technological level, it can be inferred that a third of the 50 sugar mills currently in operation in México present incorporation of novel technologies concerning juice extraction and energy use; another third with intermediate technologies, and the rest with evident technological backwardness, which has led to the production of lower quality sugars for the domestic and international market (Fig. 18.3).

18.5 Public policies in México

During the 20th century, the Mexican Government constantly issued laws, agreements, regulations, mandates, and norms for the survival of this agroindustry and to

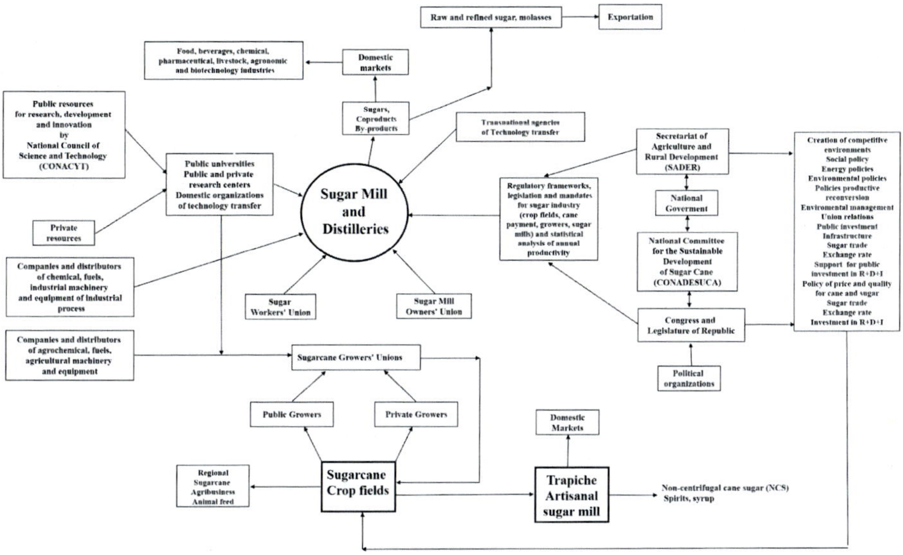

Figure 18.2 Stakeholders in the Mexican sugar industry.

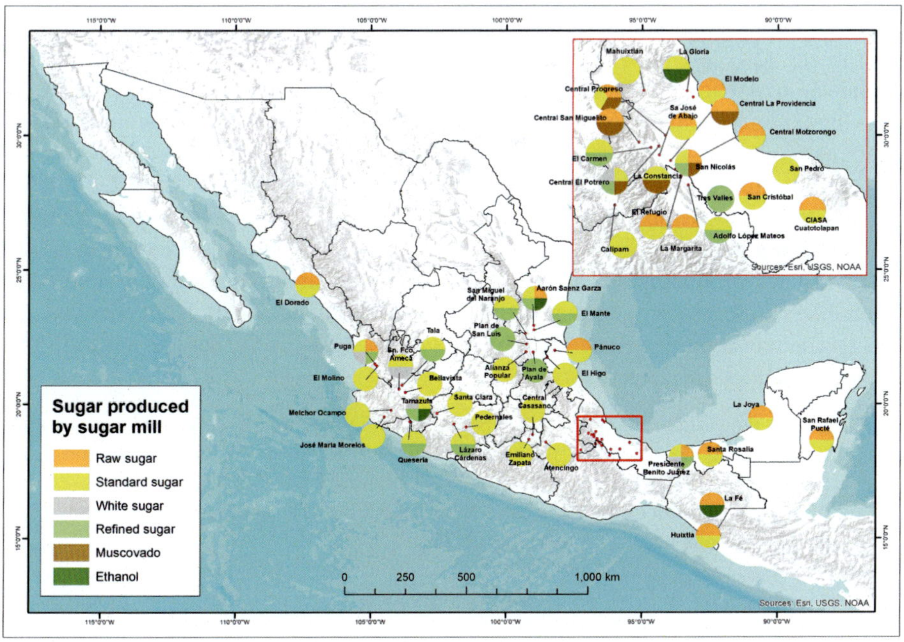

Figure 18.3 Sugar production by quality in Mexican sugar mills.
Source: From http://conadesuca.gob.mx/DocumentosExternos/7_Informe_Estadistico.pdf.

avoid social conflicts that come to structure the historical legal framework of the sugar agroindustry, from the sowing, management of the crop, harvesting, research and development priorities and technology transfer, payment of cane stalks and industrialization to the production and commercialization of sucrose, but the most important were from the 1940s (Table 18.2).

The sugar industry is subject to a series of challenges related to laws, regulations, and agreements such as environmental, labor laws, inefficiency and lack of competitiveness, public safety, tax, health problems, antitrust laws, and strict export guidelines of sugar as a result of agreements such as the United States–México–Canada Agreement (T-MEC). These are aspects that, for years, authorities and legislators have tried to implement jointly or unilaterally with growers and sugar mills without any generalized and systematic progress being made in practice. Furthermore, the political, socioeconomic, environmental, business, and academic debate of stakeholders and the technologies employed for cogeneration, as a necessary complement for the joint production of ethanol, sugar, and electric energy under the concept of biorefining in sugar mills, remain essentially unchanged since the establishment of last Sugarcane Decree (1991), the current legislation of 2005 (law of sustainable development of sugar cane) and later.

This situation causes the economic agents involved in the sugarcane agroindustry, especially the cane growers, input suppliers, industrialists, and researchers, to remain in a state of permanent legal uncertainty that does not correspond to the importance of the activity or to the circumstances that prevail in globalized markets, where to gain access in competitive conditions it is necessary, first, to have a solid legal base. Therefore, there are no conditions or interest of the current Mexican government (2018–24) to reconvert and modernize the sugarcane agroindustry, despite the existence of a solid base of renowned Mexican researchers at universities with experience in establishing a national 1G and 2G ethanol program based on sugar cane in synergy with other raw materials.

According to Aguilar-Rivera et al. (2019), a national policy will be required to launch a multidimensional approach to make the ethanol blending market competitive. Since sugarcane, as a crop, has good prospects, the biorefinery concept at the sugar mills for producing 1G and 2G ethanol and sugar production can benefit the stakeholders involved in sugarcane milling and cropping, apart from meeting the climate change commitments. To achieve ethanol competitiveness in ethanol production in México, it is necessary to:

1. Modernize the productive activity of the value chain: sugarcane crop fields, sugar mills distilleries, and marketing system (inputs, modernization, mechanization, and infrastructure).
2. Apply key best practices for the sustainable production of sugarcane under rainfed and irrigated conditions according to agroecological zoning.
3. Increase sugarcane acreage for ethanol.
4. Provide continuity in field labors with conservation farming, compost and biofertilizers, and biological pest control.
5. Ensure efficient use of water in the field.

Table 18.2 Most significant public policies of the Mexican government in the management of the sugarcane agroindustry.

Law or decree	Regulatory aspects
Decree of the Sugar and Alcohol Market January 31, 1931	Prohibition of making alcohol from any cereals, flours, and fodder, with the purpose that the only raw material to be used would be molasses from sugar mill, thus encouraging another destination for sugarcane production.
Sugar S.A. January 9, 1932	The associated mills were obliged to sell all of their sugar, uncrystallizable molasses, ethyl alcohol, and distillation residues produced to be handled in the domestic market according to prices set by the federal government.
Secretary of Agriculture and Development and the National Economy, December 23, 1941	It establishes the prices for cane and sugar. Also, it excludes cane growers from participating in the final molasses and other by-products, forcing the mills to exhaust the molasses.
Sugarcane decree March 29, 1944	Liquidation guidelines were established for the cane used to produce sugar and alcohol and manufacture noncentrifuged sugar produced in the artisanal Noncentrifuged facilities (trapiches). Sugarcane growers had 25% of their net value in final molasses and 25% in alcohol production, discounting a value for processing expenses and taxes. In the final molasses, transport costs were discounted.
Presidential decree of 1951	Establishes the prices that were in force in the various areas of the country on the different qualities of sugar.
Presidential decree November 17, 1956	The liquidation price of sugarcane was modified, giving the grower a 50% share of sales in molasses and alcohols and 50% in sugar exports.
Agreement of the Secretary of Industry and Commerce December 17, 1970	The new prices would govern the areas of the country for the different qualities of sugars and the creation of the National Commission of the Sugar Industry (CNIA) as a body to restructure the industry, increase productivity in the field and manufactures, manage the mills, are set. government sugar mills and ensure the distribution and marketing of sugar, alcohol, and molasses.
National Sugarcane Agreement UNPASA, CNC, Secretary of Agriculture and Livestock, Industry and Commerce December 9, 1972	The participation of growers in the sale of sugarcane derivatives and the promotion of the industrialization of by-products, the diversification of crops, and the integration and promotion of agricultural production with livestock programs in the supply areas were regulated.

(Continued)

Table 18.2 (Continued)

Law or decree	Regulatory aspects
Sugarcane decree October 24, 1975	The sowing, cultivation, harvest, and industrialization of sugar cane, the total payment of the cane for a single concept, and its quality derived from the sucrose content are declared of public interest. The purpose was to end the link that existed between the price of cane with that of sugar, alcohol, and molasses; give sugarcane farmers more significant incentives to grow sugarcane; modernization of production methods; Sufficient and timely financing, and above all, strengthen the State's leadership in the sugar industry.
Presidential decree December 28, 1979	Declared of public interest the sowing, cultivation, harvest, and industrialization of sugar cane as a raw material of the sugar industry, without the economic participation of growers in the other by-products of sugar cane.
Reconversion Program of the parastatal sugar industry SEMIP-Azúcar S.A. 1986	It established the need to promote research programs aimed at the development of new varieties of cane with better agroindustrial and phytosanitary characteristics and projects for the use of new products and by-products of cane in an integral way, such as livestock feed, organic fertilizers, oxalic acid, sucrochemicals, liquid sugar from molasses and juice, fuel alcohol and other results of biotechnology.
Mexican government 1990	The disappearance of the Mexican Institute of Sugar Production (IMPA) and of the public investigation related to the cultivation and processing of sugar cane and its derivatives that the National Union of Producers of sugar (UNPASA), had founded as the Experimental Fields office, which later (1956), became the Institute for the Improvement of Sugar Production (IMPA), with six regional experimental fields and one hybridization station.
Sugarcane Decree 1991 Secretary of Agriculture and Hydraulic Resources	The mills and their cane suppliers may undertake co-investments, form joint ventures or enter into contracts to increase the productivity, efficiency, and diversification of the sugarcane field as long as it is not contrary to the uniform agreement of planting, harvesting and receiving sugar cane.

(Continued)

Table 18.2 (Continued)

Law or decree	Regulatory aspects
The North American Free Trade Agreement (NAFTA) January 1, 1994	The North American Free Trade Agreement (NAFTA) establishes the gradual deregulation of the market until it is accessible in 2008 at a zero tariff rate.
Sustainable Development Law of Sugar Cane 2005	As legislation, its provisions are of public interest and social order; due to the basic and strategic nature of sugar for the national economy in terms of the Sustainable Rural Development Law, it aims to regulate the activities associated with contract farming and sustainable integration of sugarcane, sowing, cultivation, harvesting, industrialization and marketing processes, its products, by-products, co-products, and derivatives. The National Committee for the Sustainable Development of Sugarcane (CONADESUCA) is constituted and dedicated to the coordination and implementation of all activities related to the sugarcane agroindustry. The Sugarcane Scientific and Technological Research Center (CICTCAÑA) is created, which will aim to guide research and development projects to give more competitiveness and profitability to the sugarcane agroindustry.
Law for the promotion and development of bioenergetics February 1, 2008	Promote the sugar cane and corn agro-industries to produce ethanol, oilseed plants for biodiesel production, as bioenergetics, and the gasification of biomass for its use in electricity generation.
National Sugar Cane Agroindustry Program 2007–12 (PRONAC)	Provide certainty to productive activity, promote agreements, research, and technology transfer, access to financing, and encourage effective diversification of ethanol, sucrochemicals, and derivatives. Establish a payment system for sugar cane differentiated from the production of sweetener, ethanol, sucrochemicals, and derivatives at an international level and allow our country to compete successfully. Consider the sugarcane bagasse and its residual biomass as a renewable and intermittent energy source. Consolidate an attractive offer of bioenergetics that helps reduce the environmental impact and implement the urgent and necessary measures to take care of the ecosystems.

(Continued)

Table 18.2 (Continued)

Law or decree	Regulatory aspects
National Sugar Cane Agroindustry Program (PRONAC)2014–18	Promote the productive diversification of sugarcane agroindustry. Promote the linkage of the sugarcane agroindustry and biotechnology research institutions. Optimize the thermal arrangements in sugar mills to ensure energy self-sufficiency in the factory from the sugarcane bagasse. Promote investment in cogeneration plants to commercialize surplus electricity through the public grid. Promote the production of ethanol and other by-products in the mills to substitute imports. Promote projects for the sustainable production of anhydrous ethanol based on market studies. Reuse waste generated in the field and factory for composting and subsequent conversion into biofertilizers. Promote the development of a research and technology transfer agenda for the sugarcane agroindustry. Generate new varieties of high productivity, adapted to each productive region, humidity regime, and resistant to adverse factors and pests.
NOM-016-CRE-2016, Fuel Quality Specifications August 2016 (5.8% ethanol) + modification July 2017 (10% ethanol)	The Energy Regulatory Commission (CRE) establishes that the Mexican Petroleum Institute, based on the work carried out so far in México and on international experience, considers the introduction of gasoline with up to 10% ethanol (E10 mixture) in the regions considered "rest of the country" and not in the critical metropolitan areas of Monterrey, Guadalajara, Valley of México, for which laboratory studies are required to observe the behavior that would be expected with vehicles that represent the main technological strata of the parked vehicle that circulates in these areas.

(*Continued*)

Table 18.2 (Continued)

Law or decree	Regulatory aspects
Institutional Program 2020–24 of the National Committee for Sustainable Development for Sugar Cane (CONADESUCA) September 28, 2020	The diversification of the use of sugarcane for purposes other than traditional use will be promoted to stabilize the income of the sector. The production of ethanol from sugar cane can be a sustainable alternative for the diversification of sugarcane agribusiness. In addition, it has broad market potential to replace the oxygenating methyl tert-butyl ether (MTBE) used in gasoline, currently imported from the United States, with the objectives: 1. Promote the reduction of production costs in the sugarcane agroindustry at the national level. 2. Increase interinstitutional coordination for developing research, innovation, and technology transfer in the sugarcane agroindustry sector. 3. Improve the quality of information for making decisions that affect the supply of the national sugar market and promote commercial order in the sugarcane agroindustry.

S.A. Anonymous society.

6. Place more emphasis on people throughout the value chain and reinforce environmental protection.
7. Establish a differentiated price for the final use of sugarcane as a raw material for sugar or ethanol.
8. Augment factory yields.
9. Generate more electric power and decrease fossil fuel consumption.
10. Increase the capacity of existing distilleries and Implement incentives for the construction of new distilleries using cane juice, bagasse, and bagasse.
11. Create new distilleries in sugar mills and transfer technology for ethanol production from first-, and second-generation sources.
12. Establish auto assemblers with cars equipped to use various ethanol/gasoline mixtures (flex-fuel).
13. Ensuring the internal market of biofuels for gasoline in Guadalajara, Monterrey, México City, and other cities.

14. Decrease production costs.
15. Apply lessons learned from the experience of countries like Brazil, the United States, Australia, Colombia, and South Africa in México.

The problem mentioned above needs to be addressed through the following strategies:

1. Public policies development and improvement of the management in sugarcane crop fields (harvest with burning bagasse).
2. Agroecological zoning to reduce cost and inputs and increase productivity with the same land, sugar mill modernization.
3. Generation of sustainable processes.
4. Transformation of sugar mills into biorefineries.
5. Diversification of the production chain.
6. Use of ethanol in gasoline.

18.5.1 Use of ethanol in gasoline in México

México, as the sixth world producer of sugar cane, needs to diversify its energy sources based on available renewables since it is an oil exporter (primary energy) but demands to import more than half of the gasoline (secondary fuel) required by transport where the cost is a function of fluctuations in the price of crude oil (Table 18.3).

Therefore México will continue to be dependent on imports since it is not economically viable to build new refineries. However, the Mexican government currently has one under construction in Tabasco and another bought in Texas, United States of America. In the world scenario, there is an oversupply of gasoline, and demand is reducing due to the incorporation of ethanol in blends from 5% to 100% worldwide; in the short term, the supply of hybrid or electric vehicles.

Besides, Mexican gasoline uses methyl tert-butyl ether (MTBE) or TAME (tert-amyl methyl ether) as additives in 5% of any production volume to obey the official Mexican normative. Hence, MTBE and/or TAME could be substituted by ethanol 1G or 2G to reduce the environmental impact of fossil fuel consumption (Rendon-Sagardi et al., 2014).

18.5.2 Cane harvest with burning

The cane stalks are harvested on average under the cane-burning system (preburning, burning, and re-burning) in 90.7% of the total cane-producing regions, and mechanized harvesting barely reaches 17.6% during the harvest season 2019/2020, generating a national average of 11 970,508 t/y, the calorific value of 17.41 MJ/kg and average energy potential of 208.4 tJ/y for each sugarcane region (Fig. 18.4).

Burning directly influences a stagnation of sugarcane and sucrose yield productivity and increases the amount of noncrystallizable sugars due to the effects of the harvest with burning on the stems and low mechanization (Figs. 18.5 and 18.6).

Table 18.3 National gasoline balance 2014−32 (thousands of barrels per day) (https://sie.energia.gob.mx/herramientas.do?action = search).

Year	Domestic production	Refineries							Importation	Importation (%)	National consumption
		Cadereyta	Madero	Tula	Salamanca	Minatitlán	New refinery	Salina Cruz			
2014	409,21	73,37	40,13	86,53	56,32	62,13	0,00	90,72	370,05	47,54	778,41
2015	370,27	63,08	50,72	80,38	44,15	52,34	0,00	79,60	425,75	53,58	794,64
2016	333,53	52,15	34,67	73,86	56,68	42,67	0,00	73,49	476,89	58,85	810,42
2017	257,00	51,09	12,08	69,02	49,06	32,88	0,00	42,87	571,83	71,46	800,19
2018	241,05	48,35	6,29	44,58	48,97	31,08	0,00	61,79	578,62	70,59	819,67
2019	335,54	63,86	34,33	72,26	51,12	48,74	0,00	65,22	508,37	60,24	843,91
2020	361,02	48,94	46,44	62,97	54,66	61,27	0,00	86,75	495,58	57,85	856,60
2021	396,35	79,58	46,44	85,12	47,86	61,27	0,00	76,09	480,23	54,78	876,58
2022	476,15	81,87	51,02	110,42	61,22	74,48	0,00	97,15	427,71	47,32	903,86
2023	612,17	83,80	51,02	116,13	66,74	74,48	113,74	106,25	325,09	34,68	937,26
2024	619,83	88,20	51,02	122,24	77,08	74,48	120,06	86,75	340,86	35,48	960,69
2025	670,08	88,20	51,02	122,24	81,36	74,48	126,38	126,39	309,39	31,59	979,46
2026	676,72	88,20	51,02	122,24	85,64	74,48	126,38	128,75	325,51	32,48	1002,22
2027	683,50	88,20	51,02	122,24	85,64	74,48	126,38	135,53	334,68	32,87	1018,17
2028	683,50	88,20	51,02	122,24	85,64	74,48	126,38	135,53	339,11	33,16	1022,61
2029	683,50	88,20	51,02	122,24	85,64	74,48	126,38	135,53	348,41	33,76	1031,90
2030	683,50	88,20	51,02	122,24	85,64	74,48	126,38	135,53	356,00	34,25	1039,49
2031	683,50	88,20	51,02	122,24	85,64	74,48	126,38	135,53	359,99	34,50	1043,48
2032	683,50	88,20	51,02	122,24	85,64	74,48	126,38	135,53	356,54	34,30	1040,03

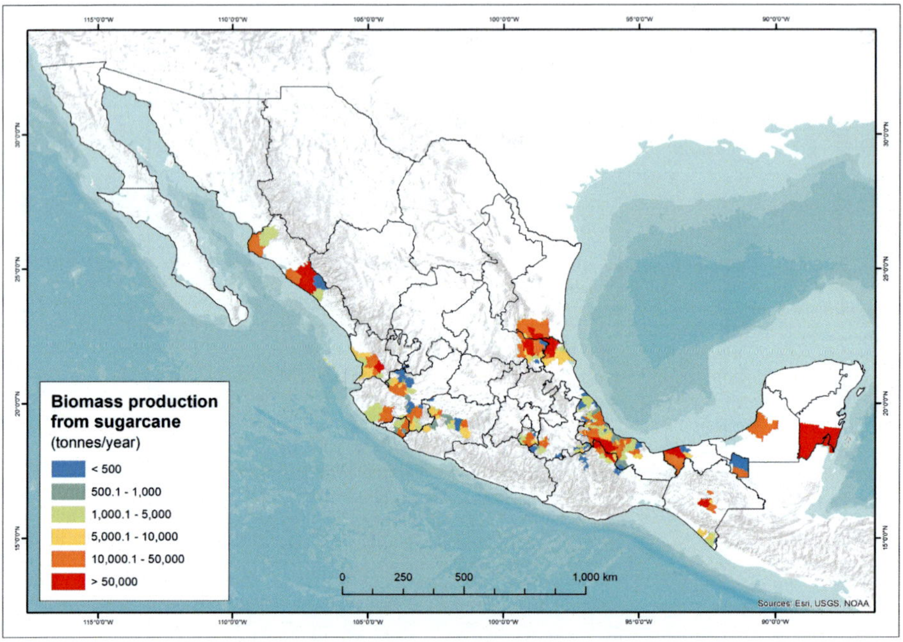

Figure 18.4 Sugarcane biomass generation by crop in México (SENER, 2018). Producción agrícola especializada para biocombustibles—Caña de azúcar. Atlas Nacional de Biomasa, ANBIO. Secretaria de Energía.
Source: From https://www.gob.mx/sener/articulos/atlas-nacional-de-biomasa.

Therefore the burning of the cane and bagasse (green leaves and tops) is mainly wasted by either burning on the field or is left on the ground to decay and not only contaminates the environment with CH_4, CO, N_2O, SO_x, and NO_x and aggravates the extreme heat situation in the harvest season, but also affects the fertility of the soils by reducing the content of organic matter and nitrogen. Therefore, it is the leading cause of forest fires and contributes to the global warming of the atmosphere.

The work on the manual cutting of sugarcane, especially of burned sugarcane, exposes workers to various risks, with different health impacts such as respiratory symptoms, inflammatory markers, oxidative stress, pulmonary disease, cardiovascular effects, renal effects, and miscellaneous infections (Leite et al., 2018). Therefore, the manual harvest of sugarcane, which was previously associated with burning to remove the bagasse and facilitate cutting, could be gradually replaced by mechanical harvest.

Nonetheless, considering the amounts of bagasse generated and that all the cellulose and hemicelluloses present could be converted into electricity by cogeneration and ethanol with the use of bagasse and molasses could significantly increase ethanol production without the need for expansion of cultivated areas. However, there is still no consensus about the optimal amount of bagasse that should be left on the

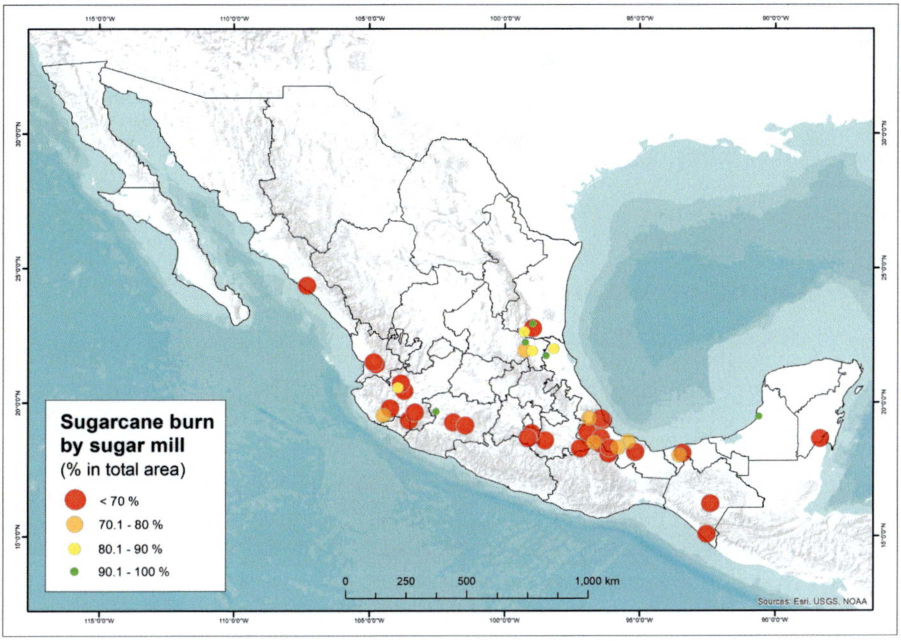

Figure 18.5 Sugar cane burning in México (%).
Source: From http://conadesuca.gob.mx/DocumentosExternos/7_Informe_Estadistico.pdf.

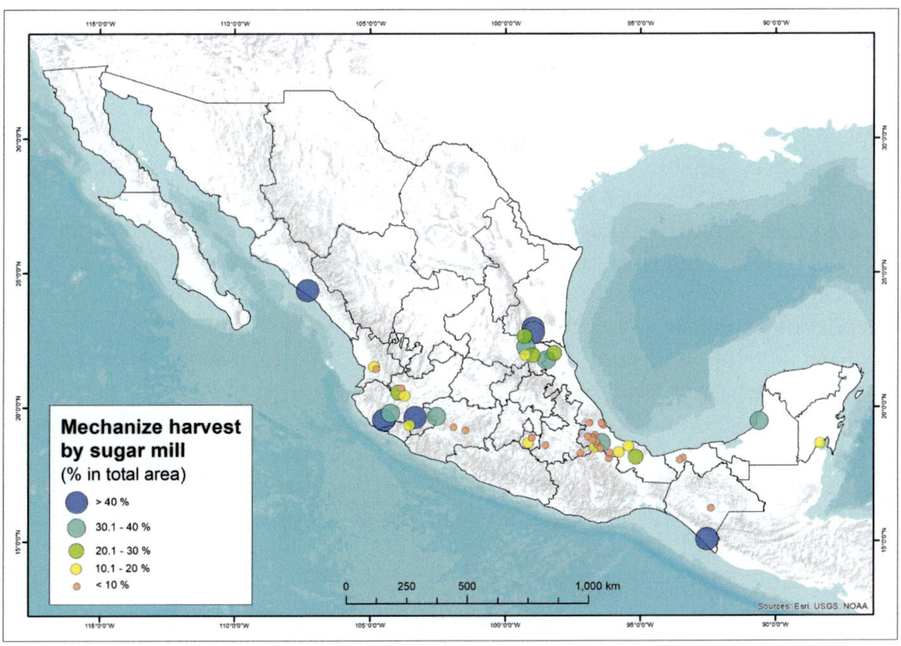

Figure 18.6 Mechanization of the sugarcane harvest in México (%) (CONADESUCA, 2021).
Source: From http://conadesuca.gob.mx/DocumentosExternos/7_Informe_Estadistico.pdf.

ground. It has been observed that 65% of the residue from sugarcane harvesting can be removed from the fields as dry leaves (straw) without causing adverse agricultural effects and the soil (Pereira et al., 2015). For this reason, it is necessary to establish and apply environmental policies articulated to the social and economic realities of each sugarcane region and other crops such as corn to use their biomass, promote sustainability, and reduce health risks.

18.6 Zoning of sugarcane and corn in México

For the zoning, the cultivation of corn was also considered since many sugarcane areas overlap with this, the largest crop in the country. That also generates a large amount of biomass that, like sugar cane, is burned after harvest.

The production of sugarcane and corn depends significantly on the environment and climatic conditions where it is developed. Recognizing these areas and their distribution is an essential part of the agroecological improvement that can be given to crops. Agricultural zoning is undoubtedly a state-of-the-art method for productive efficiency. It is defined as the surface division into units with similar characteristics related to suitability, potential production, and environmental impact (de Oliveira Aparecido et al., 2021). The zoning of both crops was carried out from the multicriteria analysis in the geographic information system, where through an analytical hierarchy process, the areas with the most suitability for the best production of the sugarcane and corn crops were identified. Saha et al. (2021) and Shivaswamy et al. (2021) assume that this method is simplified and very suitable for analyzing the suitability of the sites, which helps those responsible for formulating agricultural use plans for the development of the region.

Specific data were used for each crop, prioritizing zoning relevance in four categories (optimal, suitable, marginal, and unsuitable). The variables considered were in the climatic and edaphoecological order, giving weight to each variable and standardizing their values, viewing their importance in terms of the aptitude of the land to grow sugarcane and corn, respectively. All variables were considered with a discrete scale, so a direct classification was used to assign the aptitude value to each class (Goodwin and Wright, 2014).

For the cultivation of corn, eight variables were used for zoning, obtained from various sources of information. Each variable has a value and weight concerning the best establishment requirements, and they were obtained from Sotelo et al. (2016) (Table 18.4 and Fig. 18.7). The cartographic data were obtained from the INEGI cartography scale 1: 250,000, rasterized at a spatial resolution of 100 m.

For the cultivation of sugarcane, the same variables were used for zoning. Each variable has a value and weight concerning the best establishment requirements, and they were obtained from Aguilar Rivera et al. (2013) and Aguiar et al. (2021) (Table 18.5 and Fig. 18.8). The cartographic data were obtained from the INEGI cartography scale 1: 250,000, rasterized at a spatial resolution of 100 m.

Table 18.4 Climatic and edafoecological requirements for the zoning of the corn crop.

Features	Optimal	Suitable	Marginal	Unsuitable
Temperature (°C)	15–22	12/15 – 22/28	10/12 – 28/40	<10– >40
Rainfall (mm)	600 – 1200	500/600 – 1200/ 1600	400/ 500 – 1600/ 2000	<400– >2000
Relative humidity (%)	40 – 60	20/40 – 60/80	80 – 90	<20– >90
Elevation (m. a.s.l.)	1200 – 2500	0/1200 – 2500/ 2900	2900 – 3300	>3300
Slope (%)	<8	15 – 30	30 – 45	>5
Soil type	Andosols, Feozem, vertisols	Fluvisols, planosols, Cambisols, histosols, regosols	Acrisols, luvisol, Lepstosols	Solonchak, Gleysols
pH	6.0 – 7.0	5.5/6.0 – 6.0/7.5	5/5.5 – 7.5/8.0	<5– >8.0
Soil depth (cm)	80 – 100	50/80 – 100/170	10/50 – 170/ 250	<10

m.a.s.l, Meters above sea level.
Source: Sotelo et al. (2016) and Ruiz et al. (2013).

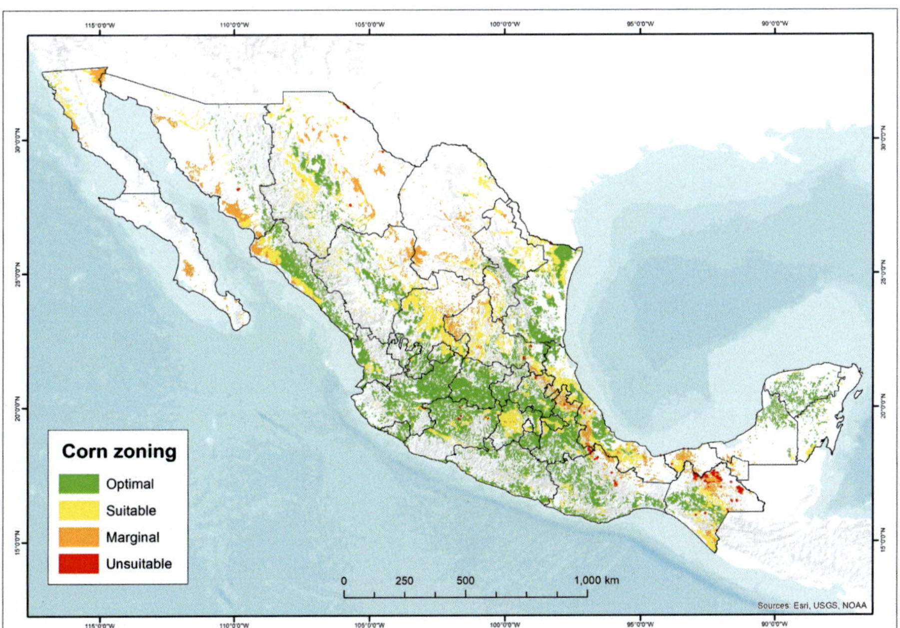

Figure 18.7 Agroecological zoning for corn production in México.

Table 18.5 Climatic and edafoecological requirements for the zoning of the sugarcane crop.

Features	Optimal	Suitable	Marginal	Unsuitable
Temperature (°C)	24 – 32	18/24 – 32/35	18/10 – 35/38	<10– >38
Rainfall (mm)	>1500	1250 – 1500	1250 – 1000	<1000
Relative humidity (%)	>70	50 – 70	10 – 50	<10
Elevation (m.a.s.l.)	<400	400–900	900–1600	>1600
Slope (%)	<8	8 – 16	16 – 30	>30
Soil type	Leptosols, luvisol	Vertisols, fluvisols	Cambisols	Flooding or saline soils
pH	6.5 – 7.5	6.0 – 8.0	4.5 – 8.5	<4.5
Soil depth (cm)	>60	50 – 60	10 – 50	<10

m.a.s.l, Meters above sea level.
Source: Aguilar-Rivera et al. (2018), Aguilar Rivera et al. (2013), Ruiz et al. (2013), Palma-López et al. (2002), and Rojas and Eldin (1983).

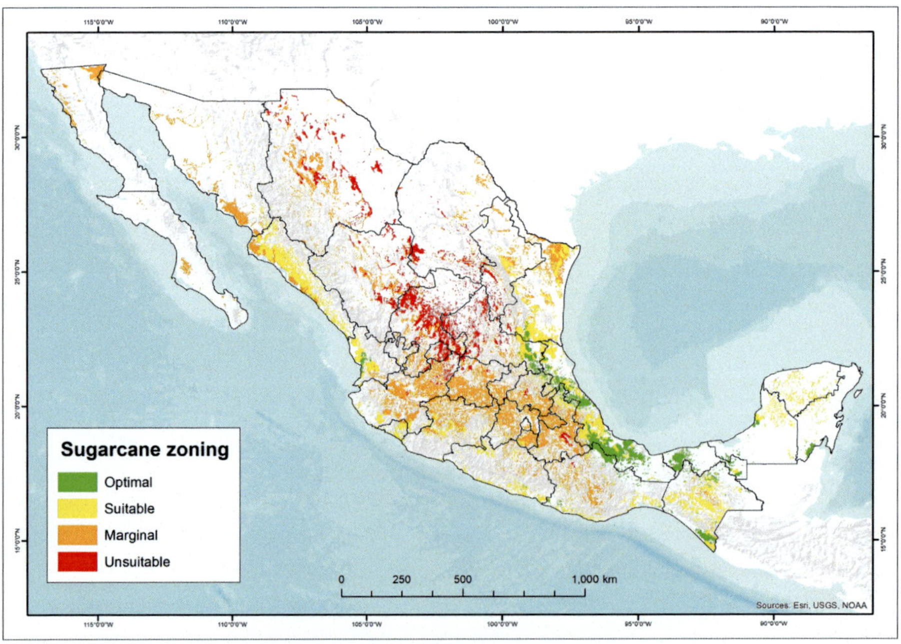

Figure 18.8 Agroecological zoning for cane production in México.

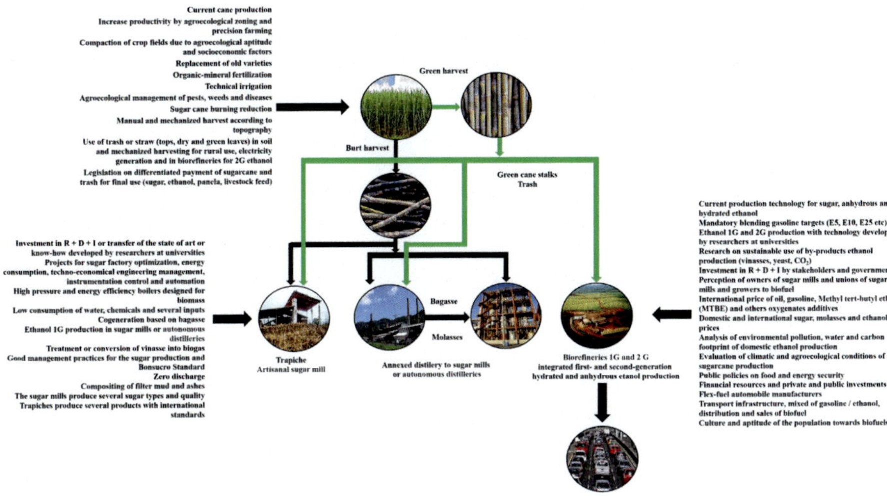

Figure 18.9 Actions to produce 1G and 2G ethanol from sugarcane.

Bioenergy derived from the conversion of bagasse and biomass can be a short- and medium-term alternative for:

1. diversify the energy matrix (ethanol 2 G),
2. reduce dependence on oil,
3. boost the rural economy,
4. open new avenues for economic development.

For this, it is essential to integrate bioenergy with public policies and stakeholders in energy planning and all sectors: electricity, biofuel transport, industrial, residential, and rural (Fig. 18.9).

Further designing a sustainable biofuels policy, multiple economic, environmental, and social aspects need to be considered in three objective groups simultaneously (Ziolkowska, 2014):

1. *Economic objectives*: Reducing biofuels production costs, increasing biofuels productivity/acreage, ensuring national food security, securing farmers' incomes or even increasing them.
2. *Environmental objectives:* Reducing GHG emissions, water usage, land use, protecting biodiversity and landscapes.
3. *Social objectives:* Increasing consumer welfare, supporting local communities, improving health and safety issues, and creating new jobs.

18.7 Conclusions and future perspectives

1G or 2G ethanol biofuel production based on the use of marginal agricultural lands, agroindustrial by-products, and technological innovations, as a function of economic, ecological, political, and social aspects are primordial factors in the pursuit of sustainable development of domestic biofuel policy to mitigate the climate change and decrease dependence on imported energy sources (gasoline, diesel, MTBE, ETBE, and petrochemicals) to achieve energy security, job creation in the rural sector and economic growth. Moreover, the work concludes that there is still a lack of knowledge and uncertainty among stakeholders, car manufacturers, the population, and users in general, about the possible beneficial or negative results regarding the socioeconomic and environmental effects of biofuels since most of the published results are technological and case studies mainly for biofuel producers such as Brazil, the United States of America and the European Union. The impact of biofuels on rural employment, food security, deforestation to increase the area of agricultural land, and environmental and socioeconomic cost are still barriers to successful promotion. Therefore, studies must be carried out for each region based on local conditions to design the best technological option, raw materials, and a national policy as a domestic biofuels program will indeed be required for launching a multidimensional approach to make the ethanol blending market competitive and the co-creating knowledge to support a national ethanol program without various inconsistencies and ambiguities. For México, it is necessary to continuity in field labors with conservation farming, continuity in the use of compost and biofertilizers, continuity in biological pest control, better yields in the factory and more electric power generation and decrease of fossil fuels consumption as initial steps in establishing a national ethanol program.

References

Aguiar, A., Milessi, T.S., Mulinari, D.R., Lopes, M.S., da Costa, S.M., Candido, R.G., 2021. Sugarcane straw as a potential second generation feedstock for biorefinery and white biotechnology applications. Biomass Bioenergy 144, 105896.

Aguilar Rivera, N., Olvera Vargas, L.A., Galindo Mendoza, G., 2013. Evaluación de aptitud de tierras al cultivo de caña de azúcar en la Huasteca potosina, México, por técnicas geomáticas. Rev. Geogr. Norte Grande 55, 141–156.

Aguilar-Rivera, N., Algara-Siller, M., Olvera-Vargas, L.A., Michel-Cuello, C., 2018. Land management in Mexican sugarcane crop fields. Land. Use Policy 78, 763–780.

Aguilar-Rivera, N., Michel-Cuello, C., Serna-Lagunes, R., de Jesús Debernardi-Vázquez, T., Trujillo-Mata, A., 2019. Ethanol production from the Mexican sugar industry: perspectives and challenges. In: Khan, M.T., Khan, I.A. (Eds.), Sugarcane Biofuels Status, Potential, and Prospects of the Sweet Crop to Fuel the World. Springer Nature Switzerland AG, p. 203.

Ajanovic, A., Haas, R., 2014. On the prospects and limits of biofuels in Brazil, the US, and EU. Appl. Energy 135, 730–737.

Ali, Z.M., Shah, S.F.A., Tunio, M.M., Abbas, M., Arshad, M., 2021. Policy implications for sustainable ethanol production. In: Arshad, M. (Ed.), Sustainable Ethanol and Climate Change. Springer Nature Switzerland AG, p. 181.

Cavalett, O., Chagas, M.F., Junqueira, T.L., Watanabe, M.D., Bonomi, A., 2017. Environmental impacts of technology learning curve for cellulosic ethanol in Brazil. Ind. Crop. Products 106, 31–39.

Chandel, A.K., Forte, M.B., Gonçalves, I.S., Milessi, T.S., Arruda, P.V., Carvalho, W., et al., 2021. Brazilian biorefineries from second generation biomass: critical insights from industry and future perspectives. Biofuels. Bioprod. Bioref. 15, 1190–1208.

de Oliveira Aparecido, L.E., de Meneses, K.C., Torsoni, G.B., e Costa, C.T.S., Mesquita, D. Z., 2021. Climate efficiency for sugarcane production in Brazil and its application in agricultural zoning. Sugar Tech. 23, 776–793.

Esparza, I., Jiménez-Moreno, N., Bimbela, F., Ancín-Azpilicueta, C., Gandía, L.M., 2020. Fruit and vegetable waste management: conventional and emerging approaches. J. Environ. Manag. 265, 110510.

FAOSTAT, 2022. Crop. Livest. Products. https://www.fao.org/faostat/en/#data/QCL.

Fava, F., Totaro, G., Diels, L., Reis, M., Duarte, J., Carioca, O.B., et al., 2015. Biowaste biorefinery in Europe: opportunities and research and development needs. New Biotechnol. 32, 100–108.

Galanopoulos, C., Giuliano, A., Barletta, D., Zondervan, E., 2020. An integrated methodology for the economic and environmental assessment of a biorefinery supply chain. Chem. Eng. Res. Des. 160, 199–215.

Galdos, M.V., Cantarella, H., Hastings, A., Hillier, J., Smith, P., 2017. Environmental sustainability aspects of second generation ethanol production from sugarcane. In: Buckeridge, M., De Souza, A. (Eds.), Advances of Basic Science for Second Generation Bioethanol from Sugarcane. Springer, Cham, p. 177.

Gilani, H., Sahebi, H., Oliveira, F., 2020. Sustainable sugarcane-to-bioethanol supply chain network design: a robust possibilistic programming model. Appl. Energy 278, 115653.

Gonçalves, F., Perna, R., Lopes, E., Maciel, R., Tovar, L., Lopes, M., 2021. Strategies to improve the environmental efficiency and the profitability of sugarcane mills. Biomass Bioenergy 148, 106052.

Goodwin, P., Wright, G., 2014. Decision Analysis for Management Judgment, 5th (Edn.). John Wiley and Sons, Chichester. New York, p. 454.

Hoque, N., Biswas, W., Mazhar, I., Howard, I., 2020. Life cycle sustainability assessment of alternative energy sources for the western Australian transport sector. Sustainability 12, 5565.

Jesus, G.K., Jugend, D., Paes, L.A. B., Siqueira, R.M., Leandrin, M.A., 2021. Barriers adoption circular economy. Braz. Sugarcane Ethanol Sect. https://assets.researchsquare.com/files/rs-318975/v1/c167c934-6749-4c93-b22c-249051633c7d.pdf.

Karp, S.G., Medina, J.D., Letti, L.A., Woiciechowski, A.L., de Carvalho, J.C., Schmitt, C.C., et al., 2021. Bioeconomy and biofuels: the case of sugarcane ethanol in Brazil. Biofuels. Bioprod. Bioref. 15, 899–912.

Leite, M.R., Zanetta, D.M.T., Trevisan, I.B., Burdmann, E.D.A., Santos, U.D.P., 2018. Sugarcane cutting work, risks, and health effects: a literature review. Rev. de. Saude Publ. 52, 80.

Maga, D., Thonemann, N., Hiebel, M., Sebastião, D., Lopes, T.F., Fonseca, C., et al., 2019. Comparative life cycle assessment of first-and second-generation ethanol from sugarcane in Brazil. Int. J. Life Cycle Assess. 24, 266–280.

Neto, A.C., Guimarães, M.J.O., Freire, E., 2018. Business models for commercial scale second-generation bioethanol production. J. Clean. Prod. 184, 168–178.

Palma-López, D.J., Salgado García, S., Obrador Olán, J.J., Trujillo Narcía, A., del C. Lagunes Espinoza, L., Zavala Cruz, J., et al., 2002. Sistema integrado para recomendar dosis de fertilización en caña de azúcar (SIRDF). TERRA Latinoamericana 20, 347–358.

Pattanaik, L., Pattnaik, F., Saxena, D.K., Naik, S.N., 2019. Biofuels from agricultural wastes. In: InL, A., Basile, Dalena Second, F. (Eds.), Third Generation of Feedstocks. Elsevier, Netherlands, p. 103.

Pereira, S.C., Maehara, L., Machado, C.M.M., Farinas, C.S., 2015. 2G ethanol from the whole sugarcane lignocellulosic biomass. Biotechnol. Biofuels 8, 1–16.

Rendon-Sagardi, M.A., Sanchez-Ramirez, C., Cortes-Robles, G., Alor-Hernandez, G., Cedillo-Campos, M.G., 2014. Dynamic analysis of feasibility in ethanol supply chain for biofuel production in Mexico. Appl. Energy 123, 358–367.

Rojas, O.E., Eldin, M., 1983. Zonificación agroecológica para el cultivo de caña de azúcar en Costa Rica. Turrialba 33, 151–160.

Ruiz, C.J.A., Medina García, G., González Acuña, I.J., Flores López, H.E., Ramírez Ojeda, G., Ortiz Trejo, C., et al., 2013. Requerimientos Agroecológicos de Cultivos. Instituto Nacional de Investigaciones Forestales Agrícolas y Pecuarias, México. 564 p.

Saha, S., Sarkar, D., Mondal, P., 2021. GIS and multi-criteria decision-making assessment of sites suitability for agriculture in an anabranching site of sooin river. India Modeling Earth Syst. Environ. 7, 571–588.

Sherpa, K.C., Rajak, R.C., Banerjee, R., Banerjee, R., 2017. Sugarcane: a potential agricultural crop for bioeconomy through biorefinery. In: Sharma, V. (Ed.), Lignocellulosic Biomass Production and Industrial Applications. Scrivener Publishing LLC, p. 171.

Shivaswamy, M., Ravikumar, A.S., Shivakumar, B.L., 2021. Crop suitability analysis for Kaberi command area using RS and GIS techniques—a multi-criteria approach. In: Narasimhan, M.C., George, V., Udayakumar, G., Kumar, A. (Eds.), Trends in Civil Engineering and Challenges for Sustainability. Springer, Singapore, p. 715.

Soleymani Angili, T., Grzesik, K., Rödl, A., Kaltschmitt, M., 2021. Life cycle assessment of bioethanol production: a review of feedstock, technology, and methodology. Energies 14, 2939.

Sotelo, Ruiz, E.D., Cruz Bello, G.M., González Hernández, A., Moreno Sánchez, F., 2016. Determining the suitability of the land for maize by multi-spatial analysis in the State of Mexico. Revista mexicana de ciencias agrícolas 7(2), 401–412. Available from: https://doi.org/10.29312/remexca.v7i2.353.

Sydney, E.B., de Carvalho, J.C., Letti, L.A.J., Magalhães Jr, A.I., Karp, S.G., Martinez-Burgos, W.J., et al., 2020. Current developments and challenges of green technologies for the valorization of liquid, solid, and gaseous wastes from sugarcane ethanol production. J. Hazard. Mater. 124059.

Tan, J., Tan, F.J., Ramakrishna, S., 2022. Transitioning to a circular economy: a systematic review of its drivers and barriers. Sustainability 14, 1757.

Vanholme, B., Desmet, T., Ronsse, F., Rabaey, K., Van Breusegem, F., De Mey, M., et al., 2013. Towards a carbon-negative sustainable bio-based economy. Front. Plant Sci. 4, 174.

Vieira, S., Barros, M.V., Sydney, A.C.N., Piekarski, C.M., de Francisco, A.C., de Souza Vandenberghe, L.P., et al., 2020. Sustainability of sugarcane lignocellulosic biomass pretreatment for the production of bioethanol. Bioresour. Technol. 299, 122635.

Ziolkowska, J.R., 2014. Optimizing biofuels production in an uncertain decision environment: conventional vs. advanced technologies. Appl. Energy 114, 366–376.

Index

Note: Page numbers followed by "*f*" and "*t*" refer to figures and tables, respectively.

Printed in the United States
by Baker & Taylor Publisher Services